Rashid Mehmood · Victor Alves · Isabel Praça ·
Jarosław Wikarek · Javier Parra-Domínguez ·
Roussanka Loukanova · Ignacio de Miguel ·
Tiago Pinto · Ricardo Nunes · Michela Ricca
Editors



# Distributed Computing and Artificial Intelligence, Special Sessions I, 20th International Conference

Springer

*Editors*
Rashid Mehmood
High Performance Computing Center
King Abdulaziz University
Jeddah, Saudi Arabia

Isabel Praça
ISEP/GECAD
Instituto Superior de Engenharia do Porto
Porto, Portugal

Javier Parra-Domínguez
BISITE
University of Salamanca
Salamanca, Spain

Ignacio de Miguel
Universidad de Valladolid
Valladolid, Spain

Ricardo Nunes
Universidade de Trás-os-Montes e Alto Douro
Vila Real, Portugal

Victor Alves
University of Minho
Braga, Portugal

Jarosław Wikarek
Kielce University of Technology
Kielce, Poland

Roussanka Loukanova
Institute of Mathematics and Informatics
Bulgarian Academy of Sciences
Sofia, Bulgaria

Tiago Pinto
INESC-TEC and GECAD/ISEP
Universidade de Trás-os-Montes e Alto Douro
Vila Real, Portugal

Michela Ricca
University of Calabria
Arcavacata, Rende CS, Cosenza, Italy

ISSN 2367-3370 ISSN 2367-3389 (electronic)
Lecture Notes in Networks and Systems
ISBN 978-3-031-38317-5 ISBN 978-3-031-38318-2 (eBook)
https://doi.org/10.1007/978-3-031-38318-2



This Springer imprint is published by the registered company Springer Nature Switzerland AG
The registered company address is: Gewerbestrasse 11, 6330 Cham, Switzerland

# Lecture Notes in Networks and Systems 741

The series "Lecture Notes in Networks and Systems" publishes the latest developments in Networks and Systems—quickly, informally and with high quality. Original research reported in proceedings and post-proceedings represents the core of LNNS.

Volumes published in LNNS embrace all aspects and subfields of, as well as new challenges in, Networks and Systems.

The series contains proceedings and edited volumes in systems and networks, spanning the areas of Cyber-Physical Systems, Autonomous Systems, Sensor Networks, Control Systems, Energy Systems, Automotive Systems, Biological Systems, Vehicular Networking and Connected Vehicles, Aerospace Systems, Automation, Manufacturing, Smart Grids, Nonlinear Systems, Power Systems, Robotics, Social Systems, Economic Systems and other. Of particular value to both the contributors and the readership are the short publication timeframe and the world-wide distribution and exposure which enable both a wide and rapid dissemination of research output.

The series covers the theory, applications, and perspectives on the state of the art and future developments relevant to systems and networks, decision making, control, complex processes and related areas, as embedded in the fields of interdisciplinary and applied sciences, engineering, computer science, physics, economics, social, and life sciences, as well as the paradigms and methodologies behind them.

Indexed by SCOPUS, INSPEC, WTI Frankfurt eG, zbMATH, SCImago.

All books published in the series are submitted for consideration in Web of Science.

For proposals from Asia please contact Aninda Bose (aninda.bose@springer.com).

# Preface

The 20th International Symposium on Distributed Computing and Artificial Intelligence 2023 (DCAI 2023) is a forum for exchange of ideas between scientists and technicians from both academic and business areas. DCAI is essential to facilitate the development of systems that meet the demands of today's society. Research on Intelligent Distributed Systems has matured during the last decade, and many effective applications are now deployed. Nowadays, technologies such as Internet of Things (IoT), Industrial Internet of Things (IIoT), Big Data, Blockchain, and Distributed Computing, in general, are changing constantly because of the large research and technical effort being undertaken in both universities and businesses. Most computing systems from personal laptops to edge/fog/cloud computing systems are available for parallel and distributed computing. This conference is a forum in which to present application of innovative techniques to complex problems in all these fields.

This year's technical program will present both high quality and diversity, with contributions in well-established and evolving areas of research. Specifically, 108 papers were submitted, by authors from 31 different countries (Algeria, Angola, Austria, Brazil, Burkina Faso, Canada, Croatia, Czechia, Denmark, Ecuador, Egypt, France, Germany, Greece, India, Israel, Italy, Japan, Moldova, Netherlands, Nigeria, Norway, Poland, Portugal, Serbia, South Africa, Spain, Tunisia, Turkey, USA, and Zambia), representing a truly wide "area network" of research activity. The DCAI'23 technical program has selected 50 full papers in the special sessions, and as in past editions, there will be special issues in ranked journals such as Electronics, Sensors, Systems, and Advances in Distributed Computing and Artificial Intelligence Journal. DCAI'23 special sessions have been a very useful tool to complement the regular program with new or emerging topics of particular interest to the participating community.

This year, DCAI 2023 has especially encouraged and welcomed contributions on: Advances in Sustainable Economics and Technology (ASET); AI-Driven Methods for Multimodal Networks and Processes Modeling 2023 (AIMPM); Artificial Intelligence for Enhanced Cybersecurity (AI4CS); Computational Linguistics, Information, Reasoning, and AI 2023 (CLIRAI); New Perspectives and Solutions in Cultural Heritage (TECTONIC); Power System Optimization and Machine Learning Models (PSO-ML); Smart Factories of the Future (SmartFoF); and the Next Generation of IoT Infrastructures, Cybersecurity, and Applications (IoTalentum).

DCAI 2023 is organized by the LASI and Centro Algoritmi of the University of Minho (Portugal). We would like to thank all the contributing authors, National Associations (AEPIA, APPIA), the sponsors (AIR Institute), the funding supporting of the project "ActiveEdgeMobile, Reference: PDC2022-133161-C31, financed by MCIN /AEI/10.13039/501100011033 and NextGeneration EU/PRTR, UE", and finally, the

Local Organization members and the Program Committee members for their hard work, which was essential for the success of DCAI'23.

June 2023

Rashid Mehmood
Victor Alves
Isabel Praça
Jarosław Wikarek
Javier Parra
Roussanka Loukanova
Ignacio de Miguel
Tiago Pinto
Ricardo Nunes
Michela Ricca

# DCAI 2023 Organization

## Honorary Chairman

| | |
|---|---|
| Sigeru Omatu | Hiroshima University, Japan |

## Program Committee Chairs

| | |
|---|---|
| Sascha Ossowski | King Juan Carlos University, Spain |
| Pawel Sitek | Kielce University of Technology, Poland |

## Special Sessions Chairs

| | |
|---|---|
| Rashid Mehmood | King Abdulaziz University, Saudi Arabia |
| Victor Alves | University of Minho, Portugal |

## Advisory Board

| | |
|---|---|
| Yuncheng Dong | Sichuan University, China |
| Francisco Herrera | University of Granada, Spain |
| Kenji Matsui | Osaka Institute of Technology, Japan |
| Tan Yigitcanlar | Queensland University of Technology, Australia |
| Tiancheng Li | Northwestern Polytechnical University, China |

## Organizing Committee Chairs

| | |
|---|---|
| Sara Rodríguez | University of Salamanca, Spain |
| Cesar Analide | University of Minho, Portugal |
| Goreti Marreiros | ISEP/GECAD, Portugal |

## Local Organizing Committee

| | |
|---|---|
| Paulo Novais (Chair) | University of Minho, Portugal |
| José Manuel Machado (Co-chair) | University of Minho, Portugal |

| | |
|---|---|
| Hugo Peixoto | University of Minho, Portugal |
| Regina Sousa | University of Minho, Portugal |
| Pedro José Oliveira | University of Minho, Portugal |
| Francisco Marcondes | University of Minho, Portugal |
| Manuel Rodrigues | University of Minho, Portugal |
| Filipe Gonçalves | University of Minho, Portugal |
| Dalila Durães | University of Minho, Portugal |
| Sérgio Gonçalves | University of Minho, Portugal |

## Organizing Committee

| | |
|---|---|
| Juan M. Corchado Rodríguez | University of Salamanca and AIR Institute, Spain |
| Fernando De la Prieta | University of Salamanca, Spain |
| Sara Rodríguez González | University of Salamanca, Spain |
| Javier Prieto Tejedor | University of Salamanca and AIR Institute, Spain |
| Ricardo S. Alonso Rincón | AIR Institute, Spain |
| Alfonso González Briones | University of Salamanca, Spain |
| Pablo Chamoso Santos | University of Salamanca, Spain |
| Javier Parra | University of Salamanca, Spain |
| Liliana Durón | University of Salamanca, Spain |
| Marta Plaza Hernández | University of Salamanca, Spain |
| Belén Pérez Lancho | University of Salamanca, Spain |
| Ana Belén Gil González | University of Salamanca, Spain |
| Ana De Luis Reboredo | University of Salamanca, Spain |
| Angélica González Arrieta | University of Salamanca, Spain |
| Angel Luis Sánchez Lázaro | University of Salamanca, Spain |
| Emilio S. Corchado Rodríguez | University of Salamanca, Spain |
| Raúl López | University of Salamanca, Spain |
| Beatriz Bellido | University of Salamanca, Spain |
| María Alonso | University of Salamanca, Spain |
| Yeray Mezquita Martín | AIR Institute, Spain |
| Sergio Márquez | AIR Institute, Spain |
| Andrea Gil | University of Salamanca, Spain |
| Albano Carrera González | AIR Institute, Spain |

## DCAI 2023 Sponsors

# Special Session on Artificial Intelligence for Enhanced Cybersecurity (AI4CS)

This special session aims to bring together high-quality research papers and reviews that focus on the application of Artificial Intelligence Techniques for Enhancing Cybersecurity, with a strong emphasis on ensuring robustness, security, and privacy. AI technologies possess unprecedented capabilities in detecting and mitigating cyberthreats; however, their effectiveness can be compromised by vulnerabilities and attacks. This session will showcase research and strategies aimed at bolstering the robustness of AI systems. Additionally, the proceedings delve into approaches that prioritize security and privacy, mitigating the risks associated with AI deployment. By doing so, this session aims to instill confidence in the reliability and resilience of AI-driven cyber-defense solutions. Valuable insights into cutting-edge methodologies, case studies, and best practices that promote a secure and privacy-aware AI ecosystem, empowering organizations to leverage AI while safeguarding sensitive information and maintaining data integrity will be covered during the session. It will encompass different perspectives on these and related topics, fostering comprehensive discussions and exploration, specifically:

- Adversarial Learning
- Authorization, Authentication, and Access Control
- Cloud Security
- Human Factors in Security and Privacy
- Identity Management
- Machine Learning for Intrusion Detection
- Privacy-Enhancing Technologies
- Robust AI
- Software Security
- Trustworthy AI
- Web Application Security and Privacy.

## Organizing Committee


| | |
|---|---|
| Isabel Praça | Instituto Superior de Engenharia do Porto, Portugal |
| Juan Manuel Corchado | University of Salamanca, Spain |
| Eva Maia | Instituto Superior de Engenharia do Porto, Portugal |
| Pablo Chamoso | University of Salamanca, Spain |
| Alfonso González-Briones | University of Salamanca, Spain |

# Special Session on AI-Driven Methods for Multimodal Networks and Processes Modeling (AIMPM)

The special session entitled AI-Driven Methods for Multimodal Networks and Processes Modeling (AIMPM 2023) is a forum that will share ideas, projects, researches results, models, experiences, applications, etc. associated with artificial intelligence solutions for different multimodal networks-born problems (arising in transportation, telecommunication, manufacturing, and other kinds of logistic systems).

Recently, a number of researchers involved in research on analysis and synthesis of multimodal networks devote their efforts to modeling different, real-life systems. The generic approaches based on the AI methods, highly developed in recent years, allow to integrate and synchronize different modes from different areas concerning: the transportation processes synchronization with concurrent manufacturing and cash ones or traffic flow congestion management in wireless mesh and ad hoc networks as well as an integration of different transportations networks (buses, rails, subway) with logistic processes of different character and nature (e.g., describing the overcrowded streams of people attending the mass sport and/or music performance events in the context of available holiday or daily traffic services routine). Due to the above-mentioned reasons, the aim of the workshop is to provide a platform for discussion about the new solutions (regarding models, methods, knowledge representations, etc.) that might be applied in that domain. There is a number of emerging issues with big potential for methods of artificial intelligence (evolutionary algorithms, artificial neural networks, constraint programming, constraint logic programming, data-driven programming, answer set programming, hybrid methods—AI/OR operation research, fuzzy sets) like:

- multimodal processes management,
- modeling and planning production flow,
- production planning and scheduling,
- stochastic models in planning and controlling,
- simulation of discrete manufacturing system,
- supply chain management,
- mesh-like data network control,
- multimodal social networks,
- intelligent transport and passenger and vehicle routing,
- security of multimodal systems,
- network knowledge modeling,
- intelligent web mining and applications,
- business multimodal processes and projects planning.

## Organizing Committee

### Chairs

| | |
|---|---|
| Paweł Sitek | Kielce University of Technology, Poland |
| Grzegorz Bocewicz | Koszalin University of Technology, Poland |
| Izabela E. Nielsen | Aalborg University, Denmark |

### Co-chairs

| | |
|---|---|
| Peter Nielsen | Aalborg University, Denmark |
| Zbigniew Banaszak | Koszalin University of Technology, Poland |
| Robert Wójcik | Wrocław University of Technology, Poland |
| Jarosław Wikarek | Kielce University of Technology, Poland |
| Arkadiusz Gola | Lublin University of Technology, Poland |
| Mukund Nilakantan Janardhanan | University of Leicester, UK |
| Tomasz Michno | Kielce University of Technology, Poland |
| Adam Krechowicz | Kielce University of Technology, Poland |

# Special Session on Advances in Sustainable Economics and Technology (ASET)

The special session, entitled Advances in Sustainable Economics and Technology (ASET 2023), is concerned with state-of-the-art ideas, projects, research results, models, experiences, applications, etc., in the field of sustainable economics. The forum focuses on examining the relationship between the latest advancements and the intersecting branches of knowledge, ranging from new technologies to new sociological and organizational developments.

The concern for sustainability transcends today's society, and it is, therefore, toady's effort that will enable sustainability among future generations. In these new times, sustainability research is closely tied to the progress of society itself. Advances in sustainability principally concern environmental, economic, and social aspects. Hence, its transcendent involvement in issues related to care for the environment or the moral order. Countless fields of technology evidence the importance of sustainability; in fact, the connection between sustainability and technological development is crucial for social development.

There are several emerging themes with great potential for economic and technological sustainability, related to the universal impact of sustainability, environmental variables, and the new technological era.

Therefore, the forum offers a space for debate on economic and technological sustainability in relation to different organizations and societies' capacity to manage the resources at their disposal so as to generate responsible long-term profitability.

## Organizing Committee

### Chairs

| | |
|---|---|
| Juan Manuel Corchado | University of Salamanca, Spain |
| Javier Parra | University of Salamanca, Spain |

### Co-chairs

| | |
|---|---|
| Tan Yigitcanlar | Queensland University of Technology, Australia |
| Paulo Leitão | Polytechnic Institute of Bragança, Portugal |

## Program Committee

| | |
|---|---|
| Sara Rodríguez González | University of Salamanca, Spain |
| Elsa Beatriz Suarez Kimura | University of Buenos Aires, Argentina |
| María Ángela Jiménez Montañés | University of Castilla La Mancha, Spain |
| Jorge Cruz Cárdenas | ESTec Research Center Universidad Tecnológica Indoamérica, Ecuador |
| María Ángeles Fernández López | University Camilo José Cela, Spain |
| Pablo Garrido Prada | University of Salamanca, Spain |
| Vicente Antonio Pérez Chamorro | University Pablo Olavide, Spain |
| Vanessa Jiménez Serranía | Universitat Oberta de Catalunya, Spain |
| Pedro José Martínez Córdoba | University of Salamanca, Spain |
| Laís Viera Trevisan | Hamburg University of Applied Sciences, HAW/Hamburg |
| Sebastián López Florez | Technological University of Pereira, Colombia |
| Lica Erhan | Academy of Economic Studies of Moldova, Moldova |
| Emiliano García Coso | University Sergio Arboleda, Colombia |
| Esteban Sánchez Hernández | University of Salamanca, Spain |
| Ricardo S. Alonso | AIR Institute, Spain |
| Yeray Mezquita Martín | AIR Institute, Spain |
| Jean-Paul Andrieu | Loyola Marymount University, USA |
| Joanna Ejdys | Bialystok University of Technology, Poland |
| Ana María López Echeverry | Technological University of Pereira, Colombia |
| Jovanny Bedoya Guapacha | Technological University of Pereira, Colombia |
| Luis Fernando Castillo Ossa | University of Caldas, Colombia |
| Jonas Queiroz | São Paulo State University, Brazil |
| Ramón J. Durán Barroso | Universidad de Valladolid, Spain |
| Ignacio de Miguel | Universidad de Valladolid, Spain |
| Raquel Guzmán Ordaz | University of Salamanca, Spain |
| Marcelo Lampkowski | ICLEI – Local Governments for Sustainability, Germany |
| Nuria G. Rabanal | University of Leon, Spain |
| Nikola Balic | University of Split, Croatia |
| Ivana Vuka | University of Split, Croatia |

## Organizing Program Committee

| | |
|---|---|
| Andrea Gil | University of Salamanca, Spain |
| Natali Chourio | AIR Institute, Spain |
| Diego Sebastián Escobar | University of Buenos Aires, Argentina |

# Special Session on Computational Linguistics, Information, Reasoning, and AI (CLIRAI)

Computational and technological developments that incorporate natural language and reasoning methods are proliferating. Adequate coverage encounters difficult problems related to the phenomena of partiality, under specification, perspectives of agents, and context dependency. These phenomena are signature features of information in nature, natural languages, and reasoning.

The session covers theoretical work, applications, approaches, and techniques for computational models of information, language (artificial, human, or natural in other ways), and reasoning. The goal is to promote computational systems and related models of thought, mental states, reasoning, and other cognitive processes.

## Organizing Committee

| | |
|---|---|
| Roussanka Loukanova | Institute of Mathematics and Informatics, Bulgarian Academy of Sciences, Sofia, Bulgaria |
| Sara Rodríguez | University of Salamanca, Salamanca, Spain |

## Program Committee

| | |
|---|---|
| Benedikt Ahrens | Delft University of Technology, The Netherlands; School of Computer Science, University of Birmingham, UK |
| Krasimir Angelov | University of Gothenburg and Chalmers University of Technology, Gothenburg, Sweden |
| Wojciech Buszkowski | Faculty of Mathematics and Computer Science, Adam Mickiewicz University, Poznan, Poland |
| Marie Duzi | VSB-Technical University of Ostrava, Ostrava, Czech Republic |
| Antonín Dvořák | University of Ostrava, Ostrava, Czech Republic |
| Annie Foret | IRISA and Univ Rennes 1, France |
| Håkon Robbestad Gylterud | University of Begen, Bergen, Norway |
| Lars Hellan | Norwegian University of Science and Technology, Trondheim, Norway |
| Ali Hürriyetoğlu | Koç University, Istanbul, Turkey |

| | |
|---|---|
| M. Dolores Jiménez López | GRLMC-Research Group on Mathematical Linguistics, Universitat Rovira i Virgili, Tarragona, Spain |
| Manfred Kerber | School of Computer Science, University of Birmingham, UK |
| Peter Koepke | Mathematisches Institut, Rheinische Friedrich-Wilhelms-Universität Bonn, Bonn, Germany |
| Stepan Kuznetsov | Steklov Mathematical Institute, Russian Academy of Sciences, Moscow, Russia |
| Kristina Liefke | Ruhr-University Bochum, Bochum, Germany |
| Inari Listenmaa | Centre for Computational Law Singapore Management University, Singapore |
| Zhaohui Luo | Royal Holloway, University of London, London, UK |
| Richard Moot | Université de Montpellier and LIRMM-CNRS, Montpellier, France |
| Petra Murinová | Institute for Research and Applications of Fuzzy Modeling, University of Ostrava, Ostrava, Czech Republic |
| Rainer Osswald | Heinrich-Heine-Universität Düsseldorf, Düsseldorf, Germany |
| Christian Retoré | Université de Montpellier and LIRMM-CNRS, Montpellier, France |
| Frank Richter | Goethe University Frankfurt am Main, Frankfurt, Germany |
| Ana Paula Rocha | University of Porto, LIACC/FEUP, Porto, Portugal |
| Sylvain Salvati | Université de Lille, INRIA, CRIStAL UMR 9189, Lille, France |
| Milena Slavcheva | Bulgarian Academy of Sciences, Sofia, Bulgaria |
| Alexander Steen | Universität Greifswald, Institut für Mathematik und Informatik, Greifswald, Germany |
| Alexey Stukachev | Sobolev Institute of Mathematics, Novosibirsk State University, Novosibirsk, Russia |
| Satoshi Tojo | School of Information Science, Japan Advanced Institute of Science and Technology (JAIST), Japan |
| Jørgen Villadsen | Technical University of Denmark, Copenhagen, Denmark |
| Marek Zawadowski | Institute of Mathematics, University of Warsaw, Warsaw, Poland |

# Special Session on the Next Generation of IoT Infrastructures, Cybersecurity, and Applications (IoTalentum)

This special session aims to present work carried out within the IoTalentum project, which addresses how to develop the next generation of the Internet of Things (IoT), producing coordinated advantages in the area of infrastructures and cybersecurity, and how to use it for the development of IoT applications and, in particular, those of smart grids. The main objectives of this special session are:

To maximize the value of research outcomes by promoting knowledge exchange, interactions, partnerships, and inclusive engagement between IoT researchers, individuals, and organizations outside the immediate research community.

To encourage the implementation and transmission of research outcomes and to communicate them and the knowledge acquired among researchers and stakeholders.

IoTalentum is funded by the European Union Horizon 2020 research and innovation program within the framework of Marie Skłodowska-Curie Actions ITN-ETN with grant number 953442.

## Organizing Committee



## Program Committee

| | |
|---|---|
| Gianluca Rizzi | Wind Tre S.p.A, Italy |
| Óscar González de Dios | Telefonica I+D, Spain |
| Daniel E. Lucani Rötter | Aarhus Universitet, Denmark |
| Zita Vale | ISEP, Portugal |
| Jessica Chaves | PH Energía, Portugal |
| Manuel Acevedo | PH Energía, Portugal |
| Marcell Feher | Chocolate Cloud ApS, Denmark |

# Special Session on Power System Optimization and Machine Learning Models (PSO-ML)

Technological developments and environmental concerns have led to an unprecedented evolution of energy systems. In the new paradigm, advanced energy resource management is needed to enable the intensive integration of distributed energy resources, taking advantage of its full potential. Furthermore, modern power systems require a satisfactory level of flexibility over different time horizons to maintain operation at a reasonable cost and satisfy restrictions imposed by the distribution system operator. Despite the improvement in forecasting techniques, it is a fact that power systems were not designed to effectively deal with the increased levels of uncertainty associated with distributed energy resources.

Due to memory requirements and limited runtimes, current works are often restricted to small cases in terms of problem size and are rarely able to consider all sources of uncertainty together, thus lacking realism. Computational intelligence approaches are proving to be very effective in dealing with large problems with nonlinearities, and new metaheuristics have been continually proposed in the literature to mitigate these problems. Moreover, recent advances in the field of machine learning are enabling reaching more accurate predictions of generation and demand, consequently reducing the uncertainty, and contributing to a more efficient power system management and operation.

PSO-ML aims at providing an advanced discussion forum on recent and innovative work in the fields of machine learning and optimization as solutions to address power system-related problems. Special relevance is indorsed to solutions involving the application of artificial intelligence approaches, including metaheuristic optimization, automated machine learning, forecasting methodologies, and agent-based systems.

## Organizing Committee

| | |
|---|---|
| João Soares | Polytechnic of Porto, Portugal |
| John Fredy Franco | Universidade Estadual Paulista, Brazil |
| Leonardo H. Macedo | Universidade Estadual Paulista, Brazil |
| Ruben Romero | Universidade Estadual Paulista, Brazil |
| Tiago Pinto | Universidade de Trás-os-Montes e Alto Douro, Portugal |
| Zita Vale | Polytechnic of Porto, Portugal |

## Program Committee

| | |
|---|---|
| Brígida Teixeira | GECAD, Instituto Superior de Engenharia do Porto, Portugal |
| Carlos Ernani Veiga | Instituto Federal de Santa Catarina, Brazil |
| Gabriel Puerta | Universidade Estadual de São Paulo, Brazil |
| Tiago Soares | INESC TEC, Portugal |
| Juan Home-Ortiz | São Paulo State University, Brazil |
| Ricardo Faia | School of Engineering, Polytechnic of Porto, Portugal |
| José Baptista | Universidade de Trás-os-Montes e Alto Douro, Portugal |

# Special Session on Smart Factories of the Future (SmartFoF)

A digital revolution in industry is underway, characterized by a rapid technological evolution related to production processes. Industries are, consequently, being driven to transform themselves into Factories of the Future, i.e., by being highly flexible to any type of process and product, quickly adaptable to external changes (such as alterations to structural technical specifications of the product), reaching a high degree of sustainability with regard to environmental impact and resource consumption, and, at the same time, bring competitive from the economic point of view. These new Factories of the Future must be intelligent and equipped with productive systems with global connectivity, allowing human intervention in repetitive tasks to be replaced by intelligent equipment and the operator's role being fundamental for optimizing the information management of the implemented technological tools, thus generating a highly efficient and effective production process. SmartFoF aims at providing an advanced discussion forum, bringing together the perspective from both Academia and the Industry, on recent and innovative work addressing the transition toward Factories of the Future. Special relevance is indorsed to solutions related to digitization and computing (exploring Industrial Internet of Things (IIoT) technologies, cloud computing, digital twin, data analytics, protocol layers, and industrial management), cybersecurity, and symbiotic Man-Machine and Man-Factory relationship.

## Organizing Committee

| | |
|---|---|
| Anabela Ribeiro | Continental Advanced Antenna, Portugal |
| Armindo Fernandes | Continental Advanced Antenna, Portugal |
| Arsénio Reis | Universidade de Trás-os-Montes e Alto Douro, Portugal |
| João Barroso | Universidade de Trás-os-Montes e Alto Douro, Portugal |
| Paulo Nogueira | Continental Advanced Antenna, Portugal |
| Ricardo Nunes | Universidade de Trás-os-Montes e Alto Douro, Portugal |
| Tiago Pinto | Universidade de Trás-os-Montes e Alto Douro, Portugal |

## Program Committee

# Special Session on New Perspectives and Solutions in Cultural Heritage (TECTONIC)

The special session entitled "New Perspectives and Solutions in Cultural Heritage" is a forum that will share ideas, projects, researches results, models, experiences, applications, etc. that focus on Preventive Conservation of Marine and Aerial Cultural Heritage. New technological proposals, many of them based on Artificial Intelligence, are currently being applied as measures and processes for the conservation of Cultural Heritage. This special session has its main objectives:

To maximize the value of research outcomes by promoting knowledge exchange, interactions, partnerships, and inclusive engagement between Cultural Heritage researchers, individuals, and organizations outside the immediate research community.

To encourage the implementation and transmission of research outcomes and to communicate them and the knowledge acquired among researchers and stakeholder sectors.

Nowadays Preventive Conservation of Cultural Heritage (CH) and, in particular, Underwater Cultural Heritage (UCH) represents a crucial issue to safeguard tangible testimonies of past human life preserved on the seabed. Solutions to preventive conservation in the UCH sector are still costly and sometimes very difficult to develop and implement. This is because the research activities in the marine environments imply greater difficulties than the aerial environment in terms of methods as well as in the use of advanced technologies. Underwater sites are generally difficult to access and more dangerous, compared with working on dry land. Also, working in the marine environment and using specific equipment require a lot of experience, highly skilled operators, and the coordination of different "technical actors", thus keeping such solutions out of the reach of "common operators".

In this context, the whole management of an underwater site, starting from the evaluation of its state of conservation until the restoration works, is often carried out by "external experts", implying enormous costs, not always manageable and applicable. In this framework, the main objective of the session is to share research solutions tailored to the technical needs of all the actors and operators working in the Preventive Conservation of Cultural Heritage, and in particular, but not limited to this, in the UCH field (i.e., underwater site manager, managing institution, managing authority, etc.).

The session tries to join experts and researchers that recently focus their work on intelligent approaches for a more sustainable management of aerial and UCHs, facilitating monitoring operations, encouraging the conservation in situ, and providing accurate guidelines for a correct maintenance. We would like to thank all the contributing authors and the funding supporting of the TECTONIC project funded by the European Union Horizon 2020 research and innovation program within the framework of Marie Skłodowska-Curie Actions ITN-ETN with grant number 873132.

## Organizing Committee

# Contents

## Session on New Perspectives and Solutions in Cultural Heritage (TECTONIC)



## Special Session on Power System Optimization and Machine Learning Models (PSO-ML)

**Special Session on the Next Generation of IoT Infrastructures, Cybersecurity, and Applications (IoTalentum)**

# Special Session on AI–driven methods for Multimodal Networks and Processes Modeling (AIMPM)

# Use of KPI Indicators to Optimize Production Process Performance – Case Study

Mariusz Piechowski[1](✉) and Artur Meller[2]

[1] Poznan School of Banking, Powstańców Wielkopolskich 5, 61-895 Poznan, Poland
mariusz.piechowski@wsb.poznan.pl
[2] Poznan University of Technology, Pl. Marii Skłodowskiej-Curie 5, 60-965 Poznań, Poland

**Abstract.** The progressive organizational transformation of manufacturing companies combined with new technologies supported by automation and digitization of production processes is caused, among other things, by the growth of diverse customer requirements. Increasing customer requirements oblige manufacturers to improve production process management methods. The basic direction of improvement is the optimization of production management processes in terms of productivity and product quality, while one of the most important tools for monitoring the course of the process is the analysis of KPIs (Key Performance Indicators). They are used to monitor strategically, operationally, and tactically relevant parameters of the production process, and thus to evaluate the company's management system. The purpose of the article is to present the possibility of using indicators to control and optimize the production process. The indirect goal of these measures is to achieve a more efficient manufacturing process by developing a decision-making model based on key performance indicators, which will enable the optimization of the decision-making process and the effective exchange of information at various decision-making levels. The first part of the article describes the most relevant features of KPIs. Also presented are the assumptions for such a decision-making model, taking into account the various levels and areas of management, and the course of action for selecting, implementing, and improving a system based on this type of indicators. The analysis conducted using the developed diagram is described, which made it possible to select the key performance indicators of the enterprise. Guidelines and design assumptions for an information system to support decision-making processes based on KPIs were created, enabling: automation of data processing procedures, optimization of the function of overseeing objectives for individual decision-making levels, and improvement of the reporting system and calculation algorithms.

**Keywords:** KPIs (Key Performance Indicators) · Production Management Processes · Support Decision-making Processes

## 1 Introduction

The unfolding global economic crisis associated with the pandemic and the war in Ukraine has caused radical changes in the long-term policies of companies associated with ensuring competitiveness in the market. As a result, manufacturing companies characterized by complex business processes are subject to even more stringent constraints


R. Mehmood et al. (Eds.): DCAI 2023, LNNS 741, pp. 3–12, 2023.
https://doi.org/10.1007/978-3-031-38318-2_1

than ever before [2]. Many businesses affected by the crisis are changing their strategic goals to ensure shorter delivery times and increased product variability, as well as expanding into new markets. New economic and political conditions and changing customer requirements are forcing the improvement of production process management methods and the adjustment of manufacturing systems to ensure greater reliability, flexibility, and productivity. In order to meet rapidly changing and diverse customer requirements, manufacturing companies must achieve and maintain high efficiency in the production process coupled with high quality products [1, 7]. Meeting these expectations involves reducing response time to changing market conditions through increased flexibility and efficiency of the management system. Such challenges were faced by Fabryka Armatur "Swarzedz" sp. z o.o. (FAS), which belongs to the small- and medium-sized enterprises sector operating in the metal industry, particularly associated with hot forging and machining of various grades of brass and steel. The company offers products for the demanding automotive market, including transmission synchronizer rings used in many leading automotive makes.

## 2 KPIs

Modern manufacturing systems integrate advanced mechanical, electronic and IT solutions. The complexity of these systems makes it necessary to consolidate knowledge from all engineering disciplines in order to fully understand the control process at all decision-making levels and enable proper evaluation of its status [11]. A key aspect from the point of view of a company's strategic goals is the proper identification of relevant areas in the production management process. For the areas identified in this way, tools should be developed to enable effective evaluation of the objectives being pursued.

Important for decision-making is to obtain useful information about the state of the process on the basis of which we can effectively eliminate or reduce the causes of disruptions. Such information can be obtained by monitoring the state of the process and its parameters such as quality, productivity, efficiency, time-consumption [26].

One of the dedicated tools to support decision-making processes are KPIs. These indicators make it possible to monitor and evaluate the state of the process in many manufacturing and support areas, as well as at many decision-making levels [23]. The interpenetration of enterprise areas and decision-making levels makes it possible to determine relationships between individual indicator characteristics and aggregate their values. Relationships between the characteristics of different indicators allow the design of advanced inference methods covering different areas of influence on the production process [10, 29, 23, 25]. A KPI-based inference model should include information that is crucial to the company's decision-making processes. The information provided to managers, engineers and operators must be conveyed in a clear, transparent, and understandable manner [23]. Indicators wherever possible and reasonable should present both absolute and relative values. This is because absolute values alone do not indicate the scale of evaluation of the state of the process, while percentages cannot be aggregated. Indicators at the level of strategic goals should be result indicators, that is, they should reflect the intended outcomes [1]. On the other hand, at the operational level, in addition to result indicators, product indicators should also be defined, which indicate the activities leading to the achievement of the goals [1]. For goals defined in this way, it is necessary to identify the right indicators to support the decision-making process [2].

## 3 Performance Indicators Attributes

Performance indicators are widely used in enterprises, but usually in a limited way, i.e. either they apply to selected aspects of management, e.g. maintenance, or only selected indicators, e.g. OEE, are used. Of the dozens of KPIs, one can choose one that enables the design of an enterprise-wide decision support system [9, 10, 25]. With such a system, it is possible to verify and evaluate assumptions, goals, actions.

One such attribute is the extent of the indicator's impact, i.e. the relevance of the indicator. Each management model has its own key (strategic) objectives that affect its effectiveness, so it is necessary to carefully analyze the current state of your company, its constraints and opportunities, so that in this way it is possible to identify a clearly defined main objective towards which the whole enterprise will aim at all levels of decision-making. Transferring the company's mission and strategy into key performance indicators, provides an opportunity to create a framework for a system for strategy measurement and management. Therefore, the indicators must be correlated with these goals, so as to direct the decisions made to the strategic goals of the company, for example: for the TPM (Total Productive Maintenance) model, the key indicator is OEE (Overall Equipment Effectiveness), which reflects the effectiveness of both individual machines and entire processes [27].

The second attribute is the precision and accuracy of the representation of the state of the process, i.e. the authoritative nature of the indicator [1, 2, 10]. Indicators should make it possible to assess the state of the process at different decision-making levels:

1. Strategic, enabling decisions that are crucial for the entire company, e.g. PMP (Percent Maintenance Planned) indicator – reflecting the effectiveness of preventive actions,
2. Tactical, where decisions are important for departments or production lines, e.g. FPY (First Past Yeld) indicator–reflecting the efficiency of the production process,
3. Operational, where decisions are made by the operator at a given position, e.g.: the effective use of working time (ECP) indicator – reflecting the normative working time at a production position.

Another attribute is the completeness of the indicator's components. Often the indicator is representative of many decision-making levels, e.g. the OEE indicator, the value of which can reflect the state of a machine, a group of machines, as well as the entire production process. The correct choice of the component parameters of the indicator allows the correct representation of the actual state of the process, and the omission of some of the parameters may result in an erroneous assessment of its state [2, 10]. Deciding, for example, to introduce corrective actions on the basis of the value of the indicator obtained, despite the fact that the actual value of the indicator does not indicate the need to take these actions, or, conversely, not taking actions despite the fact that the actual value of the indicator indicates the need to introduce them.

Another characteristic is the effect on the moment of decision-making i.e. the timeliness of the indicator value. This characteristic is also important in the decision-making process, i.e. evaluation of the state of the process on the basis of historical indicator values (from previous weeks, days or even hours) becomes useless. Making decisions based on a single indicator overrun can also introduce unnecessary distortion into the decision-making process [1, 25].

Digitization of production processes is important when it comes to improving management methods. The digitization process affects the way the manufacturing process is monitored and controlled by optimizing its resources through the use of efficient methods of collecting and processing information. Digitization also makes it possible to build intelligent production management systems and create autonomous production planning and control systems. Therefore, monitoring the value of the indicator in real time requires its digitization – a multidimensional combination of the real and virtual worlds, i.e. the physical parameters of the indicator (e.g.: abundance, temperature, time, availability) and the information system (devices, meters, algorithms, software). Project assumptions Proper selection of indicators requires adequate preparation. As a first step, it is necessary to conduct a thorough analysis of the current state of the process in relation to the defined objectives in line with the management strategy. To carry out such an analysis, a team should be formed with members representing all areas of the company. Once the analysis is done, it should be narrowed down to the most relevant aspects: areas, characteristics, and features, e.g. by using a Pareto-Lorenz diagram. Too much data can negatively affect the selection of appropriate indicators. Recording analysis data in the form of tables describing individual areas, characteristics and features allows subsequent verification of compliance with the characteristics of the selected indicators. One method that illustrates the entire structure of the indicator management model is to visualize individual aspects in the form of a hierarchical relationship diagram. The analysis carried out will enable the effective selection of indicators. Translating a company's mission and strategy into a set of data and values, using key performance indicators, provides an affordable and transparent opportunity to create and deliver a framework for a strategy measurement and management system [2, 20]. In an era where success in running a business is often determined by access to information, the ability to obtain it, process it, interpret it, and draw appropriate conclusions based on it resulting in sound decision-making by managers, methods of selecting and implementing performance indicators come to the rescue. They describe the universal mechanisms and steps that should be taken by an organization wishing to successfully control, monitor, and thus respond appropriately to the course of its processes, which is intended to lead it to ensure the effective implementation of its long-term operations. The choice of method should also be tailored to the company's strategy and objectives, taking into account its current problems.

## 4 System Model

The developed structure of the indicator management model reflects the hierarchical structure of management levels [10, 25]. We can describe each node of the structure in an object-oriented way, assigning to it parameters and attributes relevant to each management level, as well as features associated with enterprise areas. Attributes of objects make it possible to create points aggregating information at each management level and area. This structure allows to link indicators to object attributes and verify the impact of indicators on defined management objectives [24]. An additional advantage of such a solution is the detection of correlations between individual indicator parameters.

## 5 System Implementation Concept

Implementing an indicator-based system is a long-term process, so it requires a plan. The plan should be divided into stages with defined goals and objectives [23]. Each stage should have a timeframe and milestones. Each indicator should be assigned an engineer (manager, operator) responsible for its verification, optimization, periodic reviews, checking of historical data and correction of reporting algorithms.

Metrics should be developed for such selected indicators, including a description of:

1. The extent and level of impact of the indicator on the achievement of defined strategic, tactical, and operational goals,
2. The accuracy and precision of the representation of the state of the process at the various decision-making levels and impact areas,
3. Description of the individual components of the indicator affecting its effectiveness by defining the scale of assessment and the baseline and target values,
4. Precise description of the type of indicator: its name, the unit of measurement, and the sources of data at each point of monitoring the indicator.

Such a metric is the basis for the development of a position manual, especially for operators making decisions at the operational level. If an analysis of the current state of the process has been carried out and, on the basis of this analysis, appropriate indicators have been selected, then what remains to be done is building an information system that allows the management and improvement of the entire indicator structure covering selected decision-making positions. Indicator management without an information system, as experience shows, is 90% doomed to failure. The information system should include algorithms for calculating indicators (including methods for aggregating and propagating values to higher decision-making levels), a notification system (as well as warning of critical values), clear and precise visualization of the description indicator and indicator, and be available at any decision-maker's position. It should also include procedures to verify that the indicator is in line with the project assumptions, and provide validation of the indicator structure with the results of the current state analysis.

Based on the methodology thus proposed, a decisional support system model based on key performance indicators was designed. The studies were conducted at Fabryka Armatur "Swarzedz" sp. z o.o (FAS) in Rabowice, Poland. As a leader in brass consumption in Poland, the company has modern technologies in hot forming, heat treatment and machining. The extensive machine park consists of, among others, hydraulic presses for drop-free die forging, or specialized transfer machines for the cavity processing of tee-type products. The mission of FAS is to create a modern industry based on the policy of continuous improvement of product quality and precision and the policy of continuous improvement of the company's efficiency, taking into account sustainable development in the areas of personnel competence, business ethics, social responsibility, ecology, organization as well as technology and digitization. This is part of a management system aimed at striving for excellence in the execution of manufacturing processes, while aiming to completely eliminate waste, workplace accidents, defective products and breakdowns. This requires the involvement of all employees, at every level of decision-making, including top management, to move from simply reacting to occurrences to continuous improvement [17, 18]. The vision of digital development, but above all the adoption of an appropriate strategy for running the company became the starting point for the selection of key performance indicators. For this reason, it was important to set and verify the strategic direction of the analyzed industrial plant aimed at identifying, delineating and defining its strategic areas. An in-depth SWOT analysis (S – Strengths, W – Weaknesses, O – Opportunities, T – Threats) of the enterprise was used for this, where individual process managers in the company, most of whom are managers of individual organizational units, identified strengths and weaknesses, as well as opportunities and threats to the organization. The surveyed group was then asked to rate each factor in each group individually on a scale of 1–4, where 1 means an unimportant factor and 4 is the highest importance, which made it possible to draw an average for each factor classifying the company's capabilities. Thus, the following results were obtained: S – 3.30; W – 2.88; O – 3.16; T – 3.06. Their interpretation indicates a preponderance of strengths and opportunities, which in turn predetermines the definition and adoption of the company's strategy as aggressive (maxi-maxi). It involves maximizing the synergies associated with the organization's strengths, as well as opportunities from its environment. With the company's direction established, the next step was to identify the core areas for the production department, which was undertaken by an implementation team formed of managers and specialists representing each of the company's departments, and where the following were identified during the work carried out: organization, productivity, maintenance and quality. For the areas selected in this way, the main context was described, goals were set, and key actions or steps in achieving the goal were defined. Quantitative and qualitative requirements were created, defining the target state and actions – what specifically has to be done to achieve the goal. Indicators of the goal's performance were proposed, and a RASCI (Responsible, Accountable, Supported, Consulted, Informed) matrix was created – defining responsibilities, the role to be played by each person in achieving the goal. Verification of whether the goal was set correctly was done using the SMARTER method (Specific, Measurable, Attainable, Relevant, Time-bound, Exciting, Recorded).

In the next step, the work of the implementation team was based on the use of elements of the Hoshin Kanri concept [21, 22], a strategic management method relying on guiding the company in a well-defined and agreed direction, the so-called True North, which was previously defined after the SWOT analysis had been performed. One of its elements is an organizational improvement template called X-matrix. It allows the designated strategic areas, goals, key activities (projects) and indicators to be brought together in one place. Based on the definition of the strength of the influence of the various components of the matrix on each other (directly, indirectly, no influence), it was possible to determine the correlations between them. The resulting picture allowed for precise verification and delineation of irrelevant correlations, thus omitting them from further work, and, above all, identifying the critically important ones, resulting in the need to subject them to an in-depth analysis, aiming to determine potential KPIs. Moreover, the network of linkages built perfectly shows which of the initially proposed KPIs, affect more than one core area of production, making them all the more predestined for further evaluation. The determination of appropriate performance measures is an internal task of the organization and cannot be carried out by external consultants on the basis of observations made [19]. The result of using such an approach, was the identification of a set of indicators, closely integrated with the studied organization, responsive to its needs and in line with the digital transformation that the company is undergoing, contained in the four core areas of production. In such a set, ten indicators will be monitored (seven main and three additional ones that make up OEE) that fit into the multi-area support of the organization of the manufacturing process:

- MRT – Mean Reaction Time – maintenance indicator;
- MTTR – Mean Time To Repair – maintenance indicator;
- FPY – First Pass Yield – quality indicator;
- SC – Scrap Ratio – quality indicator;
- MPST – Mean planning and scheduling time – organizational indicator;
- OEE – Overall equipment effectiveness (OEE = Availability * Effectiveness * Quality ratio) – efficiency indicator.

The indicators set are a response to the current challenges faced by the analyzed company, and their implementation and monitoring are intended to assess the state of the production process at the defined decision-making levels: strategic, tactical, and operational. The proposed comprehensive approach covering important parts of business operations is designed to bring modifications and improvements to the management processes of the production organization. Any organization, in order to function efficiently, should base its operations on reliably collected and compiled data. Hence, the correct determination of indicators alone, although an absolutely key task, will not bring the desired results, without a reliable definition of the standard of indicator description. Such a metric should contain a range of information relating both to the precisely defined KPI itself, but also to conditions associated with the environment in which it is to operate. For this reason, a comprehensive sheet was created, divided into three main parts, and consisting of: basic characteristics (name of the indicator, what it relates to, definition, area of influence, owner, calculation formula), specific characteristics (method of measurement, source of origin of measured value, unit of measurement, range of values, trend, frequency of measurement) and reporting (frequency of reporting, place of reporting, alert levels, corrective actions, method of presentation). This allows any interested person to quickly get acquainted with the system of indicators operating in the organization and is also an excellent starting point for future modifications, resulting from the changing environment or simply a desire to improve its performance. The above study was the basis for defining the assumptions for the designed information system to support the decision-making process monitoring the set objectives in the indicated areas of the company's operations and at individual decision-making levels:

1. Re-modelling of production task registration forms: recording of additional information (parameters for indicators), automation of data entry (RFID, QR codes), verification of data entered by the operator, additional data access terminals (POS computers, tablets and cell phones), monitors for data visualization,

2. Expansion of the database with additional objects (tables and views): recording of KPI parameters, automatic generation of reports and statements, generation of notifications and alerts,
3. Additional modules for data collection and time recording devices: counting systems for products at workstations, time recording readers for machines (Golem, Historian) and operators (RFID cards),
4. Organizational change connected with:
   a. maintenance: handling breakdowns and maintenance,
   b. registration of tasks performed by machine operators,
   c. logistic processes: logistic in production, maintenance logistics,
   d. production planning: scheduling of production orders,
   e. quality control: first piece inspection, registration of deficiencies.

## 6 Conclusions

A modern manufacturing company wishing to compete in a dynamically changing market is obliged not only to meet the challenges of its external environment, e.g. socio-cultural (work life balance), economic-legal (frequent changes in legislation), geopolitical (unpredictable events), technological (digitalization), but even more so, it must efficiently adapt to the current situation, appropriately formulating its internal policy related to its own organization and management methods. The rapid development of information technology and the accompanying widespread digitization, leads entrepreneurs to seek their own path to control production processes, based on modern methods of doing business using rapid access to large sets of data, often obtainable in real time.

The appointment of a multifunctional implementation team also had a not insignificant impact on the shape of the system, as it often turned out that, going very deeply into the details of individual issues (objectives, key activities), despite many years of work in the company, those participating in the work did not realize what the current work in the production department looked like. Explaining these issues and describing production from the position of managers managing the process, allowed other employees to understand the mechanisms used. Implementing a system based on the developed concept was an excellent lesson in getting to know better the organization in which one works. Explaining and understanding the mechanisms used in the company's production organization management processes was an invaluable lesson for many people and opened their horizons, allowing them to interpret with greater understanding what happens every day in the manufacturing area. The creation of the right atmosphere and culture in the company's approach to the performance of daily duties has resulted in the intensification of joint efforts to seek and implement improved solutions for the existing processes [12, 13].

Further work is in progress, as the system is not yet completed. The next step is to design and build an information system in which the solutions developed earlier will be implemented and reflected. The assumption is that such action should not only serve to indiscriminately implement the transferred solutions, but also be a stimulus for further improvement, and be oriented towards achieving long-term success, which is based primarily on satisfaction, meeting the needs and expectations of the customer (external and internal), as well as benefits for the organization itself, its members and society as a whole [14–16].

**Acknowledgement.** The authors thank the Ministry of Education and Science for their financial support (the Implementation Doctorate program, Agreement No. DWD/4/22/2020 of 06.11.2020).

## References

1. Bocewicz, G., Szwarc, E., Wikarek, J., Nielsen, P., Banaszak, Z.: A competency-driven staff assignment approach to improving employee scheduling robustness. Eksploatacja i Niezawodnosc – Maint. Reliab. **23**(1), 117–131 (2021). https://doi.org/10.17531/ein.2021.1.13
2. Czerwińska, K., Pacana, A.: Kluczowe wskaźniki efektywności (KPI), Politechnika Rzeszowska (2020)
3. Saari, E.: KPI framework for maintenance management through Maintenance Development, implementation, assessment, and optimization. Luleå University of Technology (2019)
4. Legutko, S.: Trendy rozwoju utrzymania ruchu urządzeń i maszyn, Eksploatacja i niezawodność nr 2/2009 (2009)
5. Golonka, P.: Kluczowe wskaźniki efektywności – KPI. www.leanactionplan.pl
6. Hyla, A.: Optymalizacja produkcji w 5 krokach z wykorzystaniem KPI (2020). www.przemyslowy.pl
7. Slimani, H.: The Big Book of Key Performance Indicators. ACADEMIA Accelerating the World's Research (2006)
8. Kang, N., Zhao, C., Li, J., Horst, J.A.: A Hierarchical structure of key performance indicators for operation management and continuous improvement in production systems. Department of Industrial Engineering, Tsinghua University, Beijing, China; Department of Industrial and Systems Engineering, University of Wisconsin, Madison, WI, USA; Engineering Laboratory, National Institute of Standards and Technology, Gaithersburg, USA (2015)
9. Michlowicz, E.: Assessment of the modernized production system through selected TPM method indicators. Eksploatacja i Niezawodność – Maint. Reliab. **24**(4), 677–686 (2022). https://doi.org/10.17531/ein.2022.4.8
10. Król, D., Skowroński, J., Zaręba, M., Bartecki, K.: Wyznaczanie kluczowych wskaźników wydajności procesu produkcyjnego – część II: badania przemysłowe. Pomiary Automatyka Robotyka **24**(3/2020), 19–28 (2019). https://doi.org/10.14313/PAR_237/19. ISSN 1427-9126
11. Wohlers, B., Dziwok, S., Pasic, F., Lipsmeier, A., Becker, M.: Monitoring and control of production processes based on key performance indicators for mechatronic systems. Int. J. Prod. Econ. www.elsevier.com/locate/ijpe
12. Stadnicka, D. (ed.): Lean Manufacturing. Kompendium wiedzy. Oficyna Wydawnicza Politechniki Rzeszowskiej. Rzeszów (2021)
13. Vinodh, S.: Lean Manufacturing: Fundamentals, Tools, Approaches, and Industry 4.0 Integration. CRC Press. Taylor & Francis Group (2022)
14. Stapenhurst, T.: The Benchmarking Book: A How-to-Guide to Best Practice for Managers and Practitioners. Butterworth-Heinemann. elsevierdirect.com (2009)
15. Andersen, B.: Industrial benchmarking for competitive advantage. Hum. Syst. Manag. **18**(3–4), 287–296 (1999)
16. Hellsten, U., Klefsjö, B.: TQM as a management system consisting of values, techniques and tools. TQM Mag. **12**(4), 238–244 (2000)

17. Hendry, L.C.: Applying world class manufacturing to make-to-order companies: problems and solutions. Int. J. Oper. Prod. Manag. **18**(11), 1086–1100 (1998)
18. Harrison, A.: Manufacturing strategy and the concept of world class manufacturing. Int. J. Oper. Prod. Manag. **18**(4), 397–408 (1998)
19. Parmenter, D.: Key Performance Indicators – Developing, Implementing, and Using Winning KPIs, 4th edn. Wiley (2019)
20. Kaplan Robert, S., Norton David, P.: Strategiczna karta wyników. Jak przełożyć strategię na działanie. Wydawnictwo Naukowe PWN, Warszawa (2022)
21. Melander, A., Andersson, D., Elgh, F., Fjellstedt, F., Löfving, M.: Implementing Hoshin Kanri. How to Manage Strategy Through Policy Deployment and Continuous Improvement. Routledge, Taylor & Francis Group, Productivity Press Book (2021)
22. Thürer, M., Maschek, T., Fredendall, L., Gianiodis, P., Stevenson, M., Deuse, J.: On the integration of manufacturing strategy: deconstructing Hoshin Kanri. Manag. Res. Rev. **42**(3), 412–426 (2019)
23. Ishaq Bhatti, M., Awan, H.M., Razaq, Z.: The key performance indicators (KPIs) and their impact on overall organizational performance. Qual. Quant. **48**(6), 3127–3143 (2013). https://doi.org/10.1007/s11135-013-9945-y
24. Ferreira, S., Silva, F.J.G., Casais, R.B., Pereira, M.T., Ferreira, L.P.: KPI development and obsolescence management in industrial maintenance. Proc. Manuf. **38**, 1427–1435 (2019). www.sciencedirect.com
25. Kang, N., Zhao, C., Lib, J., Horst, J.A.: A Hierarchical structure of key performance indicators for operation management and continuous improvement in production systems. Int. J. Prod. Res. **54**(21), 6333–6350 (2016). https://doi.org/10.1080/00207543.2015.1136082
26. Jasiulewicz-Kaczmarek, M., Saniuk, A.: How to make maintenance processes more efficient using lean tools? In: Goossens, R.H.M. (ed.) AHFE 2017. AISC, vol. 605, pp. 9–20. Springer, Cham (2018). https://doi.org/10.1007/978-3-319-60828-0_2
27. Jasiulewicz–Kaczmarek, M., Antosz, K., Żywica, P., Mazurkiewicz, D., Sun, B., Ren, Y.: Framework of machine criticality assessment with criteria interactions. Eksploatacja i Niezawodność – Maint. Reliab. **23**(2), 207–220 (2021)

# Job Rotation for the Competencies Maintaining: A Case Study in IT Project Management

Eryk Szwarc[1(✉)], Paulina Golińska-Dawson[2], Grzegorz Bocewicz[1], and Zbigniew Banaszak[1]

[1] Faculty of Electronics and Computer Science, Koszalin University of Technology, Sniadeckich 2, 75-453 Koszalin, Poland
eryk.szwarc@tu.koszalin.pl

[2] Faculty of Engineering Management, Poznan University of Technology, ul. Jacka Rychlewskiego 2, 60-965 Poznań, Poland
paulina.golinska@put.poznan.pl

**Abstract.** This paper deals with scheduling in software development organizations. A new version of the software project scheduling problem (SPSP) is considered that takes into account the phenomenon of learning and forgetting. Their impact on maintaining the competencies level of employees is important, especially in case of occurrence of unplanned events, such as absenteeism and/or change in the priorities of projects. The paper proposes a new reference declarative SPSP model for proactive planning of assignments of multi-skilled programmers to tasks in the portfolio of a network of activities in IT projects, taking into account forgetting effects. It's implementation in the constraints programming environment, enables decision-making in project management at software companies. It helps to solve the problem of finding a feasible solution, and identifying conditions under which given expectations are met. The potential of the proposed SPSP model is illustrated by numerical examples.

**Keywords:** Job Rotation · Project Management · Competencies Maintenance · Learning-Forgetting Effect

## 1 Introduction

Job rotation (sometimes called work rotation) is a managerial practice used in organizations to reduce work monotony, boredom, fatigue and burnout [6], as well as to increase job satisfaction and motivation [18, 22]. It is used in areas such as nursing, engineering, software development, and the manufacturing industry [2, 19]. Job rotation in producing software is defined as the systematic movement of employees from job to job (J2J) or from project to project (P2P) [29]. In P2P rotation, employees are transferred between projects with similar characteristics, often performing the same technical role, which results in increased team flexibility and reduced monotony of work. Additionally, it contributes to the development of innovation and multicultural teams.


R. Mehmood et al. (Eds.): DCAI 2023, LNNS 741, pp. 13–23, 2023.
https://doi.org/10.1007/978-3-031-38318-2_2

The purpose of the paper is to review the state of research in the field of job rotation in software development organizations, taking into account the learning and forgetting effect [11]. Furthermore, the conditions are determined, that cyclic rotation of employees between different tasks to keep their competencies at a constant level in case of IT projects [28]. Competencies are defined as a set of knowledge, experience, and skills that are necessary to conduct a predefined set of tasks. The work focuses on the skills of programmers, which are also referred to as competencies. In project on software development, the level of programmers' competencies (which changes over time due to job rotation) affects the execution time of tasks. Low-skilled programmers make more mistakes (requiring time-consuming corrections) compared to high-skilled programmers.

Therefore, this article can be seen as a continuation of previous works in the field of proactive staff allocation under competence structure constraints [4, 25]. Previous studies have not proposed models of balancing multi-skilled employees with requirements resulting from both the scope of work, and the need to maintain the team's competences at a constant level.

The main contributions are as follows:

- Development of a generic model for multi-skilled workforce proactive job rotation planning.
- Proposal of constraint satisfaction problem where the desired rotation of staff is conditioned by the time of task execution, and by the competencies' validity.
- Numerical results, which confirm the possibility of using the developed method of proactive job rotation planning in online mode.

This paper is organized as follows: In Sect. 2, the main research gap based on the review of related works were identified. In Sect. 3 a declarative framework enabling the formulation of SPSP in terms of a constraint satisfaction problem were introduced. Then in Sect. 4 the concepts of competence structure robustness and maintenance of team competences at a constant level in were integrated. In Sect. 5 presented the conducted computer experiments to reveal the effectiveness of proactive job rotation schedules in SPSP. Section 6 summarized the contributions of the paper.

## 2 Related Works

In the literature, there are many definitions describing job rotation. One of the broadest definition was proposed by Wood [29], as the systematic movement of employees from job to job or from project to project within an organization during the implementation of a task. Thanks to job rotation, many goals of human resource management can be met, such as filling jobs, targeting new employees, preventing boredom or burnout, rewarding employees, and supporting career development. Rotation can be used to meet organizational goals related to resource allocation as well as individual motivational needs. In general, rotation is used in two cases:

- When a specific project requires a larger team (number of programmers) or a different set of skills (diversity of competences)

- When an employee shows a desire to change a project.

In general, research shows that the use of rotation positively affects mobility, job satisfaction, and innovation [15] and negatively affects job specialization [14].

Many of the models proposed to solve the software project scheduling problem (SPSP) are similar to the model originally presented in [1]. The set of tasks and a set of employees with specific characteristics are included: skills, remuneration, and working time limits are given. In the context of software projects, it is possible to have skills in specific technologies or tasks such as programming, database management, design, etc.

Some SPSP models assume that the skill level of programmers is constant [7, 31]. In fact, employees can improve their skills during software development; i.e., an employee may have a low level of a skill, but if he works long enough on a task that requires that skill, he raises the level of the skill thanks to the experience gained. On the other hand, when they rarely use a particular skill, they can experience a forgetfulness effect, and then the skill level decreases [20], which affects the effectiveness and working time of programmers, and thus the timely completion of projects [26].

Wright was the first to analyze the effect of the skill level of workers on the quantity of products produced in aircraft production and described the Wright Learning Curve (WLC) [30], as the dependence of mastering a given skill or acquiring knowledge as a function of time. It has been used to solve problems related to project planning [21] and in work [17] it has been proven that the use of the learning curve affects the productivity of programmers.

The ability to learn is accompanied by the forgetfulness effect, which is considered a non-minor function of the loss of a skill while it is not in use. Many models have been developed, such as variable regression to variable forgetting (VRVF) [5], variable regression to invariant forgiveness (VRIF) [8], learning-forgetting curve model (LFCM) [10] and recency model (RC) [16].

Despite many studies in this area, only a small literature is done on the learning and forgetfulness effect in SPSP [27]. In particular, its impact on the cost and duration of the project is analyzed [12]. It should be noted that attention is focused primarily on the learning effect, i.e., the repetition of programmers' work (assignment to the same tasks in the initial phase of the project (without rotation) raises the level of specific skills, which results in shortening the time of implementation of these tasks in later stages). This is done at the expense of unused skills, the level of which is decreasing.

The results of the research do not clearly show whether the lack of rotation (repetitive task assignments) and specialization of employees in certain skills gives better results than the use of job rotation. In the field of SPSP, there is also a lack of research on the impact of rotation and related learning outcomes and forgetting on maintaining the skill level of employees in constant readiness, which is significant in cases of unplanned events (disruptions) during the operation of software companies. In this context, this contribution, which constitutes a continuation of earlier research [3, 24], introduces a new generic model for multi-skilled workforce proactive job rotation planning in SPSP with disruptions.

## 3 Problem Formulation

The essence of SPSP boils down to the allocation of employees (hereinafter also referred to as: programmers) with many skills to perform a specific set of tasks $Z = \{Z_1, \ldots, Z_i, \ldots, Z_Q\}$, such as: physical design and implementation of the database, user interface design, program coding, testing, etc. A portfolio of projects $E = \{E_1, \ldots, E_j, \ldots, E_K\}$ is given, where $E_j \subseteq Z$. The projects are executed according to the sequence: $W = (w_1, .., w_l, \ldots, w_L)$, where $w_l \in E$.

A group of $\mathcal{P} = \{P_1, \ldots, P_k, \ldots, P_M\}$ programmers to carry out individual $Z_i$ tasks is given. The competence structure of employees $P$ during the execution of the $E_j$ project has the form of a matrix $G^j$:

$$G^j = \left[g^j_{k,i}\right]_{k=1...M;i=1...Q} \tag{1}$$

where: $g_{k,i} \in \{1, \ldots, 5\}$,

$$g^j_{k,i} = \begin{cases} 5 \text{ or } 4 & \text{programmer } P_k \text{ performs the task } Z_i \text{ in } t^j_i = 1 \text{ time unit} \\ 3 & \text{programmer } P_k \text{ performs the task } Z_i \text{ in } t^j_i = 1 \text{ time units} \\ 2 \text{ or } 1 & \text{programmer } P_k \text{ performs the task } Z_i \text{ in } t^j_i = 1 \text{ time units} \end{cases}$$

Each project $E_j$ from portfolio $E$ is associated with the $X^j$ assignment of programmers to tasks. Similarly to the competence structure, it takes the form of a matrix:

$$X^j = \left[x^j_{k,i}\right]_{k=1...M;i=1...Q} \tag{2}$$

where: $x^j_{k,i} \in \{0, 1\}$, $x^j_{k,i} = 1$ when the programmer $P_k$ is assigned to perform the task $Z_i$ in the project $E_j$, otherwise $x^j_{k,i} = 0$.

Assignments $X^j$ make up the sequence of assignments $X = \left(X^1, \ldots, X^j, \ldots, X^K\right)$, which determines the schedule for project execution:

$$Y = (y_1, \ldots y_j, \ldots, y_K) \tag{3}$$

where: $y_j$ specifies when the project $E_j$ starts (all project tasks are $E_j$ started simultaneously). The moment when schedule $Y$ ends is determined by the $EY$ variable.

According to the effect of learning and forgetting, the level of competence does not remain constant. Curves that model this effect are the subject of many studies [10, 13], in which different shapes are given, with different degrees of inclination. In this work, for simplicity assumed that the level increases if the programmer performs a specific task. The range of this increase depends on the level of competence, and similarly the level of competence decreases if the programmer does not perform a specific task. The classification rules are as follows:

- A competence increases by 1 for every two time units of task duration if the competency level is 1 or 2,
- A competence increases by 1 for each time unit of task duration if the competency level is 3 or 4,

- A competence decreases by 1 for each time unit when the task is not performed if the competence level is 2 or 3,
- A competence decreases by 1 every two time units of the time when the task is not performed if the competency level is 4 or 5.

In other words, the level of programmers' competences changes during the implementation of subsequent projects. It depends on the assigned sequence adopted. The elements $g_{k,i}^{l}$ of the competence structure $G^j$ for the project $E_j$ are determined as follows:

$$g_{k,i}^{j} = \varphi\left(g_{k,i}^{j-1}, x_{k,i}^{j-1}\right), \text{ dla } j > 1, \tag{4}$$

where: $\varphi$ is a function determining the level of competence of programmers for the $E_j$ project, based on the values of the elements: $g_{k,i}^{j-1}$, $x_{k,i}^{j-1}$ (according to above rules).

In addition, the problem under consideration assumes the following:

- the task can be performed by only one programmer with any level of competence;
- all tasks must be assigned to employees,
- order $W$ must complete within the given horizon $EY \leq H$.

Unplanned events (disruptions) during the execution of projects mean adding a new project ${}^{(r)}E$ to the $W$ order (usually as the last one): ${}^{(r)}W = W || {}^{(r)}E$ (where || means sequence concatenation operation). The ability of the competency structure to deal with this type of disruption can be assessed by the robustness measure (competence structure robustness (CSR)), i.e. the ratio of variants of the number of disruption ($LP$) for which the competency structure $G$ guarantees the execution of ${}^{(r)}W$ to all disruption scenarios (set of disruptions $E^*$):

$$R = \frac{LP}{|E^*|}. \tag{5}$$

The problem illustrates a simplified version of SPSP, omitting a number of limitations encountered in practice, such as working time limits, payroll costs, etc. The presented approach is limited to one learning and forgetting model (LFCM). Comparing the proposed approach to other models such as RC (Recency) and PID (Power Integration Diffusion) is the subject of future research. In the general case, the different shapes of the forgetting function characteristic of different professions, sex, age groups could be considered. Since there are no statistics available at this time to determine the shapes of the forgetting function for selected occupational groups, this paper uses approximate characteristics collected over the past 5 years in a population of 25–30 individuals.

In this form, an answer to a question is sought:

1. Is there a (and if so, what) variant of the assignment $X$ that guarantees the execution of the project's portfolio $W$ (in the given horizon $EY \leq H$)?
2. Is there an assignment $X$ that guarantees a given robustness $R$ (e.g. $R = 1$) of the completion of the tasks in project order?

For the illustration (Fig. 1) considered an order involving a portfolio of three $E = \{E_1, E_2, E_3\}$ projects that require different set of tasks: $E_1 = \{Z_1, Z_2, Z_3\}$; $E_2 = \{Z_1, Z_2, Z_4\}$; $E_3 = \{Z_2, Z_3, Z_4\}$. Projects are executed in the following order: $W = (E_1, E_2, E_3)$. Initial competence structure $G^1$ before $E_1$ project execution is given. An assignment $X$ variant is sought, that guarantees the completion of the projects on $H = 3$ (answer to question number 1). Such a variant is illustrated in Fig. 1b).

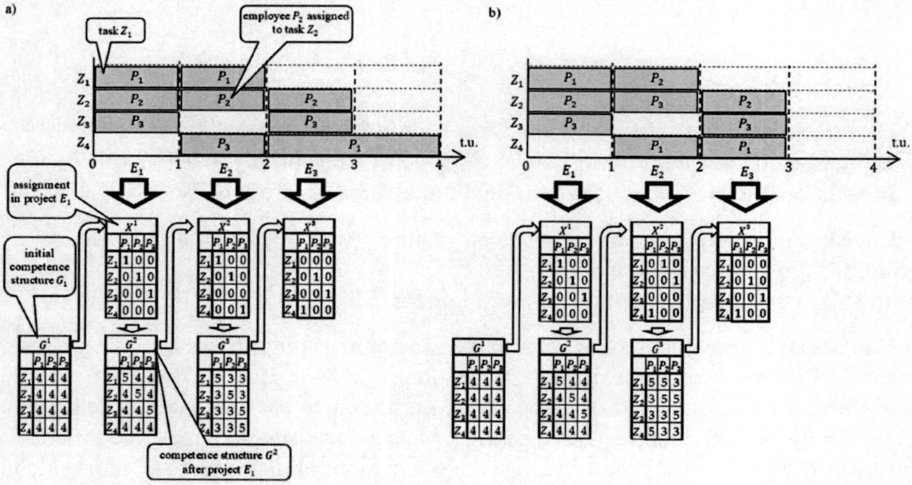


**Fig. 1.** Project execution $E_1, E_2, E_3$ in horizon $H = 4$ (a), $H = 3$ (b).

# 4 Reference Model of SPSP

The considered problem could be described using the declarative modeling paradigm.

Parameters:

$Z$: set of tasks $Z = \{Z_1, \ldots, Z_i, \ldots, Z_Q\}$,
$E$: portfolio of projects $E = \{E_1, \ldots, E_j, \ldots, E_K\}$,
$\mathcal{P}$: group of programmers $\mathcal{P} = \{P_1, \ldots, P_k, \ldots, P_M\}$,
W: sequence of projects,
$G^1$: initial competence structure,
$\varphi$: function determining the level of competence of programmers $\varphi\left(g_{k,i}^{j-1}, x_{k,i}^{j-1}\right)$,
$E^*$: set of disruptions $E^* \subseteq E$ (unplanned tasks/projects),
${}^{(r)}W$: sequence of projects with disruption ${}^{(r)}E \in E^*$: ${}^{(r)}W = W || {}^{(r)}E$,
$GR$: expected value of competence structure robustness for set of disruption $E^*$.

Decision variables:

${}^{(r)}X^j$: assignment of programmers to tasks (2), in the case of disruption ${}^{(r)}E \in E^*$,
${}^{(r)}G^j$: competence structure (1), in the case of disruption ${}^{(r)}E \in E^*$,

${}^{(r)}Y$: schedule for project execution, in the case of disruption ${}^{(r)}E \in E^*$,
${}^{(r)}\omega$: competence structure robustness in the case of disruption ${}^{(r)}E \in E^*$.

Constraints:

- In each project $E_j$ for all tasks $Z_i$ the programmers of set $\mathcal{P}$ must be assigned:

$$\sum\nolimits_{k=1}^{M} {}^{(r)}x_{k,i}^{j} = 1;\ \forall Z_i \in E_j,\ \forall^{(r)}E \in E^*, \tag{6}$$

$$\sum\nolimits_{k=1}^{M} {}^{(r)}x_{k,i}^{j} = 1;\ \forall Z_i \notin E_j,\ \forall^{(r)}E \in E^*, \tag{7}$$

$${}^{(r)}x_{k,i}^{j} =^{(q)} x_{k,i}^{j};\ q \neq r,\ E_j \neq^{(r)} E, \tag{8}$$

- In each project $E_j$ programmer $P_k$ can only be assigned to one task:

$$\sum\nolimits_{i=1}^{Q} {}^{(r)}x_{k,i}^{j} = 1;\ \forall Z_i \notin \mathcal{P},\ \forall^{(r)}E \in E^*, \tag{9}$$

- Elements ${}^{(r)}g_{k,i}^{j}$ of the competence structure ${}^{(r)}G^j$ of project $E_j$ (for $j > 1$) depend on the preceding project $E_{j-1}$:

$${}^{(r)}g_{k,i}^{j} = \varphi\left({}^{(r)}g_{k,i}^{j-1},\ {}^{(r)}x_{k,i}^{j-1}\right),\ \text{for } j > 1,\ \forall^{(r)}E \in E^*, \tag{10}$$

- The duration of a task ${}^{(r)}t_i^j$ for $E_j$ project depends on the competence of the $P_k$ employee who is performing it:

$${}^{(r)}t_i^j = \begin{cases} 1 & \text{if } {}^{(r)}x_{k,i}^{j} = 1 \text{ and } {}^{(r)}g_{k,i}^{j} \geq 4 \\ 2 & \text{if } {}^{(r)}x_{k,i}^{j} = 1 \text{ and } {}^{(r)}g_{k,i}^{j} = 3 \\ 4 & \text{if } {}^{(r)}x_{k,i}^{j} = 1 \text{ and } {}^{(r)}g_{k,i}^{j} \leq 2 \end{cases},\ \forall Z_i \in E_j,\ \forall^{(r)}E \in E^*, \tag{11}$$

- Project $E_j$ starts after project $E_{j-1}$ is finished:

$${}^{(r)}y_j =^{(r)} y_{j-1} + \max_{i=1\ldots Q}\left\{{}^{(r)}t_i^{j-1}\right\},\ \text{for j} > 1,\ {}^{(r)}y_1 = 0,\ \forall^{(r)}E \in E^*, \tag{12}$$

- If the portfolio of projects (with additional project ${}^{(r)}E$) can be completed within the given horizon $H$ then competence structure is robust for disruption ${}^{(r)}E$:

$$\left({}^{(r)}y_{K+1} + \max_{i=1\ldots Q}\left\{{}^{(r)}t_i^{K+1}\right\} \leq H\right) \Rightarrow \left({}^{(r)}\omega = 1\right), \tag{13}$$

- Competence structure robustness $R$ (5) for disruption of set $E^*$ should be at least equal to $GR$:

$$R = \frac{\sum\nolimits_{r=1}^{|E^*|} {}^{(r)}\omega}{|E^*|} \geq GR. \tag{14}$$

The presented model allows to answer the following question: Is there an assignment ${}^{(r)}X$ that guarantees a robustness $R \geq GR$ for given set of disruptions $E^*$? The above problem can be formulated as a following CSP (constraint satisfaction problem):

$$CS = ((\mathcal{V},\ \mathcal{D}),\ \mathcal{C}), \tag{15}$$

where: $\mathcal{V} = \left\{{}^{(r)}x_{k,i}^{j},\ (r)g_{k,i}^{j},\ (r)y_j | k = 1, \ldots, M;\ i = 1, \ldots, Q;\ j = 1,\ \ldots,\ \mathrm{K}+1,\ \mathrm{r} = 1 \ldots \left|E^*\right|\right\}$, a set of decision variables representing assignment ${}^{(r)}X$ schedule $(r)Y$ and competence structure ${}^{(r)}G^j$; $\mathcal{D}$ is a finite set of domains of decision variables; $\mathcal{C}$ is a set of constraints specified in inequalities (6)–(14).

To solve the problem $CS$ (15), one should determine the values of decision variables ${}^{(r)}x_{k,i}^{j}$, ${}^{(r)}g_{k,i}^{j}$, ${}^{(r)}y_j$, for which all the constraints given in the set $\mathcal{C}$ are satisfied. Solving $CS$ means determining the assignment which guarantees the execution of the portfolio of projects with given robustness for occurrence of a new project ${}^{(r)}E \in E^*$.

## 5 Computation Experiment

The model proposed in the previous section can be used to find assignment ${}^{(r)}X$ that guarantees assumed robustness $R \geq GR$ for given set of disruptions $E^*$. Finding this kind of assignment provides a solution to the problem (15) formulated in the form of constraint programming, and then implemented in the declarative programming environment IBM ILOG CPLEX (Intel Core i7-M4800MQ 2.7 GHz, 32 GB RAM).

The effectiveness of the proposed approach, orientated towards the assessment of its scalability, was verified in multiple computer experiments. The results obtained concerning a different number of programmers ($m$ = 10, 12, 15) and a different number of tasks ($n$ = 20, ..., 50) are presented in Fig. 2.

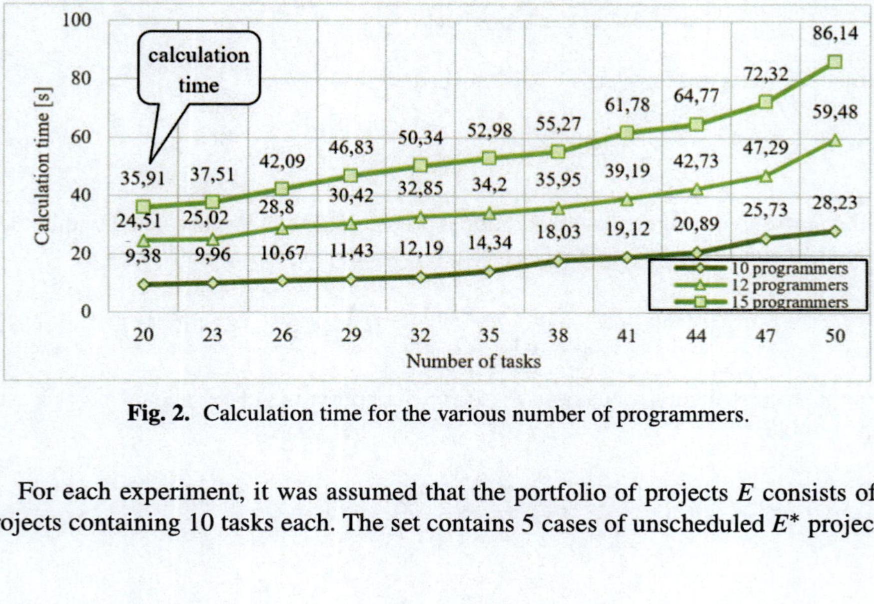


**Fig. 2.** Calculation time for the various number of programmers.

For each experiment, it was assumed that the portfolio of projects $E$ consists of 5 projects containing 10 tasks each. The set contains 5 cases of unscheduled $E^*$ projects.

Expected robustness level $R = 1$. It is easy to see that with increasing scale of the problem, the computation time increases exponentially. To illustrate the resulting consequences, let us compare the cases in which there are 50 tasks (operations). Let us note that an increase in programmers from 10 to 12 causes more than a twofold extension of the calculation time. Moreover, the increase in the size of team size from 10 to 15 programmers leads to the calculation time growing more than three times. That means the proposed approach can be used online for practical cases not exceeding 50 tasks (for the whole project portfolio) and teams of 15 programmers.

## 6 Conclusions

The paper proposed a new reference model for proactive planning of assignments of multi-skilled programmers to tasks (activities, operations) occurring in a network of activities in IT projects' portfolio, taking into account forgetting effects. To solve the related problem of maintaining the competences of employed programmers at a given level it is necessary to rotate them regularly between the tasks (activities).

The main benefit of the proposed approach is open structure of the model, which allows to take into account new relations occurring between decision variables. This is done without loss of computational efficiency (in programming environments with constraints, i.e. IBM ILOG, increasing the number of constraints reduces the time of determining the solution).

The limitations of the proposed approach result from the characteristics of the work environment (the way tasks are performed, working in groups, the amount and type of resources used), and the nature of the disruptions occurring (employee absenteeism, loss of competencies, change in organizational structure etc.). In order to overcome these limitations future research will focus on the extension of the model to cases related to the uncertainty of operation times, planned deadlines for order execution, and the level of skills (expressed in fuzzy numbers).

## References

1. Alba, E., Chicano, J.F.: Software project management with Gas. Inf. Sci. **177**(11), 2380–2401 (2007)
2. Battini, D., Berti, N., Finco, S., Zennaro, I., Das, A.: Towards industry 5.0: a multi-objective job rotation model for an inclusive workforce. Int. J. Prod. Econ. **250**, 108619 (2022)
3. Bocewicz, G., Golińska-Dawson, P., Szwarc, E., Banaszak, Z.: Preventive maintenance scheduling of a multi-skilled human resource-constrained project's portfolio. Eng. Appl. Artif. Intell. **119** (2023). https://doi.org/10.1016/j.engappai.2022.105725
4. Bocewicz, G., Szwarc, E., Wikarek, J., Nielsen, P., Banaszak, Z.: A competency-driven staff assignment approach to improving employee scheduling robustness. Eksploatacja i Niezawodnosc – Maint. Reliab. **23**(1), 117–131 (2021). https://doi.org/10.17531/ein.2021.1.13
5. Carlson, J.G., Rowe, A.J.: How much does forgetting cost? Ind. Eng. **8**(9), 40–47 (1976)
6. Coyne, P.F.: An evidence-based model of job rotation (PhD Dissetation). Middlesex University, London, UK (2011)

7. Crawford, B., Soto, R., Johnson, F., Monfroy, E., Paredes, F.: A max–min ant system algorithm to solve the software project scheduling problem. Expert Syst. Appl. **41**(15), 6634–6645 (2014)
8. Elm'Aghraby, S.E.: Economic manufacturing quantities under conditions of learning and forgetting (EMQ/LAF). Prod. Plan. Control **1**, 196–208 (1990)
9. Fægri, T.E., Dybå, T., Dingsøyr, T.: Introducing knowledge redundancy practice in software development: experiences with job rotation in support work. Inf. Softw. Technol. **52**(10), 1118–1132 (2010)
10. Glock, C.H., Grosse, E.H., Jaber, M.Y., Smunt, T.L.: Applications of learning curves in production and operations management: a systematic literature review. Comput. Ind. Eng. **131**, 422–441 (2019)
11. Glock, C.H., Jaber, M.Y.: A multi-stage production-inventory model with learning and forgetting effects, rework and scrap. Comput. Ind. Eng. **64**, 708–720 (2013)
12. Guo, Y., Ji, J., Ji, J., Gong, D., Cheng, J., Shen, X.: Firework-based software project scheduling method considering the learning and forgetting effect. Soft. Comput. **23**(13), 5019–5034 (2018). https://doi.org/10.1007/s00500-018-3165-2
13. Hoedt, S., Claeys, A., Aghezzaf, E.-H., Cottyn, J.: Real time implementation of learning-forgetting models for cycle time predictions of manual assembly tasks after a break. Sustainability **12**, 5543 (2020)
14. Hsieh, A., Chao, H.: A reassessment of the relationship between job specialization, job rotation and job burnout: example of Taiwan's high-technology industry. Int. J. Hum. Resour. Manag. **15**(6), 1108–1123 (2004)
15. Kaymaz, K.: The effects of job rotation practices on motivation: a research on managers in the automotive organizations. Bus. Econ. Res. J. **1**(3), 69–85 (2010)
16. Nembhard, D.A., Uzumeri, M.V.: Experiential learning and forgetting for manual and cognitive tasks. Int. J. Ind. Ergon. **25**(4), 315–326 (2000)
17. Ngwenyama, O., Guergachi, A., Mclaren, T.: Using the learning curve to maximize it productivity: a decision analysis model for timing software upgrades. Int. J. Prod. Econ. **105**(2), 524–535 (2007)
18. Ollo-Lopez, A., Bayo-Moriones, A., Larraza-Kintana, M.: The relationship between new work practices and employee effort. J. Ind. Relat. **52**(2) (2010)
19. Platis, C., Ilonidou, C., Stergiannis, P., Ganas, A., Intas, G.: The job rotation of nursing staff and its effects on nurses' satisfaction and occupational engagement. In: Vlamos, P. (eds.) GeNeDis 2020. AEMB, vol. 1337, pp. 159–168. Springer, Cham (2021). https://doi.org/10.1007/978-3-030-78771-4_18
20. Podolski, M., Rosłon, J., Sroka, B.: The impact of the learning and forgetting effect on the cost of a multi-unit construction project with the use of the simulated annealing algorithm. Appl. Sci. **12**, 12667 (2022)
21. Qin, S., Liu, S., Kuang, H.: Piecewise linear model for multiskilled workforce scheduling problems considering learning effect and project quality. Math. Probl. Eng. **2016**(4), 1–11 (2016)
22. Richardson, A., Douglas, M., Shuttler, R., Hagland, M.R.: Critical care staff rotation: outcomes of a survey and pilot study. Nurs. Crit. Care **8**(2), 84–89 (2003)
23. Shen, X.N., Minku, L., Bahsoon, R., Yao, X.: Dynamic software project scheduling through a proactive-rescheduling method. IEEE Trans. Softw. Eng. **1**, 658–686 (2016)
24. Szwarc, E., Bocewicz, G., Golińska-Dawson, P., Banaszak, Z.: Proactive operations management: staff allocation with competence maintenance constraints. Sustainability **15**(3), 1949 (2023). https://doi.org/10.3390/su15031949
25. Szwarc, E., Wikarek, J., Gola, A., Bocewicz, G., Banaszak, Z.: Interactive planning of competency-driven university teaching staff allocation. Appl. Sci.-Basel **10**(14) (2020). https://doi.org/10.3390/app10144894

26. Van Peteghem, V., Vanhoucke, M.: Influence of learning in resource-constrained project scheduling. Comput. Ind. Eng. **87**, 569–579 (2015)
27. Ünal, U., Tüzün, E., Gezici, T., Ahmed, A.: Investigating the Impact of Forgetting in Software Development. MSR (2022)
28. Wikarek, J., Sitek, P.: Proactive and reactive approach to employee competence configuration problem in planning and scheduling processes. Appl. Intell. **52**, 3445–3454 (2022). https://doi.org/10.1007/s10489-021-02594-x
29. Wood, R.: Human Resources Management. AHMA, Michigan (1995)
30. Wright, T.P.: Factors affecting the cost of airplanes. J. Aeronaut. Sci. **3**(4), 122–128 (1936)
31. Xiao, J., Ao, X.-T., Tang, Y.: Solving software project scheduling problems with ant colony optimization. Comput. Oper. Res. **40**(1), 33–46 (2013)

# Applying an Enterprise Architecture Approach to Modeling Production Procesess

Paweł Sitek(✉), Jarosław Wikarek, and Zbigniew Juzoń

Kielce University of Technology, Al. Tysiąclecia P.P. 7, 25-314 Kielce, Poland
{sitek,j.wikarek,zjuzon}@tu.kielce.pl

**Abstract.** One of the most important aspects of work planning at the production system level is the contract negotiation process. In this process, it is necessary to negotiate the production volume as well as lead time acceptable to both the production system and the customer. Failure to obtain sufficiently accurate forecasts regarding order volume and task completion time may result in penalties and/or loss of new orders in the future. The article proposes an innovative approach to modeling complex production planning problems based on customer orders. What distinguishes this approach is the ability to determine which orders will not be fulfilled due to which tasks. The authors proposed the use of enterprise architecture in the process of the preliminary analysis to obtain various information from the organization affecting the functioning of the production system. This approach was demonstrated with an example for which a mathematical model was built, and the obtained simulation results are stored on an enterprise data bus for possible reuse at the enterprise level.

**Keywords:** KPI · enterprise architecture · production process · meta-model · mathematical programming · decision support

## 1 Introduction

Detailed models of production optimization are characterized by a large number of decision variables, which are at least partially discrete in nature, and numerous constraints. The authors' motivation to undertake the research was the observation that the current approaches to optimizing production systems usually boil down to building a mathematical model based on data obtained directly from the technological layer without referring to the business layer, which has a significant impact on the processes functioning in the company and its environment. The authors propose to use the assumptions of enterprise architecture [1] resulting from The Open Group Architecture Framework (TOGAF) [2] to develop an architecture meta-model [3] in order to select key data and information that should be included in the modeling process and to build a detailed model of optimization of the production system. The meta-model itself is defined as an explicit model for constructing domain models [4]. The presented approach considers problems classified as discrete Constrained Optimization Problems (COPs) [5]. With such problems,


R. Mehmood et al. (Eds.): DCAI 2023, LNNS 741, pp. 24–35, 2023.
https://doi.org/10.1007/978-3-031-38318-2_3

after lengthy computations, it is often the case that a arises. Such a situation is unacceptable because it does not provide managers with any information and does not allow them to react in real time [6]. The detailed problem under consideration (Sect. 2) can be classified as a variant and extension of the Just-In-Time Job-Shop Scheduling (JIT-JSS) problem [7]. The goal of JIT-JSS is to obtain a schedule, i.e. the time of completion of the operation, with the smallest total earliness and lateness penalties [8]. In order to model the problem under consideration, it is crucial to conduct an analysis (Fig. 1) to identify information that affects the functioning of the organization.

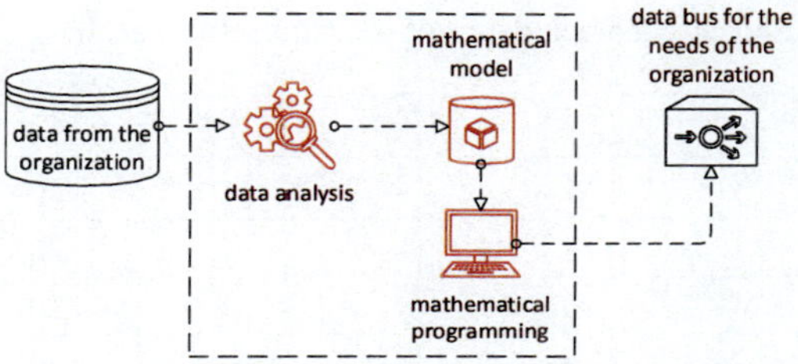


**Fig. 1.** Schema data analysis

Only the following data collected on the data bus was used, which can be called Key Performance Indicators (KPI) [9]: production cycle time, changeover time, production rate, utilization of production capacity, implementation of the production plan, percentage of planned and emergency service requests, availability, production rate, complaints from customers, quality of suppliers, customer order fulfillment indicator.

## 2 Problem Statement and Illustrative Example

In a production plant, a set of machines $M$ ($m \in M$) is given. The set of operations $O$ ($o \in O$) can be performed on machines. The company is to complete a set of products $P$ ($p \in P$) in a given planning period. To produce a product, a specific set of operations must be performed on it (*operations must be performed in a specific order*). The planning period is divided into $T$ ($t \in T$) execution periods. The demand for products can be divided into parts. Preparing the machine for the production of the product requires its retooling. Production planning should aim to meet as many requirements as possible with minimal changeovers. The most important differences between Job-Shop (classic) and problem statement are as follows: Job-Shop (classic) has fixed routes, while the routes in the problem under consideration are dynamic; Job-Shop (classic) does not have specific production batches, while in the problem under consideration the production batches have been specified. For such a problem and the data presented in Table 1 and Table 2, manual allocations of products to machines were made to fulfill orders. Table 3 and Fig. 2 and 3 present a solution in the form of process production routes for a selected case where the penalty for non-execution of orders was 1500, while the cost of execution on machines was 700. Appropriate manual allocation by the operator took about 20 min, even for such a small sample.

**Table 1.** Characteristics of operations for products (duration, order, etc.)

| $C_{p,o}$ | o1 | o2 | o3 | o4 | o5 | $W_p$ |
|---|---|---|---|---|---|---|
| p1 | 1 | 1 | 1 | - | - | 20 |
| p2 | 1 | - | 1 | 1 | 1 | 40 |

*$C_{p,o}$, $W_p$, $B_{m,o}$, $E_m$, $E_m$ - the meaning of these parameters is described in Tab. 4*

| p | First | Second |
|---|---|---|
| p1 | o1 | o2 |
| p1 | o2 | o3 |
| p2 | o1 | o3 |
| p2 | o3 | o4 |
| p2 | o4 | o5 |

**Table 2.** Characteristics of operations for machines

| Bm,o | o1 | o2 | o3 | o4 | o5 | Em | Em |
|---|---|---|---|---|---|---|---|
| m1 | 1 | 1 | – | – | 1 | 20 | 120 |
| m2 | 1 | 1 | 1 | – | – | 20 | 150 |
| m3 | – | – | 1 | 1 | 1 | 20 | 100 |
| m4 | – | – | 1 | 1 | - | 20 | 100 |

**Table 3.** The routes of the production process

| Production of products on machines | | | | |
|---|---|---|---|---|
| **p** | **o** | **m** | **t** | Quantity |
| **1** | 1 | 2 | 3 | 20 |
| **1** | 2 | 2 | 4 | 20 |
| **1** | 3 | 3 | 5 | 20 |
| **2** | 1 | 2 | 1 | 20 |
| **2** | 1 | 2 | 2 | 20 |
| **2** | 3 | 4 | 2 | 20 |
| **2** | 3 | 4 | 3 | 20 |
| **2** | 4 | 3 | 3 | 20 |
| **2** | 4 | 3 | 4 | 20 |
| **2** | 5 | 1 | 4 | 20 |

| Changeovers required | | | |
|---|---|---|---|
| **m** | **t** | **p** | **o** |
| 1 | 4 | 2 | 5 |
| 2 | 1 | 2 | 1 |
| 2 | 3 | 1 | 1 |
| 2 | 4 | 1 | 2 |
| 3 | 3 | 2 | 4 |
| 3 | 5 | 1 | 3 |

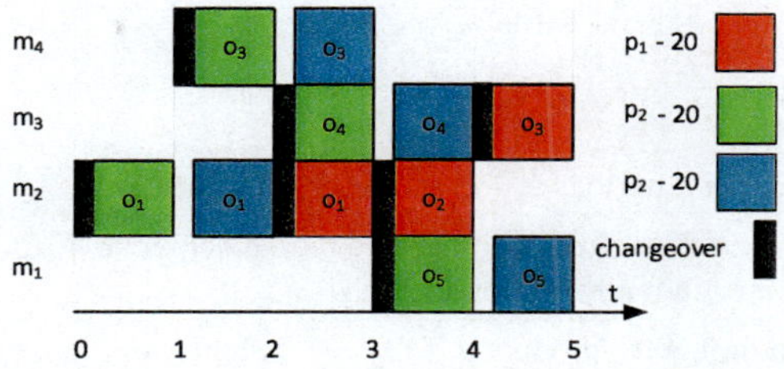


**Fig. 2.** Production process schedule (acceptable solution)

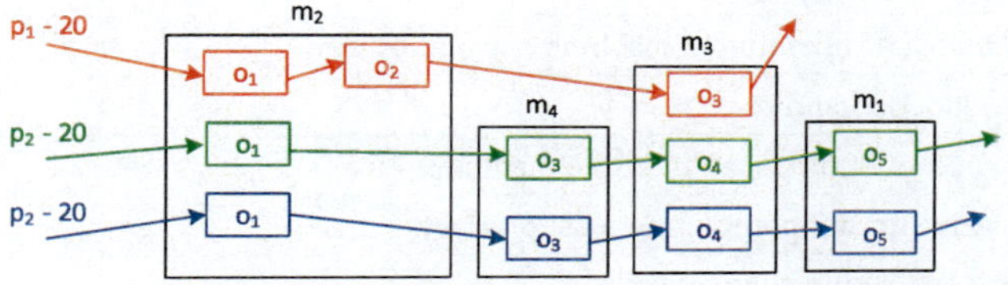


**Fig. 3.** Routes for the schedule of the production process from Fig. 2 (acceptable solution)

# 3 Mathematical Model of Production Planning and Resource Allocation for Dynamic Job-Shop Production

For the problem presented in Sect. 2, two main questions Q1 and Q2 can be posed:

*Q1: Will it be possible to fulfill the order, and if not, what will not be possible?*
*Q2: When and how many products can be made on which machine?*

In order to find answers to these questions, as well as to automate this process, a formal model of the problem of production planning and resource allocation has been proposed. This model (1)..(11) was formulated in the form of Mixed Integer Linear Programming (MILP). The description of the parameters, decision variables and constraints along with their mapping to the meta-model is presented in Table 4. The objective function (11) consists of three elements. The first element contains penalties for unfulfilled orders (Fc1). The second element covers the cost of operating the machine per unit of time (Fc2), while the third element (Fc3) of the objective function (11) includes penalties for unplanned machine downtime.

**Table 4.** Defining elements of the mathematical model

| A[a] | B[b] | Description |
|---|---|---|
| Rs | ZM | Set of machines (*m* - machine index) |
| | ZP | Product set (*p* - product index) |
| | ZO | Set of operations for the product (*o* - operation index) |

(*continued*)

**Table 4.** (*continued*)

| A[a] | B[b] | Description |
|---|---|---|
| | ZT | Set of execution periods (*t* – execution period index) |
| P | $B_{m,o}$ | Standard execution time for operation *o* on machine *m*. If $B_{m,o} = 0$ it means that machine *m* cannot perform operation *o* |
| | $C_{p,o}$ | Operation time *o* for product *p* If $C_{p,o} = 0$, then for product *p* there is no need to perform operation *o* |
| | $D_{p,o1,o2}$ | If for product *p* operation $o_1$ is performed immediately before $o_2$ then $D_{p,o1,o2} = 1$ otherwise $D_{p,o1,o2} = 0$ |
| | $F_m$ | The cost of operating a machine per unit of time |
| | $W_p$ | Product demand *p* |
| | $E_m$ | The production capacity of the machine |
| | K1 | Penalty for not completing each product |
| | K2 | Cost per single changeover |
| R | $U_p$ | The size of an order for a product *p* which cannot be fulfilled due to production capacity |
| | $Z_{p,o,m,t}$ | How much operation *o* for product *t* is performed on machine *m* in period *t* |
| | $J_{p,o,m,t}$ | If machine m was retooled in period *t* to perform operation *o* for product *p* $J_{p,o,m,t} = 1$, otherwise $J_{p,o,m,t} = 1$ |
| | $X_{p,t}$ | Part of the batch of demand for the product, the fulfillment of which began in period *t* |
| | $Y_{p,o,t}$ | How many operations *o* on product *p* are performed in period *t* |
| | $G_{p,o,m,t}$ | If operation *o* on product *t* is performed on machine *m* in period *t* then $G_{p,o,m,t} = 1$ otherwise $G_{p,o,m,t} = 0$ |
| | $H_{m,t}$ | If machine *m* worked in period *t* $H_{m,t} = 1$, otherwise $H_{m,t} = 0$ |
| | 1 | Fulfilling the entire demand or setting a correction |
| | 2 | Execution of all operations for the product |
| | 3 | Operation order |
| | 4 | Assignment of operations to machines |
| | 5 | Machine capability check |
| | 6 | Linking variables $H_{m,t}$ and $Z_{p,o,m,t}$ |
| | 7 | Linking variables $Z_{p,o,m,t}$ and $G_{p,o,m,t}$ |
| | 8 | In a given period, the machine only one type of operation |
| | 9 | Determination of the number of changeovers |
| | 10 | Binary and integer |

[a]The architecture meta-model elements.
[b]The symbol.

$$\sum_{t \in T} X_{p,t} = W_p - U_p \forall p \in P \tag{1}$$

$$\sum_{t \in T} Y_{p,o,t} = \sum_{t \in T} X_{p,t} \forall p \in P, o \in O \wedge c(p,o) \neq 0 \tag{2}$$

$$Y_{p,o1,t} = Y_{p,o2,t+1} \forall p \in P, o1, o2 \in O, t \in T \wedge D(p, o1, o2) = 1 \tag{3}$$

$$\sum_{m \in M \wedge B_{m,o} \neq 0} Z_{p,o,m,t} = Y_{p,o,t} \forall p \in P, o \in O, t \in T \tag{4}$$

$$\sum_{p \in P} \sum_{o \in O} Z_{p,o,t,m} = E_m \forall m \in M, t \in T \tag{5}$$

$$\sum_{p \in P} \sum_{o \in O} Z_{p,o,t,m} \leq H_{m,t} \cdot St \forall m \in M, t \in T \tag{6}$$

$$St \cdot G_{p,o,m,t} \geq Z_{p,o,m,t} \forall p \in P, o \in O, m \in M, t \in T \tag{7}$$

$$\sum_{p \in P} \sum_{o \in O} G_{p,o,t,m} \leq 1 \forall m \in M, t \in T \tag{8}$$

$$J_{p,o,m,t} \geq G_{p,o,m,t} - G_{p,o,m,t-1} \forall p \in P, o \in O, m \in M, t \in T \tag{9}$$

$$\begin{aligned} &X_{p,t} \in C \forall p \in P, t \in T \\ &Y_{p,o,t} \in C \forall p \in P, o \in O, t \in T \\ &Z_{p,o,m,t} \in C \forall p \in P, o \in O, m \in M, t \in T \\ &G_{p,o,m,t} \in \{0, 1\} \forall p \in P, o \in O, m \in M, t \in T \\ &H_{m,t} \in \{0, 1\} \forall p \in P, o \in O, m \in M, t \in T \end{aligned} \tag{10}$$

$$K1 \cdot \sum_{p \in P} U_p + \sum_{m \in M} \sum_{t \notin T} F_m \cdot H_{m,t} + K2 \cdot \sum_{p \in P} \sum_{o \in O} \sum_{m \in M} \sum_{t \in T} J_{p.o.m.t} \tag{11-Fc}$$

The input parameters can be divided into 2 groups. The first is constant data, which does not actually change, and the entire description of the technology is as follows: $[B_{m,o}]$, $[C_{p,o}]$, $[D_{p,o1,o2}]$, $[F_m]$. The second are extortions (these are input data that would change in subsequent cycles): $[W_p]$, $[E_m]$, $[F_m]$, [K1], [K2]. The parameters $[B_{m,o}]$, $[C_{p,o}]$, $[D_{p,o1,o2}]$, $[F_m]$ refer to the length of the execution period, which translates into production capacity (which is why it was given previously as an excitation variable). In each case of production (capacity) study, the answers to the following questions are sought: What is the cost in each case (elements of the objective function)? Is the variable $[U_p]$ what we will not be able to do? What is the load on the machines (% utilization) and how long will it take? Also, the variable $[U_p]$ answers the question: Will it be possible to fulfil the order, and if not, what will not be possible? Variables $[Z_{p,o,m,t}]$ and $[J_{p,o,m,t}]$: When and how many products can be made on which machine?

## 4 Implementation of the Model

The AMPL (A Mathematical Programming Language) language and the *Gurobi* mathematical programming [10] optimization package were used to implement the model (Sect. 3). AMPL is an algebraic modeling language for describing and solving high-complexity problems for large-scale mathematical computing. The general scheme of the implementation is shown in Fig. 4.

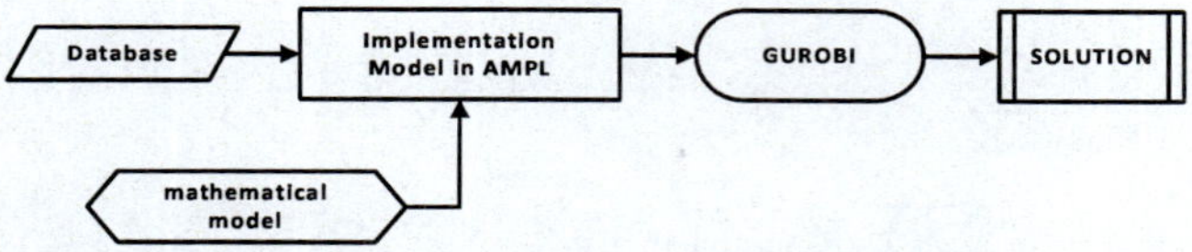


**Fig. 4.** The method of implementing the modeled problem using the AMPL and the Gurobi

Gurobi is an optimization software developed by Gurobi Optimization, LLC. It is considered one of the most diverse optimization software that can solve a wide range of problem types. These include linear programming, mixed-integer linear programming, quadratic programming, mixed-integer quadratic programming, quadratically constrained programming, and mixed-integer quadratically constrained programming.

## 5 Computational Experiments

After the implementation of the model (1)..(11) according to the concept presented in Fig. 4, a number of computational experiments were carried out. The experiments were performed in two stages. In the first stage, experiments were performed for a modeled problem of small size for identical data instances as in Sect. 2 (Table 1 and Table 2). The results obtained are shown in Table 5 and in Fig. 5 and Fig. 6. The routes of the production process (optimal) (unfulfilled cost of 1,330, cost of implementation on machines of 700) presented in Table 5 was carried out for K1 = 10,000 and K2 = 100. The computations took less than 3 s.

The data for the second stage are listed as averaged values for problems encountered in the practice of medium and small manufacturing enterprises. Experiments were carried out for the data collected in Table 6 and Table 7, Inputs part (Appendix A), using the implemented model. The results are presented in Table 7 (Output part) for K1 = 1,000 and various values of K2 and production capacity. The analysis of the obtained results enables the decision maker to make a decision regarding the scope of production, allocation methods, etc., as well as to analyze how the solution may change if the changeovers costs change. The computations took about 1,200 s per instance.

**Table 5.** The routes of the production process (optimal)

| **Production of products on machines** | | | | |
|---|---|---|---|---|
| **p** | **o** | **m** | **t** | Quantity |
| 1 | 1 | 2 | 3 | 20 |
| 1 | 2 | 1 | 4 | 20 |
| 1 | 3 | 4 | 5 | 20 |
| 2 | 1 | 2 | 1 | 20 |
| 2 | 1 | 2 | 2 | 20 |
| 2 | 3 | 3 | 2 | 20 |
| 2 | 3 | 3 | 3 | 20 |
| 2 | 4 | 4 | 3 | 20 |
| 2 | 4 | 4 | 4 | 20 |
| 2 | 5 | 3 | 4 | 20 |
| 2 | 5 | 3 | 5 | 20 |

| **Changeovers required** | | | |
|---|---|---|---|
| **m** | **t** | **p** | **o** |
| 1 | 4 | 1 | 2 |
| 2 | 1 | 2 | 1 |
| 2 | 3 | 1 | 1 |
| 3 | 1 | 2 | 3 |
| 3 | 4 | 2 | 5 |
| 4 | 3 | 2 | 4 |
| 4 | 5 | 1 | 3 |

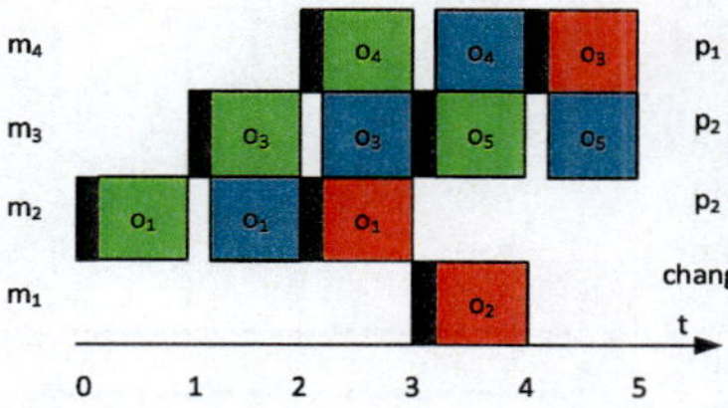


**Fig. 5.** Production process schedule *(optimal result)*

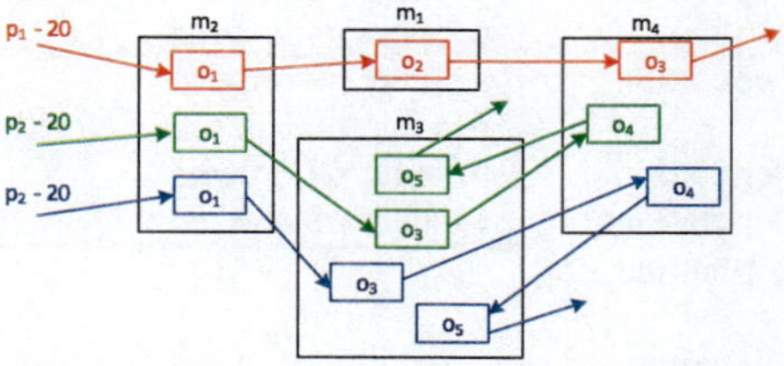


**Fig. 6.** Routes for the schedule of the production process from Fig. 2 *(optimal result)*

**Table 6.** Data for a large experiment

| Sizes | |
|---|---|
| Number of products | **8** |
| Number of machines | **10** |
| Number of operations | **6** |
| Planning horizon | **10** |
| K1 | **1000** |

| Order of operations |
|---|
| **product p1** |
| o1 operation precedes o2 operation |
| o2 operation precedes o3 operation |
| o3 operation precedes o6 operation |
| **p2 product** |
| o1 operation precedes o2 operation |
| o2 operation precedes o3 operation |
| o3 operation precedes o4 operation |
| o4 operation precedes o5 operation |
| o5 operation precedes o6 operation |
| **p3 product** |
| o3 operation precedes o4 operation |
| o4 operation precedes o5 operation |
| o5 operation precedes o6 operation |
| **p4 product** |
| o1 operation precedes o2 operation |
| o2 operation precedes o5 operation |
| o5 operation precedes o6 operation |
| **p5 product** |
| o3 operation precedes o6 operation |
| **p6 product** |
| o1 operation precedes o6 operation |
| **p7 product** |
| o3 operation precedes o4 operation |
| o4 operation precedes o5 operation |
| o5 operation precedes o6 operation |
| **p8 product** |
| o1 operation precedes o2 operation |
| o2 operation precedes o5 operation |
| o5 operation precedes o6 operation |

| **Operation for the product** | | | | | | |
|---|---|---|---|---|---|---|
| | o1 | o2 | o3 | o4 | o5 | o6 |
| p1 | **1** | **1** | **1** | **-** | **-** | **1** |
| p2 | **1** | **1** | **1** | **1** | **1** | **1** |
| p3 | **-** | **-** | **1** | **1** | **1** | **1** |
| p4 | **1** | **1** | **-** | **-** | **1** | **1** |
| p5 | **-** | **-** | **1** | **-** | **-** | **1** |
| p6 | **1** | **-** | **-** | **-** | **-** | **1** |
| p7 | **-** | **-** | **1** | **1** | **1** | **1** |
| p8 | **1** | **1** | **-** | **-** | **1** | **1** |

| **Operation on machines** | | | | | | |
|---|---|---|---|---|---|---|
| | o1 | o2 | o3 | o4 | o5 | o6 |
| m1 | **1** | **-** | **-** | **-** | **-** | **-** |
| m2 | **1** | **-** | **-** | **-** | **-** | **-** |
| m3 | **-** | **-** | **1** | **-** | **-** | **-** |
| m4 | **-** | **-** | **1** | **-** | **-** | **-** |
| m5 | **-** | **-** | **1** | **-** | **-** | **-** |
| m6 | **-** | **-** | **-** | **-** | **1** | **-** |
| m7 | **-** | **1** | **-** | **1** | **-** | **-** |
| m8 | **-** | **1** | **-** | **1** | **-** | **1** |
| m9 | **-** | **1** | **-** | **-** | **-** | **1** |
| m10 | **-** | **-** | **-** | **-** | **-** | **1** |

| Orders for products | |
|---|---|
| product p1 has been ordered | **80** |
| product p2 has been ordered | **40** |
| product p3 has been ordered | **60** |
| product p4 ordered | **50** |
| product p5 ordered | **200** |
| product p6 has been ordered | **20** |
| product p7 ordered | **40** |
| product p8 ordered | **10** |

## 6 Conclusion

Manufacturing companies continually aim at increasing the performance and effectiveness of maintenance processes. The emphasis is put on the elimination of unexpected failures which generate unnecessary costs and production losses [11]. The paper proposes a model and its implementation, as well as the use of the architecture meta-model for modeling and solving decision-making and optimization problems in the production area. On the one hand, the framework uses the enterprise architecture (TOGAF) and on the other, mathematical programming. The production and maintenance optimization model (Sect. 3) was built on the basis of the enterprise architecture meta-model. The proposed optimization model (Sect. 4) enables the optimization costs and also ensures better machine utilization. This model is only one of the few that can be built based on the proposed meta-model. These can be models related to distribution, routing, scheduling and distribution of loads, etc. The modeled problem, like most problems of this type known in the literature, is classified as a discrete constrained optimization problem [12, 13]. Such problems are characterized by high computational complexity, which results in the involvement of large resources and time to solve them. The computational experiments carried out showed the usefulness of using the model in relation to manual allocation, both in terms of computation time and the results obtained, which turned out to be better (comparison of the results presented in Fig. 2 and Fig. 3 with the results shown in Fig. 5 and Fig. 6).

The second experiment showed how to optimally execute orders (i.e. according to what schedule) while taking into account changeover costs variable that affects the load on machines. In future works, it is planned to use not only mathematical programming methods to solve the problem, but also constraint programming methods, hybrid methods and selected heuristics such as genetic algorithms, ant algorithms, etc. The use of these implementation methods will be aimed at reducing the computation time and solving larger problems. It is also planned to expand the meta-model towards the development of an ontology [14,15,16] for planning production processes using the technological and business layers.

## Appendix A. Checking the Model for the Data from the Illustrative Example

**Table 7.** Results for illustrative example

| Input | | | | | | | | | | | | |
|---|---|---|---|---|---|---|---|---|---|---|---|---|
| | Weights | | Capacity | | | | | | | | | |
| Exp | K1 | K2 | m1 | m2 | m3 | m4 | m5 | m6 | m7 | m8 | m9 | m10 |
| 1 | 1000 | 100 | 200 | 200 | 200 | 200 | 200 | 200 | 200 | 200 | 200 | 200 |
| 2 | 1000 | 100 | 100 | 100 | 100 | 100 | 100 | 100 | 100 | 100 | 100 | 100 |
| 3 | 1000 | 100 | 50 | 50 | 50 | 50 | 50 | 50 | 50 | 50 | 50 | 50 |
| 4 | 1000 | 100 | 25 | 25 | 25 | 25 | 25 | 25 | 25 | 25 | 25 | 25 |
| 5 | 1000 | 50 | 200 | 200 | 200 | 200 | 200 | 200 | 200 | 200 | 200 | 200 |
| 6 | 1000 | 50 | 100 | 100 | 100 | 100 | 100 | 100 | 100 | 100 | 100 | 100 |
| 7 | 1000 | 50 | 50 | 50 | 50 | 50 | 50 | 50 | 50 | 50 | 50 | 50 |
| 8 | 1000 | 50 | 25 | 25 | 25 | 25 | 25 | 25 | 25 | 25 | 25 | 25 |
| 9 | 1000 | 20 | 200 | 200 | 200 | 200 | 200 | 200 | 200 | 200 | 200 | 200 |
| 10 | 1000 | 20 | 100 | 100 | 100 | 100 | 100 | 100 | 100 | 100 | 100 | 100 |
| 11 | 1000 | 20 | 50 | 50 | 50 | 50 | 50 | 50 | 50 | 50 | 50 | 50 |
| 12 | 1000 | 20 | 25 | 25 | 25 | 25 | 25 | 25 | 25 | 25 | 25 | 25 |
| Output | | | | | | | | | | | | |
| | Comment | | Machine load (%) | | | | | | | | | |
| Exp | Fc | C | m1 | m2 | m3 | m4 | m5 | m6 | m7 | m8 | m9 | m10 |
| 1 | 9130 | 1200 | 2,5 | 7.5 | 6 | 2 | 13 | 10 | 8.5 | 18.5 | 11.5 | 2.5 |
| 2 | 9270 | 1200 | 19 | 1 | 14 | 8 | 20 | 20 | 18 | 13 | 28 | 23 |
| 3 | 20230 | 1200 | 14 | 24 | 52 | 16 | 16 | 38 | 36 | 34 | 38 | 52 |
| 4 | 54640 | 1200 | 28 | 40 | 80 | 46 | 30 | 66 | 70 | 64 | 70 | 86 |
| 5 | 7360 | 1200 | 9 | 1 | 21 | 0 | 0 | 10 | 14 | 24 | 1 | 2 |
| 6 | 7890 | 1200 | 4 | 16 | 24 | 14 | 4 | 20 | 20 | 24 | 26 | 12 |
| 7 | 19110 | 1200 | 40 | 0 | 52 | 0 | 30 | 38 | 38 | 18 | 30 | 74 |
| 8 | 53020 | 1200 | 58 | 10 | 76 | 50 | 30 | 66 | 70 | 54 | 80 | 86 |
| 9 | 6470 | 1200 | 10 | 0 | 16 | 2 | 3 | 10 | 5 | 6 | 17 | 13 |
| 10 | 7160 | 1200 | 12 | 8 | 22 | 16 | 4 | 20 | 20 | 21 | 31 | 10 |
| 11 | 18110 | 1200 | 14 | 24 | 54 | 0 | 30 | 38 | 28 | 0 | 58 | 74 |
| 12 | 51900 | 1200 | 60 | 8 | 56 | 30 | 70 | 66 | 66 | 76 | 70 | 78 |

| Ex | Fc1 | Fc2 | Fc3 | p1 | p2 | p3 | p4 | p5 | p6 | p7 | p8 |
|---|---|---|---|---|---|---|---|---|---|---|---|
| 1 | 0 | 6030 | 3100 | 80 | 40 | 60 | 50 | 200 | 20 | 40 | 10 |
| 2 | 0 | 6170 | 3100 | 80 | 40 | 60 | 50 | 200 | 20 | 40 | 10 |
| 3 | 10000 | 7530 | 2700 | 80 | 40 | 60 | 50 | 200 | 20 | 40 | 0 |
| 4 | 40000 | 11240 | 3400 | 75 | 25 | 50 | 50 | 200 | 20 | 40 | 0 |
| 5 | 0 | 5860 | 1500 | 80 | 40 | 60 | 50 | 200 | 20 | 40 | 10 |
| 6 | 0 | 6290 | 1600 | 80 | 40 | 60 | 50 | 200 | 20 | 40 | 10 |
| 7 | 10000 | 7410 | 1700 | 80 | 40 | 50 | 50 | 200 | 20 | 40 | 10 |
| 8 | 4000 | 11070 | 1950 | 75 | 25 | 50 | 50 | 200 | 20 | 40 | 0 |
| 9 | 0 | 5790 | 680 | 80 | 40 | 60 | 50 | 200 | 20 | 40 | 10 |
| 10 | 0 | 6500 | 660 | 80 | 40 | 60 | 50 | 200 | 20 | 40 | 10 |
| 11 | 10000 | 7430 | 680 | 80 | 40 | 60 | 50 | 200 | 20 | 40 | 0 |
| 12 | 40000 | 11120 | 780 | 75 | 25 | 50 | 50 | 200 | 20 | 40 | 0 |

## References


1. Greefhorst, D., Proper, E.: Architecture Principles - The Cornerstones of Enterprise Architecture, pp. 1867–8920 (2011). https://doi.org/10.1007/978-3-642-20279-7
2. The Open Group Architecture Framework (TOGAF) Standard (2018). https://pubs.opengroup.org/architecture/togaf9-doc/arch/index.html
3. Juzoń, Z., Wikarek, J., Sitek, P.: The use of corporate architecture in planning and automation of production processes. In: Machado, J.M., et al. (eds.) DCAI 2022. LNNS, vol. 585, pp. 21–32. Springer, Cham (2023). https://doi.org/10.1007/978-3-031-23210-7_3
4. Opekunova, L.A., Opekunov, A.N., Kamardin, I.N., Nikitina, N.V.: Modeling enterprise architecture using language ArchiMate. In: Ashmarina, S.I., Vochozka, M., Mantulenko, V.V. (eds.)

ISCDTE 2019. LNNS, vol. 84, pp. 506–513. Springer, Cham (2020). https://doi.org/10.1007/978-3-030-27015-5_61

5. Scrimieri, D., Afazov, S.M., Ratchev, S.M.: Design of a self-learning multi-agent framework for the adaptation of modular production systems. Int. J. Adv. Manuf. Technol. **115**(5–6), 1745–1761 (2021). https://doi.org/10.1007/s00170-021-07028-z
6. Mejía, G., Montoya, C., Bolívar, S., et al.: Job shop rescheduling with rework and reconditioning in Industry 4.0: an event-driven approach. Int. J. Adv. Manuf. Technol. **119**, 3729–3745 (2022). https://doi.org/10.1007/s00170-021-08163-3
7. Horng, S.C., Lin, S.S.: Apply ordinal optimization to optimize the job-shop scheduling under uncertain processing times. Arab. J. Sci. Eng. **47**, 9659–9671 (2021). https://doi.org/10.1007/s13369-021-06317-9
8. Ahmadian, M.M., Salehipour, A.: The just-in-time job-shop scheduling problem with distinct due-dates for operations. J. Heuristics **27**(1–2), 175–204 (2020). https://doi.org/10.1007/s10732-020-09458-6
9. Parmenter, D.: Key Performance Indicators (KPIs). Creation, Implementation and Use, ISBN 978-83-283-8523-8. EAN 9788328385238 (2021)
10. Gurobi (2022). https://www.gurobi.com
11. Antosz, K., Pasko, L., Gola, A.: the use of intelligent systems to support the decision-making process in lean maintenance management. IFAC PapersOnLine **52–10**, 148–153 (2019). https://doi.org/10.1016/j.ifacol.2019.10.005
12. Sitek, P., Wikarek, J.: A multi-level approach to ubiquitous modeling and solving constraints in combinatorial optimization problems in production and distribution. Appl. Intell. **48**(5), 1344–1367 (2017). https://doi.org/10.1007/s10489-017-1107-9
13. Sitek, P., Wikarek, J.: A hybrid programming framework for modeling and solving constraint satisfaction and optimization problems. Sci. Program. **2016**, 13 (2016). https://doi.org/10.1155/2016/5102616. ID 5102616
14. Thibbotuwawa, A., Bocewicz, G., Nielsen, P., Banaszak, Z.: Planning deliveries with UAV routing under weather forecast and energy consumption constraints. IFAC-PapersOnLine **52**(13), 820–825 (2019). https://doi.org/10.1016/j.ifacol.2019.11.231
15. Denkena, B., André Dittrich, M., Vibora Münch, G.: Ontology-based production planning under the consideration of system robustness **104**, 547–552 (2021). https://doi.org/10.1016/j.procir.2021.11.092
16. Hnatkowska, B., Kozierkiewicz, A., Pietranik, M.: Semi-automatic definition of attribute semantics for the purpose of ontology integration. IEEE Access **8**, 107272–107284, 9 (2020). https://doi.org/10.1109/ACCESS.2020.3000035

# Web Based Application for Probability Job Scheduling

Tomasz Michno[1,2(✉)], Roman Stanislaw Deniziak[2], and Aleksandra Michno[1]

[1] Austrian Institute of Technology, Vienna, Austria
tomasz.michno@gmail.com, aleksandra.michno90@gmail.com
[2] Kielce University of Technology, Kielce, Poland
s.deniziak@tu.kielce.pl

**Abstract.** The problem of job scheduling has been studied by various researchers since the early fifties of the 20th century. It can be defined as finding the most optimal assignment of different jobs to some resources such as machines, e.g. to find the lowest cost, which may be time. One of the most common problems in preparing the job scheduling is the precise definition of the time required to complete the task. In some situations this can be difficult because there may be many factors with uncertainties. In this paper we have focused on this problem from a practical point of view, which may be useful for e.g. project managers. For such situation, solutions using fuzzy sets or probability distributions may be relevant, but unfortunately they may be not easy to understand and use by people without advanced mathematical background. This paper presents a tool for probabilistic job scheduling. The web-based application has been prepared as well as a job scheduling approach using probability and four types of results showing the most probable total time, least probable, maximum and minimum. This can be useful for decision making by e.g. project managers. As a result, a web application for probabilistic job scheduling was created using Python, HTML, CSS and Flask framework.

**Keywords:** job scheduling · probability · web application

## 1 Introduction

The problem of job scheduling has been studied by various researchers since the early fifties of the 20th century [5]. It can be defined as finding the most optimal assignment of different jobs to some resources such as machines, e.g. to find the lowest cost, which may be a time. There have been developed many algorithms to solve this problem, e.g. using heuristic approaches such as NEH [9], genetic algorithms [7], constraint logic programming [13] or deep learning [2].

The job scheduling problem has many applications, such as job scheduling of different parts of products in a factory [1], but also job scheduling in project management [6]. Another possible use is surgical case scheduling [10]. Job scheduling


R. Mehmood et al. (Eds.): DCAI 2023, LNNS 741, pp. 36–45, 2023.
https://doi.org/10.1007/978-3-031-38318-2_4

has also many connections and similarities with graph-subgraph matching. Due to that fact, some algorithms can be applied to this area after adaptations [3].

There are many variations on the job scheduling problem, using for example fuzzy approaches [14], scheduling with spread time constraints [4], cost sharing [8] or the cyclic scheduling problem for job shops [15]. There are also some situations where there is a need to set criticality levels for jobs. In addition, there may be added probabilities of running a job [11]. Another research focuses on the probability of success or failure of a manufacturing job and its impact on other jobs [12].

One of the most common problems in preparing the job scheduling is the precise definition of the time required to complete the task. In some situations this can be difficult because there may be many factors with uncertainties. In this paper we have focused on this problem from a practical point of view, which may be useful for e.g. project managers. For such situation, solutions using fuzzy sets or probability distributions may be relevant, but unfortunately they may be not easy to understand and use by people without advanced mathematical background. Therefore, we propose to use a simplified approach that will be easy to use for users. For each job, the most probable, minimum and maximum completion times are defined with relevant probabilities. Moreover, as a result of the scheduling, four different solutions to the problem are proposed: the one that focuses on the most probable total time and the least probable situation and two for the maximum and minimum times. The concept of using probability values for each type of time can be used even if users are not able to calculate their precise values. Instead of them, some rough approximations may be used. The approach most similar to our is presented in [11], but the authors focused on different aspects of probabilities usage. Moreover, their work does not cover results presentation using different diagrams, as well as they did not provide the practical application with graphical user interface.

The paper is structured as follows: in the next section we describe the proposed approach, including the practical application. The third section is the presentation of experimental results. The last section summarizes our paper and includes a description of future work.

## 2 Proposed Approach

### 2.1 A Probabilistic Job Scheduling Problem

The main problem with many classical approaches to job scheduling is that they do not take into account the uncertainties that most often occur in real life. During scheduling, there may be different types of tasks: those that are easy to estimate fairly accurately and those that are difficult to estimate due to insufficient knowledge about them or dependencies on uncertain factors. Sometimes the people who prepare estimates may also not be sure, for example, how long it will take to complete the job. Despites this issues job scheduling and some decisions have to be made.

As a solution to such problems, the authors of this paper propose the approach based on the use of some degree of uncertainty and probability for job scheduling. To make the concept easier to read and understand, we use as a value associated with the job its completion time, but in the presented concept any type of numerical value can be used.

The main idea of the paper is to define three different completion times for each job: the most possible, the shortest and the longest. For each time, its probability should be given, e.g. for the most possible time its probability can be 0.7, for the shortest time 0.1 and for the longest time 0.2. Sometimes it can be difficult to estimate all probabilities directly, so they can be estimated roughly using the following values:

- high probability = 0.8,
- low probability = 0.1,
- moderate probability = 0.5.

The main advantage of such an approach is the possibility to provide information about the scheduling with the longest time, the shortest time, the most probable time and the least probable time, all with information about their probability. Such an information can be very important for making decisions e.g. about the final delivery of a product.

In order to provide an example of the presented idea an example may be given with the jobs and machines assignment shown in the Table 1. For these jobs, four different schedules have been computed using a greedy algorithm, as shown in the Fig. 1. Not only the total time is shown, but also the total probability of the whole schedule. In this situation it can be seen that the most probable time is 9 and all the other values are very unlikely to happen. Instead of the probabilities, there can be also seen the maximum (20) and minimum (4) times.

**Table 1.** Jobs and machines assignment.

| | | Machine 1 (M0) | Machine 2 (M1) |
|---|---|---|---|
| Job 1 (J0) | Probs: | 0.1 \| 0.8 \| 0.1 | 0.4 \| 0.5 \| 0.1 |
| | Times: | 3 \| 5 \| 10 | 10 \| 15 \| 30 |
| Job 2 (J1) | Probs: | 0.3 \| 0.6 \| 0.1 | 0.1 \| 0.8 \| 0.1 |
| | Times: | 2 \| 4 \| 5 | 1 \| 2 \| 3 |
| Job 3 (J2) | Probs: | 0.1 \| 0.8 \| 0.1 | 0.1 \| 0.8 \| 0.1 |
| | Times: | 1 \| 10 \| 11 | 2 \| 8 \| 15 |

Aforementioned approach may be extended by two types of jobs:

- dependent - where a job may be started only after finishing another one,
- optional - where a job may be executed or not.

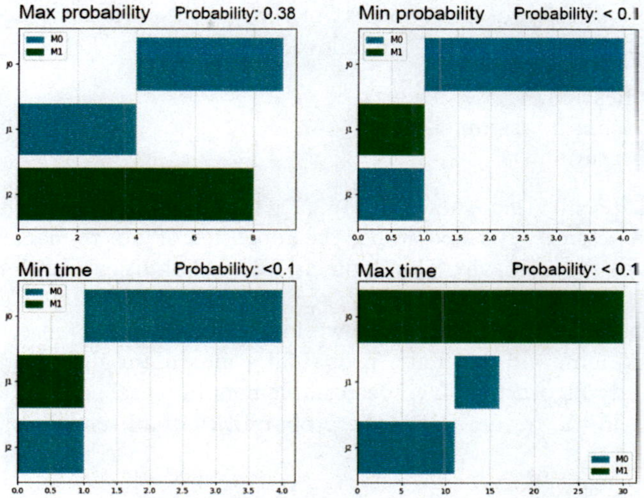


**Fig. 1.** The graphical illustration of different types of job scheduling

Dependent jobs can be defined as jobs that must wait for the completion of at least one other job. They can be run on different machines (unless defined otherwise). Optional jobs can be run and completed only if they do not increase the total time, which can reduce the total unused time of the machines. For computing all times different existing algorithms may be used, e.g.:

- brute force,
- NEH,
- greedy,
- random search.

For the most frequently repeated job schedules (e.g. a production of the same product), statistics can be made about the real times for each job and machine. This can then be used to update the probabilities and make some recommendations for future decisions. Future research will investigate the influence of the use of real data for generating the distributions and different ways to make them easily usable in the system. One of the ideas is to aggregate all times and compute probabilities to use this data to generate suggestions during query preparation. The suggestions can be provided with time constraints (e.g. last 30 days or last 6 months).

The proposed job scheduling method can be applied not only to production but also to other areas, e.g. scheduling of different work packages or tasks in different partners of a grant consortium.

### 2.2 The Web-Based Application for Job Scheduling

As another part of the research, the practical application was implemented. The application was written in Python with HTML and CSS added. In order to simplify the preparation of the web part of the tool, the Flask library has been used. The application consists of the following parts:

- initial specification,
- jobs and machines assignment,
- results presentation.

The initial specification, shown in Fig. 2, is used to define the number of jobs and available machines. The next step is the allocation of jobs to machines, shown in Fig. 3. At the current stage of development, all probabilities and times must be defined one after the other, separated by semicolons. The last step is the calculation and presentation of the results, shown in Fig. 4. Four different schedules are presented: maximum and minimum probability, maximum and minimum time. NEH and Greedy algorithms have been implemented as scheduling algorithms. The last graph in the results shows the probabilities of all schedules.

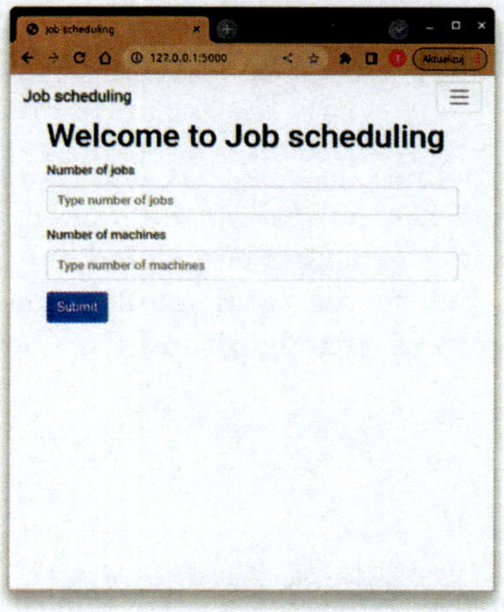


**Fig. 2.** The initial specification screen.

## 3 Experimental Results

The web application implemented during the research was tested both at the user interface level and at the level of the algorithms used. The user level tests consisted of checking that all system elements were responsive and working as designed.

The second type of testing was carried out to investigate how two implemented job scheduling algorithms, the NEH and Greedy, behave with different numbers of jobs and machines. To do this, the following sets of tests were examined:

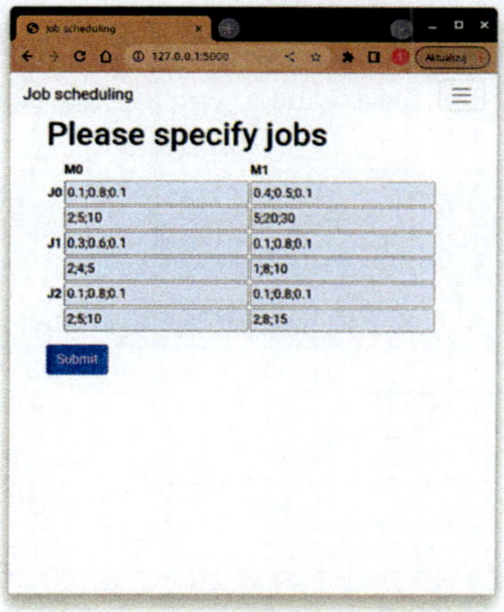


**Fig. 3.** The jobs and machines assignment screen.

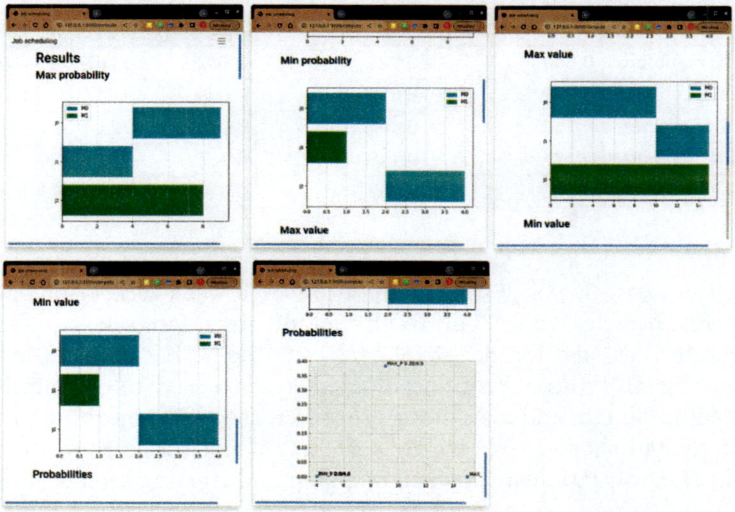


**Fig. 4.** The scheduling result.

- 3 jobs on 2 machines,
- 4 jobs on 2 machines,
- 8 jobs on 4 machines,
- 20 jobs on 4 machines,
- 30 jobs on 4 machines.

The computation times for different numbers of jobs and machines are shown in the Table 2 and 3 and in the Fig. 6. As can be seen for most of the test sets, the times for both algorithms and all result types are quite similar. Only for 20 jobs and 4 machines the Max probability result with the Greedy algorithm was much slower, but this may be due to the atypical path that had to be processed.

**Table 2.** Computation times for greedy algorithm and different number of jobs and machines in ms

| | Max probability | Min probability | Max time | Min time | Avg |
|---|---|---|---|---|---|
| 3 jobs, 2 machines | 0.171 | 0.035 | 0.035 | 0.024 | 0.066 |
| 4 jobs 2 machines | 0.032 | 0.018 | 0.05 | 0.019 | 0.03 |
| 8 jobs 4 machines | 0.05 | 0.053 | 0.027 | 0.066 | 0.049 |
| 20 jobs 4 machines | 1.126 | 0.101 | 0.090 | 0.097 | 0.353 |
| 30 jobs 4 machines | 0.381 | 0.370 | 0.183 | 0.392 | 0.332 |

**Table 3.** Computation times for NEH algorithm and different number of jobs and machines in ms

| | Max probability | Min probability | Max time | Min time | Avg |
|---|---|---|---|---|---|
| 3 jobs, 2 machines | 0.069 | 0.025 | 0.022 | 0.061 | 0.044 |
| 4 jobs 2 machines | 0.029 | 0.027 | 0.069 | 0.047 | 0.043 |
| 8 jobs 4 machines | 0.114 | 0.163 | 0.068 | 0.133 | 0.120 |
| 20 jobs 4 machines | 0.374 | 0.119 | 0.135 | 0.420 | 0.262 |
| 30 jobs 4 machines | 0.317 | 0.357 | 0.377 | 0.426 | 0.369 |

During the tests, not only the computation time was checked, but also the total time for the scheduling. The results for different test sets are shown in Fig. 5. It can be seen that for most of the test sets the NEH algorithm produced higher values for all types of results compared to the Greedy algorithm (with a greater difference in min and max times). However, the NEH algorithm produced schedules that had shorter time when a machine is not running, which can be beneficial in some situations (e.g. when stopping and starting a machine is time consuming). An example of such a schedule is shown in Fig. 7.

As a last part of the results, probabilities were examined. Their values were quite similar for all types of results and tests sets - the highest values were obtained for "Max probability" and much lower for other types of results (Fig. 6).

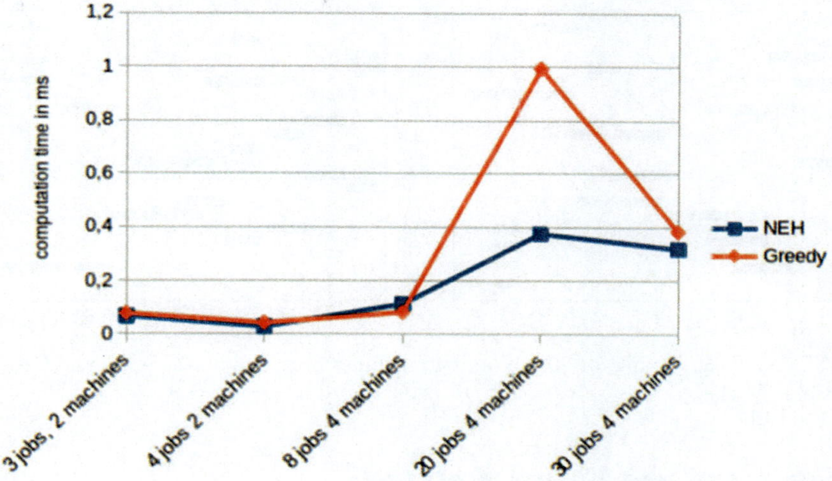


**Fig. 5.** The diagram of computation times for NEH and Greedy algorithms for maximum probability.

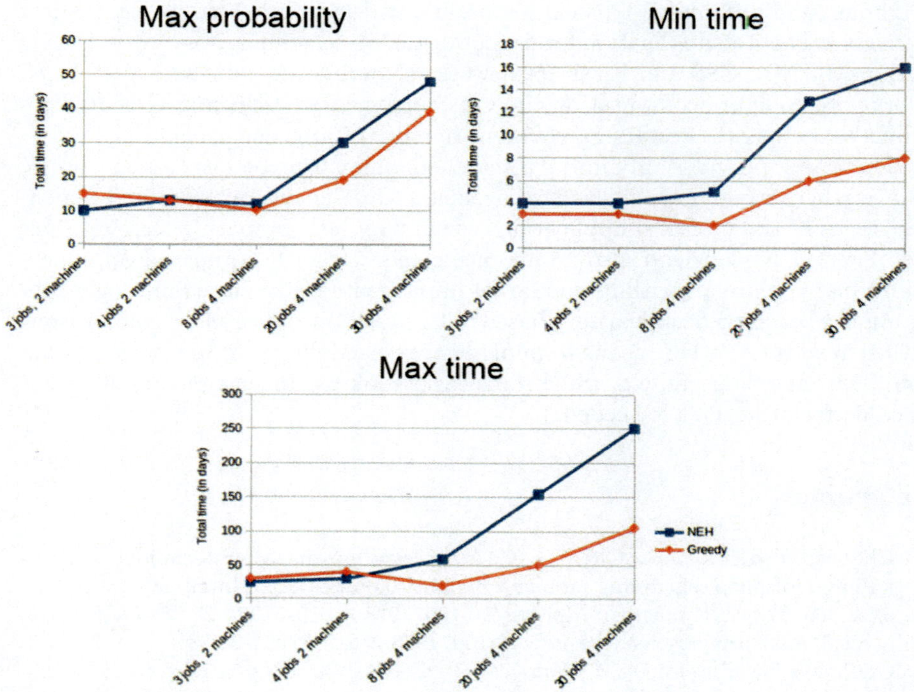


**Fig. 6.** The diagram of total times for NEH and Greedy algorithms.

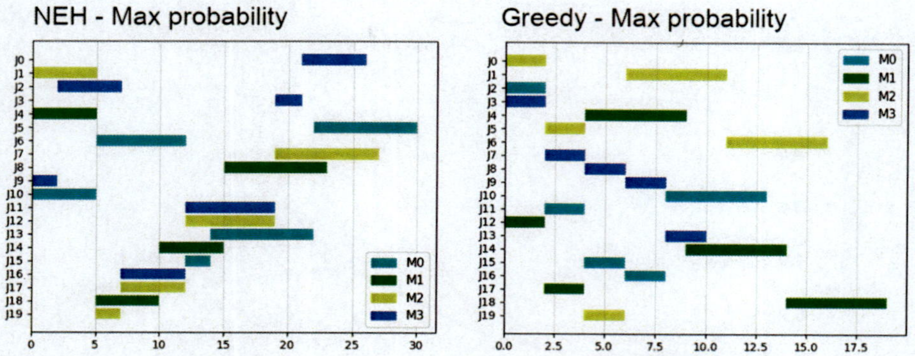


**Fig. 7.** The job scheduling for 20 jobs and 4 machines provided by the NEH and Greedy algorithms.

## 4 Summary and Future Research

This paper presents a tool for probabilistic job scheduling. The web-based application has been prepared as well as a job scheduling approach using probability and four types of results showing the most probable total time, least probable, maximum and minimum. This can be useful for decision making by e.g. project managers. The web application for probabilistic job scheduling was created using Python, HTML, CSS and Flask framework. Experiments have been performed in order to compare computation times of implemented NEH and Greedy algorithms as well as the results for different number of jobs and machines.

This paper provides only initial research and may be extended in many areas. As a further development ideas we consider to implement optional and dependent tasks in the practical application, which may help preparing tests on their usefulness. This can lead also to executing more tests on prepared approach. Additionally, more algorithms should be implemented, like constraint logic programming based or deep learning based with possibility to prepare comparisons of their results. Another idea is to implement rescheduling during active production. This may be useful e.g. when some tasks took too much time and different schedule for other jobs is needed.

## References

1. Chaouch, I., Driss, O.B., Ghedira, K.: A survey of optimization techniques for distributed job shop scheduling problems in multi-factories. Adv. Intell. Syst. Comput. **574**, 369–378 (2017). https://doi.org/10.1007/978-3-319-57264-2_38/TABLES/2, https://link.springer.com/chapter/10.1007/978-3-319-57264-2_38
2. Cunha, B., Madureira, A.M., Fonseca, B., Coelho, D.: Deep reinforcement learning as a job shop scheduling solver: a literature review. Adv. Intell. Syst. Comput. **923**, 350–359 (2020). https://doi.org/10.1007/978-3-030-14347-3_34/COVER, https://link.springer.com/chapter/10.1007/978-3-030-14347-3_34

3. Deniziak, R.S., Michno, T.: Graph of primitives matching problem in the world wide web CBIR searching using query by approximate shapes. Adv. Intell. Syst. Comput. **1004**, 77–84 (2020). https://doi.org/10.1007/978-3-030-23946-6_9/COVER, https://link.springer.com/chapter/10.1007/978-3-030-23946-6_9
4. Fischetti, M., Martello, S., Toth, P.: The fixed job schedule problem with spread-time constraints. Oper. Res. **35**, 849–858 (1987). https://doi.org/10.1287/opre.35.6.849, https://pubsonline.informs.org/doi/abs/10.1287/opre.35.6.849
5. Graham, R.L., Lawler, E.L., Lenstra, J.K., Kan, A.H.: Optimization and approximation in deterministic sequencing and scheduling: a survey. Ann. Discrete Math. **5**, 287–326 (1979). https://doi.org/10.1016/S0167-5060(08)70356-X
6. Kianpour, P., Gupta, D., Krishnan, K.K., Gopalakrishnan, B.: Automated job shop scheduling with dynamic processing times and due dates using project management and industry 4.0. J. Ind. Prod. Eng. **38**, 485–498 (2021). https://doi.org/10.1080/21681015.2021.1937725, https://www.tandfonline.com/doi/abs/10.1080/21681015.2021.1937725
7. Kundakci, N., Kulak, O.: Hybrid genetic algorithms for minimizing makespan in dynamic job shop scheduling problem. Comput. Ind. Eng. **96**, 31–51 (2016). https://doi.org/10.1016/J.CIE.2016.03.011
8. Mishra, D., Rangarajan, B.: Cost sharing in a job scheduling problem. Soc. Choice Welfare **29**, 369–382 (2007). https://doi.org/10.1007/S00355-006-0210-3/METRICS, https://link.springer.com/article/10.1007/s00355-006-0210-3
9. Nawaz, M., Enscore, E.E., Ham, I.: A heuristic algorithm for the m-machine, n-job flow-shop sequencing problem. Omega **11**, 91–95 (1983). https://doi.org/10.1016/0305-0483(83)90088-9
10. Pham, D.N., Klinkert, A.: Surgical case scheduling as a generalized job shop scheduling problem. Eur. J. Oper. Res. **185**, 1011–1025 (2008). https://doi.org/10.1016/J.EJOR.2006.03.059
11. Seddik, Y., Hanzálek, Z.: Match-up scheduling of mixed-criticality jobs: maximizing the probability of jobs execution. Eur. J. Oper. Res. **262**, 46–59 (2017). https://doi.org/10.1016/J.EJOR.2017.03.054
12. Shoval, S., Efatmaneshnik, M.: A probabilistic approach to the stochastic job-shop scheduling problem. Procedia Manuf. **21**, 533–540 (2018). https://doi.org/10.1016/J.PROMFG.2018.02.154
13. Sitek, P., Wikarek, J.: A hybrid programming framework for modeling and solving constraint satisfaction and optimization problems. Sci. Program. 2016 (2016). https://doi.org/10.1155/2016/5102616
14. Wang, G.G., Gao, D., Pedrycz, W.: Solving multiobjective fuzzy job-shop scheduling problem by a hybrid adaptive differential evolution algorithm. IEEE Trans. Ind. Inf. **18**, 8519–8528 (2022). https://doi.org/10.1109/TII.2022.3165636
15. Wójcik, R., Pempera, J., Bocewicz, G., Banaszak, Z.: The no-wait cyclic scheduling of delivery traffic in the grid distribution network. ISA Trans. **131**, 282–298 (2022). https://doi.org/10.1016/J.ISATRA.2022.05.031

# Working Time Standardization in the Assembly Process of Regenerated Elements Using MES System and Timing Method

Łukasz Wójcik[1](✉) and Arkadiusz Gola[2]

[1] BORG Automotive sp. z o.o., ul. Metalurgiczna 13B, 20-001 Lublin, Poland
lwoj@pl.borgautomotive.com
[2] Department of Production Computerisation and Robotisation, Lublin University of Technology, ul. Nadbystrzycka 36, 20-619 Lublin, Poland
a.gola@pollub.pl

**Abstract.** The issue of working time standardization is a key element in the process of planning, organization and control of production processes - the possibility of timely execution of orders for production orders depends on the compliance of standards with the actual time of task execution. As a consequence, establishing reliable time standards, as close as possible to the real times of individual tasks, becomes a significant challenge. This article deals with the issue of working time standardization in the process of regenerating brake calipers. In particular, the focus was on the assembly operation, analyzing unit times collected by the MES (Manufacturing Execution System) class system, and - for the purpose of comparative analysis - timing was carried out at assembly stations. As part of the conclusions, general guidelines were formulated on the method of time standardization in processes with different times of unit execution of production tasks.

**Keywords:** regeneration · assembly · unit time · timing · MES

## 1 Introduction

Contemporary market requirements impose on companies the requirement of timely execution of orders, which translates into the need for proper planning and scheduling of production [1]. As a consequence, the task of key importance is the correct determination of the execution times of individual technological tasks (operations) - properly established time norms are to enable both reliable determination of the dates and costs of execution of production orders and, consequently, have a direct impact on the correctness of the developed plans operational (production schedules) [2, 3]. While in the case of repetitive serial production, determining the correct standards is a relatively easy task - in the case of processes with high variability - this task is a big challenge even if you have historical data (e.g. in ERP - Enterprise Resources Planning or MES class systems) [4].

In production conditions, apart from the operation time, the factors determining productivity are [5]:

- availability of parts and their correct preparation,


R. Mehmood et al. (Eds.): DCAI 2023, LNNS 741, pp. 46–55, 2023.
https://doi.org/10.1007/978-3-031-38318-2_5

- tools and devices for efficient work,
- degree of complexity of the assembled parts,
- bug fixes,
- failures and downtime.

All of the above factors have a direct impact on the decline in productivity caused by improper use of available working time (Fig. 1) and are of interest to many researchers dealing with the optimization of production processes (see e.g. [6–10]).

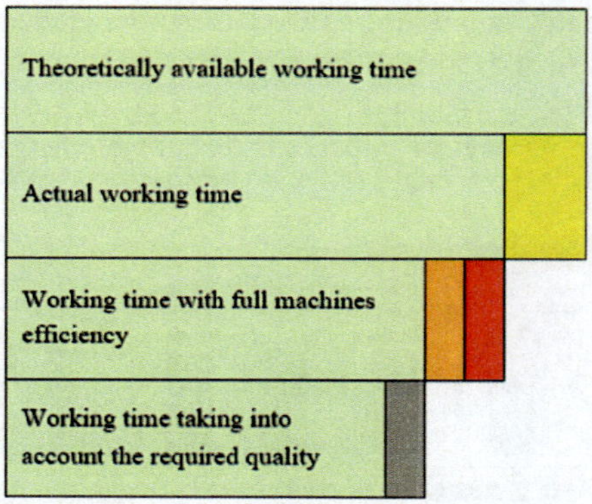


**Fig. 1.** Working time losses during production processes.

| | |
|---|---|
| | Time of planned downtime<br>• changeovers<br>• scheduled breaks (e.g. rest breaks)<br>• cleaning/cleaning<br>• other planned activities |
| | Time for unplanned downtime<br>• failures and faults<br>• staff unavailability<br>• prolonged changeovers<br>• waiting (for raw material, for an employee, for tools) |
| | Time lost due to low efficiency (hard to assemble details) |
| | Time wasted on scrap production |

As part of this article, special attention is paid to the issue of determining unit times in processes with high variability, which can be used e.g. in intelligent robust scheduling algorithms (see e.g. [11]). In particular, the compliance of the unit times recorded in the MES class system with the actual times of the assembly operation of regenerated brake calipers was analyzed. This analysis is aimed at developing a methodology for determining unit times for the needs of operational planning in the production of remanufactured parts.

## 2 The Role of MES Systems in the Management of Production Processes

Both developed and developing countries are increasingly betting on the ideas and tools of Industry 4.0 [12]. The vision of factories with intelligent solutions is becoming not only a direction of development but a kind of norm [13]. Challenges such as disruptions in supply chains and rapidly changing market trends mean that companies must be ready to implement production processes in a flexible manner [14]. As a consequence, in industrial conditions, technologies such as 3D printing, Internet of Things, work in the "cloud" (Cloud Computing), as well as more and more complex production implementation systems such as MES (Manufacturing Execution System) [15].

MES type software (Fig. 2) enables direct transfer of data from machines, production cells, operators or measuring devices to persons responsible for data analysis.

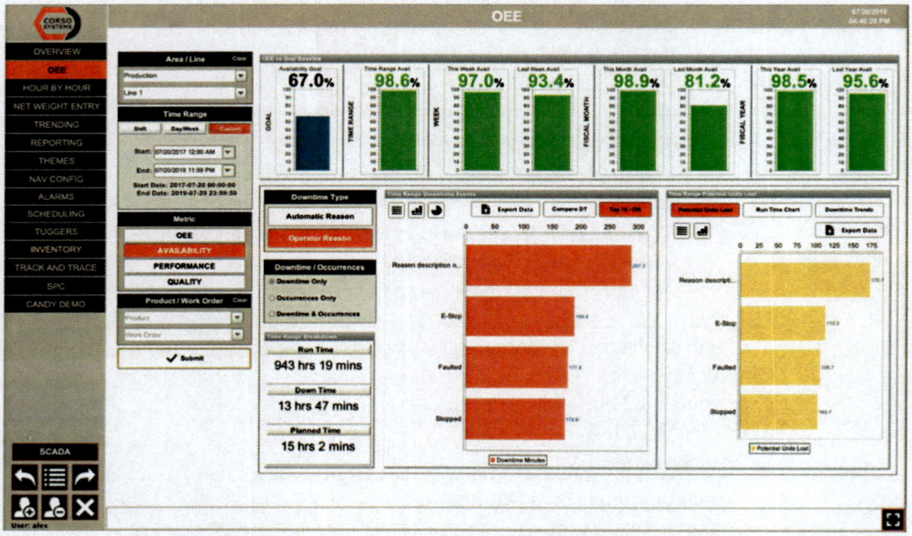


Fig. 2. An example of MES interface [16].

The implementation of the MES class system enables, among others [17]:

- determination of the current location of the semi-finished product in production,
- identification of the bottleneck resulting from the availability of machinery,
- finding the places where the most common failures or downtimes occur,
- reading the execution time of given operations and processes,
- setting production standards at individual stages of the production process,
- collecting data on product defects,
- collecting data on qualitative indicators,
- system integration at the automatic level (between machines) and manual level (entered by the operator).

In addition, using data from the MES system, it is possible:

- extracting data from individual production and assembly stations,
- identification of the exact time of entry and exit of the production batch from the workplace
- identification of the operator(s) performing the work,
- monitoring the flow of a production batch through the process.

As a consequence, having an MES class system, it is possible to collect and analyze real data on production processes, which (in principle, can be used in the process of planning and scheduling production).

## 3 Research Methodology

In order to verify the reliability of unit times of assembly operations of regenerated brake calipers (Fig. 3) generated by the MES class system, research works were carried out covering:

1. Recognition of the characteristics of the enterprise.
2. Determination of measures and indicators of productivity.
3. Performing a general audit.
4. Conducting a detailed audit.
5. Collecting data from IT systems (MES).
6. Analysis and comparison of the obtained results.

**Fig. 3.** Brake calipers after assembly process.

The first stage of the productivity survey was to discuss the directions of the company's development and expectations related to work efficiency. The talks discussed the automation and robotization of manufacturing processes, the possibility of using systems that save energy and materials, methods of developing employees' creativity (KAIZEN), as well as the creation of clear indicators for monitoring productivity.

In the second stage, productivity measures that will be used in the empirical research were indicated. It has been taken into account:

- place of measurement - assembly station,
- comprehensiveness: partial,
- the complexity of the process,
- staff qualifications.

The general audit consisted in reviewing the entire production process and selecting places where a detailed audit will be carried out. During the audit, attention was paid to:

- complexity of operations at individual positions,
- the necessary qualifications of employees required to perform the required work,
- layout of audited areas,
- work culture (adherence to instructions, standards, tests, analysis of mistakes made, efficient cooperation with quality and technology departments).

The next stage was a detailed audit for which a brake caliper assembly station was selected. Each analyzed production series included the same clamp model. The number of pieces in a production series varies and this is a factor resulting from the specificity of the regeneration process. The editors were randomly selected to test differences in productivity for several different teams.

The actual assembly times were determined by the timing method and were performed by the production technologist (Fig. 4).

**Fig. 4.** Brake caliper assembly time measurements.

## 4 Measurement Results and Analysis

Assembly time measurements were made for 24 production series of various sizes. The time of performing individual operations included in the assembly process was measured. The results of exemplary measurements for selected production series are shown in Figs. 5–6.

| | | | | |
|---|---|---|---|---|
| Shift | Morning | B | Date: | 20.12.2022 |
| Workstation | BCF (assembly) | | Observer: | - |
| Series no./caltalogue no. | POB89955 / 87-1349 | | Producer of brake caliper | LUCAS |
| Batch quantity | 36 | | Family of brake caliper | MANUAL |

| No. | Following operations | Recorded operation times in seconds: 1 | 2 | Average | assembly time [s/1piece] | Remarks |
|---|---|---|---|---|---|---|
| 1 | MES | 00:00:15 | - | 00:00:15 | 00:00:00 | - |
| 2 | Disassembling the components - unpacking the parts from the containers | 00:05:30 | - | 00:05:30 | 00:00:09 | - |
| 3 | Lubricating the pistons, putting on the pp cover, grease for the screw | 00:03:48 | - | 00:03:48 | 00:00:05 | - |
| 4 | Distributing the segers into the cuvette, lubricating the sealing rings and o-rings with oil | 00:00:59 | - | 00:00:59 | 00:00:02 | - |
| 5 | Putting o-rings on the mechanisms | 00:14:00 | - | 00:14:00 | 00:00:23 | - |
| 6 | Assembly of the mechanism, circlip, sealing rings. [removing the bodies from the container on the table] | | | | | |
| 7 | Screwing the piston | 00:11:00 | - | 00:11:00 | 00:00:18 | - |
| 8 | Flange lubrication, sealant installation, sealant lubrication | | | | | |
| 9 | Preparation of the press, lubrication of loctite nuts | 00:01:00 | - | 00:01:00 | 00:00:02 | - |
| 10 | Pressing the lever | 00:21:30 | - | 00:21:30 | 00:00:36 | - |
| 11 | Handbrake test (cable selection, marking of the correct test, installation of the spring and putty) | 00:13:00 | - | 00:13:00 | 00:00:22 | - |
| 12 | MES | 00:00:15 | - | 00:00:15 | 00:00:30 | - |
| | | | | | 00:01:59 | - |
| | | | | | 00:00:59 | *Divided by 2 because of two operators* |

**Fig. 5.** The results of the timing operation for a production run of 36 units.

| | | | |
|---|---|---|---|
| Shift | Morning B | Date: | 20.12.2022 |
| Workstation | BCF (assembly) | Observer: | - |
| Series no./caltalogue no. | POB69955 / 87-1349 | Producer of brake caliper | LUCAS |
| Batch quantity | 33 | Family of brake caliper | MANUAL |

| | | Recorded operation times in seconds | | | | |
|---|---|---|---|---|---|---|
| Lp. | Kolejne operacje | 1 | 2 | Average | Assembly time [s/pcs] | Uwagi |
| 1 | MES | 00:00:35 | - | 00:00:35 | 00:00:01 | - |
| 2 | Disassembling the components - unpacking the parts from the containers | 00:00:50 | - | 00:00:50 | 00:00:02 | - |
| 3 | Lubricating the pistons, putting on the pp cover, grease for the screw | 00:05:34 | - | 00:05:34 | 00:00:10 | - |
| 4 | Distributing the segers into the cuvette, lubricating the sealing rings and o-rings with oil | 00:00:30 | - | 00:00:30 | 00:00:01 | - |
| 5 | Putting o-rings on the mechanisms | 00:10:20 | - | 00:10:20 | 00:00:19 | - |
| 6 | Assembly of the mechanism, circlip, sealing rings. [removing the bodies from the container on the table] | | | | | |
| 7 | Screwing the piston | 00:12:55 | - | 00:12:55 | 00:00:23 | - |
| 8 | Flange lubrication, sealant installation, sealant lubrication | | | | | |
| 9 | Preparation of the press, lubrication of loctite nuts | 00:00:00 | - | 00:00:00 | 00:00:00 | - |
| 10 | Pressing the lever | 00:15:55 | - | 00:15:55 | 00:00:29 | - |
| 11 | Handbrake test (cable selection, marking of the correct test, installation of the spring and putty) | 00:11:30 | - | 00:11:30 | 00:00:21 | - |
| 12 | MES | 00:00:15 | - | 00:00:15 | 00:00:00 | - |
| | | | | | 00:01:46 | - |
| | | | | | 00:00:53 | *Divided by 2 because of two operators* |

**Fig. 6.** The results of the timing operation for a production run of 33 units.

The last stage of data collection was downloading data on the time of assembly operations for individual production orders. Generated data along with the calculation of unit times resulting from the size of the production series (Fig. 7).

| Shift | - | Date | - |
|---|---|---|---|
| Workstation | BCF (assembly) | Obserwator: | - |
| Series no./caltalogue no. | 87-1349 | Producer of the caliper: | LUCAS |
| Quantity | - | Family of brake caliper: | MANUAL |
| Start of measurement | - | | |

| Lp. | Batch quantity | Recorded operation time | Assembly time [s/1piece] | Remarks |
|---|---|---|---|---|
| 1 | 32 | 00:15:00 | 00:00:28 | deviates from average |
| 2 | 42 | 00:29:00 | 00:00:54 | |
| 3 | 27 | 00:21:00 | 00:00:39 | |
| 4 | 31 | 00:39:00 | 00:01:13 | |
| 5 | 28 | 00:26:00 | 00:00:49 | |
| 6 | 37 | 00:45:00 | 00:01:24 | |
| 7 | 23 | 00:21:00 | 00:00:39 | |
| 8 | 35 | 00:47:00 | 00:01:28 | |
| 9 | 21 | 00:52:00 | 00:01:38 | deviates from average |
| 10 | 40 | 00:32:00 | 00:01:00 | |
| 11 | 28 | 00:35:00 | 00:01:06 | |
| 12 | 31 | 00:18:00 | 00:00:34 | deviates from average |
| 13 | 35 | 00:45:00 | 00:01:24 | |
| 14 | 30 | 00:33:00 | 00:01:02 | |
| 15 | 34 | 00:43:00 | 00:01:21 | |
| 16 | 27 | 00:42:00 | 00:01:19 | |
| 17 | 35 | 00:38:00 | 00:01:11 | |
| 18 | 37 | 00:35:00 | 00:01:06 | |
| 19 | 36 | 00:43:00 | 00:01:21 | |
| 20 | 34 | 00:48:00 | 00:01:30 | |
| 21 | 27 | 00:36:00 | 00:01:07 | |
| 22 | 34 | 00:35:00 | 00:01:06 | |
| 23 | 40 | 00:44:00 | 00:01:23 | |
| 24 | 35 | 00:36:00 | 00:01:07 | |
| | Average | 00:35:45 | 00:01:07 | *Two people in workstation* |

**Fig. 7.** The results of the generated data along with the calculation of unit times resulting from the size of the production series

## 5 Summary and Conclusions

Differences in assembly times between physical readings and the results from the MES system are insignificant and amount to approx. 15%. Physical measurements are a very good method for analyzing individual assembly components. They can be the basis for mapping the value stream at this position. The data collected from the MES system, on the other hand, allow for a broader view of the assembly process. You can easily analyze the generated files. A long history of production is available, thanks to which you can check how productivity has changed with the increase in employee experience, the introduction of new devices or layout changes. It also gives a signal how much time is wasted by improvements, errors and failures.

In the future, research using the MES system will be carried out at other stations. They will be aimed at obtaining full data from the entire brake caliper regeneration line, which will allow optimal planning and management of production.

## References


1. Relich, M., Nielsen, I., Gola, A.: Reducing the total product cost at the product design stage. Appl. Sci. **12**(4), 1921 (2022)
2. Juzoń, Z., Wikarek, J., Sitek, P.: The use of corporate architecture in planning and automation of production processes. In: Machado, J.M., et al. (eds.) DCAI 2022. LNNS, vol. 585, pp. 21–32. Springer, Cham (2023). https://doi.org/10.1007/978-3-031-23210-7_3
3. Musiał, K., Balashov, A., Burduk, A., Rysińska-Wojtasik, D.: Solving manufacturing orders scheduling problem using annealing simulation. In: Machado, J., et al. (eds.) icieng 2022. LNME, pp. 279–292. Springer, Cham (2023). https://doi.org/10.1007/978-3-031-09382-1_25
4. Danilczuk, W., Gola, A.: Computer-aided material demand planning using ERP systems and business intelligence technology. Appl. Comput. Sci. **16**(3), 42–55 (2020)
5. Juzoń, Z., Wikarek, J., Sitek, P.: The concept of use of process data and enterprise architecture to optimize the production process. In: Szewczyk, R., Zieliński, C., Kaliczyńska, M., Bučinskas, V. (eds.) AUTOMATION 2023. LNNS, vol. 630, pp. 73–84. Springer, Cham (2023). https://doi.org/10.1007/978-3-031-25844-2_7
6. Wittbrodt, P., Łapuńka, I., Baytikenova, G., Gola, A., Zakimova, A.: Identification of the impact of the availability factor on the efficiency of production processes using the AHP and fuzzy AHP methods. Appl. Comput. Sci. **18**(4), 116–129 (2022)
7. Burduk, A., Łapczyńska, D., Kochańska, J., Musiał, K., Więcek, D., Kuric, I.: Waste management with the use of heuristic algorithms and Internet of Things technology. Sensors **22**(22), 8786 (2022)
8. Smutnicki, C., Pepmera, J., Bocewicz, G., Banaszak, Z.: Cyclic flow-shop scheduling with no-wait constraints and missing operations. Eur. J. Oper. Res. **302**(1), 39–49 (2022)
9. Grzelak, M., Borucka, A., Guzanek, P.: Application of linear regression for evaluation of production processes effectiveness. In: Machado, J., Soares, F., Trojanowska, J., Ivanov, V. (eds.) icieng 2021. LNME, pp. 36–47. Springer, Cham (2022). https://doi.org/10.1007/978-3-030-78170-5_4
10. Paśko, Ł., Antosz, K.: Neural model of manufacturing process as a way to improve predictability of manufacturing. In: Gapiński, B., Ciszak, O., Ivanov, V. (eds.) MANUFACTURING 2022. LNME, pp. 24–38. Springer, Cham (2022). https://doi.org/10.1007/978-3-031-00805-4_3
11. Sobaszek, Ł., Gola, A., Świć, A.: The algorithms for robust scheduling of production jobs under machine failure and variable technological operation times. In: Machado, J., Soares, F., Trojanowska, J., Ivanov, V. (eds.) icieng 2021. LNME, pp. 56–67. Springer, Cham (2022). https://doi.org/10.1007/978-3-030-78170-5_6
12. Jasiulewicz-Kaczmarek, M., Antosz, K., Zhang, C., Ivanov, V.: Industry 4.0 technologies for sustainable asset life cycle management. Sustainability **15**(7), 5833 (2023)
13. Kanski, L., Pizon, J.: The impact of selected components of Industry 4.0 on project management. J. Innov. Knowl. **8**(1), 100336 (2023)
14. Pizoń, J., Gola, A.: The meaning and directions of development of personalized production in the era of Industry 4.0 and Industry 5.0. In: Machado, J., et al. (eds.) icieng 2022. LNME, pp. 1–13. Springer, Cham (2023). https://doi.org/10.1007/978-3-031-09360-9_1
15. Kaczmarczyk, V., et al.: Revisiting the role of manufacturing execution systems in Industry 4.0. IFAC PapersOnLine **55**(4), 151–157

16. https://corsosystems.com/manufacturing-execution-systems-guide. Accessed 23 Apr 2023
17. Shojaeinasab, A., et al.: Intelligent manufacturing execution systems: a systematic review. J. Manuf. Syst. **62**, 503–522 (2022)

# Massive Natural Language Processing in Distributed Environment

Adam Krechowicz(✉)

Faculty of Electrical Engineering, Automatic Control and Computer Science, Kielce University of Technology, Kielce, Poland
a.krechowicz@tu.kielce.pl

**Abstract.** Natural Language Processing (NLP) models are one of the most promising topics nowadays. Applications like ChatGPT uncover the power of such models and their broad applications. However, developing such models requires an extensive amount of training data sets to operate on. As such data sets grow, it becomes impossible to process them on a single machine, even using parallel computations. In this paper, the problem of distributing the NLP computations is taken into consideration. The Scalable Distributed Two-Layer Data Structures were used to organize the distributed environment. Stanford Core NLP library was used to perform accurate NLP computations. The experiments focused on the efficiency of the computation in the distributed environment and the accuracy of the predictions in relation to the scale of the distributions. The results clearly show that an increase in performance can be obtained without a significant decrease in accuracy.

**Keywords:** NLP · SD2DS · BigData

## 1 Introduction

Natural Language Processing (NLP) techniques and applications are one of the most interesting topics nowadays. ChatGPT, the most recognizable example of NLP application, may become a real game changer for many aspects of our lives and work [8]. Developing such models requires huge training data sets [20] which require a great deal of resources to process.

Dealing with such massive data sets can be a very challenging task. The most typical solution to deal with such data sets is to use parallel processing with the help of a general purpose Graphics Processing Unit (GPU) [25]. Such a solution allows processing the data in an efficient way, but is still unable to process the massive data set at once. Developers of such systems often have problems with respect to good memory organization. The most common and also the most infamous problem is the out–of–memory error [6]. In many applications such kind of errors cannot be dealt with without distributing the data to process it in separate fragments.


R. Mehmood et al. (Eds.): DCAI 2023, LNNS 741, pp. 56–65, 2023.
https://doi.org/10.1007/978-3-031-38318-2_6

On the other hand, not all calculations can be performed using GPU. There are many proven implementations that provide accurate results and are not prepared in such a way that GPU acceleration is possible. In many cases developers and users expect the same results that come from such implementations but also expect performance at the specific level [24].

In this paper, the problem of distributing the NLP computations is taken into consideration. The Scalable Distributed Two-Layer Data Structures (SD2DS) were used to manage the distributed environment. The proposed experiments used a Stanford Core NLP library as a reference implementation. The paper focuses on answering research questions: 1) Is it possible to use proven implementation of NLP in a distributed environment? (RQ1); 2) Does this distribution cause an improvement in efficiency? (RQ2); 3) Does this distribution affect the accuracy of the predictions? (RQ3). To answer those questions author prepared the distributed implementation of NLP processing tasks. Next, those implementations were used to evaluate the efficiency and accuracy of the prediction with different configurations of the distributed environment.

The paper is organized as follows. The reminder of the Introduction describes the basic concepts connected with the NLP and SD2DS. The next section describes the related work. Methodology and Experimental Results are presented in two subsequent sections. The paper ends with conclusions.

### 1.1 Natural Language Processing

Natural Language Processing tasks are evolving rapidly recently. It finds applications in many different areas such as medicine [18], cybersecurity [27], or entertainment [7]. Typical NLP tasks include machine translation, text summation, and question answering [3].

Most modern NLP systems are built on top of Deep Learning models that can be efficiently run in parallel using Graphical Processing Units (GPUs). However, as the need for better accuracy arises, the number of documents increases seriously. The fast growth of documents can cause problems on many levels such as data storing, data pre–processing, data processing, and so on. In massive processing of text, a single GPU can not be sufficient because of the memory or pipeline limitations. Despite the great performance that can be ensured by GPU acceleration, many specific solutions were not developed to take advantage of these opportunities.

Even when the NLP system extensively uses GPU acceleration, some performance bottlenecks can arise during the pre–processing phase of NLP tasks. The most popular NLP pre–processing tasks consist of a series of annotations that often need to be executed in specific order. The most popular annotation technique consists of *tokenization*, *sentence splitting*, *part of speech tagging* (pos), *lemmatization*, *named entity recognition* (ner). To catch the grammatical structure, both *parsing* and *dependency parsing* are often used. In more advanced applications, some specific annotators are used, which consists of *coreference detection*, *knowledge base population* (kbp), *quotation detection*, *natural logic sentence detection* and *entity relation detection*.

### 1.2 Scalable Distributed 2–Layered Data Structures

The rapid growth of the collected data sets forced the direction of storage management systems to change. Typical Relational Databases have been displaced by more robust and efficient NoSQL and NewSQL systems. This trend is maintained despite the lack of advanced querying and strong consistency mechanisms.

In many situations distributed storage systems do not allow to perform advanced processing. In many cases, traditionally used algorithms are not appropriate for such data sets, or they are not prepared for massive data distribution. Because of that, new data models are needed. MapReduce is one of the most popular framework for such massive distributed processing. Unfortunately, such a framework often requires one to redesign the existing algorithms to fit into such an environment. Additionally, many advanced algorithms are unable to run in distributed environments because they require to use all data sets. In such cases, a trade–off between the performance time and accuracy of processing is required.

Scalable Two–Layer Datastructures (SD2DS) [17] are one of the very efficient data storage solutions that allow one to store a massive data set in a distributed environment. SD2DS proved to be used especially when using massive data sets [12,16]. The most recognizable property of SD2DS is the division of the whole system into two layers that are responsible for managing the actual data (*component bodies*) and their metadata (*component headers*) separately. The component headers and buckets are stored on separate nodes (buckets) in the cluster. The headers of the components are stored in the first layer buckets (denoted as $B_1^1, \ldots, B_n^1$) and the components bodies are stored in the second layer (denoted as $B_1^2, \ldots, B_m^2$). This separation allows to introduce many additional features that can easily equip the system to fulfill current clients needs. Such features include content search for both numerical [11,15] and image data [4] or user anonymity [10,13].

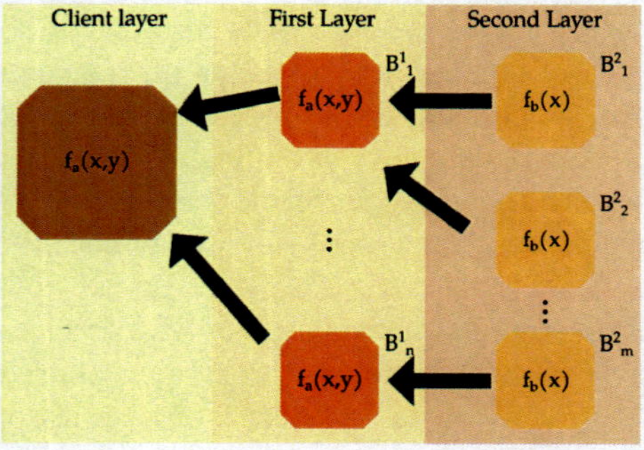


**Fig. 1.** Scalable Distributed Two–Layer Datastructure Architecture.

The division into component headers and bodies allows to introduce efficient processing model into SD2DS architecture. The idea behind the SD2DS processing model is depicted in Fig. 1. The client sends a request to the first-layer bucket. These buckets then forward those requests to the second-layer buckets that actually store the data. Second-layer buckets perform all the necessary operations depending on the nature of the processing. This operation is represented as $f_b(x)$. The results are then sent back to the first–layer buckets, which are responsible for result aggregation. This operation is represented as $f_a(x, y)$. The partially aggregated results are then sent to the client who performs the final aggregation of the results. In most cases, the client can apply the same $f_a(x)$ function to perform the final aggregation.

Consider a simple example of finding the maximum value from all buckets [14]. In such a case, the responsibility of $f_b(x)$ is to find the maximum value in the second–layer buckets. Similarly, the responsibility of $f_a(x, y)$ is to compare the two values that are found and return the highest one.

## 2 Related Work

In [2] the author presented a preliminary performance analysis of Apache Spark and Hadoop's implementation of the MapReduce programming in natural language text processing for named-entity recognition. It was found out that the evaluated distributed systems can be an attainable approach to natural language text processing with large data sets. The presented work shows that the performance of the processing tasks are strongly related to the architecture (number of cluster cores, number of processors per virtual machine etc.).

In [23] the approach for processing of large amount of text in distributed environments using MapReduce for name entity recognition was proposed. It was found out that the developed solution overcomes the limitations of NER systems connected with working slow on huge datasets. It provides better performance with fast extraction and reduction of storage cost.

In [22] an algorithm was developed for implementing natural language processing using MapReduce program. In [21] the authors proposed a distributed approach based on the Apache Hadoop and MapReduce for crawling web documents and running Natural Language Processing tasks in a parallel fashion. Its performance scalability was evaluated on a large corpus taken from web and other documents. In the framework called KOSHIK proposed in [5], a large scale-processing and querying of unstructured multilingual NL documents is enabled on the developed NLP platform. The documents are distributed on a Hadoop-based cluster. The application of the MapReduce programming (Hadoop distributed architecture) allowed to scale efficiently and easily. In [26] the authors presented GATECloud.net, which is a platform for large scale NLP processing using Amazon cloud. In this solution solution the GATE software is suited to a cloud computing environment thanks to the application of the PaaS paradigm. Behemoth 6 is a commercial tool, which is a platform enabling large scale document processing using Apache Hadoop, and NLP tools such as GATE, Tika and UIMA [19].

In [9] a library developed on Apache Spark ML was presented. It is able to use different kinds of languages and annotators that can be distributed over the Apache Spark environment.

In summary, there are many various distributed NLP solutions with different features. Each of them covers only a subset of typical annotator types. Furthermore, since they are trained on different datasets, the accuracy of their results differs significantly.

## 3 Methodology

In the experiments, the SD2DS system was used to store text documents. The documents from [1] were used in all experiments. Up to 27 cluster nodes were used to store and process the data. Each node consists of 16 Intel® Xeon® 8 core processors and 16GiB of RAM. The nodes were connected using 1GiB Ethernet.

The distribution of the documents was carried out as shown in the Fig. 2. The client was responsible for splitting the entire text into $N$ separate chunks. Then those chunks were located in the appropriate second–layer bucket by forwarding them through the first–layer bucket.

Document divisions were performed using a simple sentence splitting strategy. This is presented in the a) part in Fig. 3. The documents (of size $d$) were divided into relatively equal parts $(d_1, d_2, \dots)$ of size $s$. This strategy is suitable for most annotators which do not require relation detection between different sentences.

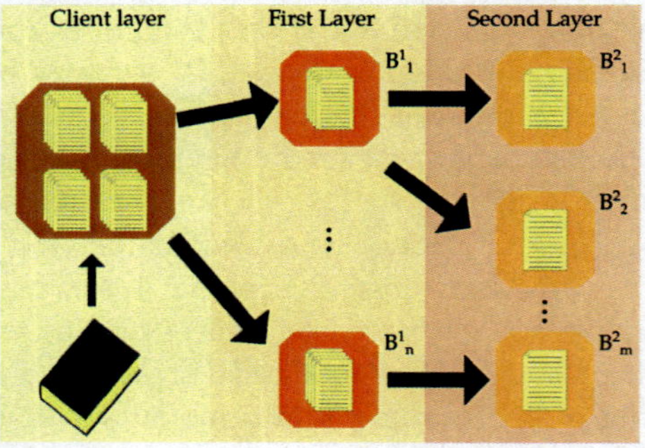


**Fig. 2.** Text distribution in SD2DS buckets.

In more advanced scenarios (e.g., when coreferences between sentences need to be detected), strategy b) from Fig. 3 can be used. In this case, the text fragments overlap, so references between fragments can be detected. The size of the overlapped text is denoted as $p$. In this scenario, the overall size of processed documents ($S$) can significantly increase. The size can be calculated as:

$$S = N * (d/N) + (N - 1) * 2p \quad (1)$$

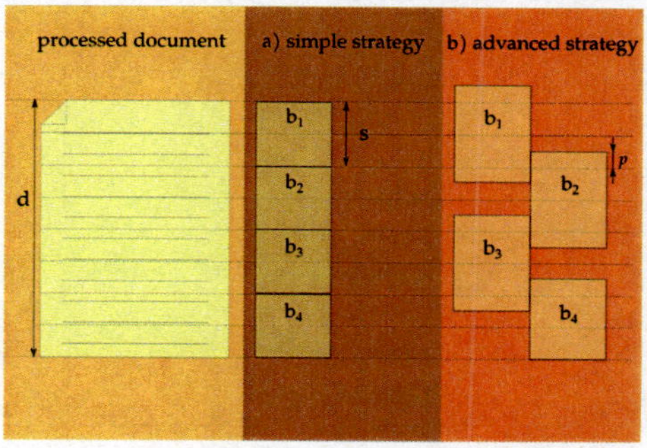


**Fig. 3.** Strategies for splitting text documents

Experiments were carried out to assess the processing time and accuracy of the results obtained. Experiments were conducted in relation to the number of second layer buckets used and with different numbers of annotators. Stanford Core NLP library was used to perform text processing.

In case of accuracy evaluation, the predicted annotations were checked. The annotations obtained from processing non–distributed documents (processed on single bucket) were considered as a reference. Each difference in the predicted annotations was considered an error. This allowed us to calculate the percentage of all errors using different distribution configurations.

## 4 Experimental Results

Figures 4a) and 4b) present the performance results in relation to the number of buckets used. In Fig. 4 a) the results of using simple annotators are shown. Figure 4b) presents the results of using more advanced annotators. As can be seen, the performance of most simple annotators does not increase when using the data distribution. Using more and more buckets results in increasing the overall execution time. Because these annotators work pretty fast, the cost of aggregation of results (function $f_a(x, y)$) seriously affects performance. On the other hand, when using more time–consuming annotators, the performance can be seriously increased with using more buckets. This trend can be visible for annotation with name entity recognition, dependency parsing, coreference detection, knowledge base population, quotation detection, natural logic detection, and entity relation detection.

This trend can be confirmed in Fig. 5. There is a significant threshold in the number of annotators that have a serious impact on performance. When the annotators use name entity recognition and subsequent annotators, the performance increases with the number of buckets used.

The results of the evaluation in terms of the accuracy of the prediction is presented in Fig. 6. In most cases, there is no visible relation between the number of buckets used and the total number of errors. In case of the most annotators

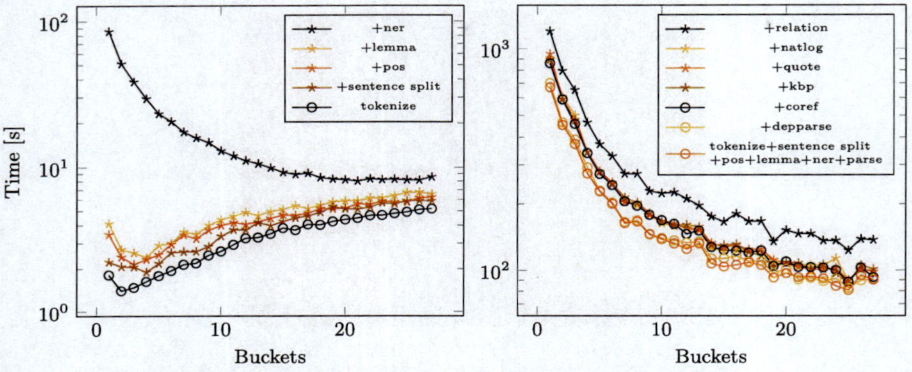


a) simple annotators b) advanced annotators

**Fig. 4.** Performance analysis in terms of the number of buckets used

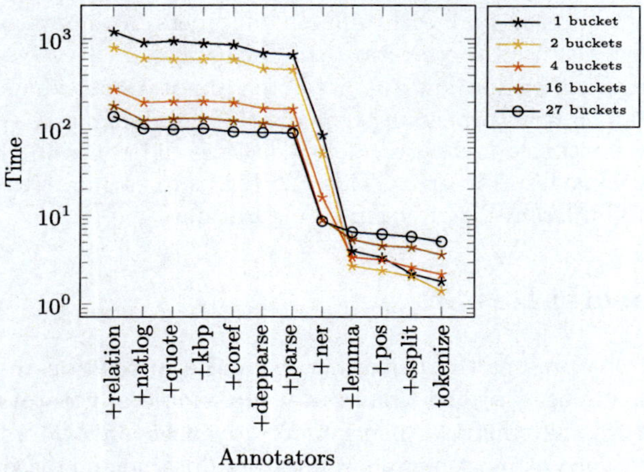


**Fig. 5.** Performance analysis in terms of the annotators used

only single errors were observed. However, the relation between the coreference annotation errors and the number of buckets can be clearly visible. Since the coreference annotators require text neighborhood to perform their work, it is an expected effect. Despite the fact that the maximum observed error slightly exceeds 1%. The highest number of errors (up to 5%) is related to the results of the dependency parser.

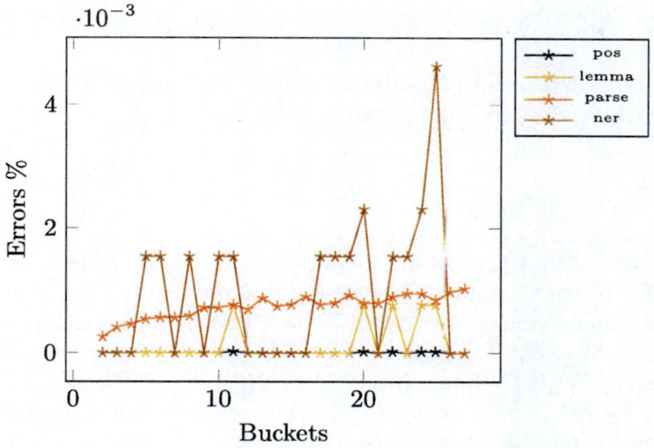


**Fig. 6.** Errors with respect to the distribution factor

## 5 Conclusions

In this paper, the problem of distributing massive natural language processing was considered. The main goal was to determine the possibility of obtaining accurate results in a distributed environment. Thanks to the conducted research, the following research questions were answered: (RQ1) Is it possible to use proven implementation of NLP in distributed environment? (RQ2) Does the distribution cause an improvement in efficiency? (RQ3) Does the distribution affect the accuracy of the prediction?

The RQ1 was addressed by proposing methods for both splitting the data into computational cluster and preparing framework for processing the text data in distributed environment. A standard and proven implementation of NLP processing (Stanford Core NLP) was used to perform the analysis. An efficient solution of the NoSQL system (SD2DS) was used to organize the data in a distributed environment.

The results obtained in the performance analysis show that it is possible to improve the performance for time-intensive computations of NLP tasks. Increasing the number of buckets allows to improve the performance of the following annotators: name entity recognition, parsing and dependency parsing, coreference detection, knowledge base population, quotation detection, natural logic detection, and entity relation detection. This allows an affirmative answer for RQ2. For simple annotators (tokenization, sentence splitting, and lemmatization), the overhead associated with the organization of the distributed environment was greater than the computation themselves. This trend allows to conclude that the distributed architecture needs to be specifically designed for the specified problem, which is consistent with the conclusions form [2]. The analysis of the impact of the accuracy of the annotator prediction allows to affirmatively answer the RQ3. The observed accuracy errors were in the accepted range (5% at most).

In future work, the problem of automatically choosing the optimal distributed architecture for the problem considered will be discussed.

## References

1. Project gutenberg. (n.d.). www.gutenberg.org. Accessed 10 Mar 2023
2. Andersson, L.: Natural language processing in a distributed environment: A comparative performance analysis of apache spark and hadoop mapreduce (2016)
3. Cambria, E., White, B.: Jumping NLP curves: a review of natural language processing research. IEEE Comput. Intell. Mag. **9**(2), 48–57 (2014)
4. Deniziak, S., Michno, T., Krechowicz, A.: The scalable distributed two-layer content based image retrieval data store. In: Ganzha, M., Maciaszek, L., Paprzycki, M. (eds.) Proceedings of The 2015 Federated Conference on Computer Science and Information Systems. ACSIS-Annals of Computer Science and Information Systems, vol. 5, pp. 827–832. IEEE, 345 E 47TH ST, NEW YORK, NY 10017 USA (2015). https://doi.org/10.15439/2015F272
5. Exner, P., Nugues, P.: Koshik-a large-scale distributed computing framework for NLP. In: ICPRAM, pp. 463–470 (2014)
6. Gao, Y., Liu, Y., Zhang, H., Li, Z., Zhu, Y., Lin, H., Yang, M.: Estimating GPU memory consumption of deep learning models. In: Proceedings of the 28th ACM Joint Meeting on European Software Engineering Conference and Symposium on the Foundations of Software Engineering, pp. 1342–1352 (2020)
7. Hamarashid, H.K., Saeed, S.A., Rashid, T.A.: A comprehensive review and evaluation on text predictive and entertainment systems. Soft Comput. **26**, 1–22 (2022)
8. Kasneci, E., et al.: ChatGPT for good? on opportunities and challenges of large language models for education. Learn. Individ. Differ. **103**, 102274 (2023)
9. Kocaman, V., Talby, D.: Spark NLP: natural language understanding at scale. Softw. Impacts **8**, 100058 (2021)
10. Krechowicz, A.: Scalable distributed two-layer datastore providing data anonymity. In: Kozielski, S., Mrozek, D., Kasprowski, P., Małysiak-Mrozek, B., Kostrzewa, D. (eds.) BDAS 2015-2016. CCIS, vol. 613, pp. 262–271. Springer, Cham (2016). https://doi.org/10.1007/978-3-319-34099-9_20
11. Krechowicz, A.: Content-aware data distribution over cluster nodes. Intell. Data Anal. **25**(4), 907–927 (2021)
12. Krechowicz, A., Chrobot, A., Deniziak, S., Łukawski, G.: SD2DS-based datastore for large files. In: Janech, J., Kostolny, J., Gratkowski, T. (eds.) SDOT 2015. AISC, vol. 511, pp. 150–168. Springer, Cham (2017). https://doi.org/10.1007/978-3-319-46535-7_13
13. Krechowicz, A., Deniziak, S.: SD2DS-based anonymous datastore for IoT solutions. DEStech Trans. Comput. Sci. Eng. (WCNE) (2016)
14. Krechowicz, A., Deniziak, S.: Business intelligence platform for big data based on scalable distributed two-layer data store. In: Communication papers of the Federated Conference on Computer Science and Information Systems. Annals of Computer Science and Information Systems, vol. 13 (2017)
15. Krechowicz, A., Deniziak, S.: Hierarchical clustering in scalable distributed two-layer datastore for big data as a service. In: 2018 Sixth International Conference on Enterprise Systems (ES), pp. 138–145. IEEE (2018)

16. Krechowicz, A., Deniziak, S., Bedla, M., Chrobot, A., Łukawski, G.: Scalable distributed two-layer block based datastore. In: Wyrzykowski, R., Deelman, E., Dongarra, J., Karczewski, K., Kitowski, J., Wiatr, K. (eds.) PPAM 2015. LNCS, vol. 9573, pp. 302–311. Springer, Cham (2016). https://doi.org/10.1007/978-3-319-32149-3_29
17. Krechowicz, A., Deniziak, S., Łukawski, G.: Highly scalable distributed architecture for NoSQL datastore supporting strong consistency. IEEE Access **9**, 69027–69043 (2021)
18. Locke, S., Bashall, A., Al-Adely, S., Moore, J., Wilson, A., Kitchen, G.B.: Natural language processing in medicine: a review. Trends Anaesth. Crit. Care **38**, 4–9 (2021)
19. Ltd, D.: Behemoth module tutorial. https://github.com/DigitalPebble/behemoth/wiki/tutorial. Accessed 03 Mar 2023
20. Mutuvi, S., Boros, E., Doucet, A., Jatowt, A., Lejeune, G., Odeo, M.: Multilingual epidemiological text classification: a comparative study, pp. 6172–6183 (2020). https://doi.org/10.18653/v1/2020.coling-main.543
21. Nesi, P., Pantaleo, G., Sanesi, G.: A hadoop based platform for natural language processing of web pages and documents. J. Vis. Lang. Comput. **31**, 130–138 (2015)
22. Nigam, J., Sahu, S.: An effective text processing approach with mapreduce. Int. J. Adv. Res. Comput. Eng. Technol. (IJARCET) **3**(12) (2014)
23. Nigam, J., Sahu, S.: Fast and effective system for name entity recognition on big data. Int. J. Comput. Sci. Eng. **3**(2), 31–35 (2015)
24. NLP, S.: How to speed up with corenlp client or get the same result with stanza on GPU? https://github.com/stanfordnlp/stanza/issues/595. Accessed 10 Mar 2023
25. Strubell, E., Ganesh, A., McCallum, A.: Energy and policy considerations for deep learning in NLP. arXiv preprint arXiv:1906.02243 (2019)
26. Tablan, V., Roberts, I., Cunningham, H., Bontcheva, K.: Gatecloud. net: a platform for large-scale, open-source text processing on the cloud. Philos. Trans. R. Soc. A: Math. Phys. Eng. Sci. **371**(1983), 20120071 (2013)
27. Ukwen, D.O., Karabatak, M.: Review of NLP-based systems in digital forensics and cybersecurity. In: 2021 9th International Symposium on Digital Forensics and Security (ISDFS), pp. 1–9 (2021). https://doi.org/10.1109/ISDFS52919.2021.9486354

# Special Session on Advances in Sustainable Economics and Technology (ASET)

# State of Sustainable Entrepreneurship in SMEs: Development of a Cross-Country Sustainability Pulse Check

Nikola Balić[1,2(✉)] and Ivana Vuka[3]

[1] Faculty of Electrical Engineering, Mechanical Engineering and Naval Architecture, University of Split, Split, Croatia
nikola@pulent.com, nbalic@pmfst.hr
[2] Faculty of Science, University of Split, Split, Croatia
[3] University of Split, Split, Croatia
ivuka@unist.hr

**Abstract.** This paper presents the development and testing process of the Sustainability Pulse Check (SPC), an instrument to understand the factors driving sustainable entrepreneurship of Small and Medium-sized Enterprises (SME) across diverse economies, industries and cultures of 37 countries of the Enterprise Europe Network (EEN). The instrument was developed and tested in two iterations resulting in the development of the novel fast survey. The survey underlies three critical factors that drive sustainable entrepreneurship in SMEs: knowledge and understanding of sustainability, organizational commitment to sustainability, and employee education and culture. The SPC instrument offers several advantages, including being relatively quick and straightforward, focused on specific aspects of sustainability management, and well-received by respondents. Furthermore, the paper explores the potential contributions of the upcoming research to our understanding of sustainable entrepreneurship in SMEs highlighting the potential utility of the instrument for informing policymakers and offering actionable recommendations to support the transition towards a more sustainable future.

**Keywords:** Sustainable entrepreneurship · SME sustainabilit · Cross-country survey · Organizational commitment · Value-chain pressures · Sustainable transition

## 1 Introduction

The global transition towards sustainable development has become increasingly important as we face the challenges of climate change, resource depletion, and social inequality [2,10,24]. The United Nations (UN) has recognized the importance of promoting sustainable development and has set the 2030 Agenda for Sustainable Development, which includes 17 Sustainable Development Goals


R. Mehmood et al. (Eds.): DCAI 2023, LNNS 741, pp. 69–78, 2023.
https://doi.org/10.1007/978-3-031-38318-2_7

(SDGs) that aim to end poverty, protect the planet, and ensure prosperity for all. The transition towards a sustainable future requires collective action and collaboration from governments, businesses, and individuals [39].

As the backbone of many economies, Small and Medium-sized Enterprises (SMEs) play a crucial role in driving this transition towards a more sustainable entrepreneurial society [8,32]. SMEs make up over 99% of European businesses and contribute to two-thirds of private sector jobs creating significant economic growth [11]. Lawrence et al. (2006) state that while SMEs have limited individual impact on social, environmental, and financial issues, their cumulative impact is significant. Previous research shows that, in comparison to larger companies, the majority of SMEs take a relatively reactive approach to environmental issues [23]. Clearly, there is significant room for improvement in the environmental performance of SMEs [33].

SMEs face new challenges as global sustainability reporting and proactive sustainability considerations increase. In response to this major external change, these companies often lack a structured strategy formulation process [35]. Policymakers and SMEs struggle to understand existing frameworks' operational mechanisms in the context of SMEs [36]. According to Sloan et al. (2013), most research on successful sustainability practises has focused on large enterprises, so SMEs' sustainability implementation dynamics need further study [36].

Over the past two decades range of sustainability tools have been designed to operationalize sustainability-oriented strategies and to coordinate activities within the SMEs some of them being sector or country specific and some of them tailored to the specific SME need [19]. Overall, sustainability tools can be a valuable resource for SMEs looking to reduce their environmental impact and promote more sustainable business practices. By using these tools, SMEs can gain a better understanding of their environmental impact and identify areas where they can make changes to become more sustainable.

However, the majority of the tools have been implemented only by a relatively small number of SMEs [19]. The main reasons for low acceptance of the tools are either internal shortcomings such as lack of awareness of sustainability issues or absence of perceived benefits, lack of knowledge and expertise and lack of human and financial resources [18]. Despite these challenges, many SMEs are demonstrating their commitment to sustainability by adopting environmentally friendly practices, engaging in social responsibility initiatives, and promoting economic development [17].

Understanding the factors that drive sustainable entrepreneurship in SMEs across various countries can provide valuable insights and inform the development of effective policies and programs to support this transition. Several studies on sustainable practices in SMEs have found barriers to implementation, the need for organized approaches in emerging markets, and the importance of technological adaptation for SME sustainability in global competition [7,18,19,30].

In addition to the detection of the challenges and barriers faced by SMEs in adopting sustainable practices, it is also important to understand the factors that drive sustainable entrepreneurship in SMEs in order to better understand the dynamics and challenges faced by SMEs as they navigate this transition as well as their contributions [4,9]. Several studies have explored the factors that drive sustainable entrepreneurship in SMEs across various countries, including entrepreneurial, economic and social factors [15,28] but also motivational, psychological and emotional factors [25].

A recent study [38] explored those factors combined and found that the primary factors that drive sustainable entrepreneurship are related to behaviour and business. In particular, ethical values and principles, as well as competitive intelligence, play a critical role in taking actions that promote sustainability. On the other side environmental factors such as environmental regulations, entrepreneurship policies, are not relevant factors when creating sustainable initiatives [38].

The Enterprise Europe Network (EEN), operating in 37 countries and with more than 3,000 EEN advisors, presents an excellent opportunity to study sustainable entrepreneurship in SMEs across diverse economies, industries, and cultures [1,12,14]. The developed fast survey instrument is planned to be implemented as a pre-assessment of clients by EEN advisors, enabling the first large-scale research of sustainability perceptions and commitments in SMEs.

In this paper, we present the development and testing process of the Sustainability Pulse Check (SPC) instrument, the main findings from the initial rounds of testing and analysis, and the potential contributions of the upcoming research to our understanding of sustainable entrepreneurship in SMEs. This study represents a crucial step in advancing our understanding of these dynamics and supporting the ongoing efforts to promote sustainable entrepreneurship across the EEN countries and beyond.

## 2 Methodology

The study developed and tested an instrument to assess sustainable entrepreneurship in SMEs. A rigorous validation and verification procedure was implemented during the development process to ensure the instrument's validity and dependability. It is important to note, however, that the survey has not yet been tested with SMEs. The purpose of this study was to establish the validity of the survey instrument, and future research will involve testing it with SMEs from a variety of industries and geographies to determine how well it captures their experiences and perceptions.

The main hypothesis was that the level of knowledge and understanding of sustainability, organizational commitment to sustainability, and employee education and culture within SMEs play significant roles in driving sustainable entrepreneurial activities. Additionally, external value-chain pressures may further influence organizational efforts towards sustainability.

The initial survey was developed during the Strategy session of the Thematic Group Sustainability and was tested with a group of 26 expert EEN Sustainability Advisors from 10 countries during a workshop held at EISMEA [13]. The obtained results were analyzed using Python with the aid of pandas, factor_analyzer, networkx, and statsmodels packages [16,29,34].

The first version consisted of 17 questions and 2 questions to evaluate the survey's length and Net Promoter Score (NPS) [3] appeal to the advisors. Participants were instructed to complete the survey for one of their actual clients. The survey was designed as a 7-point Likert scale, to facilitate quick completion. Results of the Factor analysis, Pearson correlation, and Cronbach Alpha were employed to create a new iteration of the survey [37].

The new iteration of the survey was then tested with a group of 31 individuals forming a mixed group of experienced sustainability advisors and other EEN advisors during the Regional training of Western Balkan EEN experts. The second iteration of the survey consisted of nine plus two questions.

## 3 Main Findings

The survey identified three critical factors that drive sustainable entrepreneurship in SMEs: knowledge and understanding of sustainability, organizational commitment to sustainability, and employee education and culture. The results of the analysis were used to reduce the overall number of questions from 17 to 9 and convert the survey to a 5-point Likert scale (the resulting value of Cronbach's Alpha was 0.89). Following a group discussion with experts, one additional question concerning the potential impact of external value-chain pressures on organizational sustainability efforts was added (Tables 1, 2, 3 and 4).

The Net Promoter Score (NPS) was 0, which is considered positive, but it should be noted that the group was heterogeneous, with an equal number of detractors and promoters (10 each). This means that 32% of the group is expected to be highly engaged as promoters (scores 9 and 10) of the instrument. A linear regression model was used to identify factors that are positively related to the likelihood of recommending clients to fill out the survey. The model revealed that a good understanding of the concept of sustainable development and knowledgeable employees about sustainability issues and best practices are significant predictors of the likelihood of recommendation ($P>|t| > 0.05$, $R^2$ of .411) (Figs. 1 and 2).

## 4 Discussion

The study identified three critical factors that drive sustainable entrepreneurship in SMEs: knowledge and understanding of sustainability, organizational commitment to sustainability, and employee education and culture. These findings are consistent with existing literature on sustainable entrepreneurship, which emphasizes the importance of knowledge, organizational commitment, and employee engagement in achieving sustainable business practices [6, 20, 22, 27].

Compared to other tools used for assessing sustainability management, the proposed survey appears to offer several advantages. For example, some existing tools may be overly complex or time-consuming to use, which can make it difficult for companies to accurately assess their sustainability practices [26]. Furthermore, the reduced number of questions in our survey instrument achieves an

**Table 1.** Summary Statistics of the First Survey

| Abbreviated Question | Mean | Median | SD |
|---|---|---|---|
| Q1: Awareness of dimensions of sustainability | 5.9 | 6 | 1.3 |
| Q2: Understanding of sustainable development | 5.3 | 5 | 1.5 |
| Q3: Familiarity with steps to become more sustainable | 4.6 | 4 | 1.4 |
| Q4: Organization's active steps to contribute to sustainability | 4.3 | 4 | 1.4 |
| Q5: Sustainability as a critical part of business strategy | 4.6 | 5 | 1.5 |
| Q6: Organization's clear sustainability policy | 3.7 | 3 | 1.6 |
| Q7: Employee education and training on sustainability | 3.4 | 3 | 1.8 |
| Q8: Organization's culture aligns with pro-sustainability attitudes | 4.0 | 4 | 1.6 |
| Q9: Interventions to improve organizational culture for sustainability | 4.8 | 5 | 1.5 |
| Q10: Organization's culture promotes sustainability | 4.1 | 4 | 1.4 |
| Q11: Organization's regular implementation of actions for sustainability | 4.2 | 4 | 1.6 |
| Q12: Sustainability efforts positively impact economic performance | 4.4 | 4 | 1.2 |
| Q13: Sustainability efforts positively impact long-term viability | 4.3 | 4 | 1.3 |
| Q14: Sustainability efforts positively impact employees and consumers | 4.3 | 4 | 1.6 |
| Q15: Sustainability efforts positively impact community | 4.5 | 4 | 1.5 |
| Q16: Sustainability efforts positively impact environment | 4.6 | 5 | 1.5 |
| Q17: Sustainability efforts positively impact carbon footprint reduction | 4.7 | 5 | 1.4 |
| Time taken to complete the survey (0–10) | 5.6 | 5 | 2.2 |
| Likelihood to recommend clients to fill out the survey (0–10) | 6.4 | 7 | 2.1 |

**Table 2.** Summary Statistics of the Second Survey

| Abbreviated Question | Mean | Median | SD |
|---|---|---|---|
| Q1: Well-informed on sustainability | 3.0 | 3 | 1.1 |
| Q2: Well-informed on dimensions of sustainability | 3.2 | 3 | 1.1 |
| Q3: Good understanding of sustainable development | 3.2 | 3 | 1.0 |
| Q4: Organization takes active steps for sustainability | 3.1 | 3 | 1.0 |
| Q5: Sustainability as critical part of business strategy | 3.0 | 3 | 1.0 |
| Q6: Organization has clear sustainability policy | 2.6 | 3 | 1.1 |
| Q7: Employees knowledgeable about sustainability | 2.7 | 3 | 1.1 |
| Q8: Pro-sustainability culture | 3.1 | 3 | 0.8 |
| Q9: Interventions necessary for sustainability culture change | 3.6 | 4 | 0.9 |
| Time taken to complete the survey (0–10) | 4.7 | 5 | 2.1 |
| Likelihood to recommend clients to fill out the survey (0–10) | 7.2 | 8 | 2.4 |

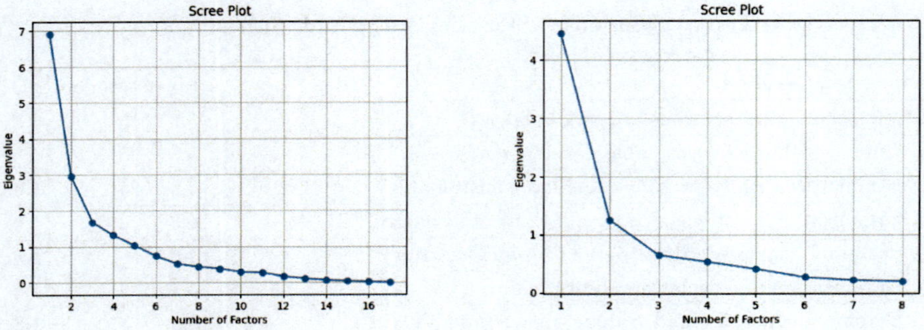


(a) The results of the first version of the survey.

(b) The results of the second version of the survey.

**Fig. 1.** Scree plots for factor analysis models on survey results.

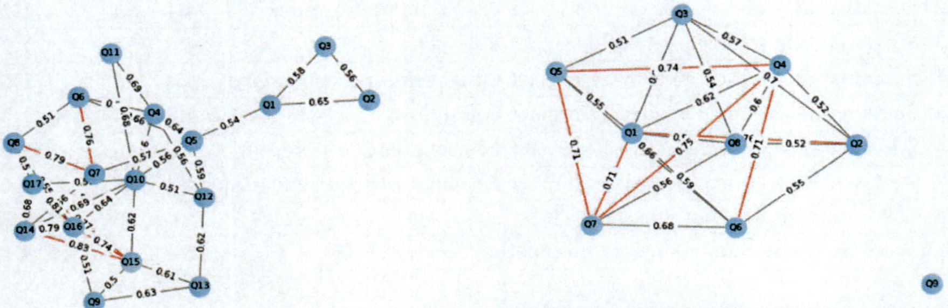

(a) The first version of the survey.

(b) The results of the second version of the survey.

**Fig. 2.** Correlation Network of Survey Questions: Thresholds at 0.5 and 0.7 for Strong Relationships.

**Table 3.** Factor Loadings on the results of the first version of the survey.

| Factor Loadings | 0 | 1 | 2 |
|---|---|---|---|
| Q1: Awareness of dimensions of sustainability | −0.171 | 0.817 | 0.016 |
| Q2: Understanding of sustainable development | −0.035 | 0.650 | 0.007 |
| Q3: Familiarity with steps to become more sustainable | 0.088 | 0.681 | 0.103 |
| Q4: Organization's active steps to contribute to sustainability | 0.165 | 0.280 | 0.788 |
| Q5: Sustainability as a critical part of business strategy | 0.279 | 0.623 | 0.526 |
| Q6: Organization's clear sustainability policy | 0.322 | 0.541 | 0.343 |
| Q7: Employee education and training on sustainability | 0.607 | 0.565 | 0.134 |
| Q8: Organization's culture aligns with pro-sustainability attitudes | 0.666 | 0.168 | 0.014 |
| Q9: Interventions to improve organizational culture for sustainability | 0.515 | 0.108 | 0.048 |
| Q10: Organization's culture promotes sustainability | 0.509 | 0.118 | 0.662 |
| Q11: Organization's regular implementation of actions for sustainability | 0.107 | −0.085 | 0.907 |
| Q12: Sustainability efforts positively impact economic performance | 0.384 | 0.306 | 0.478 |
| Q13: Sustainability efforts positively impact long-term viability | 0.590 | 0.178 | 0.268 |
| Q14: Sustainability efforts positively impact employees and consumers | 0.805 | 0.024 | 0.302 |
| Q15: Sustainability efforts positively impact community | 0.852 | −0.123 | 0.259 |
| Q16: Sustainability efforts positively impact environment | 0.866 | 0.017 | 0.170 |
| Q17: Sustainability efforts positively impact carbon footprint reduction | 0.772 | −0.211 | 0.236 |

**Table 4.** Factor Loadings on the results of the second version of the survey.

| Factor Loadings | 0 | 1 |
|---|---|---|
| Q1: Well-informed on sustainability | 0.657 | 0.573 |
| Q2: Well-informed on dimensions of sustainability | 0.515 | 0.662 |
| Q3: Good understanding of sustainable development | 0.520 | 0.596 |
| Q4: Organization takes active steps for sustainability | 0.887 | 0.134 |
| Q5: Sustainability as critical part of business strategy | 0.803 | 0.077 |
| Q6: Organization has clear sustainability policy | 0.787 | 0.145 |
| Q7: Employees knowledgeable about sustainability | 0.870 | 0.094 |
| Q8: Pro-sustainability culture | 0.534 | 0.432 |
| Q9: Interventions necessary for sustainability culture change | −0.148 | 0.400 |

optimal balance between length and depth, allowing for quick assessment without overwhelming companies and capturing essential aspects of sustainability practices. The feedback received during the validation phase indicated that the median time required to complete the survey was slightly above 3 min, which was well within acceptable limits according to the previous studies on the ideal length of surveys [31]. Previous research has also shown that shorter surveys are more reliable and have higher response and completion rates [21].

Based on the above initial testing results, the survey appears to have been well-received by the respondents that participated. Additionally, the fact that the survey questions were focused on specific aspects of sustainability management allowed for more detailed and accurate responses.

## 5 Conclusion

The SPC instrument was designed for use by EEN advisors as a pre-assessment questionnaire for their clients. It enables the first large-scale study of SME perceptions and commitments to sustainability. The SPC instrument has a number of advantages: it is quick and straightforward, it focuses on particular aspects of sustainability management, and it is well-received by those surveyed.

Research shows that organisations that fully embrace sustainability principles and innovation approaches create value by introducing new products (radical innovations) and collaborating with diverse stakeholders [5]. One of the most fundamental questions for policymakers is how a single economic entity, particularly a small business, can engage in the adoption of sustainability practises [23]. Utilising the SPC instrument could be instrumental in fostering sustainable innovation and facilitating informed decision-making among SMBs and policymakers.

Overall, this study represents a crucial step in advancing our understanding of the underlying factors driving SMEs on their sustainability journey and supporting the ongoing efforts to promote sustainable entrepreneurship across the EEN countries and beyond. By implementing the SPC instrument, EEN advisors can effectively guide SMEs and, as a result, provide policymakers with valuable guidance. In addition, the study's findings can inform the design of capacity-building programmes and policy frameworks intended to foster sustainable entrepreneurship in countries outside the EEN network.

Future research will further validate the instrument's effectiveness by conducting additional surveys with SMEs and with more extensive samples. Additionally, the survey questions could be refined or expanded to capture more nuanced aspects of sustainability management. Finally, it would be valuable to explore how the instrument can be integrated into existing sustainability programs and interventions, such as training programs, coaching sessions, or more complex sustainability assessment instruments.

In conclusion, the SPC instrument offers a simple and effective method for assessing sustainable entrepreneurship in SMEs. By providing a comprehensive picture of SMEs' sustainability practices, this instrument can help EEN advisors to support SMEs' efforts to become more sustainable and promote sustainable entrepreneurship across the European Union.

**Acknowledgements.** We thank all EEN colleagues and Thematic Group Sustainability members, especially Reyes Sansegundo Romero, for their invaluable contributions to this project. We would also like to thank Sara Lucija Dragičević for her invaluable operational support.

## References

1. Enterprise Europe Network - Sustainability
2. Adams, W.M., Jeanrenaud, S.: Transition to Sustainability: Towards a Humane and Diverse World. IUCN, Gland (2008)
3. Baehre, S., O'Dwyer, M., O'Malley, L., Story, V.M.: Customer mindset metrics: a systematic evaluation of the net promoter score (NPS) vs. alternative calculation methods. J. Bus. Res. **149**, 353–362 (2022)
4. Bapuji, H., Husted, B.W., Lu, J., Mir, R.: Value creation, appropriation, and distribution: how firms contribute to societal economic inequality. Bus. Soc. **57**(6), 983–1009 (2018)
5. Bos-Brouwers, H.E.J.: Corporate sustainability and innovation in SMEs: evidence of themes and activities in practice. Bus. Strat. Environ. **19**(7), 417–435 (2009). https://doi.org/10.1002/bse.652
6. Crals, E., Vereeck, L.: Sustainable entrepreneurship in SMEs. Theory and practice
7. Das, M., Rangarajan, K.: Corporate sustainability as a business strategy in SMEs-a literature review in the emerging market context (2017)
8. Das, M., Rangarajan, K., Dutta, G.: Corporate sustainability in small and medium-sized enterprises: a literature analysis and road ahead. J. Indian Bus. Res. **12**(2), 271–300 (2019)
9. DiVito, L., Ingen-Housz, Z.: From individual sustainability orientations to collective sustainability innovation and sustainable entrepreneurial ecosystems. Small Bus. Econ. **56**(3), 1057–1072 (2019). https://doi.org/10.1007/s11187-019-00254-6
10. Drucker, P.F.: Post-capitalist Society. Routledge, Milton Park (1993)

11. EC: Inforegio - SME competitiveness. https://ec.europa.eu
12. EEN: Enterprise Europe Network. https://een.ec.europa.eu/
13. EISMEA: European Innovation Council and SMEs Executive Agency. https://eismea.ec.europa.eu/
14. Ferraro, G., Iovanella, A.: Technology transfer in innovation networks: an empirical study of the enterprise Europe network. Int. J. Eng. Bus. Manage. **9** (2017). https://doi.org/10.1177/1847979017735748
15. Florea, L., Cheung, Y.H., Herndon, N.C.: For all good reasons: role of values in organizational sustainability. J. Bus. Ethics **114**(3), 393–408 (2013). https://doi.org/10.1007/s10551-012-1355-x
16. Hagberg, A.A., Schult, D.A., Swart, P.: Exploring network structure, dynamics, and function using networkX (2008)
17. Hansen, E.G., Schaltegger, S.: The sustainability balanced scorecard: a systematic review of architectures. J. Bus. Ethics **133**(2), 193–221 (2014). https://doi.org/10.1007/s10551-014-2340-3
18. Jaramillo, J., Sossa, J.W.Z., Mendoza, G.L.O.: Barriers to sustainability for small and medium enterprises in the framework of sustainable development-literature review. Bus. Strat. Environ. **28**, 512–524 (2018)
19. Johnson, M.P., Schaltegger, S.: Two decades of sustainability management tools for SMEs: how far have we come?*. J. Small Bus. Manage. **54**, 481–505 (2016)
20. Koe, W.L., Omar, R., Sa'ari, J.R.: Factors influencing propensity to sustainable entrepreneurship of SMEs in Malaysia. Procedia - Soc. Behav. Sci. **172**, 570–577 (2015)
21. Kost, R.G., Correa Da Rosa, J.: Impact of survey length and compensation on validity, reliability, and sample characteristics for ultrashort-, short-, and long-research participant perception surveys. J. Clin. Trans. Sci. **2**(1), 31–37 (2018). https://doi.org/10.1017/cts.2018.18
22. Kraus, S., Burtscher, J., Vallaster, C., Angerer, M.: Sustainable entrepreneurship orientation: a reflection on status-quo research on factors facilitating responsible managerial practices. Sustainability **10**, 444 (2018)
23. Lawrence, S.R., Collins, E., Pavlovich, K., Arunachalam, M.: Sustainability practices of SMEs: the case of NZ. Bus. Strat. Environ. **15**(4), 242–257 (2006). https://doi.org/10.1002/bse.533
24. Leach, M., et al.: Transforming innovation for sustainability. Ecol. Soc. **17**, 11 (2012)
25. Muñoz, P., Cohen, B.: Sustainable entrepreneurship research: taking stock and looking ahead. Bus. Strat. Environ. **27**(3), 300–322 (2018). https://doi.org/10.1002/bse.2000, https://onlinelibrary.wiley.com/doi/abs/10.1002/bse.2000
26. Ness, B., Urbel-Piirsalu, E., Anderberg, S., Olsson, L.: Categorising tools for sustainability assessment. Ecol. Econ. **60**(3), 498–508 (2007). https://doi.org/10.1016/j.ecolecon.2006.07.023, https://www.sciencedirect.com/science/article/pii/S0921800906003636
27. Parrish, B.D.: Sustainability-driven entrepreneurship: principles of organization design. J. Bus. Ventur. **25**(5), 510–523 (2010), https://EconPapers.repec.org/RePEc:eee:jbvent:v:25:y:2010:i:5:p:510-523
28. Patzelt, H., Shepherd, D.A.: Recognizing opportunities for sustainable development. Entrepreneurship Theory Pract. **35**(4), 631–652 (2011). https://doi.org/10.1111/j.1540-6520.2010.00386.x
29. Persson, I., Khojasteh, J.: Python packages for exploratory factor analysis. Struct. Equat. Model.: Multidisc. J. **28**, 983–988 (2021)

30. Prasanna, R., Jayasundara, J., Gamage, S.K.N., Ekanayake, E.M., Rajapakshe, P.S.K., Abeyrathne, G.: Sustainability of SMEs in the competition: a systemic review on technological challenges and SME performance. J. Open Innov.: Technol. Market Complex. **5**(4), 100 (2019)
31. Revilla, M., Ochoa, C.: Ideal and maximum length for a web survey. Int. J. Market Res. **59** (2017)
32. Rothwell, R., Zegveld, W.: Innovation and the small and medium sized firm: their role in employment and in economic change (1981)
33. Schaper, M.T.: The challenge of environmental responsibility and sustainable development : Implications for SME and entrepreneurship academics (2002)
34. Seabold, S., Perktold, J.: Statsmodels: econometric and statistical modeling with python (2010)
35. Shields, J., Shelleman, J.M.: Integrating sustainability into SME strategy. J. Small Bus. Strat. **25**(2), 59–78 (2015)
36. Sloan, K., Klingenberg, B., Rider, C.: Towards sustainability: examining the drivers and change process within SMEs. J. Manage. Sustain. **3**(2), p19 (2013). https://doi.org/10.5539/jms.v3n2p19
37. Tavakol, M., Dennick, R.: Making sense of Cronbach's alpha. Int. J. Med. Educ. **2**, 53–55 (2011)
38. Tur-Porcar, A., Roig-Tierno, N., Llorca Mestre, A.: Factors affecting entrepreneurship and business sustainability. Sustainability **10**(2) (2018). https://doi.org/10.3390/su10020452, https://www.mdpi.com/2071-1050/10/2/452
39. UN: Transforming our world: The 2030 Agenda for Sustainable Development — Department of Economic and Social Affairs. https://sdgs.un.org/2030agenda

# Intangible Assets in Accounting. A Bibliometric Analysis

Jonathan Bermúdez-Carvajal and Javier Parra-Domínguez(✉)

Universidad de Salamanca, Salamanca, Spain
{jcbeca,javierparra}@usal.es

**Abstract.** This scientific paper presents a bibliometric analysis of publications on the accounting treatment of intangible assets to show the evolution of literature and current trends in the field. The paper has four specific objectives: identifying the development and variations in publications related to intangible asset accounting from 1984 to 2022, establishing the type of publications made with reference to this area, determining the indexed journals that lead the publications on this topic, and identifying the most prolific authors in the area. A bibliometric analysis was performed with information from the Scopus database to answer the four research questions raised. The search equation generated 687 documents published between 1984 and 2022. The study finds that there was a significant increase in publications on this topic during the 2000s and 2010s, peaking in 2020 with 57 publications. However, the number of publications decreased in 2021 and 2022. Business, Management, and Accounting is the area where most publications on this topic are found, with the Journal of Intellectual Capital leading the publications. The study also identifies the most productive authors in the area.

**Keywords:** Intangible Assets · Accounting · Bibliometric Analysis

## 1 Introduction

### 1.1 Evolution of International Financial Reporting Standard

IFRS are a set of accounting standards issued by the International Accounting Standards Board (IASB), a private body based in the United Kingdom. These standards began to be issued in 1975 under the name of International Accounting Standards (IAS) by the International Accounting Standards Committee (IASC) which in 2001 became IASB and the standards name changed from IAS to IFRS. However, both IAS and IFRS are still in force and are part of the compendium known today as International Financial Reporting Standards (IFRS).

These standards have been widely accepted around the world, to the point that by 2010 they had been adopted by more than 150 countries around the world [1]. This convergence is supported by the improvement of the quality, transparency and comparability of information [2], although the literature is not conclusive in this respect [3]. Usually, IFRS are not applicable to every company, but only to public companies or groups of


R. Mehmood et al. (Eds.): DCAI 2023, LNNS 741, pp. 79–87, 2023.
https://doi.org/10.1007/978-3-031-38318-2_8

companies where the parent or holding company is required to prepare consolidated financial statements. In these countries, there is usually a national accounting standard for all non-publicly traded companies.

IASB also issues accounting standards aimed at use by small and medium-sized enterprises (SMEs). These standards consist of a set of requirements inspired by IFRS, but with significantly less complexity and fewer disclosures. They are known as IFRS for SMEs, which has not been as widely adopted globally as IFRS, as many countries have decided to maintain their own accounting regulations for these companies. However, in some countries, both IFRS and IFRS for SMEs have replaced the national accounting standard for most companies, as in the case of Colombia. This has not been the case of other countries, like Spain but IFRS have also had a direct influence on the determination of domestic accounting requirements, although they have not replaced the accounting standards [4].

### 1.2 Evolution of Accounting Standards in Spain

In Spain, accounting has been officially regulated since 1973, when Decree 530 containing the Plan General de Contabilidad (PGC) was issued. Since then, the PGC has been substantially modified on two occasions: in 1990 and in 2007. The 1990 reform was carried out with the purpose of adapting the internal regulations to the directives of the European Economic Community. The 2007 reform was aimed at adopting a regulation very similar to that contained in the IFRS adopted by Europe, but contemplating simpler treatments and wording, and is contained in Royal Decree 1514 of 2007.

This regulatory change is not exclusive to Spain. Variations in accounting requirements have occurred around the world and depend on different factors like political set up market capitalization [5], legal origin [6] or regulatory efficiency [7], impacting to a greater or lesser extent the financial statements of companies. In addition, technology is advancing at a dizzying pace and accounting must keep up to date, looking for innovative ways to reliably show the reality of companies [8]. Accounting standards need to be constantly up to date to respond to these changes, so it is essential to investigate how accounting is adapting to this changing reality.

This paper seeks to show how scientific publications related to intangible accounting have evolved. To this end, it is organized as follows: Sect. 2 will provide a brief introduction to intangible assets, highlighting their current importance in companies' operations. Section 3 will detail the methodology and the research questions that the paper seeks to answer. Section 4 discusses the results with reference to the evolution of publications, areas of knowledge, journals and authors. Section 5 will develop the discussions on the answers provided by the data and, finally, Sect. 6 will propose some conclusions.

## 2 Introduction to Intangible Assets

The current importance of intangible assets is undeniable [9]. These have become one of the most relevant resources within the assets of companies and are becoming increasingly relevant in their success [10]. Over the last decades, there has been a shift from industrialization-based companies to knowledge-based companies [11].

Intangible assets are economic resources that lack physical substance from which economic benefits are expected to be derived. They include, for example, software, brands, licenses, patents, copyrights, and scientific knowledge. However, not all assets without physical substance fall within the definition of intangible assets in accounting standards. Each standard may have different definitions and recognition criteria, but, in general, it is usual that in addition to intangibility, the following elements are included in the concept:

1. The asset must be a non-monetary item. To be included in the accounting concept of intangible assets, the asset must be non-monetary, i.e., it must not be received in a fixed or determinable amount of money[1] [12].
2. It must be identifiable. Identifiability is conceived as a complex characteristic that can be met through one of its two constituent elements. The asset is identifiable if either it may be separable, i.e., it may be separated from the entity to be sold, leased or assigned, or it arises from contractual or legal rights[2] [12].

In addition to the concept of intangible assets contained in the accounting standards, it is required that they meet certain criteria to be recognized in the financial statements. As is happens with the concept, it is usual for the accounting requirements to be the same ones in different accounting standards, although they may be presented with different approaches. For example, in one standard they may be presented as general requirements for all assets and in other standard they may be presented as specific requirements for specific assets such as inventories or, in the case in point, intangibles. These criteria usually are:

- Control. This is the power to receive the economic benefits derived from the asset. It usually implies the possibility of preventing other entities from receiving these benefits but does not necessarily imply that it is certain or even probable that the benefits will be received.
- Potential to produce economic benefits. Economic benefits imply that the entity can profit from it. These benefits can be received by the entity in different ways. It could be in the form of cash flows, for example, in the case of an asset being leased. It could also be used in the entity's business, as in the case of design software; or it could even be used to optimize production costs, as might be the case with the development of a new way of executing a specific process.

Probably the most argued points about intangible assets are their recognition and measurement [13–17], mainly due to the great uncertainty surrounding the profitability of some intangible assets, especially when they are in the research and development phases.

## 3 Methodology

The purpose of this paper is to perform a bibliometric analysis of publications regarding the accounting treatment of intangible assets to show the evolution of the literature and current trends in the field.

[1] Real Decreto 1514/2007. BOE 278 20/11/2007.

[2] Ibid.

Bibliometric analysis has received much attention from authors from different disciplines in recent years [18, 19] and aims to measure the scientific production on a specific area of knowledge, denoting its trends and fundamental characteristics.

Four specific objectives were targeted, which are identified with the four research questions that will be posed below. First, the identification of the development and variations in publications related to intangible asset accounting from 1984 to 2022. Second, to establish what type of publications are made with reference to this area. Third, to determine which are the indexed journals that lead the publications on this topic. And fourth, to identify the most prolific authors in the area.

In order to analyze the four questions raised about publications, a bibliometric analysis was developed with information from the Scopus database, answering the following research questions:

- RQ1: What is the trend of publications on intangible assets in the accounting field?
- RQ2: What are the characteristics of the publications?
- RQ3: Which are the journals that publish the most on the subject?
- RQ4: Who are the authors who publish the most in the area?

With the above analysis as an objective, a search equation was determined that contemplated both different terms related to accounting in general and to intangible assets. The equation is as follows:

( TITLE-ABS-KEY ( accounting) OR TITLE-ABS-KEY ( ifrs) OR TITLE-ABS-KEY ( "international financial reporting standards") OR TITLE-ABS-KEY ( niif) OR TITLE-ABS-KEY ( "normas internacionales de información financiera") OR TITLE-ABS-KEY ( "plan general de contabilidad") OR TITLE-ABS-KEY ( pgc) AND TITLE-ABS-KEY ( "intangible asset*") OR TITLE-ABS-KEY ( "activo* intangible*")).

## 4 Results

The search yielded 687 documents published between 1984 and 2022. In the first instance, to answer the first research question, the evolution of publications over time was analyzed. Initially, during the 1980s and 1990s, very few publications were made, with a maximum of only 8 publications in 1998 and an average of just two publications per year. However, in the following decades of 2000 and 2010, the average number of publications increased significantly, reaching 18.8 and 32.8 documents respectively, indicating a substantial increase (Fig. 1).

This upward trend reached its peak in 2020, with 57 publications on the topic. However, in the years 2021 and 2022, the number of publications has decreased significantly to 45 and 36 respectively. Nevertheless, this decline does not necessarily indicate a loss of interest in the subject, as similar decreases have occurred in the past, only for the number of publications to subsequently increase again.

It is worth noting that the decline in publications in recent years might be attributable to a variety of factors such as changes in funding or research priorities, shifts in the academic landscape, or the impact of the ongoing pandemic. Further analysis and research may be required to determine the underlying reasons for this decrease and to anticipate future trends in this field (Fig. 2).

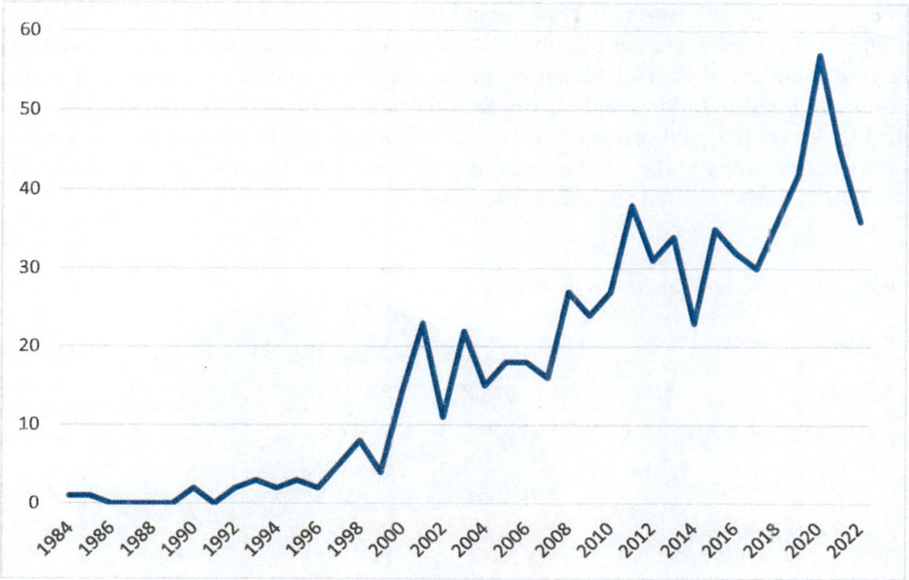


**Fig. 1.** Evolution of publications over time

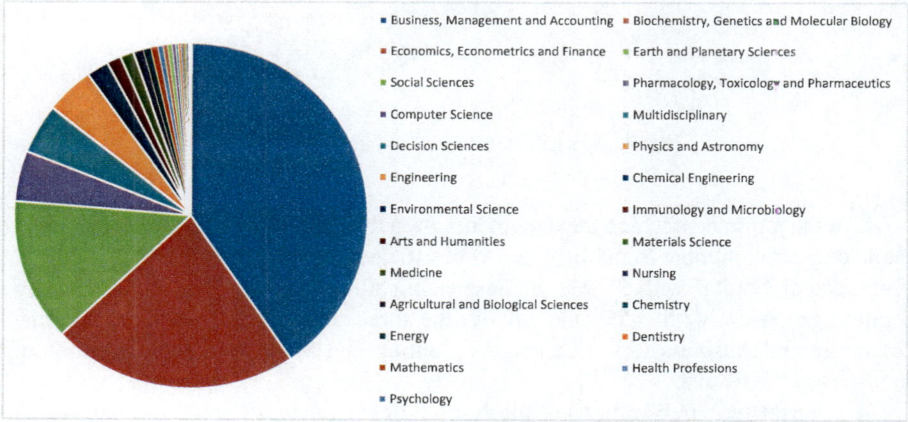


**Fig. 2.** Subject areas of publications

Another interesting issue to analyze is in which areas of knowledge the publications are developed. The study of intangible assets crosses several disciplines, since it is a complex phenomenon that encompasses, for example, administration and accounting for its management, law for its regulation, environmental sciences to analyze its impact and, in most cases, engineering, computer science and the arts for its design and implementation.

Only three disciplines exceed 10% of publications and between them, they account for 76.3% of the total number of publications. Business, Management and Accounting leads the publications with 40.4% corresponding to 496 documents, Economics, Econometrics and Finance follows with 22.6% and 277 publications and then Social Sciences with 13.3% and 163 publications. The sum of the other areas of knowledge amounts to 23.7% and individually they do not exceed 5%. This result is not surprising, since the search was clearly oriented towards accounting.

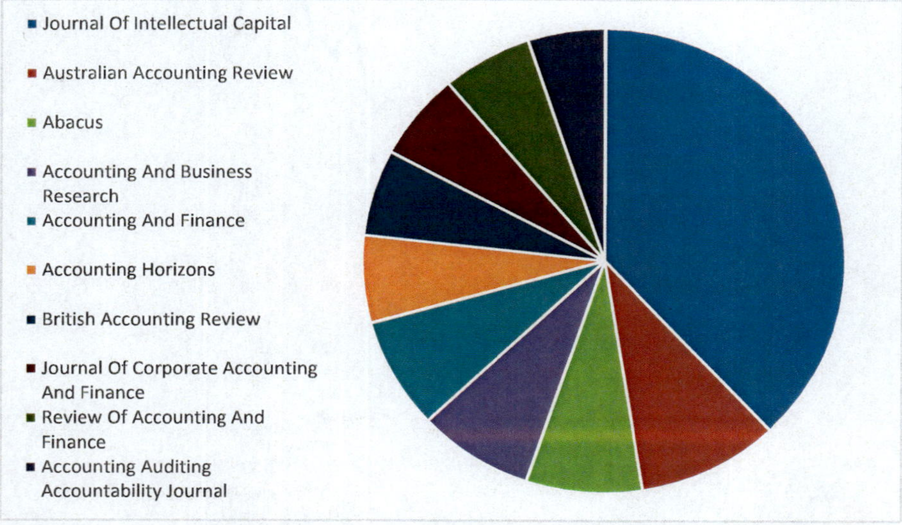


**Fig. 3.** Journals with the most publications

As for the journals in which the documents are published, as we can see in Fig. 3, the top ten journals in number of publications total 116 documents. The leader is the Journal of Intellectual Capital with 37.9% of these publications, followed by the Australian Accounting Review with 9.5% and finally, the three with 7.75%, we have Abacus, Accounting and Business Research and Accounting and Finance to complete the top five places in the results.

The superiority in publications of the Journal of Intellectual Capital is clear, which quadruples the number of papers published by the journal in second position. Thereafter, the numbers shown by the journals are quite similar (Fig. 4).

Finally, the most productive authors in the area are analyzed. In this analysis, no differentiation was made between authors who publish individually or collaboratively. Among the ten authors who have published the most, we found five with five or more publications and the remaining five with four or less. Leading the ranking is T. Tollington from Brunel University London with ten publications, followed by A. Wyatt from The University of Queensland with seven papers and the next three are M. Giuliani, B. Lev from Università Politecnica delle Marche and Popescu C.R.G.H. from Bucharest University of Economic Studies both with five. Although each author is identified with

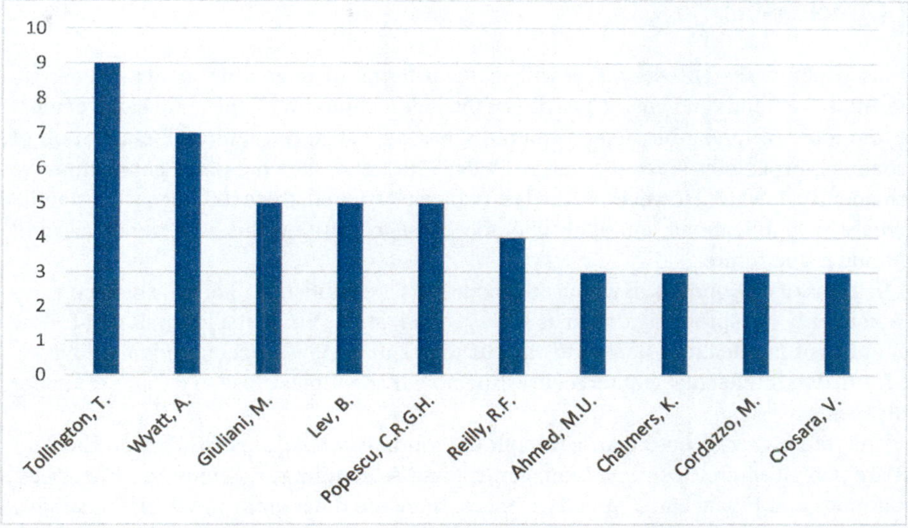


**Fig. 4.** Most prolific authors

the university with which he or she is affiliated, this analysis did not seek to determine which universities have the most impact on publications, but only the authors, regardless of the institution with which they are affiliated.

## 5 Discussion

Publications are highly concentrated in a single journal, the journal of intellectual property. The other journals are far behind in the number of publications.

It should be noted that the analysis conducted has several limitations. First, although the use of Scopus usually adequately reflects the tendencies of the scientific community, the fact that other databases have not been included may mean that some relevant publications in the area have not been considered. Secondly, since this was a basic exploratory analysis and we used information from all the publications that met the search criteria, it was possible to include publications that were less directly related to the topic in question. Finally, the search equation was generally oriented to intangible assets, so it was not intended to cover specific assets that are part of this group.

For future works that seek to continue what has been done in this paper, it could focus more on specific subtopics of the treatment of intangible assets, such as their measurement or disclosure. Likewise, a natural deepening would be the study of the accounting treatment of specific intangible assets, such as patents or trademarks. Other analyses that could be useful to deepen the understanding of publications in the area would be with respect to the universities and countries that have the greatest impact on publications.

## 6 Conclusions

In this paper, a search was made with general terms of intangible assets in order to determine the trends and characteristics of the publications. With this, it became evident that the topic of intangible asset accounting has increasingly captured the attention of academics, especially since the early 2000s. The trend shown by the publications is permanently upward. However, in the last two years studied, there has been a substantial decrease in publications, although it is too early to say that this is a trend that will continue in the future.

In terms of the journals in which the documents are published, there is a high concentration in only one, probably due to its high specialization. While the journals that follow in number of publications deal with various topics in business, accounting and finance, the Journal of intellectual capital focuses specifically on topics close to the subject under analysis.

It is interesting to note that, although the journals where most studies on the subject are published are Business, Management and Accounting, Economics, Economics, Economics and Finance and Social Sciences, there are other areas in which the subject arouses interest, such as Computer Science, Decision Sciences, Engineering and Environmental Science. Therefore, this is an area where interdisciplinarity could be used to great advantage.

In addition to the results presented, it is important to emphasize the great relevance of the relationship between intangible assets and disruptive technologies such as artificial intelligence, machine learning, blockchain or nanotechnology. Faced with a business environment that is advancing at a rapid pace, disruptive technologies are increasingly likely to impact the creation, management, valuation, and exploitation of intangible assets, thereby directly influencing the competitiveness of companies and their valuation in the market. Hence, it is essential this interaction be studied in detail, in order to clarify how these two elements, affect each other and their impact on the performance of the company in particular and the market in general. In addition, due to the multidisciplinary nature of the subject, it is necessary that the research be conducted collaboratively by professionals from different areas such as engineering, accounting, law, management, information technology, among others. This, considering that the influence of certain disruptive technologies on the overall consumer population is still in its early stages. In the future, the expansion of this influence will lead to unsuspected relationships and even to the creation of new intangible assets that today do not even cross the imagination of most individuals.

## References

1. Brown, P.: International financial reporting standards: what are the benefits? Account. Bus. Res. **41**(3), 269–285 (2011)
2. Florou, A., Pope, P.F.: Mandatory IFRS adoption and institutional investment decisions. Account. Rev. **87**(6), 1993–2025 (2012)
3. Ahmed, A.S., Neel, M., Wang, D.: Does mandatory adoption of IFRS improve accounting quality? Preliminary evidence. Contemp. Account. Res. **30**(4), 1344–1372 (2013)

4. Monclús Guitart, R., Vidal Blasco, M.A.: Los activos y pasivos financieros en la nueva economía. Adaptación contable en el caso español. Contaduría Universidad de Antioquia **53**, 81–101 (2009)
5. Nobes, C.: Towards a general model of the reasons for international differences in financial reporting. Abacus **34**(2), 162–187 (1998)
6. Dima, B., Dima (Cristea), S., Claudia Moldovan, N., Gabriel Pirtead, M.: National legislative systems and foreign standards and regulations: the case of International Financial Reporting Standards' adoption. Econ. Res.-Ekonomska Istraživanja **26**(3), 15–30 (2013)
7. Zahid, R.M.A., Simga-Mugan, C.: An analysis of IFRS and SME-IFRS adoption determinants: a worldwide study. Emerg. Mark. Financ. Trade **55**(2), 391–408 (2019)
8. Barker, R., Lennard, A., Penman, S., Teixeira, A.: Accounting for intangible assets: suggested solutions. Account. Bus. Res. **52**(6), 601–630 (2022)
9. Nakamura, L.I.: Intangible assets and national income accounting. Rev. Income Wealth **56**, S135–S155 (2010)
10. Cañibano, L.: Accounting and intangibles. Revista de Contabilidad **21**(1), 1–6 (2018)
11. Haskel, J., Westlake, S.: Capitalism Without Capital: The Rise of the Intangible Economy. Princeton University Press (2018)
12. International Accounting Standards Board (IASB). International Accounting Standard 38 (2001)
13. Chalmers, K., Clinch, G., Godfrey, J.M.: Adoption of international financial reporting standards: impact on the value relevance of intangible assets: adoption of international financial reporting standards. Aust. Account. Rev. **18**(3), 237–247 (2008)
14. Lev, B.: Remarks on the Measurement, Valuation, and Reporting of Intangible Assets (2003)
15. Lev, B.: The deteriorating usefulness of financial report information and how to reverse it. Account. Bus. Res. **48**(5), 465–493 (2018)
16. Zadorozhnyi, Z.-M., Yasyshena, V.: Intangible assets accounting and reporting issue. Mark. Manag. Innov. **4**, 182–193 (2019)
17. Van Criekingen, K., Bloch, C., Eklund, C.: Measuring intangible assets—A review of the state of the art. J. Econ. Surv. **36**(5), 1539–1558 (2022)
18. Donthu, N., Kumar, S., Pattnaik, D.: Forty-five years of Journal of Business Research: a bibliometric analysis. J. Bus. Res. **109**, 1–14 (2020)
19. Manzano, S., Parra-Domínguez, J., Pinto, F., González-Briones, A., Hernández, G.: The communicational universe of cryptocurrencies. An approach to the current scientific importance. In: Prieto, J., Benítez Martínez, F.L., Ferretti, S., Arroyo Guardeño, D., Tomás Nevado-Batalla, P. (eds.) BLOCKCHAIN 2022. LNNS, vol. 595, pp. 380–387. Springer, Cham (2023). https://doi.org/10.1007/978-3-031-21229-1_35

# Sustainability as an Informative Business Strategy: New EU Directive and the Principle of Double Materiality

Mª Angela Jiménez-Montañes(✉) and Susana Villaluenga de Gracia

University of Castilla La Mancha, Cobertizo de San Pedro Mártir, S/N 45071 Toledo, Spain
{angela.jimenez,Susana.villaluenga}@uclm.es

**Abstract.** Sustainability has become a relevant competitive strategy in business performance, which has traditionally been analyzed from three dimensions: economic, social, and environmental. From the economic point of view, it can be understood as those actions that aim to favor social and environmental activities and that help improve the financial situation of the company. From the environmental point of view, it focuses on actions that try to achieve a balance of the environment through practices that favor its quality. From the social point of view, it is the voluntary aid expressed in economic resources granted by companies to external projects of socioeconomic development. Given the impact of sustainability on business activity, we must add *the dimension of sustainable information*. At the end of 2022, the Council of the EU approved the Sustainability Reporting Directive, requiring companies to provide concrete detailed information, which strengthens the accountability of the company, avoiding divergences between the different existing rules and facilitating a transition towards a sustainable economy, aligned with the Sustainable Development Goals of the 2030 Agenda.

**Keywords:** Sustainability · Non-Financial Information · Double Materiality

## 1 Introduction

Current economic growth has proven to be incompatible with available resources, which has led to a rethinking of thinking and decision-making in terms of sustainability, a concept introduced in 1987 by the Brundtland Report [1], which defined Sustainable Development (SD) as a way of meeting the needs of the present without compromising the ability of future generations to meet their own needs. So far, much has been said about economic sustainability, mostly in connection with the environmental sustainability of the economy. The basis of sustainability is viability, which is maintained if the system can react adequately to changes in its environment [2]. A project is viable when the cost is higher than the benefits it generates. Compared to the previous concept, it can be said that, if wealth measures affluence, income can be the benchmark for determining

Member of the research group "Responsible Enterprise Information System: Accounting and Non-Financial Reporting" (SIER) of the UCLM.


R. Mehmood et al. (Eds.): DCAI 2023, LNNS 741, pp. 88–98, 2023.
https://doi.org/10.1007/978-3-031-38318-2_9

sustainability, so that income is what can be spent while leaving the asset base intact [3]. Expanding on this idea, Anand and Sen [4] associate income with the amount a person can consume without becoming poorer, so that those who consume below their income intend to save to improve their future and those who live above their income will become worse off. The concept of sustainability has been considered from three important perspectives, the economic, social and environmental perspective, but given the impact of sustainability on business activity, we must add the dimension of sustainable information. At the end of 2022 the EU Council has approved the Directive on sustainability reporting [5], requiring concrete detailed information from companies, strengthening corporate accountability, avoiding divergences between the different existing standards and facilitating a transition towards a sustainable economy, aligned with the Sustainable Development Goals of the 2030 Agenda within the framework of the European Green Pact and the Sustainable Finance Agenda.

From an economic point of view, it can be understood as those actions that aim to favour social and environmental activities at the same time, and that in turn help to improve the financial situation of the company [6]. From an environmental perspective, it is the balance generated by the harmonious relationship between human beings and the nature that surrounds them and of which they are a part, and that this relationship remains sustainable over time. From a social point of view, it is the voluntary aid expressed in economic resources granted by companies to external socio-economic development projects. Corporate social sustainability consists of identifying and managing business impacts, both positive and negative, on the company's people, on the supply chain and on the community in which the company operates. This involvement in the activity of all types of organizations leads us to incorporate the dimension of sustainable information, the point of development of this research, based on the international requirements of sustainability applied to business information to guarantee its transparency. This leads to the research questions that we have posed in the development of the work,

- RQ 1 How does the Sustainability Directive affect corporate reporting?
- RQ 2 What differences arise in the information to be prepared with the Non-Financial Reporting Statement?
- RQ 3 What advantages does it bring in terms of information transparency for stakeholders?

The methodology used is typical of the Social Sciences. According to [7], the classification of the possible lines of research in Social Sciences identifies three main orientations: empirical, interpretative, and normative-critical. Almost all classifications of research methods tend to make two major groupings, with categories, which basically have to do with the quantitative paradigm and the qualitative paradigm. Scientific research begins with a concrete problem, not with the choice of a particular methodology. We focus on non-experimental research, following [8, 9] since the researcher cannot have direct control over the independent variables with which he is conducting his research, [10–12]. For this reason, we establish the term 'sequential selection strategy' whereby progressive theoretical construction determines data collection. As the Sustainability Directive has been approved in 2022 and will not be implemented until the financial year ending 2024, we do not have quantitative data for an analysis of the data on compliance with the requirements and results of the Directive.

## 2 European Sustainability Directive: informative Dimension

In this section we answer research questions RQ1 and RQ 2.

### 2.1 Non-financial Information as a Prelude to Corporate Sustainability Reporting and its Limitations

The starting point for the analysis of the European Directive, is based on the concept of "sustainable development", it is within this framework that the latest actions of the European authorities have been developed, with the aim of creating and defining a business action model in relation to Sustainability and Corporate Social Responsibility and the Circular Economy. These actions and pronouncements relate to the application of good management and production practices, also extending to good waste management practices. It encompasses efficient management at all points in the value chain. The most innovative idea in this topic would be the approach that provides a practical framework for creating products and industrial systems in a positive relationship between ecological health and abundance and long-term economic growth [13]. The values and principles of the cooperative movement and the social economy, such as links with the local environment, inter-cooperation, and solidarity, are decisive pillars for guaranteeing sustainable development processes in their triple dimension: environmental, economic, and social, which since 2018 have been reflected in the Statement of Non-Financial Information, which accompanies the Financial Statements of certain entities with specific characteristics. It is prepared in accordance with the GRI (Global Reporting Initiative) Standards [14]. The benefits that the preparation of this report brings to the entities can be summarized as follows [15]:

- Transparency and trust in stakeholders and Global information on the organization.
- Reliability of the information through the external independent verification process.
- Generation of useful information for the decision-making process.
- Improved organizational performance.
- Fulfilment of Sustainable Development Goal (SDG) 16.

One of the most relevant aspects in the preparation of this Statement is the determination of "materiality", also referred to as non-financial or environmental and social "significance" [15]. Material information is defined as information whose omission or misstatement is reasonably considered likely to influence decisions that users make based on the company's financial statements, and therefore to the extent necessary to understand the impact of the company's activity. Within the concept of materiality, the report should address topics that [16]: A) Reflect the significant economic, environmental, and social impacts of the organization, or B) Substantially influence the judgements and decisions of stakeholders. All non-financial reporting should comply with the following fundamental principles [17]: a) Disclosure of material information, b) Fair, balanced, and understandable information, c) Complete but concise information, d) Strategic and forward-looking information, e) Stake older-oriented information, f) consistent and systematic information.

Although companies started to prepare this non-financial information from 2018 onwards, it has been considered to suffer from the following limitations and restrictions:

- *It is not sufficient factual.* Not sufficiently comparable between companies. The existence of a gap between the sustainability information reported by companies and the information needs of the information users about risks for companies to make their investment decisions.
- At present, companies reporting on the EINF face *problems arising from the different frameworks,* private standards and requirements that can be used in the presentation of non-financial information. Increased costs in the preparation of non-financial information.
- *Insufficient reporting on assets* and other intangible factors, including internally generated intangible resources, which impedes the proper assessment of the company's development, performance, and position, as well as the monitoring of investments.
- *To enable investors to better understand* the growing gap between the book value of many companies and their market valuation, which is observed in many sectors of the economy, adequate disclosure of information on intangible resources should be required for all large companies and for all companies, except for micro-enterprises, whose securities are admitted to trading on a regulated market in the Union.
- *They do not specify whether the information* to be reported should be forward-looking or backward-looking. There is currently a deficit of forward-looking information. The sustainability information presented should include forward-looking and backward-looking information, both qualitative and quantitative.
- *Sustainability reporting should also consider short*, medium, and long-term time horizons and contain information on the company's entire value chain, including its own operations, its products and services, its business relationships and its supply chain.
- *Companies are not obliged to refer to other information* in the management report or to add supplementary explanations of such information. Therefore, there is currently a lack of consistency between the non-financial information presented and the other information published in the management report.

## 2.2 Sustainability Information and CSDR

The general objective of the new Directive (CSDR) focuses on promoting the development of the preparation of non-financial information by organizations, within the European Green Deal [18] and its 2020 work program, the Sustainable Finance Agenda [19] and the Sustainable Development Goals (SDGs). The relevant point has been the change from "non-financial information" to "sustainability information" to equate both types of information with financial information, reflecting it as a complement to business activity. The most relevant aspect of sustainability reporting focuses on the requirement to present not only the information necessary to understand the company's performance, results, and position, but also the information necessary to understand the impact of the company's activities on environmental, social and personnel issues, respect for human rights and the fight against corruption and bribery. Companies are therefore obliged to report both on the impact of their activities on people and the environment and on how sustainability issues affect them. This is known as the dual significance or materiality perspective, where the risks to the company and the impact of the company represent a significance perspective. This will ensure that the characteristics of sustainability reporting are as follows: comparable, secure, relevant, and easy to find by digital means. The CSDR states that sustainability information should be captured in the broad components: environmental, social, human rights and governance factors, and incorporate the definition of the terms 'sustainability factors' set out in Regulation (EU) 2019/2088, focusing on what is termed 'sustainability risk risk' an environmental, social or governance event or condition that, if it were to occur, could result in a material negative impact on the value of the investment [20] (Table 1).

## 3 The Principle of Double Materiality

Materiality should be understood as the criteria for the inclusion of specific information in corporate reports. The basic points are [22]:

- the importance of the information in relation to the phenomenon it is intended to describe or explain,
- its ability to meet the needs and expectations of a company's stakeholders and of the company itself, enabling appropriate decision-making,
- the transparency needs corresponding to the public interest.

But what is relevant is dual materiality, which is considered as a concept that provides criteria for determining whether a sustainability issue or information should be included in the company's sustainability report. Dual materiality is the union (in mathematical terms, i.e., the union of two sets, not the intersection) of *impact materiality (in-out)* and *financial materiality (out-in)*. A sustainability issue or information therefore meets the criteria for dual materiality if it is material from either an impact perspective or a financial perspective or both. The dual materiality approach provides information on how sustainability issues affect the company (financial materiality) as well as how the company impacts the environment and society (impact materiality). Dual materiality involves the assessment of risks and opportunities and impacts in all three dimensions of sustainability – economic, social and environmental. A sustainability issue is a specific dimension of sustainability by virtue of which the company affects people or the environment (impact materiality) or which gives rise to risks or opportunities for the company that affect or may affect its financial development, performance and position (financial materiality) [23]. In materiality assessment processes, equal weight should be given to impact materiality and financial materiality. Both processes should be conducted in parallel. The basic concepts to consider in the application of the dual principle are as follows:

1. *Financial materiality*: A sustainability topic is material from a financial perspective if it triggers financial effects on companies, i.e. it generates risks or opportunities that are likely to influence future cash flows and thus the value of the company in the short, medium or long term, but which are not captured in the financial information at the reporting date,
2. *Risks and opportunities of financial materiality*: Effects on future cash flows in respect of assets and liabilities already recognized in the financial information as a result

**Table 1.** Content of Sustainability Information CSDR (Substantive Aspects).

| Factors | General points. Substantive aspects | Specific information |
|---|---|---|
| *Environmental factors* | climate, soil, water, biodiversity | Reference is made to a paper presented by the authors [21], |
| *Social factors* | - labour conditions,<br>-the involvement of the social partners,<br>-collective bargaining<br>-equality non-discrimination, diversity and inclusion,<br>-human rights<br>-entrepreneurship capacity building<br>-fundamental freedoms<br>- democratic principles | -including the impact of business on people, in particular workers and on human health<br>- data on the existence of forced and child labour in their value chains<br>- Potential risks and trends with regard to employment and income<br>- equal opportunities for all and working conditions and democratic freedoms. Wage gap, Accessibility measures for people with disabilities<br>- the proportion and breakdown of workers participating in training. Existence of works councils and collective agreements and the proportion of workers covered by such agreements<br>- employee participation in administrative and supervisory bodies, gender diversity in senior management. The number of members of the under-represented sex on their boards |
| *Governance factors* | - the role of management and governing bodies in corporate sustainability<br>-Business ethics<br>-political linkages<br>-internal control system, | - anticorruption activities and compromise<br>-Corporate lobbying activities and engagement – Lobbying activities and engagement<br>-quality of relationships with customers and suppliers, delivery dates or deadlines of suppliers<br>-rate of interest on late payment in relations with suppliers and customers, compensation for collection costs, |

of past events or assets and liabilities that may be recognised as a result of future events, or factors creating enterprise value that do not meet the accounting definition of an asset (liability) and/or the related recognition criteria, but contribute to the creation/maintenance of enterprise value.

3. *Impact materiality.* A sustainability topic or information is material from an impact perspective if the company is associated with significant actual or potential impacts on people or the environment related to the sustainability topic in the short, medium, or long term.
4. *Impacts.* The effect that the company has or could have on the economy, the environment and people, including their human rights, which in turn may indicate its contribution (negative or positive) to sustainable development. Impacts can be actual or potential, negative or positive, short or long term, intended or unintended, and reversible or irreversible.

Financial *materiality* (*out-in materiality*) is concerned with analyzing and assessing aspects linked to the development, performance, and position of the company and which depend on externalities that affect its value, such as global warming, pollution, etc. Within the stakeholders, the group that most requires information in this respect are investors. The *materiality of impact (in-out materiality)* is related to aspects that affect the environment, where they carry out their activity, as well as the environment and people. Within the different stakeholder groups, those in charge of demanding this information are citizens, consumers, employees, partners, the communities where the activity is carried out and civil society in general. About material sustainability impacts, risks, and opportunities of impact, it should be considered that the relative importance of a sustainability issue shall be assessed along the value chain and over the short-, medium- and long-term time horizon. It is important from an impact perspective whether it refers to significant actual or potential impacts directly caused or contributed to by the company, as well as impacts that are otherwise directly related to the company's entire upstream and downstream value chain, irrespective of the company's contribution to it. For this reason, in all aspects of sustainability, the company should reflect the following information:

- Description of the underlying actual and potential negative and positive impacts on the economy, environment, and people, including whether the company is implicated in the negative impacts through its activities or as a result of its business relationships, and describe the activities or business relationships.
- Description of the underlying financial risks and opportunities related to sustainability, including how:

  - Significant sustainability-related risks and opportunities that have affected the company's financial performance, position, and cash flows.
  - The company's analysis of the potential change in financial performance, position and cash flows from the effects of material sustainability-related risks and opportunities.
  - How the company is exposed to risk or how it intends to take advantage of the opportunity presented by the sustainability analysis in its three dimensions.

As relevant aspects of the dual materiality principle is the information provided in relation to Intangible assets, referred to by EFRAG as "key intangible resources", being considered as resources without physical substance on which the company's business model fundamentally depends and which are a source of value creation for the company and which are intrinsic to sustainability such as information on skills, competences, experience, loyalty to the company and motivation of employees to improve procedures, products and services is sustainability information related to social issues which could also be considered as intangible resource information, information on skills, competences, experience, loyalty to the company and motivation of employees to improve processes, products and services is sustainability information related to societal issues that could also be considered as information on intangible resources. Similarly, information on the quality of relationships between the company and its stakeholders, in particular customers, suppliers and communities affected by the company's activities, is sustainability information relevant to social or governance issues that could also be considered intangible resource information.

In the first dimension of the dual materiality matrix, the analysis should be carried out in four steps:

- Identification and updating of the key issues indicated in the ESG elements and sustainability aspects, which are material to the entity.
- Evaluation and prioritization to obtain evidence regarding the impact of society on the environment and its impact on the organization's value.
- Engagement with stakeholders, considering their analysis of the key issues identified.
- Monitoring and validation of the process.

The second dimension will allow us, based on the business model, the corporate culture and the organization's own culture, to classify them into three levels:

- strategic issues (first level), related to the business model and which must be incorporated into the business strategy and objectives of the company,
- those that have a significant impact (second level). They relate to policies, target setting, local project management, risk management,
- those, which, although considered relevant, have a more limited impact (third level). They may include aspects related to national requirements or intentional social, environmental, or human rights policies (Table 2).

**Table 2.** Relevant points of the dual materiality principle.

| First level | Second level | Third level |
|---|---|---|
| - Cybersecurity<br>- Competitiveness<br>- Economic and regulatory environment, Energy and emissions management<br>- Employee health and safety<br>- Customer privacy in a digital environment<br>- Supply chain management<br>- Workplace practices and business ethics | - Human rights<br>- Diversity, inclusion and employee engagement<br>- Waste management<br>- recycling and hazardous materials<br>- responsible marketing<br>- Ecological impacts<br>- Business model resilience, inclusive growth and sustainable finance | - Air quality and noise<br>-Water management<br>- Product safety |

Source: own elaboration based on the CSDR Directive

## 4 General Considerations and Conclusions

In response to RQ 3, from the studies carried out and the documents analyzed, the answers to our research questions elaborated for the development of this work, the following conclusions are drawn about the sustainability information proposed in the CSDR Directive:

- It seeks to satisfy the information needs of stakeholders, as opposed to the mere accountability of the manager to the owner. It has moved from the stakeholder reporting model to the shareholder reporting model promulgated by Agency Theory [24],
- For SMEs and less complex entities, sustainability information becomes a key instrument of competitiveness, with advantages for accessing markets, obtaining financing, contacts with customers, suppliers, investors and capital managers, national growth agreements and international expansion. In the specific case of SMEs trying to access a public tender, sustainability reporting becomes a relevant element for accessing the tender.
- Improve sustainability reporting at the lowest possible cost, in order to contribute and encourage markets to contribute globally to include sustainability in the economic and financial system, based on the Green Pact and the UN Sustainable Development Goals.
- To help reduce systemic risks in the economy, to focus financial capital and companies' activities on social housing, health and environmental problems. It will be a way for companies to be more aware of their impacts on the population and the environment, building a relationship between the company and society.
- Reducing unnecessary costs in the production of sustainability information and producing it in an efficient way.
- Sustainability reporting should be prepared in a specific and clearly identifiable section of the Directors' Report, on the same date as the financial reporting, to give it the same relevance.
- A progressive approach should be considered to improve the level of verification required for sustainability reporting, starting with a requirement for the statutory auditor or audit firm to issue an opinion on the compliance of sustainability reporting with Union requirements based on a limited assurance engagement.
- The auditor should also assess whether the company's reporting complies with the reporting requirements of Article 8 of Regulation (EU) 2020/852. To ensure a common understanding and expectations as to what would constitute a reasonable assurance engagement, the statutory auditor or audit firm should therefore be required to issue an opinion based on a reasonable assurance engagement on the compliance of the sustainability reporting with Union requirements,
- Statutory auditors or audit firms already verify the financial statements and management report. Verification of sustainability reporting by statutory auditors or audit firms would help to ensure linkage and consistency between financial and sustainability reporting, which is particularly important for users of sustainability information.

- The analysis of dual materiality will allow for greater transparency of organisations' commitment to the environment, society and the environment, enabling the alignment of corporate policy with the fulfilment of the SDGs, in compliance with the 2030 Agenda.
- Sustainability reporting will ensure a step forward for the comprehensive involvement of the planet, in its various aspects. CSDR will imply, in the medium term, a paradigm shifts in business performance, through the generation of sustainable added value and the possibility of accessing less costly financing using sustainable finance promoted by the EU and Next Generation funds.

## References


1. ONU: Informe Brundtland, p. 23. ONU Publishing, New York (1987)
2. Bossel, H.: Indictors for Sustainable Development. Theory, Method, Applications. International Institute for Sustainable Development, Winnipeg (1999)
3. Hicks, J.R.: Value and Capital, 2nd edn. Clarendon Press, Oxford (1939–1946)
4. Anand, S., Sen, A.: Human development and economic sustainability. World Dev. **28**(12), 2029–2049 (2000)
5. Parlamento Europeo y Consejo: «Directiva (UE) 2022/2464 del Parlamento Europeo y del Consejo de 14 de diciembre de 2022 por la que se modifican el Reglamento (UE) n.º 537/2014, la Directiva 2004/109/CE, la Directiva 2006/43/CE y la Directiva 2013/34/UE, por lo que respecta a la present,» Diario Oficial de la Unión Europea (2022)
6. Kiesnere, A., Baumgartner, R.: Sustainability management in practice: organizational change for sustainability in smaller large-sized companies in Austria. Sustainability **11**(3), 572 (2019)
7. Soltis, J.: On the nature of educational research. Educ. Res. **13**(10), 5–10 (1984)
8. Kerlinger, F.: Investigación del comportamiento: técnicas y metodología. McGraw Hill Interamericana, Mexico (1975)
9. Kerlinger, F.: Enfoque conceptual de la investigación en el comportamiento. Mc Graw Hill Internacional, México (1979)
10. Miles, M., Huberman, A.: Qualitative Data Analysis. A Source Book of New Methods. Sage, Beverly Hills (1984)
11. Miles, M., Huberman, A.: Data management and analysis methods. In: de Handbook of Cualitative Research. Sage Publication, Londres (1994)
12. Goetz, J., Lecompte, M.: Etnografia y diseño cualitativo en investigación cualitativa. Morata, Madrid (1988)
13. Bollinger, A., Braungart, M., Mcdonough, W.: Cradle-to-cradle desing: Creating healthy emissions. A strategy for eco-effective product and system desing. J. Soc.-Econ. **15**(13), 1337–1348 (2007)
14. GRI, Pacto Mundial de las Naciones Unidas: Business Reporting on the SDGs. https://www.pactomundial.org/biblioteca/business-reporting-on-the-sdgs-un-analisis-de-las-metas-y-los-objetivos-de-desarrollo-sostenible/ (2017). Accessed 1 marzo 2022
15. Comisión Europea: Directrices sobre la presentación de la información no financiera. Metodología para la presentación de la información no financiera. Diario Oficial de la Unión Europea (2017)
16. Global Reporting Iniciative: GRI 101. Fundamentos 2016, p. 30. GRI, Amsterdan (2018)
17. Comisión Europea: Comunicación de la Comisión. Directrices sobre la presentación de informes no financieros (Metodología para la presentación de información no financiera). https://eur-lex.europa.eu/legal-content/ES/TXT/PDF/?uri=CELEX:52017XC0705(01)&from=ES (2017). Último acceso: 26 noviembre 2021

18. European Comission: COM_2023_62_2_EN_ACT_A Green Deal Industrial Plan for the Net-Zero Age. Brussels (2023)
19. Pacto Mundial (Red Española): Finanzas Sostenibles y Agenda 2030. White Paper nº 1 (2021)
20. Parlamento Europeo y del Consejo: Reglamento (UE) 2019/2088 del Parlamento Europeo y del Consejo. Diario Oficial de la Unión Europea (2019)
21. Jiménez Montañés, M.: Indicadores de desempeño ambiental en el marco de la ISO 26000 RS. Pecvnia: Revista de la Facultad de Ciencias Económicas y Empresariales. Universidad de León **0**(9), 111–128 (2009)
22. EFRAG: [Draft] European Sustainability Reporting Guidelines 1. [Draft] ESRG 1 Double materiality conceptual guidelines for standard-setting (2022)
23. EFRAG: [Draft] European Sustainability Reporting Standard 4 Sustainability material impacts, risks and opportunities (2022)
24. Il Park, B., Chidlow, A., Choi, J.: Corporate social responsibility: stakeholders influence on MNEs' activities. Int. Bus. Rev. **23**(5), 966–980 (2014)

# A Quantitative Methodology to Support Local Governments in Climate Change Adaptation and Mitigation Actions

Nikos Dimitropoulos[1](✉), Elissaios Sarmas[1], Marcelo Lampkowski[2], and Vangelis Marinakis[1]

[1] Decision Support Systems Laboratory, School of Electrical and Computer Engineering, National Technical University of Athens, Athens, Greece
ndimitropoulos@epu.ntua.gr

[2] ICLEI European Secretariat, Leopoldring 3, 79098 Freiburg im Breisgau, Germany

**Abstract.** Tackling climate change is a multidisciplinary process, which affects all aspects of government, on a local, regional and national scale. Local governments are publishing and using data related to climate change response and evidence-based actions, enabling the use of transparent and data-driven solutions. It is essential to provide user-friendly data-driven methodologies and tools in support of local governments to address the challenges they are facing due to climate change. This work provides a robust framework and a unique ranking methodology to assess local governments' performance in tackling climate change. Utilising standardised annually reported questionnaires, the methodology calculates climate change related indicators, with the view of better understanding their performance, monitoring their activities and supporting their decision-making process. 7 different key metrics were created, based on the standardised questionnaire, spanning from governance related information to GHG-emissions city-wide levels. Results from this methodology are presented under the scope of supporting local governments to draw conclusions, compare and analyse actions other cities have taken when facing similar challenges, as the yearly reporting process encourages cities and regions to better understand and reflect on the climate risks they face.

**Keywords:** Climate change · decision support · data-driven · policymaking · indicators · ranking methodology

## 1 Introduction

Tackling climate change is a multidisciplinary process, which affects all aspects of government, on a local, regional and national scale. The abundance of data available can offer invaluable support to decision makers and governing bodies, as they bring a wealth of information and can be utilised for data-driven and evidence-based actions. However, the availability of data has its own challenges, from capturing the data to the actual value that data-driven services can offer given the multidisciplinary context. As the trend toward data-driven urban climate governance incentivises local city governments to prioritise narrowed metrics and external interests, it inhibits the broader transformations required to realise climate change goals [1].


R. Mehmood et al. (Eds.): DCAI 2023, LNNS 741, pp. 99–108, 2023.
https://doi.org/10.1007/978-3-031-38318-2_10

Actions to address climate change are now widespread as decision-makers prepare for its impacts on areas such as transportation, water supply, public health and infrastructure, utilising data-driven approaches [2–4]. At the same time, the demand for climate data and services that can contribute to further understanding of the situation and finding the most effective measures to deal. Local governments are now publishing and using data (open or private) to support their decision-making processes, with the assessment of their actions produce tangible results from researchers and public organisations for effective and transparent climate change response action [5–8].

The interrelationship between adaptation and other agendas at the level of both policy-making and practical implementation of actions is apparent and the need to develop collective responses to climate change impacts integrating sectors, spatial scales and temporal scales is emerging [9]. A keyway of monitoring and identifying the causes and effects of climate change is using indicators, as they can provide aggregated information to policymakers, and the importance of dynamic assessment and modification over time, as new adaptation targets are set and/or as intervention actions are monitored and evaluated [10]. There is a strong tendency to use indicators to assess a community's ability to adapt to climate change. In this way, progress can be made in the evaluation of decision-making and therefore in the design of new climate change adaptation policies.

However, a major problem is that some indicators do not show short-term changes that could immediately raise awareness and mobilise governments and citizens. Another problem is the lack of experience in collecting and managing climate data from the agencies involved. These two issues need to be addressed by implementing forecasts and setting long-term goals at an international level and appropriate staffing and training in data analysis technologies respectively. Below, a selected set of methodologies utilising indicators for climate change related actions is presented.

European Green City Index [11], is a study index in which a comparison of 30 major European cities from the same number of countries in terms of their environmental performance and politics is performed. The overall index consists of 8 categories and 30 sub-indicators, of which 16 are quantitative and measure how the city performs environmentally in practice, while the other 14 are qualitative and evaluate the cities' environmental goals.

Integrated Sustainability Monitoring [12] aims is to investigate the sustainable performance between European Union cities and to detect possible interactions between the three aspects of sustainability: economy, ecology and socio-cultural issues of a city. Each aspect is made up of themes such as land, air, education, health, work and infrastructure. Each theme consists of a multitude of indicators. Through a questionnaire and external sources such as the European Observation Network for Territorial Development and Cohesion (ESPON) and cities' websites, data were collected on 87 indicators.

Green City Conceptual Framework [13] presents a green city conceptual framework (IHS-GCCF) and a harmonised method for measuring green city performance over time (GGCPI). The final tool called the global green city performance index (IHS-GGCPI) consists of 25 sub-indicators in the three main categories, society, economy and environment. It is argued that by using the above framework green performance can be measured, targets set, and achievements tracked. The 25 indicators were divided into 8 categories (1 socio-economic and 7 environmental) and 10 experts, members of academia assigned the weights of the parameters that comprised each of them.

It is therefore essential to provide simple and easy tools in support of local governments to address the challenges they are facing due to climate change. In this paper, a unique methodology based on indicators is presented. A transparent way to gather data, through the use of standardised questionnaires, the ability to track the progress each year and the vast network of cities responding to the questions, make available to create a compact framework of participation and draw conclusions from data-driven services based on a large number of cities.

Apart from the introduction, the paper is organised as follows. Section 2 presented the methodology implemented, Sect. 3 the results and discussion, and in Sect. 4 the conclusion and the next steps for future research purposes.

# 2 Methodology

The methodology for ranking cities aims to assess their ability to address climate change by collecting and normalising data across seven categories: Governance, Climate Change Adaptation, Renewable Energy, Transportation, GHG Emissions, Waste Management, and Water Security. Cities' answers are used to calculate an initial score for each category, which is then normalised to rank cities in comparison to one another. The ranking process aims to encourage data sharing and disclosure to support better decision-making for tackling climate change.

However, the methodology acknowledges that a low score does not necessarily mean a city is not taking sufficient action, as it is determined by the overall number of cities that responded to the questionnaire and considers the information provided in the questionnaires; not the actions themselves. The indices and scores for each category are created using a combination of quantitative and qualitative data.

A key aspect of the methodology is to raise awareness among cities and local governments that open data initiatives can help optimise resource allocation for addressing climate change. The methodology also provides a simple and effective measurement by processing complex data to produce a single score for each category.

## 2.1 Data Availability and Processing

The main data source used for this methodology is derived from the CDP-ICLEI Track [14], a platform for city climate reporting based on annual questionnaires. ICLEI – Local Governments for Sustainability – is an international, non-governmental organisation that promotes urban sustainable development. Its aim is to support local and regional governments in achieving their sustainability goals. In 2022, over 2500 cities, counties and associations of local and regional governments from 126 different countries are part of the ICLEI network.

The data used for this paper comes in the form of the annual CDP-ICLEI Unified Common Reporting System questionnaire, which includes information related to several climate change adaptation and mitigation action at the local government level. The data are based on a considerable number of cities in comparison with the other methodologies cited (e.g., the European Green City Index encompasses 30 major European cities while the CDP-ICLEI Track is based on data reported from more than 250 municipalities in 33 countries between 2019 and 2021).The questionnaire is also adapted annually, with new questions added while others are removed, according to the current reality (e.g. in 2021, questions on the impact of the covid-19 pandemic on cities' actions to combat climate change were included in the questionnaire). Furthermore, the questionnaire follows the Common Reporting Framework for GHG Emissions, and its guidelines established by the UNFCCC [15] aiming to facilitate the collection, analysis, and transparency of emissions data, thereby supporting global efforts to mitigate climate change and assess progress towards climate goals.

The questionnaire is structured as follows. For 2021, 133 questions split into 15 different categories are given to local governments. The following table presents the categories, the number of questions included in each one, and the number of questions used for the metrics creation. Apart from the introduction, these are (Table 1):

**Table 1.** Questions per category and questions used for the metrics

| Category | # questions | # questions included | Category | # questions | # questions included |
|---|---|---|---|---|---|
| Waste | 7 | 3 | Local Government Emissions | 14 | 8 |
| Governance and Data Management | 9 | 3 | Climate Hazards and Vulnerability | 12 | 7 |
| Energy | 8 | 5 | Buildings | 2 | 2 |
| City-wide Emissions | 21 | 11 | Emissions Reduction | 15 | 5 |
| Adaptation | 9 | 6 | Urban Planning | 2 | 1 |
| Transport | 13 | 4 | Water Security | 7 | 4 |
| Opportunities | 15 | 4 | Food | 8 | 2 |

The questions are either multiple-choice or open-ended. In the first case the answers are alphanumerical and come from a limited set of possible answers while in the second they are either numerical or alphanumeric. The use of data from open-ended questions that required a written, lengthy response has been avoided in the methodology and emphasis is placed on numerical responses and responses to multiple-choice questions. It is observed, after all, that respondents tend to answer a greater percentage of multiple-choice questions. Each of the questions can have one or more sub-questions to complete.

The datasets with the local governments' responses to the questionnaire are available online [16], with a sample size of 258 European cities reporting between 2019 and 2021 used for this paper analysis.

A sample of the questions included in the questionnaire are "*Has a climate change risk and vulnerability assessment been undertaken for your city*", "*Please provide a breakdown of your GHG emissions by scope. Where values are not available, indicate the reason why.*", "*Please indicate the source mix of thermal energy (heating and cooling) consumed in your city.*", "*Does your city collaborate in partnership with businesses and/or industries in your city on sustainability projects?*".

## 2.2 Metrics

The selection of metrics to create the methodology is based on the questionnaires' structure and its intricacies. The questionnaires evolve over the years, as new information is being requested or questions are being retracted. This leads to the need to process the information within and combine questions for the methodology to be comparable each year as much as possible. Furthermore, it has been noticed that a lot of questions are not being answered (providing answers is not mandatory by the local governments), and this affects the representation that each city will have in the methodology. Based on this constraints, seven different metrics were considered, in line with the questionnaire structure, in order to provide familiarity to its intended audience (i.e., local governments member, filling the questionnaire). As the questionnaires are evolving, further categories can be considered as they seem fit.

The selection of the questions that would be used as indicators and would make up the metrics based on both the categorisation given by ICLEI questionnaires and considered critical, i.e., the possibility of elimination from future questionnaires is minimal. These are questions that provide essential information regarding the present situation in which a local government is, and the awareness it shows on climate issues through its activities and the attempt to take appropriate measures to deal with them.

In each metric, a combination of questions was used in order to provide a single score. The scores in each metric are normalised in the range [0.1], as the intend is to demonstrate the relative performance between local governments, rather than an objective score. This was chosen, given the constraints that assessing performance on absolute levels is not measurable (i.e., there is no perfect score in policy-making, but achievements of successful policies) and that this ranking process aimed to provide a comparison analysis between local governments, for an eventual knowledge-transfer, successful policies implementation and good practices replications.

The metrics considered, along with a description of the information contained by the questions, are the following (based on 2021 questionnaires, as 2022 results are not available at the time of writing):

- **Governance metric**, using indices that measure a city's activities and abilities to address climate change through internal infrastructure and resource allocation. It considers factors such as strategic decisions for sustainability, joint initiatives, opportunities for local governments to combat climate change, and the support and cooperation of local companies.
- **Climate change adaptation metric,** using indices that measure cities' activities and abilities to address climate hazards they are or will be facing. This metric takes into account the identification of these hazards, corresponding adaptation actions,

risk assessment for health issues, and factors that challenge or support the ability to adapt.

- **Renewable energy metric,** using indices that assess cities' activities and abilities to promote clean energy within their boundaries. This metric considers relevant actions taken to support the increase of renewable energy capacity, as well as the percentage of clean energy use within their electricity mix in current or future target years.
- **Transportation metric,** using indices that measure cities' activities and abilities to promote the use of public transportation and low-emission transportation systems. This metric considers factors such as the amount of transit used each year for different means of transportation, availability of charging spots, and air quality information.
- **GHG emissions metric,** using indices which evaluate city response to effective greenhouse gas (GHG) emissions measuring, monitoring, reporting and verification processes. It evaluates the emissions inventory quality, under defined methodologies and covering representative sources of emissions. In addition, it assesses the relevance and impact of cities GHG emissions reduction targets in alignment to the Paris Agreement.
- **Waste management metric,** using indices that assess cities' activities and abilities to ensure safe and secure waste disposal. This metric considers factors such as per capita waste disposal, regulations, initiatives and legislation followed for safe disposal, the type of disposal activities, and the amount of recycling material.
- **Water security metric,** using indices that measure cities' activities and abilities to assess the access and security of clean water for their inhabitants. This metric considers factors such as clean water access and strategic plans for water management, factors and hazards affecting water security in the future, and corresponding adaptation actions.

## 2.3 Ranking Process

A final cumulative score is given to each city, based on the metrics as defined above, to provide the ability of a unified score considering all the metrics. Since each metric contains a different number of questions to calculate their score, the weighted average of each metric is selected. This also provides the ability to choose the importance of each metric by changing the weights, to consider the varying degrees of importance of each metric, if needed.

Each metric is normalised in the [0,1] range by applying the following formula, as suggested from similar approaches [17, 18].

$$X_{normalised} = \frac{X_{actual} - X_{min}}{X_{max} - X_{min}} \tag{1}$$

The total metric is then calculated as the weighted, normalised sum of the individual metrics, where $w$ is the given weight to each metric.

$$X_{total\_metric} = \sum w_i * x_{normalised} \tag{2}$$

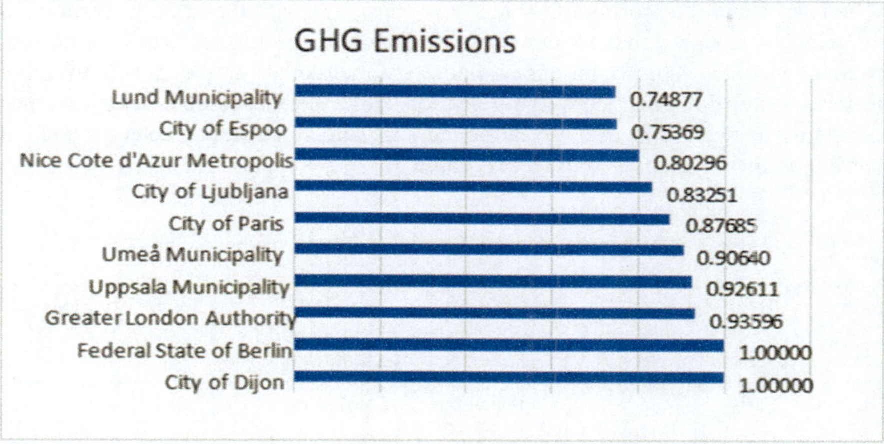


**Fig. 1.** Top 10 cities in GHG emissions metric.

## 3 Results and Discussion

Applying the methodology for the 2021 questionnaire, the following figure provide the highest ranked cities for the GHG emissions metric as an example (Fig. 1).

The results for the top 10 cities for each metric in descending order are displayed in the following Table 2:

**Table 2.** Top 10 cities per metric

| Emissions | Climat. Adapt | Renewable | Water mgmt | Waste mgmt | Transportation | Governance |
|---|---|---|---|---|---|---|
| Berlin | Nice | Copenhagen | Oslo | Rotterdam | Copenhagen | Guimarães |
| Dijon | Torino | Oslo | West Midlands | Barcelona | Berlin | Mannheim |
| London | Paris | Turku | Viseu | Paris | Heidelberg | Águeda |
| Uppsala | Águeda | Uppsala | Paris | Lisbon | Stockholm | Paris |
| Umeå | Athens | Malmö | Örebro | Milan | Barcelona | Barcelona |
| Paris | Milan | Somerset | Nice | London | Paris | Bristol |
| Ljubljana | Braga | Elsinore | Zaragoza | Berlin | Lisbon | Braga |
| Nice | Gladsaxe | Nyon | Trelleborg | Copenhagen | Rotterdam | Athens |
| Espoo | Barcelona | Paris | Barcelona | Venice | Venice | Espoo |
| Lund | Ferrara | Stockholm | Cascais | Stockholm | Helsinki | Nottingham |

The results and the scoring of each city is consistent with the findings from other methodologies, as shown in the following table. Below, the results are presented considering equal weights, $w_i$ of (2), for our methodology, in order to compare it with the other indicators, as they don't include weight selection. However, the weights can be changed according to the need of its user, given the fact that some metrics can be considered of higher importance to compared to others (Table 3).

**Table 3.** Comparison between different indicators methodologies

| Ranking | Total Metric | Green Index | Integrated Sustainability Monitoring | Smart City Index |
|---|---|---|---|---|
| 1 | Paris | Copenhagen | Stockholm | Helsinki |
| 2 | Barcelona | Stockholm | Munich | Zurich |
| 3 | Nice Cote d'Azur | Oslo | Espoo | Oslo |
| 4 | Copenhagen | Vienna | Helsinki | Copenhagen |
| 5 | Berlin | Amsterdam | Copenhagen | Geneva |
| 6 | Oslo | Zurich | Tampere | Amsterdam |
| 7 | London | Helsinki | Amsterdam | Munich |
| 8 | Elsinore | Berlin | Vienna | Dusseldorf |
| 9 | Stockholm | Brussels | Frankfurt | London |
| 10 | Milan | Paris | Nijmegen | Stockholm |

To draw concrete conclusions from the comparison between the different indices, all the proposed methodologies should have included the same cities as samples. As this is not the case, it can be noticed that all the methodologies have identified as the top cities in the respective indices, cities from north-western Europe. This can be due to the fact that local governments that are located in countries with high GDP can usually be more advanced in the climate change tackling actions, due to higher resources to combat both climate change, but also in the reporting process.

## 4 Conclusions and Next Steps

The present work provides a framework and a unique methodology to assess local governments' performance in tackling climate change. The framework derives from the standardised reporting system managed by ICLEI in combination with CDP, in which annually local governments members of ICLEI network, respond to the questionnaire. The yearly reporting process encourages cities and regions to better understand and reflect on the climate risks they face. Reporting also helps them to embed climate action planning in policymaking and spending, and motivates them to create comprehensive action plans, as well as effective monitoring them.

Utilising this questionnaire, 7 different key metrics were created, by combining a set of selected questions, spanning from Governance to GHG-emissions. The score in each metric was normalised between 0 and 1, to provide the ability of direct comparisons and to create a total metric, incorporating each unique metric, using weighted averages. The adoption of this formula was decided, due to the fact that cities are not required to provide complete responses to the questionnaires, thus not allowing to have questions answered.

The results of the methodology are in agreement with the literature and are predicted to be even more effective if the number of participants increases, as well as the quality of the data. The small number of participants is a problem that can be easier to mitigate, as by providing added value from the use of their data, local governments could be more willing to put effort in the process. The problem behind the unanswered questions cannot be ascertained with certainty. One possibility is that participants find the questionnaire bulky and skip those questions requiring extensive analysis. Replacing all "free-text" questions with multiple choice questions, would make it easier for participants to respond, but this require good planning of the questionnaire. The more likely scenario, however, is that most of the local governments have neither the information nor the means required to provide the requested data. This in turn indicates either a lack of interest in some areas or a lack of resources to strengthen the human and technical potential required to answer the questions validly and fully.

Next steps of this work will consider the further development of the methodology, incorporating annually the new responses in the questionnaires, as well as putting into action the methodology for more direct services in "production" mode, i.e , the use of it by local governments to support them in decision-making processes. This can happen with enhanced data-driven services, such as utilising the methodology and the ranking process to find similarities between cities, in both challenges and solutions, and recommendation of best practices. By having the ability to be informed in a user-friendly way of other local governments' successful actions to tackle climate change, it will motivate decision makers to adopt similar measures and/or identify gaps or opportunities in their own impact area.

## References


1. Hughes, S., Giest, S., Tozer, L.: Accountability and data-driven urban climate governance. Nat. Clim. Chang. **10**, 1085–1090 (2020). https://doi.org/10.1038/s41558-020-00953-z
2. Ma, R., Yu, K., Xiao, S., Liu, S., Ciais, P., Zou, J.: Data-driven estimates of fertilizer-induced soil NH3, NO and N2O emissions from croplands in China and their climate change impacts. Global Change Biol. **28**, 1008–1022 (2021). https://doi.org/10.1111/gcb.15975
3. Talari, G., Cummins, E., McNamara, C., O'Brien, J.: State of the art review of Big Data and web-based Decision Support Systems (DSS) for food safety risk assessment with respect to climate change. Trends Food Sci. Technol. **126**, 192–204 (2022). https://doi.org/10.1016/j.tifs.2021.08.032
4. Hussain, S., et al.: Spatiotemporal variation in land use land cover in the response to local climate change using multispectral remote sensing data. Land **11**, 595 (2022). https://doi.org/10.3390/land11050595

5. Salvia, M., et al.: Will climate mitigation ambitions lead to carbon neutrality? An analysis of the local-level plans of 327 cities in the EU. Renew. Sustain. Energy Rev. **135**, 110253 (2021). https://doi.org/10.1016/j.rser.2020.110253
6. Rivas, S., Urraca, R., Palermo, V., Bertoldi, P.: Covenant of Mayors 2020: drivers and barriers for monitoring climate action plans. J. Cleaner Prod. **332**, 130029 (2022). https://doi.org/10.1016/j.jclepro.2021.130029
7. Reckien, D., et al.: How are cities planning to respond to climate change? Assessment of local climate plans from 885 cities in the EU-28. J. Cleaner Prod. **191**, 207–219 (2018). https://doi.org/10.1016/j.jclepro.2018.03.220
8. Carter, J.G.: Climate change adaptation in European cities. Curr. Opinion Env. Sustain. **3**, 193–198 (2011). https://doi.org/10.1016/j.cosust.2010.12.015
9. Boyd, D., Pathak, M., van Diemen, R., Skea, J.: Mitigation co-benefits of climate change adaptation: a case-study analysis of eight cities. Sustain. Cities Soc **77**, 103563 (2022). https://doi.org/10.1016/j.scs.2021.103563
10. Pearce-Higgins, J.W., et al.: A framework for climate change adaptation indicators for the natural environment. Ecol. Ind. **136**, 108690 (2022). https://doi.org/10.1016/j.ecolind.2022.108690
11. Shields, K., et al.: European Green City Index. Assessing the environmental impact of Europe's major cities (2009)
12. Zoeteman, B.C.J., Van der Zande, M., Smeets, R.: Integrated Sustainability Monitoring of 58 EU-Cities: A Study of European Green Capital Award Applicant Cities (2015)
13. Brilhante, O., Klaas, J.: Green city concept and a method to measure green city performance over time applied to fifty cities globally: influence of GDP. Population Size Energy Effi. **10**, 2031 (2018). https://doi.org/10.3390/su10062031
14. CDP Website. https://www.cdp.net/en/cities. Last accessed 17 Apr 2023
15. UNFCC Website: https://unfccc.int/process-and-meetings/transparency-and-reporting/reporting-and-review-under-the-convention/greenhouse-gas-inventories-annex-i-parties/reporting-requirements. Last accessed 18 May 2023
16. CDP website: https://data.cdp.net/Governance/2021-Full-Cities-Dataset/6dea-3rud. Last accessed 18 May 2023
17. Razmjoo, AA, Sumper, A, Davarpanah, A: Development of sustainable energy indexes by the utilization of new indicators: a comparative study. Energy Reports **5**, 375–383 (2019)
18. Chen, C., Noble, I., Hellmann, J., Coffee, J., Murillo, M., Chawla, N.: University of Notre Dame Global Adaptation Index, Country Index Technical Report (2015)

# Ontologies and Accounting in the Field of Sustainability and Reporting

Javier Parra-Domínguez[1](✉), Sara Rodríguez-González[1], Alfonso González-Briones[1], and Juan Manuel Corchado[1,2]

[1] BISITE Research Group, University of Salamanca, Salamanca, Spain
javierparra@usal.es
[2] AIR Institute, Valladolid, Spain

**Abstract.** Nowadays, we live immersed in realities that take us to the accurate grasp of being to incorporate, among others, transparency. It is essential to contemplate ontology as a framework for approaching work in conceiving the company as an organisational unit. In the paper, we expose the necessary connections between accounting and ontology. We do so by considering upcoming developments, such as incorporating non-financial information, to contemplate the company in its true essence. We expose a concrete case where we can observe the reality of the work on this non-financial information by several organisations and, in this way, contemplate the necessary work in a sense that is none other than the one that best characterises and with the most reliable image of the organisation's practices. The use case we are working on focuses on the treatment of the figure of the person in charge that organisations have to incorporate in terms of sustainability.

**Keywords:** Ontology · Accounting · Sustainability

## 1 Introduction

Nowadays, the concern for the total and complete conception of entities and organisations is based on an observation of the same through information reports that attempt to capture it. The parallel with the principles of ontology is evident. It makes us reflect, in a world dominated by different metrics, measurements and institutions concerned with progress, on whether all the variables genuinely reflect the most immediate reality of the organisation under study. It is no coincidence that, in addition to what is known as financial information, which, to date, contains information that transparently provides us with a complete conception of the economic state of a company, new regulations are emerging that attempt to reflect the company's being based on more variables that investigate the actual reality of what an organisation is. With this bibliographic study and with a clear theoretical connotation, we intend to invite reflection on the measures for capturing information about a company that, based on ontological principles, aims solely to transmit the most reliable and complete image of an entity. We will do so by investigating the science of accounting as a robust information system that is aimed at making society aware of what is happening and which favours its progress in the business


R. Mehmood et al. (Eds.): DCAI 2023, LNNS 741, pp. 109–118, 2023.
https://doi.org/10.1007/978-3-031-38318-2_11

world. To this end, in the following subsections of the introduction, we will delve into the idea of ontology and accounting; in a second section, we will link it to sustainability as a capacity to create a complete and obligatory awareness of the business reality, investigating a bibliographic review of the main studies carried out to date; in a third section we will incorporate a system of reflection on the incorporation of information, and we will end with a discussion and conclusions.

## 1.1 Main Linkage Between Ontologies and Accounting

In an age full of standards to belong to and to live up to, it is essential to rethink information and the motivation for it in essence. One step towards this is ontological studies. Ontology is an area of philosophy that studies the nature of reality, i.e. the study of the fundamental principles that define what is and what is not [1]. When we want to approach and deepen on new concepts such as sustainability in its link with the business report, this top quality is necessary. As a philosophical area, ontology focuses on the categories and general principles that govern the being, knowing, acting and thinking of human beings; in essence, it is the study of the concept of "*being*" and the nature of existence [2].

Without attempting to delve deeper into the philosophical concept, beyond the reasoning behind its definition, it is our intention to establish the basic points of action of an advance prior to the contemplation of the advance as a whole, to provide robustness and determination. The work will be carried out on accounting and, specifically, in the field of sustainability, in more detail on sustainability reporting.

Before incorporating the importance of accounting as an information system by generating ontologies, it is good to dwell on sustainability reporting. Sustainability reporting is indeed a crucial aspect of corporate reporting that focuses on the disclosure of an organization's environmental, social, and governance (ESG) performance. It provides stakeholders with information about a company's activities and impacts related to sustainable development and responsible business practices. By reporting on sustainability, organizations can demonstrate their commitment to sustainable practices, transparency, and accountability. It enables stakeholders, including investors, customers, employees, and regulators, to evaluate a company's environmental and social impacts and make informed decisions. When incorporating the importance of accounting as an information system by generating ontologies, sustainability reporting can serve as a valuable source of data. It provides non-financial information that complements traditional financial data, allowing for a more comprehensive and holistic understanding of an organization's performance. By integrating sustainability reporting into accounting information systems and generating ontologies, organizations can improve the collection, analysis, and reporting of sustainability data. This integration enables better decision-making, both internally within the organization and externally with stakeholders, by providing a more complete picture of the company's financial and non-financial performance.

Overall, sustainability reporting plays a vital role in enhancing transparency, accountability, and stakeholder engagement. It aligns with the growing emphasis on sustainability and responsible business practices and serves as a foundation for generating ontologies and incorporating sustainability information into accounting systems.

By reporting information, we aim to make ourselves known and for society to contemplate our "being", our nature of existence [3]. Concerning accounting science, its definition essentially incorporates the need to inform and report on a fundamental knowledge base. Historically, accounting arose with the first records on agriculture and animal husbandry, widely used by traders in its early stages. Even the Egyptians developed systems that we can understand as accounting to administer certain taxes [4]. The most crucial milestone came with the Italian merchants who developed explicit accounting systems to keep track of their commercial transactions. In the 15th century, Luca Pacioli, an Italian Franciscan friar, was the first to address a treatise on modern accounting in an undoubtedly mathematical book [5]. In the centuries that followed, accounting systems developed to help report on businesses that became more complex, evolving to meet the demand for financial information. Today's reliable management of companies can be mapped and evaluated thanks to accounting systems providing this information [6]. To answer increasingly complex questions, the involvement of new technologies such as computers and information technology has advanced, driving accounting through electronic and digital environments. The information circle is completed with the reporting of accurate and timely financial information to executives and shareholders, among others [7].

It is precisely from this last point that the importance of knowing, not only for the stakeholders indicated, the reality of the organisation that is reporting, i.e. the report has consequences such as the verification of fundamental knowledge of ontology on which it is possible to work on what is and what is not.

## 1.2 Towards Complete Knowledge of Organisations

It is important to develop a complete understanding of an organisation by different stakeholders. To do this, information needs to be as complete as possible in order to develop the complex reality of the organisation and to make decisions about it. Information systems, by definition, are a set of interacting components that work together to collect, process, store and distribute information to support decision making and information processing itself. These systems include hardware, software, data, people and processes. It is a fact that information systems play an important role for most modern organisations, allowing users to easily access the data and processes needed to carry out their jobs, improving the efficiency of an organisation by providing more accurate and better-quality information. Another complementary view is that of a set of elements for receiving, processing, and sending data and information that fits perfectly to a set of problems [8], such as those underlying the economic science of the study of the processes of production, exchange and consumption of goods and services to meet needs that are satisfied through the economy itself.

Knowing an organisation's complexity and breadth must be pursued by any institution or private entity in the form of stakeholders [9]. The complex knowledge is transferred through sciences such as accounting, which goes beyond the economic sense to give meaning and response to social ambitions concerning organisational development. As we have already advanced, accounting reflects an essential function for the company. The main branches, such as financial, cost and management accounting, and the preparation of financial statements, work towards transparency and complete knowledge. Nowadays, it is essential to note that it is no longer only financial information on important issues that are passed on to stakeholders, but also non-financial information is reported, including information related to sustainability.

Hence, we return to the concept of ontology, in its ambition for the reality of "*being*" and whose ambition can be completed by incorporating more descriptive variables of an organisation.

## 2 Approaching Research on Ontologies and Accounting More

In an age full of standards to belong to and to live up to, it is essential to rethink information and the motivation for it in essence. One step towards this is ontological studies. Ontology is an area of philosophy that studies the nature of reality, i.e. the study of the fundamental principles that define what is and what is not. When we want to approach and deepen on new concepts such as sustainability in its link with the business report, this top quality is necessary. As a philosophical area, ontology focuses on the categories and general principles that govern the being, knowing, acting and thinking of human beings; in essence, it is the study of the concept of "*being*" and the nature of existence.

Without attempting to delve deeper into the philosophical concept beyond the reasoning behind its definition, we intend to establish the essential points of action of an advance before the new requirements of the information provided by the different organisations with the ambition of transparency to Society and the manifestation of reality.

As part of a literature review, and for this approach, following an initial bibliometric analysis in which we have sought to know the advance of science concerning the linking of accounting, ontology, and business and economics, we have identified previous lines of scientific concern in the linking of ontology and accounting, being the authors by most cited publications. For the analysis we used the Scopus database because of its breadth of study and its coverage of all areas of knowledge in the social sciences linked to economics and business. Table 1 shows the studies by authors and the number of citations up to 2022.

As can be seen, the vast majority of the most cited studies focus on accounting and business, with clear verticals of the need for information through reporting, or directly to key concepts such as sustainability.

In their study, Debreceny and Gray (2001) [10] already analysed the problems faced by stakeholders when trying to access certain financial information on the web. They proposed the use of intelligent software agents to automate the mechanics of searching the web and hinted at the potential of Extensible Markup Language (XML) to tag financial information and improve the automation of information location and retrieval. What is relevant to the study is the presentation of XBRL (eX-tensible Business Reporting Language) as the initiative upon which rich new research opportunities are generated in areas such as new taxonomies, database accounting, financial statement assurance, intelligent agents, human/computer interfaces, standards development process, adoption incentives, global adoption and, most importantly for the purposes of this study, the generation of formal ontologies. Another of the most cited works is Andon, Baxter and

**Table 1.** Articles by number of citations

| Title | Authors | Cites |
|---|---|---|
| The production and use of semantically rich accounting reports on the Internet: XML and XBRL [10] | Debreceny, R., Gray, G.L | 167 |
| Accounting change as relational drifting: A field study of experiments with performance measurement [11] | Andon, P., Baxter, J., Chua, W.F | 138 |
| A geometrically exact beam element based on the absolute nodal coordinate formulation [12] | Gerstmayr, J., Matikainen, M.K., Mikkola, A.M | 109 |
| Embodied energy and emergy evaluation of a typical biodiesel production chain in China [13] | Ju, L.P., Chen, B | 56 |
| On data reliability assessment in accounting information systems [14] | Krishnan, R., Peters, J., Padman, R., Kaplan, D | 52 |
| An ontology modelling perspective on business reporting [15] | Spies, M | 49 |
| The Socio-Technological Lives of Bitcoin [16] | Hayes, A | 43 |
| Ontological considerations about the representation of events and endurants in business models [17] | Guizzardi, G., Guarino, N., Almeida, J.P.A | 42 |
| The art of accounting for science: A prerequisite for sustainable development? [18] | Birkin, F | 38 |
| Cost ontology for enterprise modelling [19] | Tham, K.D., Fox, Mark S., Gruninger, Michael | 37 |

Chua (2007) [11], which works on an application to an Australasian telecommunications organisation. They explore the challenges for an organisation of accurate measurement through the concept of "relational drifting" by highlighting the experimental means by which accounting entries are taken into account and fabricated. On the other hand, the work of Gerstmayr, Matikainen and Mikkola (2008) [12] is certainly more distant from accounting knowledge due to its eminently technical character; even so, the article does include numerical examples and comparisons with analytical solutions and commercial application codes. Ju and Chen (2011) [13], for their part, do not propose an excessive ambition for accounting and work in the same field of knowledge but do suggest a more comprehensive understanding of the ecological costs and benefits of a production process necessary in their perhaps ontological knowledge for effective environmental management and planning, which is important for our study when we talk about sustainability. A further step in the direction of our current study is taken by Krishnan, Peters,

Padman and Kaplan (2005) [14] who go on to propose the development of a formal approach to assessing the reliability of accounting information systems data by creating an ontological metamodel that represents the key concepts needed to assess data reliability. They even delve into the development of a set of algorithms that can process instances of the ontological model to support decision making by auditors.

A further step is taken by Spies (2011) [15] with their ontological representation of the data structures and metadata of business reports defined in the XBRL language standard. They even delve into accounting taxonomies motivating the potential improvement of integrated analytical applications based on quantitative reporting data combined with structured and unstructured data from additional sources. Furthermore, they assert and conclude that such applications will enable significant improvements in compliance management, as they enable business analytics combined with inference engines for statistical and logical inferences. As Hayes (2019) [16] contributes, ontologies in their accounting ramifications could reach cryptocurrencies beyond their importance and blockchains as objects of social research, proposing three alternative ontologies for blockchains and the traceability of Bitcoin development as a socio-technological whole. Guizzardi, Guarino and Almeida (2016) [17], for their part, insist on the purpose of modelling to capture what they call events in structural conceptual models and thus analyse in a purely technical sense. Even with fewer citations, no less important is the article by Birkin (2000) [18] when already in 2000, the marginalisation of environmental accounting was raised due to its addition to existing economic and accounting concepts. At this point, it is essential to highlight the proposal to examine ontological assumptions to develop an accounting practice that assists sustainable business activities. It is based on an ontology of related events. Finally, the work of Tham, Fox and Gruninger (1994) [19] is essential as it reflects how for almost 20 years, they have been working on basic cost ontologies and micro-theories of costing for business modelling that can be used for effective activity-based cost management in any business, irrespective of sector. The ontology is created in the context of the TOVE project.

## 3 Progress in the Bibliographic Analysis

After observing the main lines of the most cited authors based on Scopus in works related to accounting, ontologies and business economics, we have to widen the focus and see if there has indeed been an evolution in this type of publication and who are the most prolific authors.

As can be seen in Fig. 1, the number of publications that relate ontology and accounting within the field of business and economics has suffered ups and downs, with an increase in publications after the economic crisis of 2007. Specifically, 2012 is the year with the highest number of publications.

In addition to the previous review of the most cited papers carried out on the same search basis as in Figs. 1 and 2, other authors have stood out in terms of the number of publications among the top 5, including Walter Schwaiger, from the Vienna University of Technology and linked to the area of accounting; Hans Weigand, from the Uni-versity of Tilburg and linked to the area of business; Christian Fischer-Pauzenberger, also from the University of Vienna and in this case linked to engineering and law; Christian Huemer,

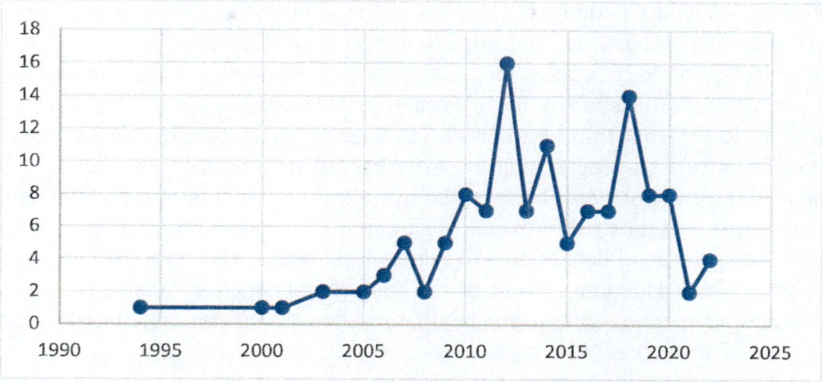


**Fig. 1.** Evolution over time by number of publications.

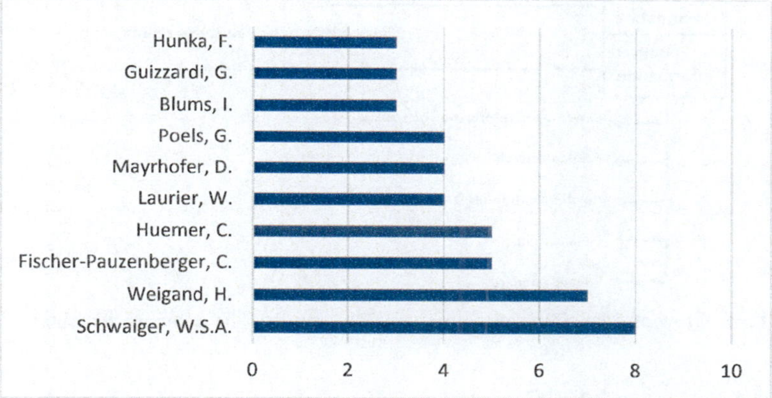


**Fig. 2.** Most prolific authors in terms of publications in the field.

another member of the Vienna University of Technology, linked to the area of computer science and Dominik Mayrhofer, from the University of Graz, specifically from the area of engineering. From all of the above, we can see the significant role of Austrian authors in the world of ontology and its link with accounting, being from areas such as accounting or business but also from engineering.

## 4 Approach to the Representation of the Main Accounting Ontological Relations

As we have previously observed, ontology is a framework used to structure knowledge and represent it logically, coherently and structured way. This knowledge structure is often used in computer science, informatics, engineering, medicine, finance, biology, etc. The ontology consists of a set of concepts and their relationships and rules that are applied to the concepts to describe their semantics. This helps developers understand concepts' meaning and their relationships in a specific subject area.

The concrete ontology is completed in the specific accounting framework by incorporating non-financial sustainability information. This non-accounting information is derived from the relationship between the accounting entity and the environment. It refers to the entity's position in relation to stakeholders and business activities. This non-financial information contributes to the overall understanding of the sustainability of the entity and its future performance. The ontology of accounting, in this sense, is enriched by non-financial data related to the environment, social responsibility, governance, human rights and ethics. These data, like financial data, are collected and organised according to specific criteria. By facilitating the comparison of different entities, the data contributes to market transparency.

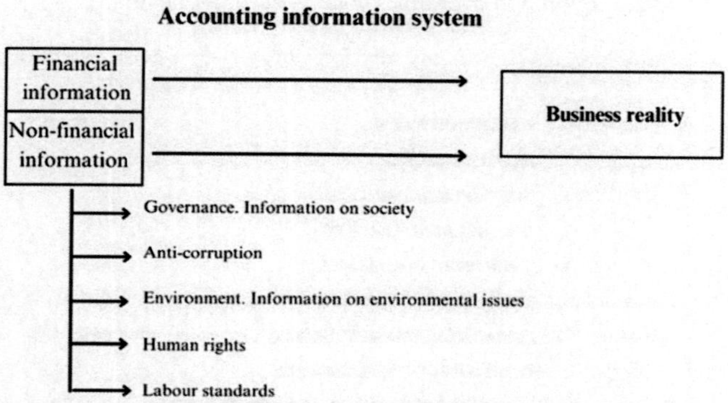


**Fig. 3.** Flow of financial information towards capturing the company's reality.

As shown in Fig. 3, the flow towards the complete vision of the business reality, thanks to the accounting information systems, is completed by incorporating non-financial information based on sustainability. A practical case of the concern for the most immediate reality is illustrated by the following correspondence:

Example case: On the designation of a person or group responsible for sustainability issues.

- Application of the ten principles of the United Nations Global Compact
- Principles 1, 7 and 10
- Mapping to the SDG targets: 5.2, 5.5, 8.1, 8.3, 8.5, 8.7, 8.8, 16.1, 16.2, 16.3, 16.4, 16.5, 16.6, 16.7
- Standards linkage: GRI Disclosures 2–13 (2021)
- European Sustainability Reporting Standards (ESRS): ESRS 2- GOV-1: 18 (b), 20 (c) and AR 2 (a).
- Corporate Sustainability Reporting Directive (CSRD):
  - Article 1. Amendments to Directive 2013/34/EU. 4) Article 19a(2)(c)
  - Article 1. Amendments to Directive 2013/34/EU. 7). Article 29a. 2. c)

## 5 Conclusions

We are currently immersed in a concern for capturing information on sustainability that provides a complete knowledge of the business reality while maintaining fundamental principles that coincide with those pursued by accounting science, such as transparency and rigour in giving information.

Through the study of ontology, the current work raises the necessary concern of the accounting community when reporting and participating in preparing non-financial information regulations. These should follow logical ontologies that show the company's or organisation's most immediate reality.

The field of ontology and its application to the essence of accounting has been studied with its ups and downs in recent years, with Austrian professors linked to engineering and accounting being the most prolific. The different works have maintained theoretical and practical contributions, even working on specific practical cases and already in the early 2000 s, coinciding with studying the true meaning of information in sustainability.

With the current research and exposing a concrete case of the use of different criteria of several organisations concerned with non-financial information, we propose the true sense of progress in regulations and their disposition, which has to be none other than the ontological one, participating in the actual reality of the being of the organisational unit or company.

**Acknowledgments.** This research has been supported by the project "COordinated intelligent Services for Adaptive Smart areaS (COSASS), Reference: PID2021-123673OB-C33, financed by MCIN/AEI/10. 13039/501100011033/FEDER, UE.

## References

1. Marietto, M.L., Ribeiro, I., Serra, F.A.R.: Strategy as practice in the structurationist perspective: what it is and why it is?: toward an ontology of practice of strategy in organizations. Ágora Revista de Divulgação Científica **21**(1), 135–143 (2016)
2. Alistair, M.: Paul Tillich: An Essay on the Role of Ontology in his Philosophical Theology (2017)
3. McCartney, S., Arnold, A.J.: Financial capitalism, incorporation and the emergence of financial reporting information. Account. Auditing Accountability J. **25**(8), 1290–1316 (2012)
4. Carmona, S., Ezzamel, M.: Accounting and accountability in ancient civilizations: mesopotamia and ancient Egypt. Account. Auditing Accountability J. **20**(2), 177–209 (2007)
5. Najemy, J.M. (ed.): Italy in the Age of the Renaissance, pp. 1300–1550. Oxford University Press (2004)
6. Adams, S., Simnett, R.: Integrated reporting: an opportunity for Australia's not-for-profit sector. Aust. Account. Rev. **21**(3), 292–301 (2011)
7. Dahal, R.K.: Changing role of management accounting in 21st Century. Rev. Public Adm. Manag. **7**(3), 1–8 (2019)
8. Stair, R., Reynolds, G.: Principles of Information Systems. Cengage Learning (2020)
9. Gordon, S.N., et al.: People, projects, organizations, and products: designing a knowledge graph to support multi-stakeholder environmental planning and design. ISPRS Int. J. Geo Inf. **10**(12), 823 (2021)

10. Debreceny, R., Gray, G.L.: The production and use of semantically rich accounting reports on the Internet: XML and XBRL. Int. J. Account. Inf. Syst. **2**(1), 47–74 (2001)
11. Andon, P., Baxter, J., Chua, W.F.: Accounting change as relational drifting: a field study of experiments with performance measurement. Manag. Account. Res. **18**(2), 273–308 (2007)
12. Gerstmayr, J., Matikainen, M.K., Mikkola, A.M.: A geometrically exact beam element based on the absolute nodal coordinate formulation. Multibody Sys. Dyn. **20**, 359–384 (2008)
13. Ju, L.P., Chen, B.: Embodied energy and emergy evaluation of a typical biodiesel production chain in China. Ecol. Model. **222**(14), 2385–2392 (2011)
14. Krishnan, R., Peters, J., Padman, R., Kaplan, D.: On data reliability assessment in accounting information systems. Inf. Syst. Res. **16**(3), 307–326 (2005)
15. Spies, M.: An ontology modelling perspective on business reporting. Inf. Syst. **35**(4), 404–416 (2010)
16. Hayes, A.: The socio-technological lives of bitcoin. Theory Cult. Soc. **36**(4), 49–72 (2019)
17. Guizzardi, G., Guarino, N., Almeida, J.P.A.: Ontological considerations about the representation of events and Endurants in business models. In: La Rosa, M., Loos, P., Pastor, O. (eds.) BPM 2016. LNCS, vol. 9850, pp. 20–36. Springer, Cham (2016). https://doi.org/10.1007/978-3-319-45348-4_2
18. Birkin, F.: The art of accounting for science: a prerequisite for sustainable development? Crit. Perspect. Account. **11**(3), 289–309 (2000)
19. Tham, K.D., Fox, M.S., Gruninger, M.: A cost ontology for enterprise modelling. In: Proceedings of 3rd IEEE Workshop on Enabling Technologies: Infrastructure for Collaborative Enterprises, pp. 197–210. IEEE (1994)

# Organizational Culture and Sustainability in the Context of Higher Education Institutions: A Bibliometric Study

Jorge Cruz-Cárdenas[1](✉) and Javier Parra-Domínguez[2]

[1] Research Center in Business, Society, and Technology, ESTec, Universidad Indoamérica, Quito, Ecuador
jorgecruz@uti.edu.ec

[2] BISITE Research Group, University of Salamanca, Salamanca, Spain
javierparra@usal.es

**Abstract.** Sustainability is of great importance for current and future societies, and higher education institutions (HEIs) play a leading role in sustainability efforts. Because organizational culture is a determining force in the success or failure of such efforts, this study examines the relationship between HEIs' organizational cultures and their sustainability efforts. The research took a bibliometric approach and selected 594 relevant documents from the Scopus database. The descriptive analysis reveals a rapid growth in the rate of document publication that accelerates from the year 2015. At the level of both research and financial institutions that fund related research, most are linked to developed and emerging countries. An analysis of the co-occurrence of terms yielded four thematic clusters: (1) teaching and sustainability, (2) organizational and cultural change, (3) impacts of sustainability efforts, and (4) sustainability planning. From a dynamic perspective, the research emphasis has shifted from plans and projects to actors in university communities and their skills and knowledge.

**Keywords:** Organizational Culture · Sustainability · Higher Education Institutions (HEIs) · Universities · Bibliometric Analysis

## 1 Introduction

Sustainability is a development philosophy that seeks to balance the satisfaction of present needs with their satisfaction in the future. In 2015, the United Nations (UN) proposed an agenda for sustainable development with a compliance horizon extending to 2030. It is organized under 17 Sustainable Development Goals (SDGs), including poverty elimination, health and well-being, quality education, economic growth, innovation and infrastructure, reduction of inequalities, and responsible production and consumption [1]. These SDGs quickly mobilized the interest of important sectors of society, and, shortly afterward, higher education institutions (HEIs) emerged as key actors for promoting compliance with the SDGs.


R. Mehmood et al. (Eds.): DCAI 2023, LNNS 741, pp. 119–127, 2023.
https://doi.org/10.1007/978-3-031-38318-2_12

HEIs have a great capacity to promote sustainable development thanks to their strengths in leadership, governance, research, and teaching [2, 3], and they may also benefit from playing an active role in sustainability. These benefits include attracting demand for sustainability education, generating a positive organizational image, demonstrating social impact, accessing financing, and building collaborative networks [2]. However, incorporating sustainability in the lives and actions of HEIs demands a change in their organizational cultures [4, 5]. This study examines the relationship between sustainability efforts and the organizational culture of HEIs.

As in many other long-term efforts and initiatives of HEIs, organizational culture has emerged as a key success factor. It is defined as the set of attitudes, norms, and practices that are shared within an organization [6]. Culture can be a powerful force that drives an organization toward meeting objectives, or it can be an obstacle to change [7]. In the particular case of HEIs, their organizational culture has been described as persistent and conservative [8].

An organizational culture aligned with sustainability is reflected in the beliefs and behaviors of the organization's members, who recognize the importance of balancing equity, environmental protection, and the efficient use of resources [9]. Therefore, successful efforts to transform an organizational cultural into a culture of sustainability must rest on the participation and involvement of the members and stakeholders of HEIs [10].

Universities pursue their sustainability efforts through various strategies, such as changes in their campuses and operations (e.g., consumption of water and electricity, consumption of resources, and waste management) [5]. Other sustainability support strategies employed by HEIs include integrating sustainability education into the curriculum and university extension [11], forming collaborative networks [3], and supporting sustainability research and disseminating its results.

Due to the great importance of sustainability and the leading role of HEIs in efforts to promote it, the relationship between sustainability and HEIs' organizational cultures is attracting growing scientific and academic interest to generate relevant knowledge [12, 13]. These rapidly growing areas of knowledge require systematization to structure the produced research, however, so that research efforts will not be undertaken in isolation or replicated inefficiently [14]. Because of these considerations, the present research adopted the following main research objective:

- Identify the descriptive characteristics and interrelationships among the body of existing publications on organizational culture and sustainability within the contexts of HEIs.

## 2 Method

This study's method conforms to that recommended for research taking a systematic approach to the state of research in a given area [8, 15, 16]. The present study followed three stages: (1) formulating the research objectives, (2) defining and implementing documentary search strategies, and (3) analyzing the selected body of documents and presenting the results. The Introduction describes the central objective of this research. This section describes stage 2 of the investigation, and the Results and Discussion sections describe stage 3.

A first decision regarding the search strategy was which database to use. The Scopus database was selected, as it offers a good balance between breadth of coverage and inclusion criteria quality [16]. After the database was selected, the next step was defining the wording of the search string. The research objective guided this step, as it required seeking two types of content in a given context; thus, the search string had to find content on organizational culture and sustainability in the context of higher education. This requirement yielded the following search string: culture AND sustainability AND ("higher education" OR universit*).

Other search strategies were related to search delimitations. The first delimitation was to select only articles published in journals to ensure that the content had gone through editorial and peer review [8, 16], and the second was that the documents be published in English. No time limit was set for the date of publication.

Once the search strategies were defined, their implementation followed several steps. The Scopus database was accessed on March 14, 2023, and the search string was entered into the search engine and the delimitations applied, yielding 594 relevant documents. Finally, this database was downloaded in a comma-separated values (CSV) format.

## 3 Results

This section presents the results of two types of analysis. The first aimed to describe the body of documents obtained and was carried out using the analysis tools incorporated into the Scopus database. The second, using VOSviewer 1.6.18 software (Van Eck & Waltman, 2010), identified the various interrelationships between the documents' contents.

An initial descriptive analysis sought to determine the evolution of the number of documents published over time. This indicator approximates the academic and scientific community's level of interest in the topic of organizational culture and sustainability in HEIs. As shown in Fig. 1, there are three well-differentiated stages. In the first period, up to the year 2000, a very small number of publications is observed. The second period, from 2001 to 2015, is characterized by growth in the number of publications. Finally, the third period, from 2016 to the present, presents a steep growth curve. In this period, the annual rate of publications leaps from 24 documents annually in 2016 to 85 annually in 2022 (the last full year analyzed).

These results indicate that the academic and scientific areas of organizational culture and sustainability in the context of HEIs are attracting growing interest. These results comport with the perceptions of other authors [12, 13].

Another topic of great interest is the disciplines in which the literature on organizational culture and sustainability in the context of HEIs is being generated. Table 1 presents the areas that had more than 50 associated documents. (The percentages in the table add up to more than 100%, as the same document may be assigned to several areas).

As Table 1 shows, the social sciences and environmental sciences stand out from other areas. Despite the predominance of these two areas, a tendency toward multidisciplinarity can be seen in the bodies of the documents. Strikingly, only 16.3% of the documents are associated with business, management, and accounting, an area that has highly developed tools for the analysis of organizations. This may be due to the fact that the dominant tendency of academics and researchers is to conceive of HEIs as nonprofit institutions.

**Fig. 1.** Evolution of the number of documents published (source: Scopus)

**Table 1.** Disciplines associated with publications

| Subject Area | Documents | % |
|---|---|---|
| Social sciences | 381 | 64.1% |
| Environmental science | 169 | 28.5% |
| Energy | 107 | 18.0% |
| Business, management, and accounting | 97 | 16.3% |
| Engineering | 97 | 16.3% |
| Medicine | 72 | 12.1% |
| Computer science | 51 | 8.6% |

To identify the most important publication routes or journals, only those that published more than 10 documents in the time frame of the analysis were considered, and only three journals met this criterion: *Sustainability*, with 66 documents (11.1%), *International Journal of Sustainability in Higher Education*, with 43 documents (7.2%), and *Journal of Cleaner Production*, with 19 documents (3.2%). Beyond these three journals, many others published relevant documents, but none published more than five.

The institutional affiliations of the documents' authors are broadly dispersed, with only three institutions exceeding five documents: The University of São Paulo (Brazil), with seven, the University of British Columbia (Canada), also with seven, and Queensland University of Technology (Australia), with six.

A similarly great dispersion is observed in the institutions that financed the published research on organizational culture and sustainability in the context of HEIs. Table 2 shows the institutions that financed five or more investigations or documents, including institutions in China, the United States, the European Union, and Brazil. These institutions are thus linked only to developed and emerging countries.

**Table 2.** Main funding institutions

| Institution | Funded documents |
|---|---|
| National Natural Science Foundation of China | 7 |
| National Science Foundation (USA) | 7 |
| European Commission | 6 |
| Fundação para a Ciência e a Tecnologia (Portugal) | 6 |
| Conselho Nacional de Desenvolvimento Científico e Tecnológico (Brazil) | 5 |
| National Institutes of Health (USA) | 5 |

The second stage of the analysis used VOSviewer 1.6.18 software [17] to identify at a deeper level the interrelationships in the content of the documents. An initial analysis established a cluster of terms based on the content of the documents' abstracts. Considering the large number of documents included in the analysis, the minimum frequency of a term's occurrence was set at 15 [8]. Subsequently, terms with a vague or very general meanings were eliminated, such as data, article, methodology, author, type, and item, yielding 103 retained terms and four clusters. Figure 2 shows the four clusters, each represented by a different color. The importance of a term (frequency) is represented in the size of its node.

The composition of the clusters is described below.

*Cluster 1 (red, 32 items).* Terms such as *project*, *teaching*, *course*, *community*, *pedagogy*, *faculty*, and *learning* associate this cluster with the teaching component of HEIs in the context of their sustainability efforts.
*Cluster 2 (green, 27 items).* Terms such as *organizational culture*, *improvement*, *transition*, *engagement*, *communication*, *team*, *involvement*, and *success* associate this cluster with organizational and cultural change in the context of integrating sustainability in HEIs.
*Cluster 3 (blue, 27 items).* Terms such as *effect*, *business*, *government*, *infrastructure*, *performance*, *response*, *environmental sustainability*, *energy*, *quality*, *benefit*, *life*, and *attitude* associate this cluster with the impacts of HEIs' sustainability efforts.
*Cluster 4 (yellow, 17 items).* Terms such as *higher education institution*, *mission*, *leadership*, *sustainable development goal*, *transformation*, *organization*, and *staff* associate this cluster with HEIs' planning efforts and the incorporation of sustainability goals.

In this same stage of deeper analysis, a second analysis was performed, called overlay visualization. Figure 3 shows the results, with the most influential terms colored according to their seniority. Those that were prominence several years ago are presented in purple, those currently prominent in yellow, and those between those extremes in green. The importance of a term is represented by the size of its node. As can be seen

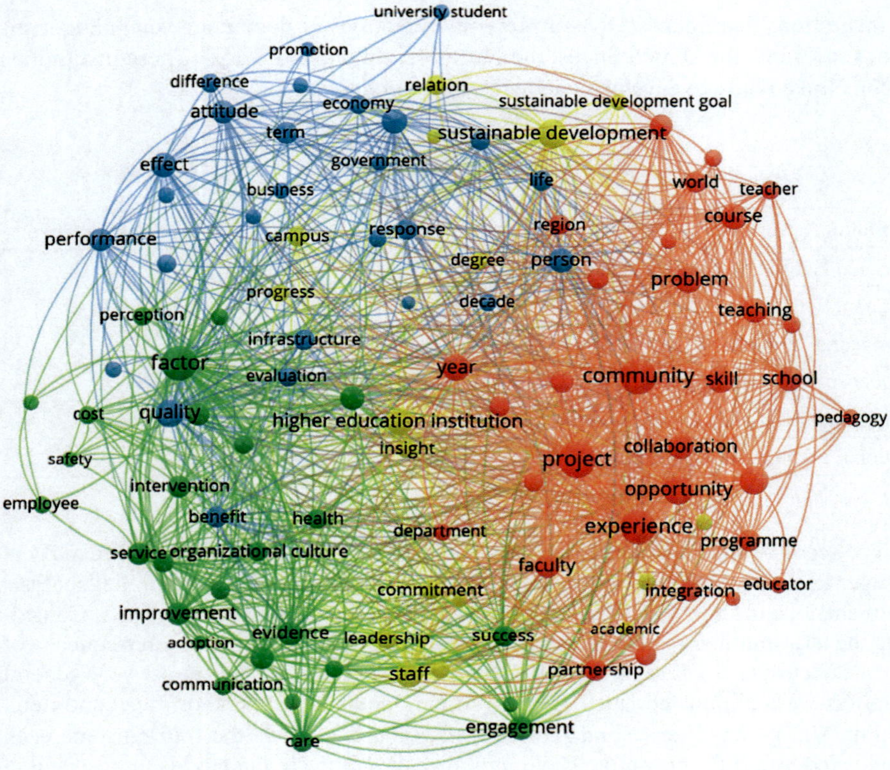


**Fig. 2.** Analysis of the co-occurrence of terms

in Fig. 3, the terms *project*, *progress*, *performance, term*, *success*, and *educator* were influential several years ago (approximately 2015), possibly because they are related to the work of planning HEIs' sustainability efforts. In the current era, *higher education institution*, *skill*, *staff*, *university student* and *communication* become more common. In short, there is a transition in importance from planning organizational culture and sustainability toward the actors in sustainability efforts.

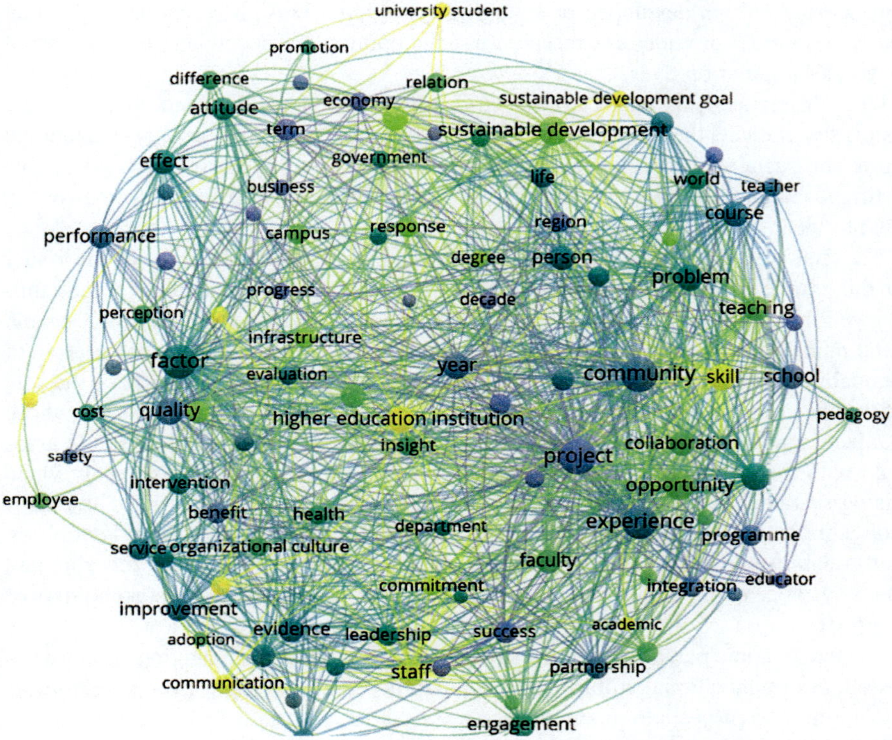


Fig. 3. Overlay visualization of terms

## 4 Discussion

Previous authors [12, 13] have described the sustainability efforts and organizational cultures of HEIs as an area of growing research interest. The rapid growth in publications observed since 2016 may be attributed to, among other factors, the UN's formulation of the SDGs in 2015. Another factor is the leading role of HEIs in sustainability efforts.

Sustainability is an urgent concern of great importance in today's societies. HEIs, as institutions that produce and transmit knowledge, are essential in the sustainability efforts of all societies. Within HEIs, organizational culture is a powerful force that can facilitate or hinder the sustainability efforts of these institutions.

This study analyzed 594 documents on organizational culture and sustainability in the context of HEIs. A first relevant finding is that this is a rapidly growing academic and research area, confirming the perceptions of other authors [12, 13]. Although this growth had been ongoing since 2001, its rate accelerated rapidly beginning in 2015. This change in the rate of growth is most likely related to the formulation of the SDGs [1].

Other relevant descriptive data from the analyzed body of documents indicate that the social sciences and environmental science contribute the most to this academic and scientific production. A broad dispersion is seen in the institutions most active in publication, such as financial institutions, but there is a discernible pattern of the institutions being located only in developed or emerging countries. Hence, it is recommended that greater support be provided for research in developing countries, where a large part of the world's population lives.

The cluster analysis based on the co-occurrence of terms yielded four thematic clusters that indicate the most prominent research areas [16] in the case of organizational culture and sustainability in the context of HEIs. Recognizing these clusters will allow existing research to be better structured and give future research efforts a framework in which to locate themselves.

The first cluster relates to the teaching of sustainability in HEIs. It is not surprising that this is one of the more important ways in which HEIs contribute to sustainability, given their mission of creating and disseminating knowledge [11, 18]. The second cluster embraces the organizational and cultural change demanded by the integration of sustainability in the life of HEIs. This topic is of great importance in light of the role that organizational culture plays in the success of new initiatives in HEIs [7], whose organizational culture has been described as conservative [8]. The third cluster is associated with the impact of HEIs' sustainability efforts, which is crucial not only when revising sustainability plans but also in building an image and demonstrating the institution's importance [2]. Finally, the fourth cluster relates to the planning and definition of sustainability goals by HEIs. To succeed, sustainability efforts must be incorporated at the highest level of planning and have the support of all members and stakeholders of HEIs [10].

From a dynamic perspective, the results of this study show the changing emphasis of research on organizational culture and sustainability over the years, from an emphasis on planning and projects to an emphasis on the actors in sustainability efforts and the skills they need. These findings inspire a final recommendation for future studies and for managers of sustainability projects: it is vital to recognize the importance of the involvement of the members and stakeholders of the university community and the knowledge and skills they need to successfully carry out sustainability efforts.

The present study has some limitations associated with its methodological design. By using the Scopus database for the documentary search, this study left out documents published in journals that are not included in that database. Additionally, no books or book chapters were included, but these decisions were justified by the need to balance scope and quality when searching for relevant content [16].

## References

1. United Nations: Sustainable Development. https://sdgs.un.org/goals (2015)
2. Kestin, T., van den Belt, M., Denby, L., Ross, K., Thwaltes, J., Hawkes, M.: Getting Started with the SDGS in Universities. https://ap-unsdsn.org/wp-content/uploads/University-SDG-Guide_web.pdf (2017)
3. Ramísio, P.J., Costa Pinto, L.M., Gouveia, N., Costa, H., Arezes, D.: Sustainability strategy in higher education institutions: lessons learned from a nine-year case study. J. Clean. Prod. **222**, 300–309 (2019)
4. Adams, R., Martin, S., Boom, K.: University culture and sustainability: designing and implementing an enabling framework. J. Clean. Prod. **171**, 434–445 (2018)

5. Filho, W.L., Amaro, N., Avila, L.V., Brandli, L., Damke, L.I., et al.: Mapping sustainability initiatives in higher education institutions in Latin America. J. Clean. Prod. **315**, 128093 (2021)
6. Ravasi, D., Schultz, M.: Responding to organizational identity threats: exploring the role of organizational culture. Acad. Manage. J. **49**(3), 433–458 (2006)
7. Hemerling, J., Kilmann, J., Danoesastro, M., Stutts, L., Ahern, C.: It's Not a Digital Transformation without a Digital Culture. https://www.bcg.com/publications/2018/not-digital-transformation-without-digital-culture (2018)
8. Cruz-Cárdenas, J., Ramos-Galarza, C., Guadalupe-Lanas, J., Palacio-Fierro, A., Galarraga-Carvajal, M.: Bibliometric analysis of existing knowledge on digital transformation in higher education. In: Meiselwitz, G., et al. (eds.) HCI International 2022—Late Breaking Papers. Interaction in New Media, Learning and Games. HCII 2022. Lecture Notes in Computer Science, vol. 13517, pp. 231–240. Springer, Cham (2022)
9. Bertels, S., Papania, L., Papania, D.: Embedding sustainability in organizational culture: a systematic review of the body of knowledge. https://nbs.net/systematic-review-organizational-culture/ (2010)
10. Disterheft, A., Azeitero, U.M., Filho, W.L., Caeiro, S.: Participatory processes in sustainable universities—what to assess? Int. J. Sustain. High. Educ. **16**(5), 748–771 (2015)
11. Manfredi, L.R., Stokoe, M., Kelly, R., Lee, S.: Teaching sustainable responsibility through informal undergraduate design education. Sustainability **13**(5), 8378 (2021)
12. Ross, N., Sassen, R., Guenther, E.: Sustainability governance toward an organizational sustainability culture at German higher education institutions. Int. J. Sustain. High. Educ. **24**(3), 553–583 (2023)
13. Žalėnienė, I., Pereira, P.: Higher education for sustainability: a global perspective. Geogr. Sustain. **2**(2), 99–106 (2021)
14. Sidney, H.: Literature review as a research methodology: an overview and guidelines. J. Bus. Res. **104**, 333–339 (2019)
15. Osobajo, O.A., Moore, D.: Methodological choices in relationship quality (RQ) research 1987 to 2015: a systematic literature review. J. Relatsh. Mark. **16**(1), 40–81 (2017)
16. Cruz-Cárdenas, J., Zabelina, E., Guadalupe-Lanas, J., Palacio-Fierro, A., Ramos-Galarza, C.: COVID-19, consumer behavior, technology, and society: a literature review and bibliometric analysis. Technol. Forecast. Soc. Change **173**, 121179 (2021)
17. Van Eck, N.J., Waltman, L.: Software survey: vosviewer, a computer program for bibliometric mapping. Scientometrics **84**(2), 523–538 (2010)
18. McKeown, R., Down, L.: A critical reflection on online teaching for sustainability. Sustainability. **14**(21), 13905 (2022)

# Micro, Small and Medium Sized Enterprises in Times of Overlapping Crises

Lica Erhan(✉) and Valentina Paladi

Academy of Economic Studies of Moldova, Chisinau, Republic of Moldova
{lica.erhan,vspaladi}@ase.md

**Abstract.** This work considers the effects of the overlapping crises on micro, small and medium-sized enterprises (MSMEs). In particular, attention will mainly focus on the MSMEs within the Republic of Moldova, analyzing the way these enterprises were affected by two most recent external shocks: Covid-19 pandemic and the war launched by Russian Federation against Ukraine, in February 2022. Methodologically, this study use secondary data analysis provided by: the National Bureau of Statistics (NBS), the Organization for Entrepreneurship Development (ODA), the World Bank Reports, the Expert-Group State of the Country Reports and Moldova Economic Growth Analysis as well as the most recent references in this field of research. In this work, we will first analyze the impact of the Covid-19 pandemic on MSMEs development from Moldova; subsequently we will analyze MSMEs in the context of the war in the neighboring country. Furthermore, the impact of the Covid-19 pandemic and Russia's war in Ukraine has emphasized the need to support and increase the resilience of MSMEs. In this regard, policy recommendations for strengthening MSMEs resilience from Moldova will be drawn up.

**Keywords:** Small business · Overlapping crises · Resilience

## 1 Introduction

In most economies, micro, small and medium-sized enterprises (hereafter, MSMEs) play essential roles in driving growth, creating jobs, and opening new markets for sustainable development through domestic trade liberalization [1].

Although, MSMEs are well known for their important role in economy, these enterprises are most seriously harmed in crises due to their lower knowledge levels, more susceptibility, greater reliance on government and local authorities, and greater financial reliance on owners [2]. Moreover, MSMEs usually suffer from high losses, reduced sales volume, inability of meeting contract terms, reduction in staff numbers, and even close down of the business during or after crises [3]. Therefore, protecting small business from economic shocks is the prime concern for policymakers because the output degradation in MSMEs has spillover effects on other segments in the economy [4].

In this regard, this work analyzes the effects of overlapping crises on MSMEs. In particular, attention will mainly focus on MSMEs within the Republic of Moldova. A


R. Mehmood et al. (Eds.): DCAI 2023, LNNS 741, pp. 128–137, 2023.
https://doi.org/10.1007/978-3-031-38318-2_13

country that in June 2022, has been granted the European Union candidate status together with Ukraine. According, to the State of the Country Report (2021), Moldova's economy has been repeatedly impacted by a range of negative shocks, which have fundamentally undermined the dynamics of economic development and the transition to a functioning market economy, based on a sustainable model of economic growth.

All these shocks underlined the vulnerabilities of MSMEs slowing their development in time. An analysis of data provided by the National Bureau of Statistics (NBS) shows that, on average, only 18 enterprises of every 100 enterprises registered since the Republic of Moldova got her independence 33 years ago, have remained active until now. The business creation rate in 2020 was two times lower compared to the liquidation rate. The survival rate of start-ups from Moldova after the first year of activity is around 46%, the same rate in the European Union is around 80%. This alarming numbers shows the need for policymakers to support the entrepreneurship in Moldova especially during these difficult times of overlapping crises. The aim of this paper is to analyze the way MSMEs within the Republic of Moldova were affected by two most recent external shocks: Covid-19 pandemic and the war launched by Russian Federation against Ukraine, in February 2022. Furthermore, policy recommendations for strengthening MSMEs resilience from Moldova against the backdrop of these crises will be drawn up. In pursuance of this aim, this study use secondary data analysis provided by the National Bureau of Statistics (NBS), the Organization for Entrepreneurship Development (ODA), the World Bank Reports, the Expert-Group State of the Country Reports, Moldova Economic Growth Analysis and the most recent references in this field of research.

The Covid-19 pandemic has started as a health crisis, but it has evolved to be much more, becoming an unprecedented socio-economic crisis that affected the whole world [5]. Similar, the impacts of the war go far beyond Ukraine. The Russian military aggression against Ukraine has deeply affected the Eastern European region, including neighboring Moldova, which shares experiences of Russian aggression in the past [6]. The war has deepened the challenges Moldova is facing in many dimensions. The Russian war in Ukraine has triggered a major economic and social crisis, which has affected the vast majority of the population from Moldova. Moldovan economy faced an unprecedented influx of refugees, trade disruptions, major energetic and security crisis. However, at the same time, Moldova was granted European Union candidate status alongside Ukraine, in June 2022.

The economic crises of the past 20 years have led to a proliferation of studies investigating the impact of a crisis on entrepreneurship [9]. Although these studies have highly increased, our general understanding of entrepreneurial best practices during and after an economic crisis, the ongoing Covid-19 pandemic, the Russian's war in Ukraine, energetic and security crisis we are currently facing is of an entirely different order and its impact on entrepreneurship remains to be studied. In this regard, we will first analyze the impact of the Covid-19 pandemic on MSMEs development; subsequently we will analyze MSMEs in the context of the war in a neighboring country. These increasing numbers of crises and external shocks underline the need for the business and MSMEs in particular to become more resilient. In this case, resilience is the ability to adapt to changing situations, rather than bouncing back to a previous equilibrium [7]. In this regard, policy recommendations for increasing MSMEs resilience will be presented in this paper.

## 2 Impact of the Covid-19 Pandemic on MSMEs Development

Small and Medium-Sized Enterprises play a crucial part in a country's economic growth and sustainable development [8]. However, Covid-19 pandemic has changed the rules of the game in our society; many firms struggled to survive while others even were forced into liquidation [9]. No country in the world was prepared for a shock of the magnitude, duration and impact of the Covid-19 pandemic. The Covid-19 pandemic starkly exposed the economic vulnerabilities of Moldova, which had been classified by the World Bank as the fourth poorest country in Europe in 2021 by GDP per capita 5,230.7 USD [10].

Based on the data of the National Bureau of Statistics in 2021, MSMEs represented about 98.6% of the total number of enterprises operating in the Republic of Moldova. Specifically, of the 98.6% the largest share of 85.4% were micro-enterprises, 10.9% were small enterprises, and only 2.3% were medium-sized enterprises. According to the State of the Country Report (2020), prepared by the think-tank Expert-Group, the vast majority of MSMEs from Moldova were caught during Covid-19 pandemic, with very limited liquidity, without internal governance tools, with a low level of technological endowment, with limited digital instrument and with insufficient diversification of suppliers and customers.

This led to the fact that in the period 2019–2020, the rate of small and medium business decreased with 3.4% [12]. The Covid-19 pandemic has led to a slowdown of Moldova's economy and a decrease in GDP. The share of MSMEs to GDP formation decreased from 50.9% in 2019 to 48% in 2020 [12]. The negative effects of Covid-19 pandemic also led to a significant decrease of people employed in MSMEs of 316.8 thousand in 2020, which is 19 thousand employees or 5.7% less than in 2019 (table 1).

**Table 1.** Share of employees in MSMEs during the years 2019–2021, (thousand persons)

| *Enterprise size* | *2019* | *2020* | *2021* | *2020/2019,%* |
|---|---|---|---|---|
| *Total MSMEs* | *335.8* | *316.8* | *351.0* | ***-5.7*** |
| *medium-sized* | *102.9* | *99.3* | *95.9* | ***-3.5*** |
| *small* | *118.6* | *114.0* | *114.4* | ***-3.9*** |
| *micro* | *114.3* | *103.5* | *104.7* | ***-9.4*** |

Source: developed by authors based on the official data from NBS

Even when businesses remain open, employees are experiencing financial cuts from lost employment or fewer hours worked. Moreover, any time a business closes or has to lay off workers, it affects entire communities of people who rely on income from jobs to support themselves, and in turn, support other local businesses and organizations [3]. The reduction of the employee's number in 2020 is observed in all categories of MSMEs. The most significant decrease was registered by the micro enterprises with 9.45% or by

10.8 thousand people. The reduction is explained by the fact that in the conditions of the pandemic crisis and especially of the lockdown period, the most affected were the micro enterprises. In order to survive, during the pandemic, they had to reduce the number of people employed. According to NBS, in 2020 most jobs decreased in trade, construction, accommodation activities, public food and agriculture. In response to this decrease, the government has approved several measures to support employees and MSMEs, but these proved to be quite modest: only 1.2% GDP, while most neighboring countries have implemented financial support packages amounting to over 2–3% GDP [13]. However, in 2021 the situation improved in the category of small and micro enterprises, attesting that MSMEs had the resilience to recover from the pandemic crises.

The pandemic also has reduced the turnover of all MSMEs, regardless of size (table 2). Inequalities have increased between MSMEs, most of which were poorly prepared for such a shock. Nonetheless, every enterprise from the MSMEs group has been affected, but deeper effects are observed in micro enterprises. It is worth mentioning that even though micro enterprises represents the largest share of entities in the MSMEs sector, their contribution to employment, and to turnover is quite modest. At the same time, micro enterprises are most vulnerable to crisis due to their reduce financial potential, which worsened substantially against the background of the pandemic. However, MSMES found sufficient forces to recover in 2021, attesting an impressive increase in compression with 2019.

**Table 2.** Evolution of the turnover in MSMEs during the years 2019–2021, (mil. MDL)

| Enterprise size | *2019* | *2020* | *2021* | *2020/2019, %* |
|---|---|---|---|---|
| *Total MSMEs* | *157. 336* | *150.112* | *183.570* | ***-4.6%*** |
| *medium-sized* | *50.422* | *48.304* | *57.124* | ***-4.2%*** |
| *small* | *63.516* | *61.125* | *76.340* | ***-3.8%*** |
| *micro* | *43.398* | *40.683* | *50.106* | ***-6.3%*** |

Source: developed by authors based on the official data from NBS

In addition, for some MSMEs, the Covid-19 pandemic represented bankruptcy or survival with the last resources, here we are referring to the food sector, the transport of goods and passengers, tourism, traditional retail, entertainment, but we also note that we have a part of the winners who benefited from the pandemic. There are those who have provided essential services: the medical sector, telecommunications, IT, water distribution utilities, construction, etc. Another affected activity is agriculture, because of two overlapping crises: Covid-19 pandemic and drought. A more detailed forecast is presented below in table 3. The secondary data analysis results show that in 2020 MSMEs from Moldova were strongly affected by Covid-19 pandemic, operating with reduced turnover, jobs lost and reduce financial resources. To support these businesses, the policymakers adopted o series of supporting measures. The main measures to support the business were: increasing the unemployment fund; reimbursement, in the amount of 100% or 60%, of taxes and fees on salaries; reducing the VAT rate from 20% to 15% for entities in the hotel, restaurant, and cafe sector. It is necessary to consider that all these measures were addressed to the business environment as a whole without underlining in particular the category of MSMEs. However, in 2021 compared to 2020, there was a positive trend of 22.29% increase in turnover recorded on all types of activities carried out by MSMEs.

**Table 3.** Evolution of the turnover in MSMEs from different types of activity during the years 2019–2021 (mil. MDL)

| *Types of activity* | *2019* | *2020* | *2021* | *2020/2019,%* |
|---|---|---|---|---|
| *Total* | *157 335.5* | *150 112.1* | *183 569.7* | ***-4.6*** |
| *agriculture, hunting and forestry* | *13 937.0* | *12 663.4* | *19 161.4* | ***-9.1*** |
| *processing industry* | *15 843.8* | *16 248.6* | *18 132.2* | ***+ 2.6*** |
| *electric energy, gas and water* | *466.8* | *310.2* | *393.6* | ***-33.5*** |
| *water distribution, sanitation* | *860.1* | *1 133.0* | *1 190.8* | ***+ 31.7*** |
| *constructions* | *11 915.2* | *11 950.8* | *12 895.0* | ***+ 0.3*** |
| *whole sale and retail trade* | *73 529.2* | *70 926.4* | *83 510.4* | ***-3.5*** |
| *transport and stoking* | *11 629.1* | *10 195.2* | *12 360.1* | ***-12.3*** |
| *public alimentation* | *3 544.0* | *2 145.7* | *3 497.1* | ***-39.4*** |
| *communication* | *4 897.2* | *5 385.3* | *7 115.4* | ***+ 9.7*** |
| *real estate business* | *3 864.7* | *4 175.7* | *4 961.7* | ***+ 8.0*** |
| *scientific and professional activity* | *4 142.9* | *4 165.4* | *5 276.1* | ***+ 0.5*** |
| *other types of activity* | *12 705.5* | *10 812.3* | *15 075.9* | ***-14.9*** |

Source: developed by author based on the official data from NBS

In 2021, Covid-19 pandemic gradually fade away, allowing MSMEs from Moldova "to breathe easier" and become more resilient. Still, the Russian military invasion in Ukraine at the beginning of 2022 was an unpredictable external shock, which affected hard Moldova's economy. In this regard, we will continue to analyze MSMEs activity in the context of war in the neighboring Ukraine.

## 3 The Business Environment and MSMEs in the Context of the War in Neighboring Ukraine

Economic groups such as the MSMEs have interregional interaction, meaning the economic condition in a certain region can affect the nearest regions [15]. Given the geopolitical situation and the war's major effects especially on the neighboring countries, the economic challenges Moldova economy is facing right now are not few. Two of the crises – the Covid-19 pandemic and the energy crisis – continued from 2021. The other crises – the refugee crisis and the security crisis, started in 2022 and arose due to Russia's war in Ukraine. Since the war in Ukraine began, nearly half a million people have fled across the border into Moldova. One from ten of these refugees ultimately stay within its borders. Moldova, with a population of only 2.6 million people, hosts the largest per capita number of Ukrainian refugees in the world. As well as the economic conditions, the population dynamics in a region can affect the development of the regions nearby, particularly the vulnerability of businesses there [15].

Another consequence of the war is massive trade disruptions, which affected MSMEs. As Moldova relied heavily on imports from Ukraine and Russia for its food and energy needs, the war disrupted imports, which in turn led to increased prices and higher economic vulnerability of the country's population [14]. Consequently, the condition of MSMEs also relies on the surrounding area, where they usually get production goods from the neighboring region [16]. This way Moldova, among the poorest countries in the Europe before the war, is now coping with missing imports and limited exports – all while supporting neighbors in need [11].

Moreover, the Russian aggression in Ukraine led to a severe energy crisis in Moldova. The energy infrastructure in Ukraine and Moldova is interlinked, so when Russia attacked Ukraine's energy grid, Moldovans experienced blackouts. In 2022, Gazprom, Russia's state-owned natural gas supplier cut the gas supply to Moldova by 60%, which increased the price of natural gas that almost doubled from December 2021 to December 2022. Consequently, the country's inflation rate reached an ultimate high of 34.6% in October 2022, four times higher than in October 2021, 8.8%. However, the main inflationary factors are external, being mainly determined by the increase in global prices, especially for energy and food resources. These phenomena have led either to a direct increase in the prices of consumer products, or to an increase in production costs.

The presence of Russian troops in Transnistria (a territory of the Republic of Moldova) is often mentioned as the main potential source of destabilization, as well as hybrid threats and the role of Kremlin proxies, political parties. The Government of Moldova is aware of potential threats to stability and is working on building resilience in the security arena. Russian missiles targeting Ukraine had crossed Moldova airspace several times. On several occasions, fragments of rockets fell on the territory of the Republic of Moldova. The security threats showed how vulnerable Moldova is, without an anti-aircraft defense and with an army of only six thousand people.

According to Moldova Economic Growth Analysis (MEGA) the effects of Russian aggression in Ukraine, the energy crisis, persistent inflation and drought have put MSMEs from Moldova under severe cost and competitiveness pressures. The main challenges for MSMEs were related to loss of markets and suppliers in Russia, Ukraine and Belarus; increase in transport and logistics costs; higher costs for energy resources and lower sales on the domestic market.

However, at the same time, a new window has opened for the country. In June 2022, Moldova was granted European Union candidate status alongside Ukraine. To increase energy security, Moldova has connected to the European Union's power grid together with Ukraine, sourced natural gas from the market, and stored it in Romania and Ukraine, with support from the European Union and other international partners.

In order to support the business climate a mechanism of compensation to cover part of the energy tariffs for households was introduced; alternative sources of natural gas and electricity were used and natural gas has been stored as measures to increase energy security; primarily, from external sources, which were available due to the generous support from development partners. In addition, thanks to external support, this year the Government from Moldova did not increase the fiscal pressure, and manage to survive through these overlapping crises.

All these, supporting measures and funds led to positive results in economy. The business environment from Moldova remains resilient against the backdrop of the current crises. According to the State of the Country Report, important structural changes in the field of entrepreneurship have been observed: while in the period 2017–2020 the number of established businesses was constantly lower than the number of closed ones, in the period 2021–2022 a reversal of the situation is observed. In particular, the number of newly created enterprises (6,800) in 2021 slightly exceeded the number of closed ones (6,600), while the results for the first eight months of the 2022 indicate that this trend has continued. Thus, the positive increase in the number of newly created companies versus closed ones was 1.000 units. The year 2021 also saw the best returns in the last five-year period, with an average return on assets (ROA) of 7.4% and a return on equity (ROE) of 16.5%. Another feature of the last two years has been an increase in the share of company stocks in total current assets, from 32.9% in 2019 to 41.6% in 2021 [14].

In the context of economic, inflationary, energy and security crises that MSMEs from Moldova are operating, such a structural change means an improvement of the business environment and creates a solid basis for economic recovery. This trend may also be the result of the new support programs in the field of entrepreneurship, launched by the Organization for Development of Entrepreneurship (ODA), but it may also be a result of a qualitative change of perception on the part of the business world regarding the business environment in the country [14]. It is worth to mention that The European Association of Guarantee Institutions (AECM) was delighted to welcome in September 2022 its new member from Moldova, the Organization for Entrepreneurship Development. In addition, in order to stand by the entrepreneurs in these difficult times a Fund for Entrepreneurship and Economic Growth (FACEM) was created. This Fund is expected to provide entrepreneurs with access to low-cost financing. In 2022, 70 million MDL were allocated to finance the FACEM Fund, some of which are also planned to be used this year. For 2023–2025, the size of the Fund will be about 2.8 billion MDL.

According to Moldova Economic Growth Analysis (MEGA) Moldova's European Union candidate status obtained in 2022 and the prospect of starting accession negotiations by the end of 2023 create significant growth opportunities. It is supposed to anchor Moldova more deeply into the European civilizational area being no longer treated as part of the grey zone that previously separated Russia from the European Union. As a result, the Republic of Moldova could receive access to additional financial resources. The materialization of these opportunities can create an enormous synergy effect for attracting investment and modernizing the country [19].

## 4 Discussion and Policy Recommendations

Small business survival is critical for economic sustainability due to the multifaceted role of the economy [17]. Therefore, protecting MSMEs from unforeseen events and economic shocks is the prime concern for policymakers because the output degradation in MSMEs has spillover effects on other segments in the economy [4].

The overlapping crises has made the Moldovan economy oscillate at extreme values. After the effects of the Covid-19 pandemic and severe drought from 2020, the year 2022 brought a mix of new crises materialized by the effects of Russian military aggression on Ukraine, the energy crisis and persistent inflation. These challenges have put pressure on the costs and competitiveness of MSMEs. The overlapping crises in the 2020–2022 period have revealed, in addition to the limited business support resources, the fact that the intervention tools from the Government are limited [14]. On the other hand, the last two years have also brought a positive change in the trend with a positive increase between the number of new businesses registered and the number of businesses closed. These figures may suggest both higher adaptability to crises and higher business confidence.

Based on what has been considered so far, in this section, we briefly discuss some policies or actions that policymakers could implement to build the resilience of MSMEs and in this way minimize the impact of ongoing and future crises on these enterprises.

In this regard, we consider policies targeted to MSMEs that aims to reduce from their vulnerability and support these enterprises especially at their start-up or survival stage. Here we referred to: establishing of a simplified activity regime for micro and family businesses, which is justified both by the high share of companies that perceive the tax administration system as complicated, and by the lack of dedicated personnel in this area within companies, to which is also added the low level of entrepreneurial culture and education. Another policy recommendation aims to reduce the amount of tax and number of mandatory documents for launching a business. There are still many administrative-bureaucratic barriers that impose unjustified additional costs for local entrepreneurs who want to start or expand their business.

In order to facilitate the access of MSMEs to finance resources we recommend the expansion of the financial support programs addressed to micro and family business. This proposal is linked with the expansion of loan guarantee schemes. The data for the last two years clearly suggest an orientation and concern of local business for investments, in order to increase the efficiency and competitiveness of businesses. The capitalization of FACEM Fund and the Loan Guarantee Fund (LGF) – both from the State Budget and by attracting concessional external financing – side-by-side with ensuring appropriate governance, should be a priority for facilitating access to finance [14].

The lack of information and advisory assistance infrastructure and the development of entrepreneurial potential, especially at regional level, insufficient staff qualification, low level of training and culture of entrepreneurship in MSMEs, continue to be an obstacle to business development [18]. Therefore, expanding consulting and training services for micro and family business is an important policy in reducing MSMEs vulnerabilities. Furthermore, it is necessary to fostering cooperation between Universities, research centers and local economic actors capable of increasing innovation of MSMEs in the entire Republic.

The two external shocks Covid-19 and the Russian war in Ukraine, underlined the need to think about additional investments in infrastructure and energy security. As we saw, some branches were affected more in times of this overlapping crisis, especially the agriculture sector that was already affected by the increasing incidence of drought. Therefore, efforts need to be made to adapt to climate change and not just to compensate for its consequences with public money, but also to stimulate supporting research in agriculture, and introducing controls on annual crop rotation within the subsidy framework [14]. Regarding the energy field, the financing of energy efficiency measures and the stimulation of investments in renewable energy resources must remain a priority. Energy efficiency is one of the key factors for sustainable development. The policy implications outlined in terms of innovation and technology are highly appreciated in this field. The recent legislative changes made to simplify and speed up the procedure for connecting, installing and operating photovoltaic panels for domestic consumers is a good example in this regard [14]. In addition, public investments in infrastructure should be prioritize. As a result, of the Russian war in Ukraine, Moldovan infrastructure is under enormous pressure. This situation has also increased the infrastructure deficit, while transport and logistics costs faced by producers have increased substantially. In this context, efforts should be focused on improving infrastructure, especially since this is also an imperative for state intervention in times of recession. Priorities include accelerating existing interconnection projects with Romania, across all dimensions, examining the possibility of building more floating bridges and making the necessary investments in the railway sector [14].

The swift implementation of these measures can maintain MSMEs confidence in economy against the backdrop of inflation and multiple uncertainties. According to Moldova Economic Growth Analysis Report the credibility that the European Union candidate country status offers, will further mobilize financial assistance. The Moldovan economy faced an unprecedented mix of crises in 2022, but thanks to proper crisis management and external support it entered 2023 stronger and with good prospects for a speedy recovery. For the coming years, the main assumptions for growth are the stabilization of the security situation in the region, improved energy and mobility connectivity with Romania and the start of European Union accession negotiations.

*This research was financed from the Moldova-Turkey Bilateral Cooperation Projects, entitled "Measuring the performance of SMEs active in the agricultural sector and/or the food industry during the health crisis COVID-19: a study in the Gaziantep and Chisinau region", project number: 22.80013.0807.3TR.*

## References

1. Puriwat, W., Tripopsakul, S.: Customer engagement with digital social responsibility in social media: a case study of COVID-19 situation in Thailand. J. Asian Financ. Econ. Bus. **8**, 475–483 (2021)
2. Runyan, R.C.: Small business in the face of crisis: Identifying barriers to recovery from a natural disaster 1. J. Contingencies Crisis Manag. **14**, 12–26 (2006)
3. Engidaw, A.E.: Small businesses and their challenges during COVID-19 pandemic in developing countries: in the case of Ethiopia. J. Innov. Entrepreneurship **11**(1), 1–14 (2022). https://doi.org/10.1186/s13731-021-00191-3
4. Miocevic, D.: Investigating strategic responses of SMEs during COVID-19 pandemic: A cognitive appraisal perspective. BRQ Bus. Res. Q. **12**, 1–40 (2021)
5. Report UNDP Moldova (2020) Needs Assessment and Evaluation of COVID-19 Impact on MSMEs
6. Nino Chanadiri: Georgian Foundation for strategic and international studies: Moldova challenges alongside the war in Ukraine. https://gfsis.org.ge/blog/view/1543 (2023)

7. Böhme, K., et al.: Research for REGI Committee – The impact of the COVID-19 pandemic and the war in Ukraine on EU cohesion, Part II: Overview and outlook European Parliament. Policy Department for Structural and Cohesion Policies, Brussels (2022)
8. Fu, Q., Sial, M.S., Arshad, M.Z., Comite, U., Thu, P.A., Popp, J.: The Inter-Relationship between Innovation Capability and SME Performance: The Moderating Role of the External Environment. Sustainability **13**, 9132 (2021). https://doi.org/10.3390/su13169132
9. Schepers, J., Vandekerkhof, P., Dillen, Y.: The Impact of the COVID-19 Crisis on Growth-Oriented SMEs: Building Entrepreneurial Resilience. Sustainability **13**, 9296 (2021). https://doi.org/10.3390/su13169296
10. The World Bank Report: GDP per capita – Europe and Central Asia (2021)
11. The World Bank Report: The World Bank in Moldova Country Overview (2022)
12. National Bureau of Statistics of the Republic of Moldova, https://statistica.gov.md/index.php?l=en. Last accessed 21 April 2023
13. Expert Group Report (2020) available at https://www.expert-grup.org/ Last accessed 19 April 2023
14. Expert Group Report (2022) available at https://www.expert-grup.org/ Last accessed 19 April 2023
15. Caraka, R.E., et al.: Micro, Small, and Medium Enterprises' Business Vulnerability Cluster in Indonesia: An Analysis Using Optimized Fuzzy Geodemographic Clustering. Sustainability **13**, 7807 (2021). https://doi.org/10.3390/su13147807
16. Foreman-Peck, J., Nicholls, T.: Interregional mobility of entrepreneurial SMEs. Ann. Reg. Sci. **54**, 57–87 (2015)
17. Pu, G., Qamruzzaman, M., Mehta, A.M., Naqvi, F.N., Karim, S.: Innovative Finance, Technological Adaptation and SMEs Sustainability: The Mediating Role of Government Support during COVID-19 Pandemic. Sustainability **13**, 9218 (2021). https://doi.org/10.3390/su13169218
18. Erhan, L., Gaugaş, T.: Policies for Small and Medium Sized Enterprises' Development in the Republic of Moldova: Results and Challenges. J. Innov. Bus. Best Pract. **2022** (2022), Article ID 706025, https://doi.org/10.5171/2022.706025
19. Moldova Economic Growth Analysis (MEGA): Economic recovery of 2023 forecast, risks and ways forward. Quarter I, 2023, Expert-Group, Independent Think-Tank, Issue No. 24, April

# Economic Management of the Local Budget to Achieve the Sustainable Development Goals

Pedro-José Martínez-Córdoba[1](✉), José-Luis Zafra-Gómez[2], and Bernardino Benito[3]

[1] IME, Universidad de Salamanca, P.º Francisco Tomás y Valiente, 37007 Salamanca, Spain
pedrojose.martinez@usal.es
[2] Department of Accounting and Finance, Faculty of Economics and Business, University of Granada, 18071 Granada, Spain
jlzafra@ugr.es
[3] Department of Accounting and Finance, Faculty of Economics and Business, Regional Campus of International Excellence "Campus Mare Nostrum", University of Murcia, 30100 Espinardo, Murcia, Spain
benitobl@um.es

**Abstract.** Making progress on the Sustainable Development Goals (SDGs) requires the involvement of citizens, with the public sector, and in particular local governments, being a key agent in their achievement. Currently, the political and media commitment to the SDGs is developing with notable success, but the indicators reflect the need to reorient the actions implemented in view of the poor achievement of the goals established. In this sense, local budgets represent the most important tool for meeting the SDGs. Therefore, the objective of our research is to contribute to the promotion of the so-called 2030 Agenda, identifying the connection between the 17 SDGs and the municipal budget, using a chord diagram. To understand the level of optimisation of these sources, the budget efficiency is then calculated in terms of their relation to the SDGs. The following questions will be answered: Which SDGs are most related to the municipal budget? What is the level of efficiency for the SDGs most linked to the municipal budget? Are there changes in the orientation of municipal budget policies to achieve the SDGs?

The results confirm high levels of efficiency, although there is a decline in the efficient management of budget expenditure in 2020 compared to 2018. Consequently, local governments need to improve their budget management if they aim to achieve the SDGs by 2030.

**Keywords:** local governments · efficiency · chord diagram

## 1 Introduction

The Sustainable Development Goals (SDGs) represent the commitment made by the United Nations member countries to improve people's lives and protect the planet. This agreement, signed in 2015, establishes 2030 as the time horizon for achieving the SDGs. Although progress is positive, the policies implemented must evolve at the necessary


R. Mehmood et al. (Eds.): DCAI 2023, LNNS 741, pp. 138–146, 2023.
https://doi.org/10.1007/978-3-031-38318-2_14

speed to reach the end, which has worsened recently with the Covid-19 pandemic, armed conflicts and climate change [1]. In this sense, the United Nations decided to promote the SDGs with the so-called Decade of Action (2020–2030), assuming a more significant mobilisation of financing, improving the implementation of the public policies involved and strengthening the participation mechanisms without leaving anyone behind.

This agreement on the Decade of Action calls on all sectors of society to mobilise for the SDGs, with particular reference to the local level, pointing to the need to reorient the policies, budgets, institutions and regulatory frameworks of governments, cities and local authorities. In this context, the SDGs allow local governments to review their organisations, offering a broad perspective for implementing sustainability [2]. Moreover, these governments represent vital institutions in delivering essential services to citizens, which, coupled with their proximity and geographical permeability, provides an ideal framework for implementing sustainability initiatives.

The authors of this manuscript aim to contribute with this research to the knowledge of the fulfilment of the SDGs. To this end, a group of Spanish cities with large populations have been selected and, using a chord diagram, budget expenditure has been related to the 17 SDGs to find out which are being met to a greater extent. Once these SDGs (4–5) per municipality have been established, it has been studied whether they are being efficient in the fulfilment of these SDGs, i.e., whether the expenditure made by the local government is moving towards greater fulfilment of the SDGs. In this way, budget management in relation to the SDGs is being analysed by answering the following questions:

- Which SDGs are most related to the municipal budget?
- What is the level of efficiency in relation to the SDGs most closely linked to the local budget?
- Are there changes in the orientation of local budget policies to achieve the SDGs?

## 2 Theoretical Approach

Budgets are the most effective tool for achieving policy objectives in the public sector. Indeed, resource allocation will determine a project's success, conditional on efficient project management. Indeed, synergies between SDGs and budget programmes have been demonstrated, where progress on one SDG or investment in a given policy can contribute to achieving other SDGs [3]. In this regard, integrating the SDGs into budget processes to deliver on commitments can improve the budget performance assessment system and serve as a management tool in resource allocation [4]. On the other hand, if the current pace of investment continues and financing weaknesses are not quickly addressed, some SDG targets will be far from being met [5].

An important aspect to consider when addressing the SDGs is efficiency, identifying whether governments will deliver on their commitments while maintaining their current strategies [6]. In this vein, research has addressed the SDGs from an efficiency perspective, pointing to the importance of optimising resources to achieve the SDGs. For example, increasing the efficient use of land in underdeveloped economic environments would facilitate the achievement of the SDGs [7]. Similarly, in the case of SDG 6, efficient management is a critical factor in achieving the targets by 2030 [8].

Although the SDGs were designed and agreed upon from a national perspective, local governments are crucial to their implementation [9]. Local actors play a critical role in achieving the 2030 Agenda, where fostering the local dimension of the SDGs or linking local resources to these goals can support local governments' sustainability efforts [10]. In addition, cross-sectoral cooperation, political will, coherent governance or partnerships between public, private and civil society actors are some keys to achieving the SDGs [11].

Achieving the SDGs means a better world, enabling all people to thrive as part of a more resilient society, where building an overarching framework for SDG adoption would foster progress [12]. In this way, the benefits of achieving the SDGs can be a source of survival for countries in a globalised world, where most of these goals are based on the well-being of society [13].

# 3 Methodology

## 3.1 Sample

To the research objective, Spanish municipal governments represent the units of analysis. The available information conditions the selection of the sample, being reduced to cities with a population of over 80,000 inhabitants, a limitation that is determined by the SDG indicators, and the measurement of the fulfilment of these, which have been developed by the Spanish Network for Sustainable Development only for these municipalities. These local governments represent 1.14% of the total; however, they administer 45.30% of the Spanish population. Likewise, the sample includes all provincial capitals, administrative capitals of Autonomous Communities and Autonomous Cities, even if they do not exceed 80,000 inhabitants. The period of analysis is limited to 2018 and 2020, since SDG data are only available for those years. A full panel was constructed, which finally, after filtering out units of analysis with incomplete data, consists of 95 local governments.

## 3.2 Variables

In line with the research objectives, authors selected the total expenditure of the budget settlement to find out the relationship between municipal public spending and the SDGs. In addition, this variable represents the input in the calculation of efficiency, i.e., the total resources allocated to fulfil the SDGs. To quantify the 17 SDGs, the indicators developed by the Spanish Network for Sustainable Development in its 2018 and 2020 reports were chosen.

## 3.3 Method

To identify the SDGs most closely related to municipal public spending, authors used a chord diagram, characterised by a circular visualisation of the relationship between objects through links based on a list or adjacency matrix (see Figs. 1 and 2). Before making this estimate, the following assumptions are made ceteris paribus:

- Authors consider that the total expenditure budget contributes to, is oriented towards or impacts the SDGs as a whole.

– That the actions of the other public administrations in the SDGs are homogeneous and do not represent significant differences.

With the results obtained in Figs. 1 and 2 (see Sect. 4), the SDGs most relevant to the municipal budget were identified, which we selected as outputs for calculating efficiency. In Table 1, authors describe the efficiency models used for the 2018 and 2020 data and the Convexity and Returns test. It can be concluded that our data are convex and that returns to scale (VRS) represent the best choice of analysis.

**Table 1.** Description of models and Convexity and Returns test.

| | | Model 1 | | Model 2 | | Model 3 | | Model 4 | |
|---|---|---|---|---|---|---|---|---|---|
| Input | | Total budget expenditure 2018 | | Total budget expenditure 2018 | | Total budget expenditure 2020 | | Total budget expenditure 2020 | |
| Outputs | | SDG3, SDG6, SDG13, SDG16 | | SDG3, SDG6, SDG8, SDG16 | | SDG3, SDG6, SDG8, SDG16 | | SDG3, SDG6, SDG13, SDG16 | |
| Test Convexity | tau | -1.103 | 0.245 | -0.478 | 0.311 | -0.251 | 0.102 | -1.441 | 0.392 |
| | pval | 0.845 | 0.584 | 0.571 | 0.330 | 0.514 | 0.979 | 0.910 | 0.249 |
| Test Rts | tau | 8.647 | 1.000 | 9.669 | 1.000 | 11.934 | 1.000 | 12.043 | 1.000 |
| | pval | 0.000 | 0.000 | 0.000 | 0.000 | 0.000 | 0.000 | 0.000 | 0.000 |

Concerning the calculation of efficiency, there are different parametric and non- parametric methods, with the characteristics of the research and its objectives determining the choice of the most appropriate technique. Among the non-parametric methods, Data Envelopment Analysis (DEA) stands out. In contrast, in parametric methods, Stochastic Frontier Analysis (SFA) stands out, with DEA mainly implemented in operational research and SFA in economic research [14]. Consequently, considering the literature analysing budget efficiency at the local level alongside the SDGs (see Sect. 2), authors opt for non-parametric methods as the most appropriate and used way [15]. Moreover, there are others within the non-parametric methods similar to DEA. Such as the Free Disposal Hull (FDH), the non-convex version of DEA, or the Order-m. Therefore, to find out which technique or techniques are more suitable for our research, authors first check the convexity of the data with the Rts test of Simar and Wilson [16]. Table 1 shows the trial results, confirming the data's convexity. Consequently, the FDH technique was discarded and DEA and Order-m techniques were used to make the results more robust.

## 4 Results and Discussion

The results of applying the chord diagram to the 2018 and 2020 data to determine the relationship between total budget expenditure and the SDGs are shown in Figs. 1 and 2, respectively. Total budget expenditure in 2018 can be confirmed to be mainly related to SDGs 3, 6, 13 and 16 (see Fig. 1). In proportion to the budget implemented.

These SDGs receive more attention, thus increasing their levels of achievement. In this sense, people's health and well-being (SDG 3), sustainable management of safe drinking water and sanitation, ensuring universal access (SDG 6), taking action to combat climate change (SDG 13), and promoting a just, peaceful and inclusive society (SDG 16) are policy priorities of local governments in their budgets expressed through the SDGs.

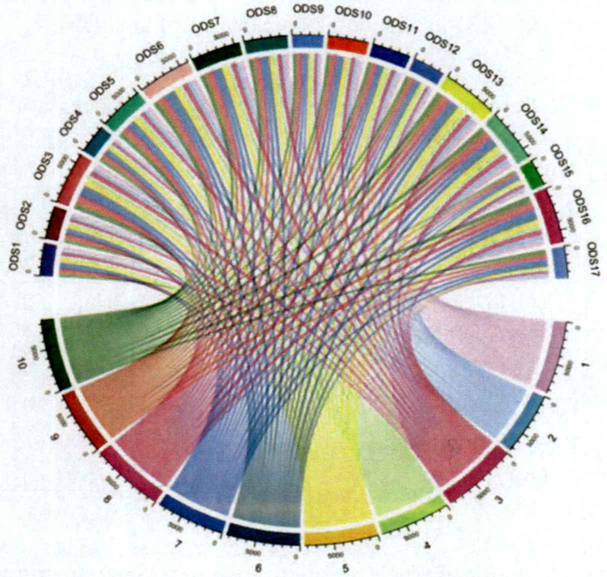


**Fig. 1.** Chord diagram 2018

In this line, 2020 budget spending (Fig. 2) maintains its orientation towards SDGs 3, 6 and 16. However, there is a variation concerning SDG 13, replaced by the need to promote inclusive and sustainable economic growth (SDG 8).

After identifying municipal governments' SDG budget priorities, efficiency levels were calculated according to the models in Table 1 (see Sect. 3). For this purpose, the input is the total budget expenditure and the outputs are the SDGs most related to this expenditure, according to the results of the chord diagrams (Figs. 1 and 2). Considering the differences between 2018 and 2020 concerning the SDGs (13 and 8, respectively) most linked to budget spending, complementary efficiency has been calculated using the 2020 results in 2018 and vice versa (Model 2 and 4, respectively). The distribution of efficiency levels for each Model and technique used is plotted below (DEA and Order-m).

Regarding the efficiency levels of budget spending in 2018, the DEA estimates (see Figs. 3 and 4) present similar results. In the case of using the outputs most related to the 2018 budget (Model 1), the levels increase slightly, providing a higher number of efficient units. In the case of the calculation with the Order-m technique (see Figs. 5 and 6), the results are similar, although in this case, the model using the SDGs most closely linked to the 2020 budget (Model 2) obtains slightly higher values. In any case, the average efficiency levels of SDG-oriented budget spending in 2018 are high.

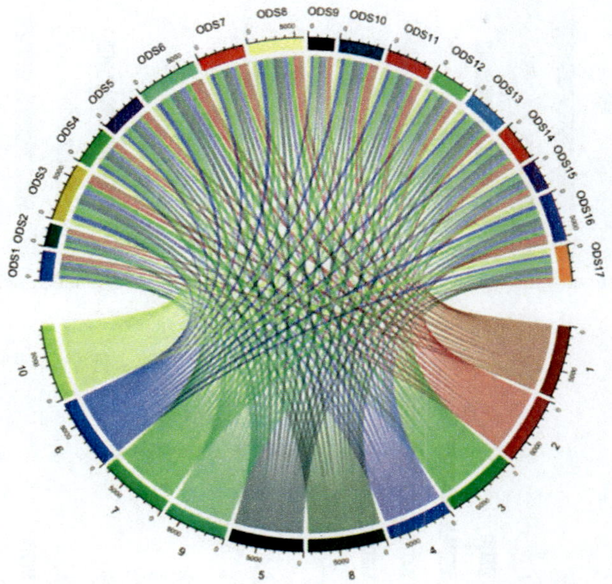


Fig. 2. Chord diagram 2020

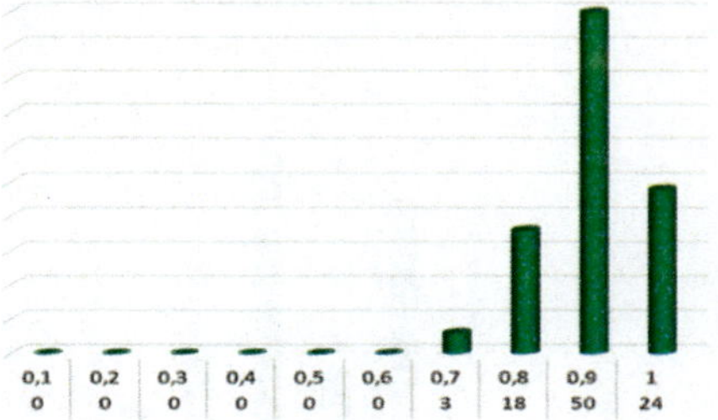


Fig. 3. DEA efficiency levels Model 1

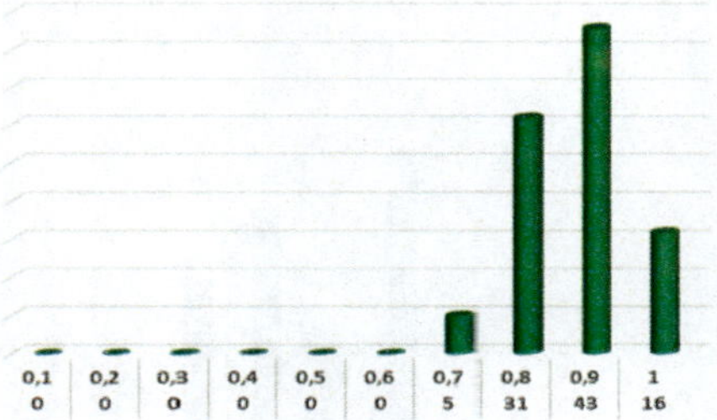


Fig. 4. DEA efficiency levels Model 2

In the case of 2020 budget spending, the trend remains the same as in 2018 (Figs. 7 and 8 for DEA and Figs. 9 and 10 for Order-m). Model 4, estimated with DEA and Model 3, with Order-m (SDGs more linked to the 2020 and 2018 budget, respectively). Present slightly higher efficiency levels. These results confirm a reduction in the average efficiency levels calculated with DEA when using the most linked SDGs in their respective periods. However, there is a slight improvement in the case of the Order-m estimates.

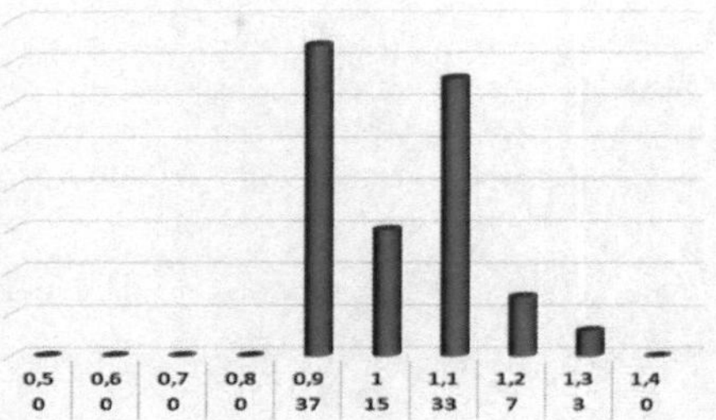


**Fig. 5.** Order-m efficiency levels Model 1

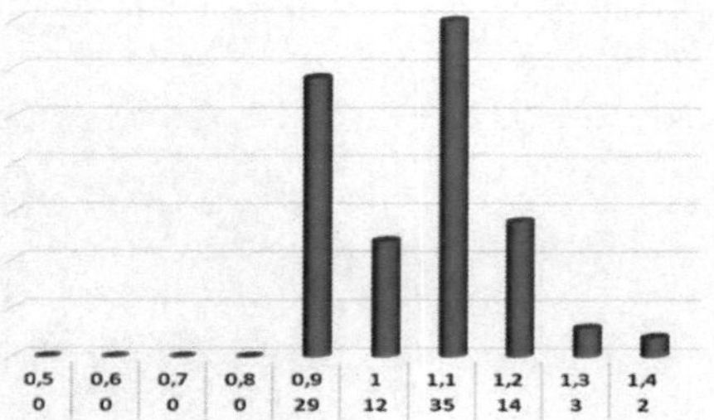


**Fig. 6.** Order-m efficiency levels Model 2

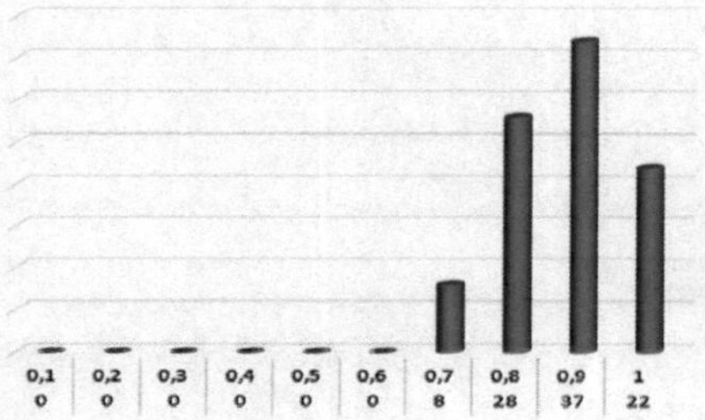


**Fig. 7.** DEA efficiency levels Model 3

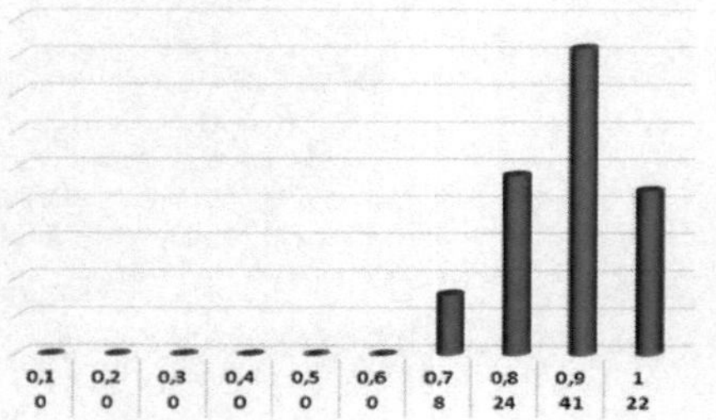


**Fig. 8.** DEA efficiency levels Model 4

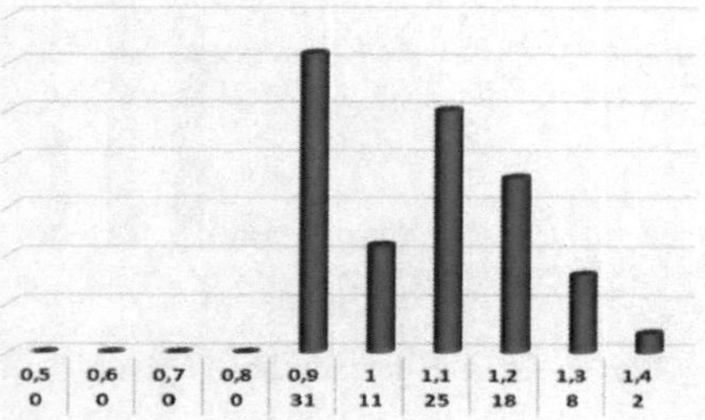


**Fig. 9.** Order-m efficiency levels Model 3

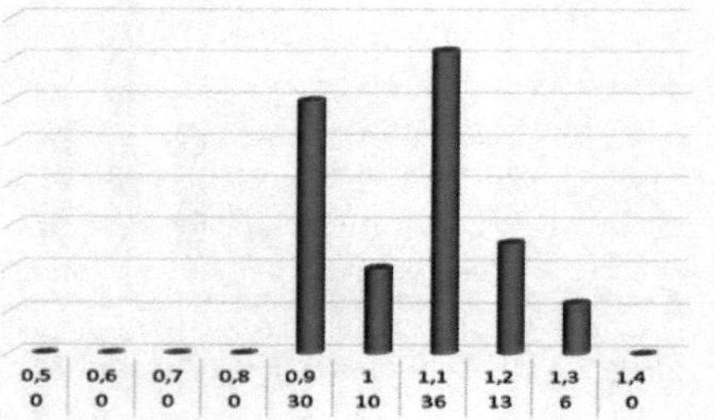


**Fig. 10.** Order-m efficiency levels Model 4

## 5 Conclusions

The poor performance of the SDGs led the United Nations to push for an improvement plan called the Decade of Action (2020–2030). This period aims to increase financing and improve the implementation of public policies to achieve the targets set for 2030. Governments must understand investments in the SDGs as a commitment to the present and future of society, which, together with efficient management, can be a very profitable economic strategy.

Given the local nature of the SDGs, this research identifies the link between the SDGs and local government budget expenditure. Using a chord diagram, it is confirmed that people's health and well-being (SDG 3), sustainable management of drinking water and sanitation (SDG 6), the need to promote inclusive and sustainable economic growth (SDG 8), taking action to combat climate change (SDG 13) and promoting a just, peaceful and inclusive society (SDG 16) are policy priorities of local governments expressed in their budgets. In addition, the efficiency of these budget allocations was calculated in terms of whether the SDGs identified as most met were reached to a greater or lesser extent, with high levels being recorded, albeit with a decreasing trend over the years studied.

The implications of this study are important. Political leaders and officials must find a balance between achieving the SDGs and maintaining sound municipal finances. Local governments, with the capacities and powers they have, can contribute to meeting the SDGs through a wide range of policy tools, such as urban development plans, traffic and public transport regulations, waste collection ordinances and procurement policies. However, if resources are scarce, municipalities should focus on the goals they consider most relevant to their municipalities. Local SDG strategies should therefore be based on the specific needs of the municipality. In this regard, we believe that local governments should develop a strategic plan for implementing the SDGs, including a section detailing how these initiatives will be financed and their impact on the financial health of the municipality. This plan could be published by local governments to gain more legitimacy and sensitise public opinion on the importance of balancing public resources and SDG implementation.

Finally, the need for more information available for a broader set of local governments and more years is a limitation of this research. In this sense, future research could extend the sample and propose a second-stage regression model to determine the factors that could improve efficiency levels.

## References

1. United Nations: The Sustainable Development Goals Report (2022). https://unstats.un.org/sdgs/report/2022/The-Sustainable-Development-Goals-Report-2022.pdf. Last accessed 26 April 2023
2. Krantz, V., Gustafsson, S.: Localizing the sustainable development goals through an integrated approach in municipalities: Early experiences from a Swedish forerunner. J. Environ. Planning Manage. **64**(14), 2641–2660 (2021). https://doi.org/10.1080/09640568.2021.1877642
3. Sisto, R., et al.: Quantitative analysis of the impact of public policies on the sustainable development goals through budget allocation and indicators. Sustainability **12**(24), 10583 (2020). https://doi.org/10.3390/su122410583
4. Hegem, E., Brimont, L., Pagnon, F.: Sustainable development goals and indicators: can they be tools to make national budgets more sustainable? Public Sector Economics **43**(4), 423–444 (2019). Retrieved from https://hrcak.srce.hr/ojs/index.php/pse/article/view/9949
5. Kherbache, N., Oukaci, K.: Assessment of capital expenditure in achieving sanitation-related MDG targets and the uncertainties of the SDG targets in Algeria. World Dev. Perspect. **19**, 100236 (2020). https://doi.org/10.1016/j.wdp.2020.100236
6. Issever Grochová, L., Litzman, M.: The efficiency in meeting measurable sustainable development goals. Int. J. Sust. Dev. World **28**(8), 709–719 (2021). https://doi.org/10.1080/13504509.2021.1882606

7. Song, W., Cao, S., Du, M., Lu, L.: Distinctive roles of land-use efficiency in sustainable development goals: An investigation of trade-offs and synergies in China. J. Clean. Prod. **382**, 134889 (2023). https://doi.org/10.1016/j.jclepro.2022.134889
8. Martínez-Córdoba, P.J., Raimo, N., Vitolla, F., Benito, B.: Achieving sustainable development goals. Efficiency in the Spanish clean water and sanitation sector. Sustainability **12**(7), 3015 (2020). https://doi.org/10.3390/su12073015
9. Fenton, P., Gustafsson, S.: Moving from high-level words to local action governance for urban sustainability in municipalities. Curr. Opin. Environ. Sustain. **26**, 129–133 (2017). https://doi.org/10.1016/j.cosust.2017.07.009
10. Masuda, H., et al.: SDGs mainstreaming at the local level: case studies from Japan. Sustain. Sci. **16**, 1539–1562 (2021). https://doi.org/10.1007/s11625-021-00977-0
11. Valencia, S.C., et al.: Adapting the Sustainable Development Goals and the New Urban Agenda to the city level: Initial reflections from a comparative research project. Int. J. Urban Sustai. Develop. **11**(1), 4–23 (2019)
12. Berrone, P., Rousseau, H.E., Ricart, J.E., Brito, E., Giuliodori, A.: How can research contribute to the implementation of sustainable development goals? an interpretive review of SDG literature in management. Int. J. Manag. Rev. **25**(2), 318–339 (2023). https://doi.org/10.1111/ijmr.12331
13. Sadiq, M., Ngo, T.Q., Pantamee, A.A., Khudoykulov, K., Thi Ngan, T., Tan, L.P.: The role of environmental social and governance in achieving sustainable development goals: evidence from ASEAN countries. Economic research-Ekonomska istraživanja **36**(1), 170–190 (2023). https://doi.org/10.1080/1331677X.2022.2072357
14. Lampe, H.W., Hilgers, D.: Trajectories of efficiency measurement: a bibliometric analysis of DEA and SFA. Eur. J. Oper. Res. **240**(1), 1–21 (2015). https://doi.org/10.1016/j.ejor.2014.04.041
15. Sant´ Ana, T.D., Lopes, A.V., Miranda, R.F.D.A., Bermejo, P.H.D.S., Demo. G.: Scientific research on the efficiency of public expenditures: how and where is it going? Int. J. Public Adm. **43**(11), 926-936 (2020). https://doi.org/10.1080/01900692.2019.1660995
16. Simar, L., Wilson, P.W.: Hypothesis testing in nonparametric models of production using multiple sample splits. J. Prod. Anal. **53**(3), 287–303 (2020). https://doi.org/10.1007/s11123-020-00574-w

# Special Session on Computational Linguistics, Information, Reasoning, and AI (CLIRAI)

# Investigating the Gap Between Single-Hop and Multi-Hop Questions in Closed-Book Question Answering via Question Decomposition

Tareq Alkhaldi(✉), Chenhui Chu, and Sadao Kurohashi

Kyoto University, Kyoto, Japan
tareq.alkhaldi@gmail.com, {chu,kuro}@i.kyoto-u.ac.jp

**Abstract.** Transformer-based language models (LMs) have been shown to perform question answering (QA) competitively even when removing context and using only questions as input (called closed-book QA). Previous work that studied closed-book has mainly used simple questions that require a single reasoning step (i.e., single-hop questions). In this study, we find that using multi-hop questions requiring multiple reasoning steps drastically drops the performance. We investigate how to close this gap using two methods: fine-tuning with explicit question decomposition using three decomposition systems, or few-shot learning with chain-of-thoughts (CoT) for implicit question decomposition. We experiment on three multi-hop datasets, considering different multi-hop question types (i.e., compositional, comparison, etc.). We demonstrate when the methods fail and identify future directions that are most promising to closing the gap between single-hop and multi-hop closed-book QA. We release the code: https://github.com/talkhaldi/mh_cbqa.

**Keywords:** Question answering · Multi-hop questions · Closed-book QA

## 1 Introduction

Closed-book QA is a setting where only questions, without accessing context or external knowledge, are used to obtain answers [14]. Models must answer using the knowledge they stored in their parameters during pre-training. Recent work shows that LMs in a closed-book setting achieve competitive performance on single-hop datasets (questions requiring a single reasoning step) compared to open-book (i.e., using context) [14]. For multi-hop questions (requiring multiple reasoning steps), previous work experiments in an open-book setting ([1,5]). However, it is unclear if multi-hop questions have a similar trend to single-hop questions in a closed-book setting.

This study aims to investigate the gap between single-hop and multi-hop questions in a closed-book setting using three representative multi-hop datasets. As single-hop performance is found to be considerably higher, we attempt to


R. Mehmood et al. (Eds.): DCAI 2023, LNNS 741, pp. 149–158, 2023.
https://doi.org/10.1007/978-3-031-38318-2_15

improve the accuracy for multi-hop questions by decomposing into single-hop sub-questions (SQ) explicitly, or implicitly via generating CoT before answering; this should facilitate finding which reasoning steps to perform. Our contributions are three-fold:

- To the best of our knowledge, this is the first study that attempts to improve the accuracy of multi-hop questions in a closed-book QA setting.
- We investigate closing the gap between single-hop and multi-hop questions in a closed-book setting by explicitly including the reasoning steps with the question in the shape of single-hop decompositions using a proposed fine-tuning method, or CoT using few-shot learning method, and compare results on three datasets.
- We identify promising future directions for improving closed-book multi-hop accuracy.

## 2 Related Work

Roberts et al. [14] evaluated T5 models on single-hop questions in a closed-book setting to estimate the amount of knowledge stored in LMs. They applied additional pre-training by masking named entities and dates. Compared to other competing models, they showed that LMs attain competitive results on open-domain QA benchmarks without accessing context or external knowledge. Ye et al. [18] attempted to improve closed-book performance by learning a masking policy for the additional pre-training.

Jiang et al. [7] investigated multi-hop reasoning via decomposition in a closed-book setting; however, different from our work, they used golden decompositions for training and dev splits and focused on T5. We experiment with three decomposers in addition to gold SQs and their answers. We also experiment on GPT-3 in addition to T5. Also, they fine-tuned on concatenated SQs to show that they are a good approximation to fine-tuning on multi-hop questions. The improvement they attempted is on a *single-hop* baseline, while we aim to improve the *multi-hop* baseline.

Previous work on CoT [10,16] experimented on datasets that mainly do not require external factual knowledge to achieve answers (e.g., math or multiple-choice datasets).

## 3 Task

We do fine-tuning and inference on multi-hop questions without context (i.e., closed-book). When a multi-hop question is given as input to an LM, it generates an answer text. We use sequence-to-sequence (Seq2Seq) LMs whose inputs and outputs are text, as opposed to LMs without a decoder like BERT [4]. Seq2Seq models are necessary for a closed-book QA setting because BERT-like models assume the availability of context from which to select text spans as output, while Seq2Seq models can generate text. As no context is provided, Seq2Seq LMs generate answers by looking only at questions.

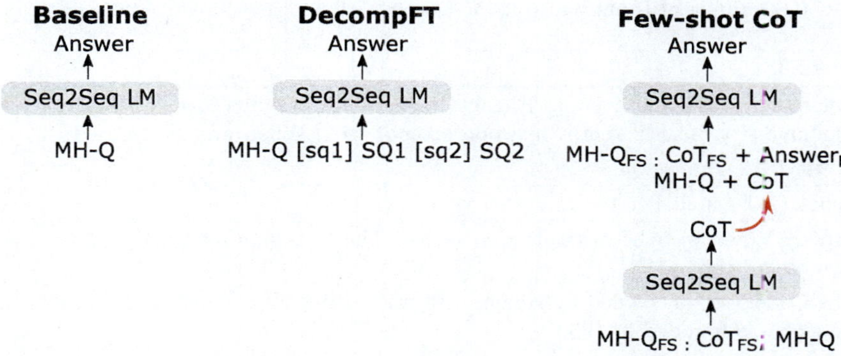


**Fig. 1.** Methods overview. MH-Q is a multi-hop question. SQ1 and SQ2 are the decomposed SQs, `[sq1]` and `[sq2]` are special tokens.

## 4 Method

An overview of the baseline, our proposed decomposition fine-tuning method and a few-shot CoT method is in Fig. 1.

### 4.1 Baseline

We use multi-hop questions as input to the LM, and answers as targets for fine-tuning and evaluating $\text{Baseline}_{\text{FT}}$, and for prefixing (question, answer) pairs as few-shot examples for evaluating another $\text{Baseline}_{\text{FS}}$.[1] We do not use context or external knowledge. This is the basic evaluation setting for closed-book QA.

### 4.2 Decomposition Fine-Tuning (DecompFT)

In the pre-processing step, we decompose the multi-hop question into two single-hop SQs: SQ1 and SQ2, using one of the following decomposers:

*DecompRC.* Introduced by Min et al. [12], this method relies on selecting text spans from the question to form SQs. They train span prediction models using 400 annotations and show that it achieves reasonable performance. They use trained models for CL and I types, while heuristic rules for C type.

*MuSiQue-Based Decomposer.* We train a T5-large[2] [13] to learn the gold decompositions provided by MuSiQue [15]. The input is the full question, while the target is the two SQs. Example input: `What year saw the creation of the region where the county of Hertfordshire is located?` Target:

[1] We tried zero-shot mode, but without few-shot examples, the LM gave a full sentence as the answer, mismatching the target. Therefore, we exclude zero-shot experiments.

[2] We also tried T5-11b but achieved similar accuracy.

**Table 1.** Example of Input/Output for DecompFT and few-shot CoT methods.

| |
|---|
| DecompFT |
| **Input**: When was the university Elizabeth Harwood attended formed? [sq1] What university did Elizabeth Harwood attend? [sq2] When was #1 formed? |
| **Output**: 1972 |
| Few-shot CoT - step 1 |
| **Input:** Q: When was the institute that owned The Collegian founded? Let's think step by step<br>A: The Collegian was owned by Houston Baptist University. Houston Baptist University was founded in 1960<br>...<br>Q: Who is the spouse of the Rabbit Hole's producer? Let's think step by step<br>A: |
| **Output**: The producer of Rabbit Hole is Nicole Kidman. Nicole Kidman's spouse is Keith Urban. |

`[sq1] In which state is Hertfordshire located? [sq2] When was #1 birthed?` The accuracy on the held-out split is 79.6 F1. This decomposer only supports CL type as MuSiQue only contains CL questions. We also compare against using gold MuSiQue decompositions in fine-tuning and inference and using intermediate gold answers to replace the answer placeholder.

*ModularQA.* Introduced by Khot et al. [9], it is a next-question generation (NQG) system that sequentially generates SQs given the full question and previous SQs with their predicted answers. They first train a model to generate questions given context, the answer, and hints that are keywords from the question. Then using this model, they generate decompositions for multi-hop questions that could lead to likely intermediate answers. They use these SQ sequences to train the NQG. At inference time, the NQG first generates SQ1, then a single-hop trained sub-model answers it, then, SQ2 is produced. In compositional question types, SQ2 includes the answer of SQ1, so we find the answer and replace it with #1 placeholder. This decomposer is not limited to any question type.

Then both SQs are concatenated with the full question to form the input for the LM. An example is in the upper part of Table 1.

### 4.3 Few-Shot CoT In-Context Learning (Few-Shot CoT)

We prefix multi-hop questions with few-shot examples of (MH-$Q_{FS}$ : $CoT_{FS}$) pairs, similar to [16], in the following format: {Q: MH-$Q_{FS}$\nA:$CoT_{FS}$...}. However, as our target is a free-form text, models had difficulty constantly generating parsable short answers.[3] Therefore, we follow Kojima et al. [10] in applying a

[3] Previous work experimented on datasets whose answers were numerical, yes/no, or multiple choice.

two-step prompting. One for generating CoT with the multi-hop question prefixed with few-shot examples as explained above, and another for generating the answer with each few-shot example appended its answer, and the CoT generated in the first step appended to the multi-hop question, as shown in Fig. 1. An example is in the bottom part of Table 1.

## 5 Experiments

### 5.1 Datasets

We use three multi-hop datasets: HotpotQA [17], 2WikiMultiHopQA [6] (2Wiki), and MuSiQue [15], which, unlike the other datasets, has 2–4 hop questions. We only use 2-hop questions for a fair comparison. MuSiQue also includes gold decompositions and intermediate answers for SQ1. We show stats in Table 2.

**Table 2.** Data statistics for all question types and datasets. (*) As HotpotQA only has "Bridge" and "Comparison" labels, we use the approximate percentage of labels "Bridging" and "Intersect" (both under "Bridge" label in HotpotQA) provided by Min et al. [12] to calculate the number of training questions. For the "Bridge" dev split, we use the labels classified by Min et al. excluding questions classified as other types.

| Question Type | HotpotQA | | 2Wiki | | MuSiQue | |
|---|---|---|---|---|---|---|
| | train | dev | train | dev | train | dev |
| Compositional/Bridging | (*) ~49,008 | (*) 2,509 | 76,481 | 5,236 | 14,376 | 1,252 |
| Comparison | 17,456 | 1,487 | 51,963 | 3,040 | – | – |
| Bridge-Comparison | – | – | 34,631 | 2,751 | – | – |
| Intersection | (*) ~23,983 | (*) 1,239 | – | – | – | – |
| Inference | – | – | 4,379 | 1,549 | – | – |
| Total | 90,447 | (*) 5,235 | 167,454 | 12,576 | 14,376 | 1,252 |

### 5.2 Question Types

Datasets have different multi-hop question types, with some common types across datasets. We list the types.

*Compositional/Bridging (CL).* It is composed of two questions connected by a bridge entity. Finding the entity is required to solve the question. Example: *Where was the director of film Ronnie Rocket born?* MuSiQue only has this type.

*Comparison (C).* Requires comparing two entities, where the two entities are mentioned explicitly in the question. Example: *Who died first, Fleetwood Sheppard or George William Whitaker?*

*Bridge-Comparison (BC).* Similar to the previous type, but requires finding the entities first before comparing them. Example: *Which film has the director born first, Once A Gentleman or The Girl In White?* This type only exists in the 2Wiki dataset. We consider this to be in the same category as "Comparison" because of similarity.

*Intersection (I).* Requires finding an entity that satisfies different conditions. Example: *Which college football team for the University of Oregon did Enoka Lucas play for?* This type only exists in HotpotQA. Min et al. [12] identified this type and only classified the dev split accordingly. As both "Bridging" and "Intersection" types are labeled as "Bridge" in HotpotQA, we fine-tune on them together but evaluate CL and I separately.

*Inference (Inf).* Requires inferring the bridge entity using common-sense and logical rules. It is different than the CL type in that it is not explicit to find the bridge entity. Example: *Who is the paternal grandmother of King Kang Of Zhou?* This type only exists in the 2Wiki dataset.

### 5.3 Experimental Settings

The used models are: a pre-trained T5-large (T5-L) with 770M parameters, and two GPT-3 versions [3]; Curie and Davinci with 13B and 175B parameters, respectively. We only fine-tune T5-L and apply few-shot prompting to GPT-3. We evaluate both models on three datasets, and we use two accuracy metrics: EM = exact match, and F1 = partial match.

### 5.4 Results

Roberts et al. [14] evaluated three single-hop datasets in a closed-book setting: NaturalQuestions [11], WebQuestions [2], and TriviaQA [8]. T5-L achieved 28.5, 30.6, and 28.7 EM respectively. We consider these as a reference for single-hop closed-book performance.

From Table 3, we see an expected large gap in performance between multi-hop and single-hop baselines in types CL and Inf for all models, and in type I for T5-L and Curie models, demonstrating the difficulty of multi-hop questions. An exception is the C type; the answer is present in the question itself, therefore the LM can generate it with ease.

*Summary of Findings.* We find that even though GPT-3 is more than two orders of magnitude larger than T5-L, fine-tuning T5-L is superior under two conditions: enough training data is provided, and the questions are truly multi-hop and not guessable from one-hop like type I questions. In addition, we find that predicting a correct bridging entity/answer to SQ1 (we call it SA1) is more important than correct SQs. We explain more in detail.

*Analysis of Baselines.* Davinci's high performance for type I is because of the nature of questions that allow the answers to be guessed without other reasoning hops (like single-hop questions). For example: *Based on a True Story is an album by which country music star, with the single My Eyes?*. The T5-L performance on MuSiQue is especially low due to small training data. The 2Wiki Inf split is smaller, however, upon investigation, most question formats are: *Who is the grandparent of X?* and a large portion of the answers overlap greatly with the

**Table 3.** Comparing the baselines against DecompFT/Few-shot CoT methods. For DecompFT, we used the highest-performing decomposition method. All results are shown in EM/F1 format. We group the question type splits of the datasets together.

| Data Split | Model | Baseline | | Proposed Method | |
|---|---|---|---|---|---|
| HotpotQA (CL) | T5-L | (Baseline$_{FT}$) | 8.45/15.94 | (DecompFT) | 8.65/16.51 |
| | Curie | (Baseline$_{FS}$) | 7.25/14.49 | (Few-shot CoT) | 4.94/10.50 |
| | Dvnc | (Baseline$_{FS}$) | 17.90/27.32 | (Few-shot CoT) | **22.52/34.28** |
| 2Wiki (CL) | T5-L | (Baseline$_{FT}$) | **16.31/18.62** | (DecompFT) | **16.31**/18.56 |
| | Curie | (Baseline$_{FS}$) | 2.90/5.48 | (Few-shot CoT) | 1.99/3.97 |
| | Dvnc | (Baseline$_{FS}$) | 6.46/10.65 | (Few-shot CoT) | 9.19/15.82 |
| MuSiQue (CL) | T5-L | (Baseline$_{FT}$) | 2.40/8.48 | (DecompFT) | 3.22/9.02 |
| | Curie | (Baseline$_{FS}$) | 2.80/9.56 | (Few-shot CoT) | 2.24/6.84 |
| | Dvnc | (Baseline$_{FS}$) | 6.56/15.31 | (Few-shot CoT) | **14.48/25.05** |
| HotpotQA (C) | T5-L | (Baseline$_{FT}$) | 53.87/60.38 | (DecompFT) | 52.80/58.77 |
| | Curie | (Baseline$_{FS}$) | 36.32/43.57 | (Few-shot CoT) | 35.64/40.57 |
| | Dvnc | (Baseline$_{FS}$) | 51.11/58.73 | (Few-shot CoT) | **60.93/68.94** |
| 2Wiki (C+BC) | T5-L | (Baseline$_{FT}$) | 50.56/52.05 | (DecompFT) | 50.26/51.68 |
| | Curie | (Baseline$_{FS}$) | 43.31/44.31 | (Few-shot CoT) | 42.86/44.11 |
| | Dvnc | (Baseline$_{FS}$) | 44.78/45.22 | (Few-shot CoT) | **58.95/59.31** |
| HotpotQA (I) | T5-L | (Baseline$_{FT}$) | 11.22/19.63 | (DecompFT) | 10.82/19.54 |
| | Curie | (Baseline$_{FS}$) | 10.57/18.97 | (Few-shot CoT) | 7.022/14.44 |
| | Dvnc | (Baseline$_{FS}$) | 25.02/37.35 | (Few-shot CoT) | **25.75/38.82** |
| 2Wiki (Inf) | T5-L | (Baseline$_{FT}$) | **10.85/30.99** | (DecompFT) | 10.40/29.38 |
| | Curie | (Baseline$_{FS}$) | 1.81/21.50 | (Few-shot CoT) | 1.55/21.08 |
| | Dvnc | (Baseline$_{FS}$) | 5.62/27.47 | (Few-shot CoT) | 8.65/30.45 |

question. This explains the high F1 score compared to other splits. The high T5-L performance for 2Wiki CL split is because of the sufficiently large training data. The high performance of GPT-3 on the HotpotQA CL split is likely due to wrongly included type I questions because we used the type prediction by Min et al. [12], as mentioned in Sect. 5.2. For types C and Inf, the T5-L baseline outperforms GPT-3 because of fine-tuning.

*Analysis of DecompFT/CoT.* We discuss adding SQs or CoT to help the model explicitly do multi-hop reasoning. Although three decomposers were experimented with, we found no or marginal changes; suggesting that the quality of predicted SQs is not sufficient. However, when using gold decompositions, we see a boost, albeit small, in Table 4. To test DecompFT with predicted SA1, we train a single-hop T5-L and T5-11b using gold SQ1 and SQ2 scoring 6.15/14.07 and 16.65/25.51, respectively. Table 4 shows that plugging low-quality predicted SA1 to the gold decompositions hurts the performance, while T5-11b's better

**Table 4.** Comparison with using gold decompositions on MuSiQue (CL), and between using predicted and gold SA1, all in both fine-tuning and inference.

| Setting | T5-L | T5-11b |
|---|---|---|
| No decompose | 2.40/8.48 | 4.31/12.20 |
| + gold decompose | 3.35/9.64 | 5.03/12.66 |
| + pred SA1 | 2.56/8.33 | 5.91/14.34 |
| + gold SA1 | **7.19/16.39** | **18.53/29.80** |

quality SA1 improves it. Gold SA1 expectedly provides remarkable improvement where the scores outperform the single-hop accuracy, confirming that DecompFT method is viable, and the importance of SA1.

*Impact of CoT's Quality.* CoT, by default, includes the equivalent of SA1 (i.e., the bridging entity). Curie's generated CoT degrades its performance. Upon examining the quality of the generated CoT, we found that the problem is mainly not with the logic, as the case with the previously experimented decomposers, but rather with the bridging entity. Predicting the wrong entity harshly harms the performance, as seen in Curie with CoT results in Table 3. However, while the phrasing of Davinci's CoT is similar to that of Curie's, the bridging entities are different. Predicting the correct entity essentially transforms the original question into a single-hop one, increasing the performance. Nonetheless, Curie occasionally makes logical mistakes in comparison steps and fails to decompose the question.

## 6 Limitations

Our hyper-parameter tuning for T5 was manual, and there could be better hyper-parameters that produce better results. Still, we chose the best parameters after some trials to run all experiments. Our few-shot examples for GPT-3 prompts were selected from the beginning of the training data of each dataset split while considering making them diverse in terms of the named entity type of the target. Although we did experiment with using a different set of few-shot examples depending on the question type, we got no obvious improvements. We did not experiment with further prompt sampling techniques or shuffling of the few-shot examples.

## 7 Conclusion

We revealed a large gap between single-hop and multi-hop questions in the closed-book QA setting. We proposed two methods, fine-tuning question decompositions or few-shot generation of CoT to facilitate teaching LMs to "hop" and find answers. We demonstrated that decomposing multi-hop questions without or wrongly predicting the bridging entity does not improve performance, while correctly predicting it boosts the performance. We also found that when enough data is provided, fine-tuning a small model like T5-L is enough to outperform few-shot GPT-3 on real multi-hop questions. Additionally, compared to Davinci, Curie's low-quality CoT is largely due to mispredicting the bridging entity rather than the logic. In the future, we will investigate more effective ways of selecting few-shot examples for the few-shot CoT method.

## References

1. Alkhaldi, T., Chu, C., Kurohashi, S.: Flexibly focusing on supporting facts, using bridge links, and jointly training specialized modules for multi-hop question answering. IEEE/ACM Trans. Audio, Speech, Lang. Process. **29**, 3216–3225 (2021). https://doi.org/10.1109/TASLP.2021.3120643
2. Berant, J., Chou, A., Frostig, R., Liang, P.: Semantic parsing on Freebase from question-answer pairs. In: Proceedings of the 2013 Conference on Empirical Methods in Natural Language Processing, pp. 1533–1544. Association for Computational Linguistics, Seattle, Washington, USA (2013). https://aclanthology.org/D13-1160
3. Brown, T., et al.: Language models are few-shot learners. In: Larochelle, H., Ranzato, M., Hadsell, R., Balcan, M., Lin, H. (eds.) Advances in Neural Information Processing Systems, vol. 33, pp. 1877–1901. Curran Associates, Inc. (2020). https://proceedings.neurips.cc/paper/2020/file/1457c0d6bfcb4967418bfb8ac142f64a-Paper.pdf
4. Devlin, J., Chang, M.W., Lee, K., Toutanova, K.: BERT: pre-training of deep bidirectional transformers for language understanding. In: Proceedings of the 2019 Conference of the North American Chapter of the Association for Computational Linguistics: Human Language Technologies, vol. 1 (Long and Short Papers), pp. 4171–4186. Association for Computational Linguistics, Minneapolis, Minnesota (2019). https://doi.org/10.18653/v1/N19-1423, https://aclanthology.org/N19-1423
5. Fang, Y., Sun, S., Gan, Z., Pillai, R., Wang, S., Liu, J.: Hierarchical graph network for multi-hop question answering. In: Proceedings of the 2020 Conference on Empirical Methods in Natural Language Processing (EMNLP), pp. 8823–8838. Association for Computational Linguistics, Online (2020). https://doi.org/10.18653/v1/2020.emnlp-main.710, https://aclanthology.org/2020.emnlp-main.710
6. Ho, X., Duong Nguyen, A.K., Sugawara, S., Aizawa, A.: Constructing a multi-hop QA dataset for comprehensive evaluation of reasoning steps. In: Proceedings of the 28th International Conference on Computational Linguistics, pp. 6609–6625. International Committee on Computational Linguistics, Barcelona, Spain (Online) (2020). https://doi.org/10.18653/v1/2020.coling-main.580, https://aclanthology.org/2020.coling-main.580
7. Jiang, Z., Araki, J., Ding, H., Neubig, G.: Understanding and improving zero-shot multi-hop reasoning in generative question answering. In: International Conference on Computational Linguistics (COLING). Gyeongju, Korea (2022)
8. Joshi, M., Choi, E., Weld, D., Zettlemoyer, L.: TriviaQA: a large scale distantly supervised challenge dataset for reading comprehension. In: Proceedings of the 55th Annual Meeting of the Association for Computational Linguistics (vol.1: Long Papers), pp. 1601–1611. Association for Computational Linguistics, Vancouver,

Canada (2017). https://doi.org/10.18653/v1/P17-1147, https://aclanthology.org/P17-1147

9. Khot, T., Khashabi, D., Richardson, K., Clark, P., Sabharwal, A.: Text modular networks: learning to decompose tasks in the language of existing models. In: Proceedings of the 2021 Conference of the North American Chapter of the Association for Computational Linguistics: Human Language Technologies, pp. 1264–1279. Association for Computational Linguistics, Online (2021). https://doi.org/10.18653/v1/2021.naacl-main.99, https://aclanthology.org/2021.naacl-main.99
10. Kojima, T., Gu, S.S., Reid, M., Matsuo, Y., Iwasawa, Y.: Large language models are zero-shot reasoners. In: ICML 2022 Workshop on Knowledge Retrieval and Language Models (2022). https://openreview.net/forum?id=6p3AuaHAFiN
11. Kwiatkowski, T., et al.: Natural questions: a benchmark for question answering research. Trans. Assoc. Comput. Linguist. **7**, 452–466 (2019). https://doi.org/10.1162/tacl_a_00276, https://aclanthology.org/Q19-1026
12. Min, S., Zhong, V., Zettlemoyer, L., Hajishirzi, H.: Multi-hop reading comprehension through question decomposition and rescoring. In: Proceedings of the 57th Annual Meeting of the Association for Computational Linguistics, pp. 6097–6109. Association for Computational Linguistics, Florence, Italy (2019). https://doi.org/10.18653/v1/P19-1613, https://aclanthology.org/P19-1613
13. Raffel, C., et al.: Exploring the limits of transfer learning with a unified text-to-text transformer. J. Mach. Learn. Res. **21**(140), 1–67 (2020), http://jmlr.org/papers/v21/20-074.html
14. Roberts, A., Raffel, C., Shazeer, N.: How much knowledge can you pack into the parameters of a language model? In: Proceedings of the 2020 Conference on Empirical Methods in Natural Language Processing (EMNLP), pp. 5418–5426. Association for Computational Linguistics, Online (2020). https://doi.org/10.18653/v1/2020.emnlp-main.437, https://aclanthology.org/2020.emnlp-main.437
15. Trivedi, H., Balasubramanian, N., Khot, T., Sabharwal, A.: MuSiQue: multihop questions via single-hop question composition. Trans. Assoc. Comput. Linguist. **10**, 539–554 (2022). https://doi.org/10.1162/tacl_a_00475, https://aclanthology.org/2022.tacl-1.31
16. Wei, J., et al.: Chain of thought prompting elicits reasoning in large language models. arXiv preprint arXiv:2201.11903 (2022)
17. Yang, Z., Qi, P., Zhang, S., Bengio, Y., Cohen, W., Salakhutdinov, R., Manning, C.D.: HotpotQA: a dataset for diverse, explainable multi-hop question answering. In: Proceedings of the 2018 Conference on Empirical Methods in Natural Language Processing, pp. 2369–2380. Association for Computational Linguistics, Brussels, Belgium (2018). https://doi.org/10.18653/v1/D18-1259, https://aclanthology.org/D18-1259
18. Ye, Q., et al.: Studying strategically: learning to mask for closed-book QA. arXiv preprint arXiv:2012.15856 (2020)

# Evaluating Transformers' Sensitivity to Syntactic Embedding Depth

Hailin Hao(✉)

University of Southern California, Los Angeles, CA 90089, USA
hailinha@usc.edu

**Abstract.** Various probing studies have investigated the ability of neural language models trained only on word prediction over large corpora to process hierarchical structures, a hallmark of human linguistic abilities. For instance, it has been shown that Long Short-Term Memory (LSTM), a type of Recurrent Neural Networks (RNNs), can capture long-distance subject-verb agreement patterns and attraction effects found in human sentence processing. However, although it's found in human experiments that attractors that are syntactically closer to but linearly farther from the verb elicit greater attraction effects than those that are linearly closer but syntactically farther, LSTM shows the opposite pattern, suggesting that it lacks knowledge regarding syntactic distance and is more sensitive to local information. The current article investigates whether state-of-the-art Generative Pre-trained Transformers (GPTs) can capture the prominence of syntactic distance. We experimented with various versions of GPT-2 and GPT-3 and found that all of them succeeded at the task. We conclude that GPT may be able to better model human linguistic cognition than LSTM, corroborating previous research, and that further investigating what mechanisms enable GPT to do so may inform research on human syntactic processing.

**Keywords:** Syntactic hierarchies · syntactic distance · agreement attraction · cognitive modeling

## 1 Introduction

Recent advancements in Natural Language Processing have largely been attributed to the introduction of neural language models [e.g., 1–3]. However, despite their striking performance in a wide range of applied NLP tasks, the nature of the representations learned by these models remains largely unknown, posing challenges for engineers, accountability, and theoretical understanding. To address this issue, previous research has investigated how language models perform on controlled sentences designed to reveal their representation of syntactic hierarchies in human behavioral or neural experiments [e.g., 4–7]. It has shown that these models are capable of learning various types of syntactic phenomena, such as number agreement, filler-gap dependencies, and islands. In this article, we focus on the so-called *syntactic embedding depth,* or *syntactic distance.* It is well-known that in human syntactic processing, hierarchical structures can be constructed over long


R. Mehmood et al. (Eds.): DCAI 2023, LNNS 741, pp. 159–166, 2023.
https://doi.org/10.1007/978-3-031-38318-2_16

distances, which is a hallmark of human linguistic abilities [8]. Therefore, other than the linear distance between words, humans also represent a syntactic distance that does not necessarily correspond to the linear distance. Syntactic distance or embedding depth can be measured as the number of intervening nodes.

We will illustrate the effect of syntactic (and linear) distance on human language processing using agreement attraction [9]. Subject-verb agreement attraction arises when the verb erroneously agrees with a more local noun (i.e., the attractor) instead of with the grammatical subject (i.e., the target), as shown in (1a). It has been found that the preamble "The key to the doors" in (1a) elicits more production of "are" than the preamble "The key to the door" in (1b) and is read faster at the verb region [10, 11]. Such effects are accounted for by either similarity-based interference that arises in memory retrieval [12], or feature percolation in memory distortion [13].

(1) a. The key to the doors are rusty.
    b. The key to the door are rusty.

It's been hypothesized that attractors that are closer to the verb should elicit more attraction errors. For example, [14] shows that when linear distance is controlled, syntactically closer attractors elicit a larger number of production errors. As shown in (2), in (2a) the attractor "popular rock singers" is embedded in a Prepositional Phrase (PP), while in (2b) the attractor is embedded in a Relative Clause (RC). Structurally, the attractor in RCs is more deeply embedded, and has to travel across more nodes to get to the verb. The difference in terms of the ratio of attraction errors is thus explained via syntactic distance. Linear distance also plays a role, where linearly closer attractors may have a bigger impact on attraction [15]. Moreover, it's been found that syntactically closer attractor exhibits bigger influences than linearly closer ones, suggesting the prominence of syntactic hierarchy in human language processing [16].

(2) a. The demo tape from the popular rock singers ...
    b. The demo tape that promoted the rock singers ...

The effect of syntactic distance thus becomes a very good testing ground for the linguistic abilities of neural language models, since it requires the models to have knowledge of latent syntactic hierarchical relations. [17] tested the ability of Long Short-Term Memory [18] to identify syntactic distance using the agreement task. They found that LSTM indeed shows more sensitivity to linearly closer attractors, when syntactic distance is controlled, and that when linear distance is controlled, syntactically closer attractors have a bigger impact on agreement attraction. However, contra human behavioral results, where syntactically distance influences attraction more strongly than linear distance, LSTM shows more sensitivity to linear distance. This revealed a key difference in syntactic processing in humans and LSTM. In this article, we look at the ability of Generative Pre-trained Transformers (GPTs) [19] to capture the prominence of syntactic distance. GPT-2 and GPT-3 have surpassed many other neural language models on various NLP tasks [e.g., 2] and may potentially better model human sentence processing [20, 21]. More specifically, we performed the experiments on four versions of GPT-2 and GPT-3, with varying numbers of parameters and different sizes of training corpora, as detailed in Table 1.

**Table 1.** Models used and their size of parameters and training corpus

| Model | Parameters | Corpus Size |
|---|---|---|
| GPT-2 Base | 117M | 8B |
| GPT-2 Medium | 345M | 8B |
| GPT-2 Large | 774M | 8B |
| GPT3-Davinci | 175B | 300B |

In the next section, we will detail each experiment we performed, and then present the results obtained from the language models. We briefly discuss the implications and conclude in Sect. 3.

## 2 Experiments

In Experiment 1, we investigate whether in GPT-2 and GPT-3 linearly closer attractors elicit greater attraction effects, controlled for syntactic distance. In Experiment 2, we test whether in GPT-2 and GPT-3 syntactically closer attractors elicit greater attraction effects, controlled for linear distance. LSTM succeeded in both tasks. In Experiment 3, we test whether in GPT-2 and GPT-3 syntactically closer but linearly farther attractors elicit greater attraction effects than linearly closer but syntactically father ones. This is the task that LSTM failed.

Following [17], we feed the preamble together with the verb (singular and plural forms of *be*) into the model, which outputs the probabilities assigned to the singular and plural forms of *be*. The error rates are calculated according to the formula in (3).

(3) for singular targets: $P(err) = P(are)/(P(is) + P(are))$
for plural targets: $P(err) = P(is)/(P(is) + P(are))$

For all statistical analysis, we fitted maximal Bayesian logistic regressions were fitted on the binary response variable using the brms package [22] in R. All predictor variables are sum coded $(1, -1)$. The same non-informative priors were chosen for all the models. We report the posterior estimate of the coefficients $\beta$, and their 95% credible intervals.

### 2.1 Linear Effects

Experiment 1 follows the behavioral experiment conducted by [15]. They found that human participants are more likely to produce plurally marked verbs (such as *are* and *were*) when the linearly closer attractor is plural, as in (4a), than when the linearly farther attractor is plural, as in (4b). We have in total 32 sets of items in the form of (4).

(4) a. Can you ask Brenda if the boy or the girls...
b. Can you ask Brenda if the boys or the girl...

**GPT Results.** The results of the four models are plotted in Fig. 1. All four models show more errors when the linearly closer attractor is plural (GPT-2 Base: $\beta = 0.41$, CrI =

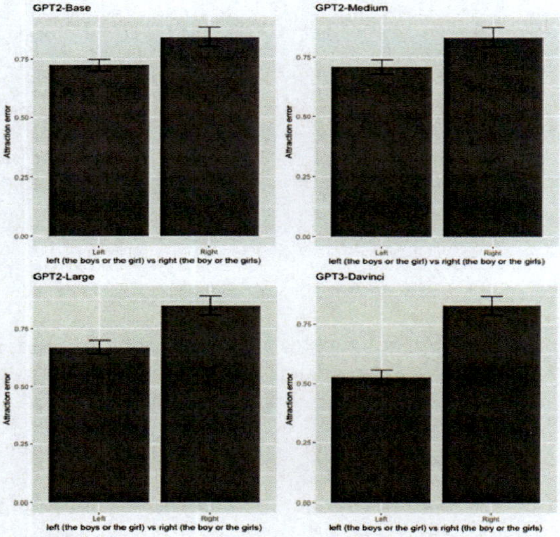


**Fig. 1.** Error rates across models by condition (linearly closer vs. farther attractor)

[0.05, 0.76]; GPT-2 Medium: β = 0.39, CrI = [0.02, 0.76]; GPT-2 Large: β = 0.53, CrI = [0.16, 0.89]; GPT-3 Davinci: β = 0.77, CrI = [0.47, 1.07).

## 2.2 Syntactic Effects

Experiment 2 follows the behavioral experiment conducted by [14]. They found that human participants are more likely to produce plurally marked verbs (such as are and were) when the attractor is nested in an RC, as in (5a), than when nested in a PP, as in (5b). What they also found is the so-called *markedness* effects [13], according to which there is an asymmetry in the attraction effects induced by singular and plural attractors, in that plural attractors in singular target conditions elicit attraction, but singular attractors in plural target conditions elicit no attraction. Therefore, there should be a difference in terms of attraction rates between (5a) and (5b), but not between (5c) and (5d). We have in total 32 sets of items in the form of (5).

(5) a. (PP; plural attractor) The demo **tape** from the popular rock **singers**
b. (RC; plural attractor) The demo **tape** that promoted the rock **singers**
c. (PP; singular attractor) The demo **tapes** from the popular rock **singer**
d. (RC; singular attractor) The demo **tapes** that promoted the rock **singer**

**GPT Results.** The results of the four models are plotted in Fig. 2. As can be seen, all models show more errors in PPs than in RCs. Our statistical analysis suggests that the effect is reliable across all models, with the exception of GPT-2 Large (GPT-2 Base: $\beta = 0.28$, CrI = [0.07, 0.49]; GPT-2 Medium: $\beta = 0.52$, CrI = [0.16, 0.87]; GPT-2 Large: $\beta = 0.01$, CrI = [−0.26, 0.28]; GPT-3 Davinci: $\beta = 0.64$, CrI = [0.20, 1.07]). However, although there's a trend that supports the markedness effects, we didn't find strong evidence for markedness effects across models (GPT-2 Base: $\beta = -0.28$, CrI = $[-0.74, 0.20]$; GPT-2 Medium: $\beta = -0.39$, CrI = $[-0.89, 0.13]$; GPT-2 Large: $\beta = -0.15$, CrI = $[-0.63, 0.33]$; GPT-3 Davinci: $\beta = -0.16$, CrI = $[-0.78, 0.46]$). We suspect that this is due to the lack of power, since we only have one estimate for each of the stimuli in 32 sets. A much larger scale study testing GPT's performance in long-distance subject-verb agreement elicited the expected markedness effects [23].

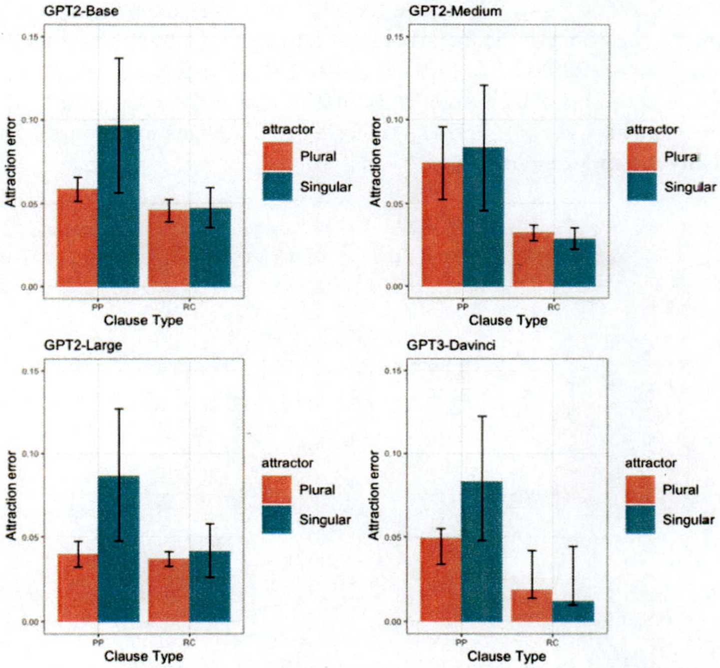


**Fig. 2.** Error rates across models by condition

## 2.3 Linear vs. Syntactic Effects

Experiment 2 follows the behavioral experiment conducted by [16]. They found that human participants are more likely to produce plurally marked verbs (such as are and were) when the linearly closer attractor is plural, as in (6b), and when the syntactically closer attractor is plural, as in (6c). However, the key finding that we hope to replicate is that the effect of syntactically closer attractors should be stronger than that of linearly closer attractors. We have in total 32 sets of items in the form of (6).

(6) a. (SSS) The **threat** to the **president** of the **company**...
b. (SSP) The **threat** to the **president** of the **companies**...
c. (SPS) The **threat** to the **presidents** of the **company**...

d. (SPP) The **threat** to the **presidents** of the **companies**…

**GPT Results.** The results of the four models are plotted in Fig. 3. For all models, the syntactic distance (GPT-2 Base: $\beta = 0.694$, CrI = [0.526, 0.858]; GPT-2 Medium: $\beta = 0.644$, CrI = [0.450, 0.835]; GPT2-Large: $\beta = 0.702$, CrI = [0.539, 0.863]; GPT3-Davinci: $\beta = 0.580$, CrI = [0.439, 0.722]) has a larger effect on attraction errors than linear distance (GPT-2 Base: $\beta = 0.481$, CrI = [0.316, 0.643]; GPT-2 Medium: $\beta = 0.461$, CrI = [0.268, 0.652]; GPT-2 Large: $\beta = 0.441$, CrI = [0.278, 0.604]; GPT-3 Davinci: $\beta = 0.314$, CrI = [0.175, 0.454]). Moreover, for all models, when the syntactically closer attractor is plural, the effect of the linearly closer attractor on attraction errors decreases (GPT-2 Base: $\beta = -0.289$, CrI = [−0.454, −0.125]; GPT-2 Medium: $\beta = -0.279$, CrI = [0.471, − 0.087]; GPT-2 Large: $\beta$ = -0.269, CrI = [−0.432, −0.103]; GPT-3 Davinci: $\beta = -0.234$, CrI = [−0.374, −0.094]), again suggesting the prominence of the syntactically closer attractor.

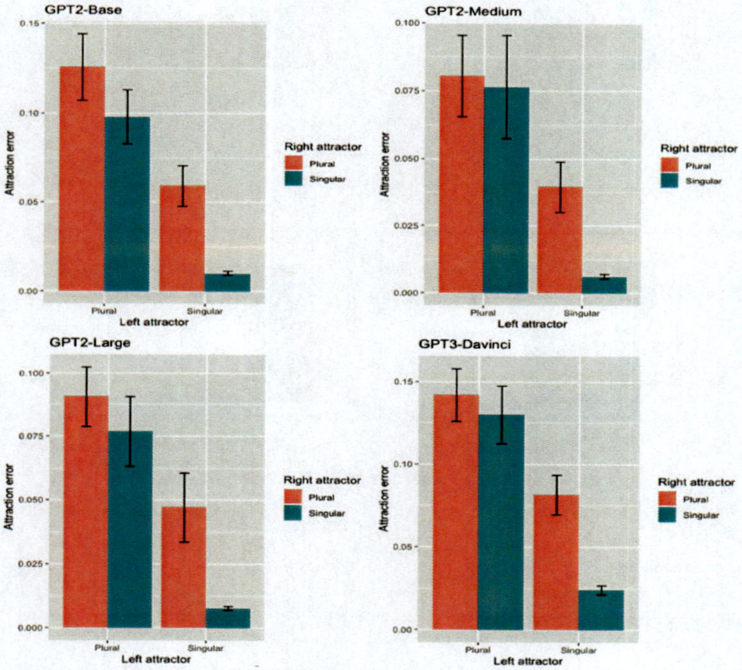


**Fig. 3.** Error rates across models by condition

## 3 Discussion and Conclusion

In order to test GPT's ability to capture the prominence of syntactic distance, we replicated three human experiments on four versions of GPT. We find that all models successfully replicated linear effects, syntactic effects, and the relatively bigger impact of syntactic effects found in human experiments. The only exception is that GPT-2 Large failed to replicate the syntactic effects, but the results are still in the expected direction. We additionally test whether GPT exhibits markedness effects. Although we do not find statistically reliable evidence for markedness effects, we argue that this might be due to the small size of our stimuli.

The fact that GPT language models can recover non-surface syntactic distance is quite striking, especially considering that it's not explicitly trained for this task. Our experiments provide another piece of evidence regarding what types of syntactic constraints can be acquired from domain-general sequence processing mechanisms [7, 23, 24], whereas those that do not may demand additional mechanisms to explain them. In addition, it provides evidence that Transformer-based language models may be able to better model human linguistic cognition, corroborating [20, 21]. RNN-based language models handle sequential data through recurrence, which means that each new piece of input is taken in, processed, and then combined with the current hidden state. The Transformer architecture is different from RNN-based models in that it doesn't rely on recurrence. Instead, the Transformer cell [25], which is made up of self-attention layers [26] and a linear feed-forward layer, allows the model to 'attend' to specific parts of previous input directly. This key difference between the models may be able to explain their divergent behavior on distinguishing syntactic and linear distance effects. Further investigating the mechanisms enabling Transformers to process syntax can also be beneficial for understanding human syntactic processing. For example, [20] argued that the attention mechanism whereby previous inputs are weighted according to how similar they are to the current input is similar to the cue-based retrieval theory in human language processing, where items in past input can be quickly retrieved based on their similarity to a given cue. This may allow the GPT language models to place more weight on words that are linearly farther but syntactically more prominent.

To conclude, through three replications of human behavioral experiments, we show that GPT language models have access to abstract syntactic hierarchies and are able to give syntactically closer elements more attention, which LSTM failed. We provide additional evidence regarding what types of syntactic constraints can be acquired via unsupervised learning and the cognitive plausibility of Transformer-based language models, and we argue that it can also inform research in psycholinguistics and human syntactic processing.

## References

1. Goldberg, Y.: Neural network methods for natural language processing. Synthesis Lect. Hum. Lang. Technol. **10**(1), 1–309 (2017)
2. Devlin, J., Chang, M.-W., Lee, K., Toutanova, K.: BERT: pre-training of deep bidirectional transformers for language understanding. arXiv preprint arXiv:1810.04805 (2018)
3. Peters, M.E., et al.: Deep contextualized word representations. In: Proceedings of NAACL (2018)
4. Linzen, T., Dupoux, E., Goldberg, Y.: Assessing the ability of LSTMs to learn syntax-sensitive dependencies. Trans. Assoc. Comput. Linguist. **4**, 521–535 (2016)
5. Gulordava, K., Bojanowski, P., Grave, E., Linzen, T., Baroni, M.: Colorless green recurrent networks dream hierarchically. In: Proceedings of NAACL (2018)

6. Wilcox, E., Levy, R., Morita, T., Futrell, R.: What do RNN language models learn about filler-gap dependencies? In: Proceedings of the 2018 EMNLP Workshop BlackboxNLP: Analyzing and Interpreting Neural Networks for NLP, vol. 2, no. 5, pp. 211–221 (2018)
7. Wilcox, E., Gauthier, J., Hu, J., Qian, P., Levy, R.P.: Learning syntactic structures from string input. In: Lappin, S. (ed.) Algebraic Systems and the Representation of Linguistic Knowledge, vol. 2, no. 5 (2022)
8. Chomsky, N.: Syntactic Structures. Walter de Gruyter, Berlin (1957)
9. Jäger, L.A., Engelmann, F., Vasishth, S.: Similarity-based interference in sentence comprehension: literature review and Bayesian meta-analysis. J. Mem. Lang. **94**, 316–339 (2017)
10. Bock, K., Miller, C.A.: Broken agreement. Cogn. Psychol. **23**(1), 45–93 (1991)
11. Wagers, M., Lau, E.F., Phillips, C.: Agreement attraction in comprehension: representations and processes. J. Mem. Lang. **61**, 206–237 (2009)
12. Lewis, R.L., Vasishth, S.: An activation-based model of sentence processing as skilled memory retrieval. Cogn. Sci. **29**(3), 375–419 (2005)
13. Eberhard, K.: The marked effect of number on subject–verb agreement. J. Mem. Lang. **36**, 147–164 (1997)
14. Bock, K., Cutting, J.C.: Regulating mental energy: performance units in language production. J. Mem. Lang. **31**, 99–127 (1992)
15. Haskell, T.R., MacDonald, M.C.: Constituent structure and linear order in language production: evidence from subject-verb agreement. J. Exp. Psychol. Learn. Mem. Cogn. **31**(5), 891–904 (2005)
16. Franck, J., Vigliocco, G., Nicol, J.: Subject–verb agreement errors in French and English: the role of syntactic hierarchy. Lang. Cognit. Process. **17**, 371–404 (2002)
17. Arehalli, S., Linzen, T.: Neural language models capture some, but not all, agreement attraction effects. Trans. Assoc. Comput. Linguist. **8**, 166–180 (2020)
18. Hochreiter, S., Schmidhuber, J.: Long short-term memory. Neural Comput. **9**(8), 1735–1780 (1997)
19. Radford, A., Narasimhan, K., Salimans, T., Sutskever, I., et al.: Improving language understanding by generative pre-training. OpenAI (2018)
20. Merkx, D., Frank, S. L.: Human sentence processing: recurrence or attention? In: Proceedings of the 2021 Workshop on Cognitive Modeling and Computational Linguistics, pp. 1–11. Association for Computational Linguistics, Mexico City (2021)
21. Michaelov, J.A., Coulson, S., Bergen, B.K.: So Cloze yet so far: N400 amplitude is better predicted by distributional information than human predictability judgements. arXiv:2109.01226 [Cs, Math] (2021)
22. Bürkner, P.-C.: brms: An R package for Bayesian multilevel models using Stan. J. Stat. Softw. **80**(1), 1–28 (2017)
23. Contreras Kallens, P., Kristensen-McLachlan, R.D., Christiansen, M.H.: Large language models demonstrate the potential of statistical learning in language. Cogn. Sci. **47**(3) (2023)
24. Wilcox, E.G., Futrell, R., Levy, R.: Using computational models to test syntactic learnability. Linguist. Inq. (2022)
25. Vaswani, A., et al.: Attention is all you need. In: Proceedings of the 31st Conference on Neural Information Processing Systems, pp. 6000–6010 (2017)
26. Luong, M.-T., Pham, H., Manning, C.-D.: Effective approaches to attention based neural machine translation. In: Proceedings of the 2015 Conference on Empirical Methods in Natural Language Processing, pp. 1412–1421 (2015)

# Parsing IsiZulu Text Using Grammatical Framework

Laurette Marais[1(✉)] and Laurette Pretorius[2]

[1] Voice Computing, NextGen Enterprises and Institutions, CSIR, Pretoria, South Africa
lmarais@csir.co.za

[2] Division of Computer Science, Department of Mathematical Sciences, Stellenbosch University, Stellenbosch, South Africa
lpretorius@sun.ac.za

**Abstract.** While Grammatical Framework (GF) resource grammars are primarily intended as linguistic software libraries, they can also be employed to enable syntax parsing via the GF runtime system. The isiZulu resource grammar (ZRG) in conjunction with the GF runtime system constitutes the first openly available parser for isiZulu. A key feature of the implementation of the ZRG is to model aspects of morphology as equivalent to syntax. In the absence of a baseline, we evaluate the parser at the token level against a morphological analyser, before characterising its phrase level analysis in order to establish a baseline for parsing of isiZulu text.

**Keywords:** chunk parsing · morphology · syntax · morphosyntax

## 1 Introduction

IsiZulu is an under-resourced language for which basic natural language processing core technologies remain an active area of research. One such effort is the development of an isiZulu GF resource grammar (ZRG). The GF runtime system, when enabled by a resource grammar, can be employed as a syntax parser [2], and the ZRG constitutes the first available parser for isiZulu.

GF-based parsers can be somewhat brittle, and a standard technique to mitigate this is to develop an extension that enables chunk parsing [17]. There have also been efforts to connect the GF project with UD via the *gf2ud* and *ud2gf* conversion tools [7,18]. It has been shown that GF treebanks can be converted to UD treebanks via *gf2ud* in order that UD parsers may be trained.

The development of an isiZulu parser via other means has not yet been achieved. IsiZulu has an agglutinating morphology with a conjunctive orthography, and hence a consensus has emerged that it is best modelled at the subword

---

This work has been funded by the South African Centre for Digital Language Resources (https://sadilar.org/).


R. Mehmood et al. (Eds.): DCAI 2023, LNNS 741, pp. 167–177, 2023.
https://doi.org/10.1007/978-3-031-38318-2_17

level. Consequently, past and recent research has focused on morphological segmentation [8,9]. On the other hand, for part-of-speech tagging, which operates on the token level, tag sets with differing levels of granularity have been experimented with [5,10]. From a machine learning perspective, segmentation and part-of-speech tagging are prerequisites to parsing, and so the ability to train an isiZulu parser is still some way away.

In this paper, we present the ZRG as an enabler for an isiZulu syntax parser. The morphological complexity and orthography of the language present some challenges, which we show can be handled in a principled way by the ZRG-enabled GF runtime system. Development of the ZRG forms part of an effort to include isiZulu in multilingual projects such as GF and UD.

## 2 The IsiZulu GF Resource Grammar

GF resource grammars are primarily intended as linguistic software libraries, where the morphology and syntax are implemented once and for all so that developers of application grammars need not be concerned with these details [16]. Instead, high level functions for combining syntax categories can be used to model the semantics of a specific domain.

Any GF grammar, however, whether it is an application grammar or a resource grammar, can be used by the GF runtime system to linearise trees defined by the grammar, as well as to parse strings in the language of the grammar. Hence, provided that the required linguistic constructions have been implemented by a resource grammar, the GF runtime system can be employed as a syntax parser for a language. This was an important reason why development of the ZRG was undertaken. In addition, a large lexicon of almost 17 000 entries was developed.

### 2.1 IsiZulu Morphosyntax and Orthography

IsiZulu is a Nguni language in the Southern Bantu family. It has an agglutinating morphology and a conjunctive orthography which exhibits morphophonological alternation, including a high degree of morpheme fusion. Like other Bantu languages, it is characterised by nominal classification and concordial agreement. IsiZulu has 18 noun classes. The concords (agreement morphemes) are linked to these noun classes and observed in all parts of an utterance, including verbs, pronouns, adjectives, relatives, possessives, etc. [11].

The combination of morphological complexity, a conjunctive orthography and morphophonological alternation creates challenges for modelling the underlying structure of words and the corresponding surface forms at the same time. This has been successfully achieved by ZulMorph, a finite-state morphological analyser [12] which remains the state of the art in isiZulu morphological analysis [4]. It transduces between surface forms and forms that represent tagged abstract morphophonemic strings, as shown below for the token *sinezinkomazi* (we have cows).[1].

```
si[SC][1pp]na[AdvPre]i[NPrePre][10]zin[BPre][10]khomazi[NStem][9-10]
si[SC][7]na[AdvPre]i[NPrePre][10]zin[BPre][10]khomazi[NStem][9-10]
```

The ZRG does not implement the morphology of isiZulu in the same way, with reference to the abstract morphophonemics, and instead uses a technique described in the next section.

## 2.2 Implementation of the IsiZulu Resource Grammar

The GF resource grammar library (RGL) is a multilingual grammar in which typical syntax categories and functions are defined via its abstract syntax, with each concrete syntax implementing the corresponding linearisation categories and functions [15]. During implementation of the ZRG, it was found that some significant deviations from the common abstract syntax of the GF RGL was necessary in order for the grammar to remain feasible in terms of compilation time, compiled size, ambiguity and overgeneration, as well as linguistic transparency and integrity.

A key feature of the implementation of the ZRG was a reliance on the orthography engineering features of the GF runtime system [1]. In particular, the `BIND` token is used to allow for the construction of tokens, or surface forms, at runtime. Typically, tokens, which constitute the atomic strings of a GF grammar, are constructed at compile time, and they can be concatenated at runtime only via the `++` operator, which inserts a space in between. This requires that all tokens of the grammar be constructed at compile time. For morphologically complex languages, this can render the grammar size intractable [14]. An extension to the GF runtime system allows the use of the `BIND` token to perform direct concatenation (without space separation) during runtime, which has been employed in many other resource grammars, especially Rukiga, another Bantu language [3]. In languages such as isiZulu, which exhibit morpheme fusion, a parameterised selection of surface strings must be implemented in order to ensure that such concatenation at runtime results in grammatically correct surface tokens. In short, in this workaround, aspects of morphology are essentially treated as syntax, and morphophonological alternation rules are treated as a form of agreement.

While in some respects this constitutes a departure from the generally accepted linguistic descriptions of isiZulu, we show in this paper that it allows the resulting parser to give a highly detailed morphosyntactic description of a sentence.

[1] A detailed discussion of this example with reference to the ZulMorph tagset is beyond the scope of this paper. A full description of the tagset is available at https://portal.sadilar.org/FiniteState/demo/zulmorph/doc.html#tagset.

### 2.3 Extending the ZRG for Chunk Parsing

Chunking extensions to the GF RGL typically add two new categories on top of the RGL, namely chunks and lists of chunks [17]: "The parser tries to build chunks of maximal lengths. Its output is a list of chunks, which are independent of each other." Chunks can contain trees of any category type, from high level categories like `S` (sentences), to uninterpreted string literals as `Symb`. This improves the parser's ability to deal with grammatically incorrect input, as well as allowing it to recover from encountering tokens that fall outside of its definition.

A new chunking extension module was implemented for the ZRG that added another layer to the usual approach, namely by mirroring each targeted category with its own chunk category (e.g. `Chunk_NP` for `NP`). This was done to simplify the analysis of chunk-based trees by making it easier to inspect the syntactic types of chunks.

As an aside, we note that the functions of the chunking extension also liberally exploit the fact that isiZulu is a pro-drop language in order to "remember" agreement information in the abstract syntax trees. This is done by including a pronoun as argument, which in the linearisation function supplies the agreement information, even if the string it contributes to the linearisation is empty.

## 3 Experiment

To our knowledge, the ZRG (together wth the GF runtime system) constitutes the first syntax parser for isiZulu, and certainly the first openly available one. As such, no baseline for parsing exists against which to evaluate it. The only sufficiently reliable existing tool which could provide a baseline for morphosyntactic analysis is ZulMorph. ZulMorph accepts a token and produces all possible morphological analyses for the token, although it is known to overgenerate analyses to some extent [13].

As a morphological analyser, ZulMorph does not disambiguate between possible analyses for a given token in context - for this, phrase level information is required. Even then, ambiguities may remain that can only be addressed at the semantic or pragmatic level and hence the latter kind of disambiguation falls outside the scope of this work. We therefore make no attempt to disambiguate between different parses returned by the GF runtime system, simply choosing the first parse for the purpose of evaluation. (The GF parsing algorithm produces a statistical ranking, allowing user-defined probabilities for functions [2]. The only material guidance we provide our parser is to strongly favour non-symbol based chunks, which we discuss in more detail in Sect. 3.2.) Given that the ZRG has been built to be correct, our *evaluation* of its ability to enable *parsing* will focus on *coverage*.

In this section, therefore, we first evaluate the parser at token level by using ZulMorph as a baseline. Then, we characterise the parser's ability to perform phrase level parsing, and in so doing establish a baseline for syntactic parsing of isiZulu.

**Table 1.** Token counts

| Measure | Count | % |
|---|---|---|
| Number of sentences | 381 | |
| Number of tokens | 2975 | |
| Number of unique tokens | 2030 | 68 |
| Avg tokens per sentence | 5.6 | |

### 3.1 Corpus

The choice of input deserves some consideration. By design, the grammar is not intended to model grammatically incorrect text (although the chunk extension aims to overcome it to some extent), and hence in order to evaluate the ZRG for what it is, an evaluation corpus should ideally consist of high quality, grammatically correct texts, since it reduces the need to identify false negatives. In order to further reduce errors caused by spelling mistakes and out-of-vocabulary words (often loan words), we filtered the corpus to only include sentences where ZulMorph is able to provide at least one analysis for each token.

The corpus we selected is a set of usage examples developed as part of the African Wordnet project [6] (381 sentences). We give some preliminary token counts for it in Table 1. We consider this corpus to be a good choice for evaluating the parser, since the sentences have been hand-crafted to exhibit typical usage of lexical entries in the isiZulu wordnet. The average token count must be considered in light of the agglutinating nature of isiZulu. To provide some perspective, the average number of tokens per sentence for the English translations of the same sentences is 9.27. As indicated in Table 1, the percentage of unique tokens is high, as would be expected considering the origin of the corpus.

### 3.2 Token-Level Evaluation

The ZulMorph morphological analyser serves as the basis of our evaluation of the parser in terms of token-level morphosyntactic coverage. We first quantify the parser's coverage by counting tokens successfully parsed. Recall that we consider only the first parse returned by the GF runtime system for each sentence. Our evaluation is based on identifying failures as so-called symbol chunks: chunks that are of type `Chunk_Symb`, which is used to include unparsed tokens in a list of chunks. Our methodology to determine token level coverage is as follows:

1. For each (first) parse, extract all chunks that are not of type `Chunk_Symb`. These chunks represent parsed phrases of one or more tokens. We refer to these as non-symbol chunks.
2. We calculate the percentage of successfully parsed tokens relative to the total number of tokens, both in terms of absolute tokens and unique tokens.
3. We also calculate the number of sentences in which all tokens were parsed.

**Table 2.** Evaluation of parser at token level

| | Successes | Total | % |
|---|---|---|---|
| Parsed tokens (absolute) | 2880 | 2975 | 97 |
| Parsed tokens (unique) | 1940 | 2030 | 96 |
| Fully parsed sentences | 299 | 381 | 79 |

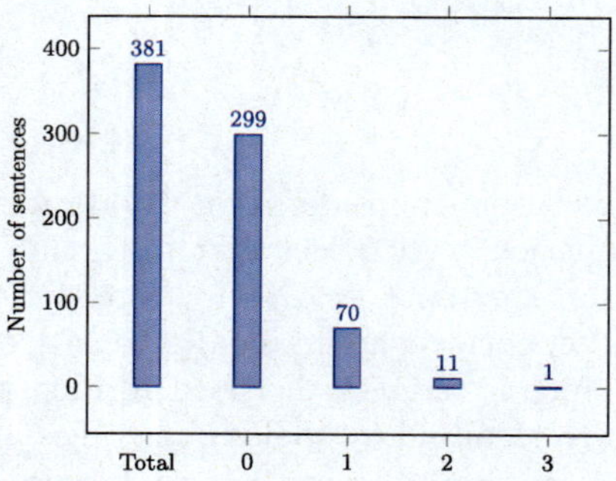


**Fig. 1.** Number of unparsed tokens

4. Finally, we investigate the nature of the token failures (parsed as `Chunk_Symb`) by searching for morpheme tags within the tokens' possible morphological analyses.

The results of performing steps 1 to 3 are shown in Table 2. We note that 79% of sentences could be fully parsed without reverting to symbol chunks. Figure 1 shows the breakdown of number of unparsed tokens per sentence.

To provide insight into the kinds of tokens left unparsed, the ZulMorph morphological analyser was utilised. All possible morphological analyses of the 95 failed tokens were inspected, and trends of morpheme tags present in these analyses were identified. Then, a count was performed to determine how many tokens exhibited certain tags (or substrings) within at least one of its analyses, shown in Table 3. These results provide insight into prioritising further development and refinement of linguistic constructions in the ZRG.

### 3.3 Phrase-Level Characterisation

We turn now to characterising the successful chunks found by the parser. As a general characterisation, the length of the average non-symbol chunk was found to be 2.71 tokens. Firstly, we consider the sentence level chunks, namely those representing well-formed isiZulu sentences (such as indicative main clauses, imperatives, questions etc.). When parsing a sentence from the corpus, it is possible that it is parsed as a complete sentence (a single "chunk"), or as a list of chunks, some of which may themselves represent well-formed sentences.

**Table 3.** ZulMorph analysis substring counts for unparsed tokens

| Substring | Description | Number of tokens | % |
|---|---|---|---|
| se[AuxVStem] | Auxiliary verb stem *se* | 11 | 12 |
| Conj | Conjunctions | 5 | 5 |
| Dem | Demonstrative pronouns | 6 | 6 |
| Pot | Potential aspect marker | 5 | 5 |
| AdvPre | Adverb prefix | 15 | 16 |
| Interrog | Interrogative | 7 | 7 |
| SCSubj | Subjunctive agreement morpheme | 21 | 22 |
| LocPre | Locative prefix | 11 | 12 |

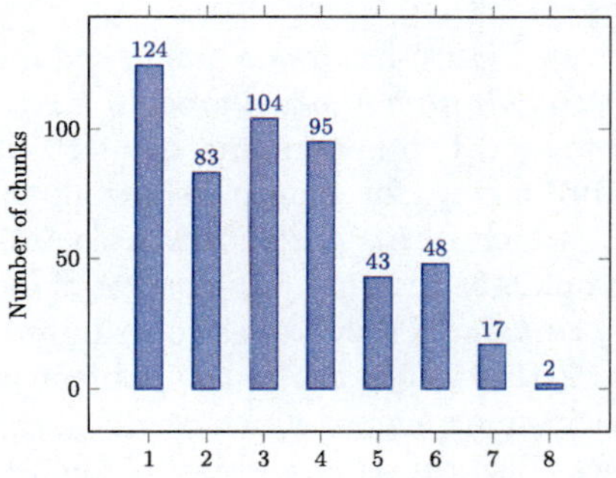


**Fig. 2.** Sentence level chunks by token length

Figure 2 depicts the token length of such sentence level chunks. It is not surprising to find a large number of single token sentence level chunks, given that isiZulu is a pro-drop language in which a single verb token can include its subject and object via agreement morphemes, constituting a well-formed sentence. The total number of sentence level chunks found is 518 and average token length of sentence level chunks is 3.2. In total, of the 2975 tokens in the corpus, 1638 were parsed as part of sentence level chunks, i.e. 55%.

Secondly, we consider verb phrase and noun phrase level chunks parsed separately from sentence level chunks (chunks cannot include other chunks). Figure 3 shows the number of such chunks according to token length. In total, 106 separate verb phrase level chunks were found with an average token length of 1.14, while 157 separate noun phrase level chunks were found with an average token length of 1.2.

### 3.4 IsiZulu Morphosyntax Using GF

So far we have focused on tokens and the token length of chunk parses, but it is also useful to consider the parses at a subword level, since the grammar

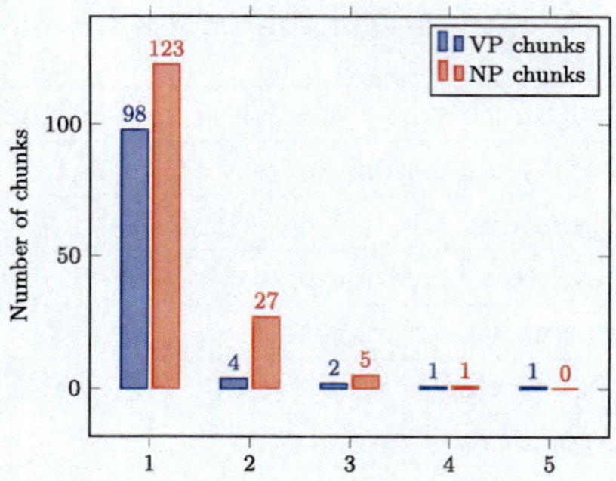


**Fig. 3.** Verb phrase and noun phrase level chunks by token length

models isiZulu at this level. Figure 4 shows a parse spanning 8 tokens, namely *epulazini sinezinkomazi ezimbili namathokazi amabili kanye nenkabi eyodwa* (on the farm we have two cows and two heifers and one ox). The tree indicates the provenance of each subword string, for example showing that the subject agreement morpheme *si-* (we), which forms part of the associative copulative prefix, was contributed by the `CopNPAssoc` function, whereas the copulative base noun (without its pre-prefix *i-*) *zinkomazi* (cows) was contributed by its corresponding lexical function `khomazi_9_10_N`. In between the subject agreement morpheme and the noun, the associative morpheme, which is *-na-* (has/have) in its abstract morphophonemic form, was also contributed by the `CopNPAssoc` function. However, due to morpheme fusion with the pre-prefix of the noun, it was rendered as *-ne-* by the function. A parameter within the noun phrase that keeps track of the abstract morphophonemic form of the pre-prefix is used to select the correct form of the associative morpheme. In this way, morphophonological changes are handled in essentially the same way as syntactic agreement in GF.

Note also that while the surface token length of the sentence is 8, the subword length is 18, showing the hidden complexity of relations between subwords in isiZulu. In fact, while the average number of tokens per non-symbol chunk was 2.71, the average number of subwords was found to be 5.96.

## 4 Conclusion

We have presented the isiZulu GF resource grammar as an enabler for the GF runtime system to perform syntax parsing of isiZulu. The ZRG exploits the ability of the runtime system to treat morphology as equivalent to syntax via its orthography engineering approach. In this way, it deals effectively with the typical language modelling challenges presented by the morphology, orthography and morphophonological alternation of the language, with 79% of sentences fully parsed without reverting to symbol chunks.

The parser has also been evaluated for coverage at the token level, where it approaches the state-of-the-art, but with the addition of phrase-level parsing.

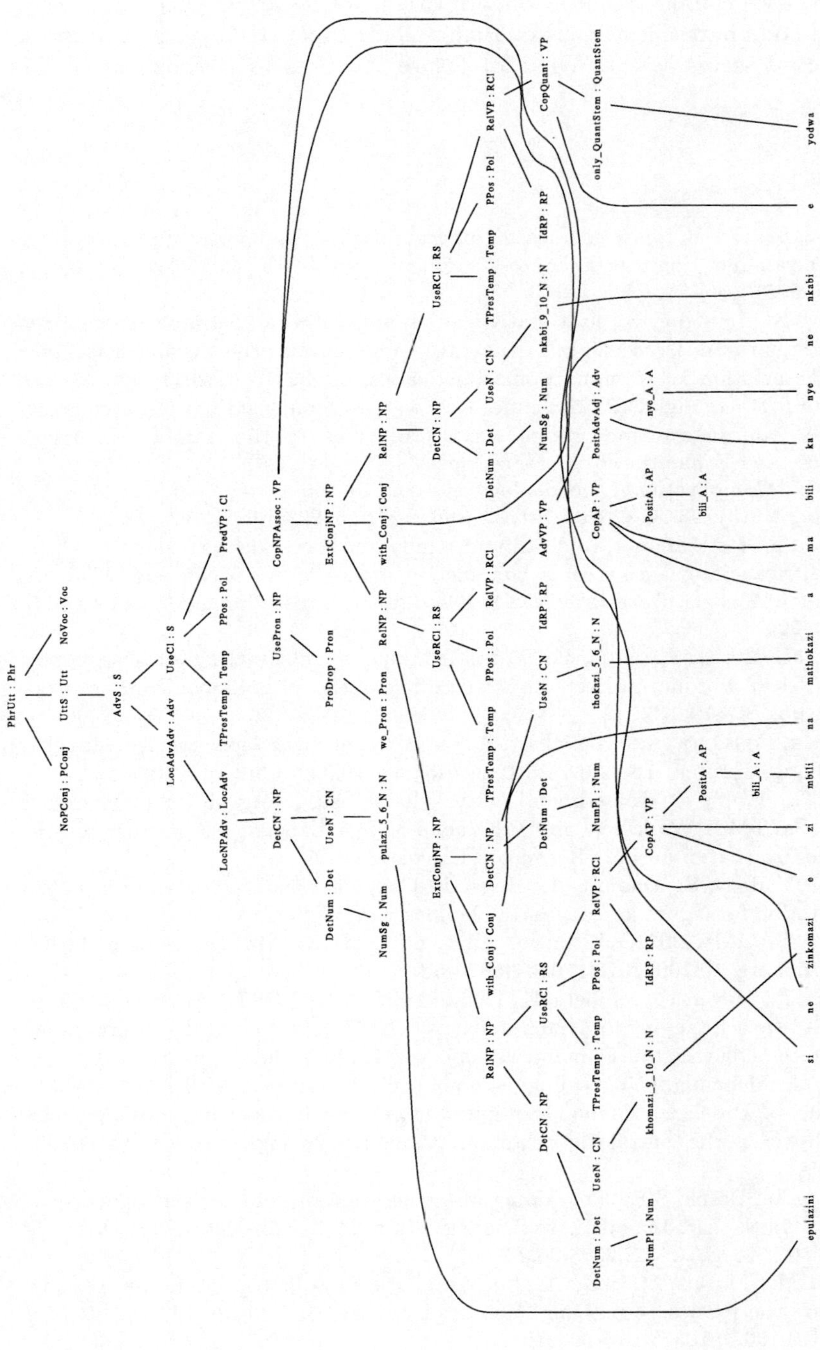


**Fig. 4.** Parse tree for *epulazini sinezinkomazi ezimbili namathokazi amabili kanye nenkabi eyodwa* (on the farm we have two cows and two heifers and one ox)

We have presented a characterisation of the phrase-level parsing with emphasis on sentence level chunks (covering 55% of tokens in the corpus), as well as verb phrase and noun phrase level chunks. Being the first for the language, the parser presented here serves as a baseline for future attempts to develop parsers for isiZulu.

## References

1. Angelov, K.: Orthography engineering in grammatical framework. In: Proceedings of the Grammar Engineering Across Frameworks (GEAF) 2015 Workshop, pp. 33–40 (2015)
2. Angelov, K., Ljunglöf, P.: Fast statistical parsing with parallel multiple context-free grammars. In: Proceedings of the 14th Conference of the European Chapter of the Association for Computational Linguistics, pp. 368–376 (2014)
3. Bamutura, D., Ljunglöf, P., Nebende, P.: Towards computational resource grammars for runyankore and rukiga. In: Proceedings of the Twelfth Language Resources and Evaluation Conference, pp. 2846–2854 (2020)
4. Bosch, S.: Computational morphology systems for Zulu-a comparison. Nord. J. Afr. Stud. **29**(3), 28–28 (2020). https://doi.org/10.53228/njas.v29i3.548
5. Gaustad, T., Puttkammer, M.J.: Linguistically annotated dataset for four official South African languages with a conjunctive orthography: IsiNdebele, isiXhosa, isiZulu, and Siswati. Data Brief **41**, 107994 (2022). https://doi.org/10.1016/j.dib.2022.107994
6. Griesel, M., Bosch, S., Mojapelo, M.L.: Thinking globally, acting locally-progress in the African wordnet project. In: Proceedings of the 10th Global Wordnet Conference, pp. 191–196 (2019)
7. Kolachina, P., Ranta, A.: From abstract syntax to universal dependencies. Linguist. Issues Lang. Technol. **13** (2016). https://doi.org/10.33011/lilt.v13i.1391
8. Kotzé, G., Wolff, F.: Experiments with syllable-based English-Zulu alignment. In: 9th SaLTMiL Workshop on "Free/open-Source Language Resources for the Machine Translation of Less-Resourced Languages" (2014)
9. Moeng, T., Reay, S., Daniels, A., Buys, J.: Canonical and surface morphological segmentation for Nguni languages. In: Jembere, E., Gerber, A.J., Viriri, S., Pillay, A. (eds.) SACAIR 2021. CCIS, vol. 1551, pp. 125–139. Springer, Cham (2022). https://doi.org/10.1007/978-3-030-95070-5_9
10. Pannach, F., Meyer, F., Jembere, E., Dlamini, S.Z.: NLAPOST2021 1st shared task on part-of-speech tagging for Nguni languages. In: Proceedings of the International Conference of the Digital Humanities Association of Southern Africa (2021)
11. Poulos, G., Msimang, C.T.: A linguistic analysis of Zulu. Via Afrika Ltd. (1998)
12. Pretorius, L., Bosch, S.: Enabling computer interaction in the indigenous languages of South Africa: the central role of computational morphology. Interactions **10**, 56–63 (2003)
13. Pretorius, L., Bosch, S.E.: Containing overgeneration in Zulu computational morphology. South. Afr. Linguist. Appl. Lang. Stud. **26**(2), 209–216 (2008). https://doi.org/10.2989/SALALS.2008.26.2.3.567
14. Pretorius, L., Marais, L., Berg, A.: A GF miniature resource grammar for Tswana: modelling the proper verb. Lang. Resour. Eval. **51**(1), 159–189 (2016). https://doi.org/10.1007/s10579-016-9341-z

15. Ranta, A.: The GF resource grammar library. Linguist. Issues Lang. Technol. **2** (2009)
16. Ranta, A.: Grammatical Framework: Programming with Multilingual Grammars. CSLI, Stanford, California (2011)
17. Ranta, A., Angelov, K., Gruzitis, N., Kolachina, P.: Abstract syntax as interlingua: scaling up the grammatical framework from controlled languages to robust pipelines. Comput. Linguist. **46**(2), 425–486 (2020). https://doi.org/10.1162/coli_a_00378
18. Ranta, A., Kolachina, P.: From universal dependencies to abstract syntax. In: Proceedings of the NoDaLiDa 2017 Workshop on Universal Dependencies (UDW 2017), pp. 107–116 (2017)

# Can Large Language Model Surprisal Capture the Informativity Bias in Human Language Processing?

Hailin Hao(✉) and Muxuan He

University of Southern California, Los Angeles, CA 90089, USA
{hailinha,muxuanhe}@usc.edu

**Abstract.** Surprisal as a general theory of human sentence processing proposes that the processing difficulty of a word is inversely proportional to its contextual probability. However, it is not always the case that more predictable material is easier to process. Recent studies suggest that humans also have an informativity bias, whereby they prefer more informative or newsworthy contents. If a word is too predictable given the context, it provides virtually no new information and may cause extra processing difficulty. Despite many existing studies showing that the surprisal value obtained from large language models (LLMs) is a good fit for human reading time data, it has not been investigated whether they can also explain the informativity bias in humans, or they will only favor more probable information. We answer this question by comparing the results of several state-of-the-art autoregressive LLMs and two recent human reading experiments, and find that contra humans, LLMs do not show a bias towards more informative material. We discuss the implications of the current study on Natural Language Processing and human linguistic cognition.

**Keywords:** Surprisal · informativity bias · human sentence processing · cognitive modelling

## 1 Introduction

It has been a central question in the study of human language cognition what determines the processing difficulty of a word in context. A prominent theory in this field is the so-called Surprisal theory [1, 2], under which the processing difficulty of a word is proportional to its surprisal value, the negative of the log of the word's conditional probability given previous input, the context *c*, as formulated below:

(1) surprisal $\propto -\log P(w|c)$.

Surprisal theory has gained wide support in various empirical studies of human language processing, as indexed by longer reading time in self-paced reading or eye-tracking experiments or more negative N400 components in EEG experiments [e.g., [3–6]]. However, there have been deputes regarding how conditional probability is calculated [7]. Previous research mostly employed the *cloze task* [8], where human


R. Mehmood et al. (Eds.): DCAI 2023, LNNS 741, pp. 178–185, 2023.
https://doi.org/10.1007/978-3-031-38318-2_18

participants fill in the blanks after a sentence preamble, as illustrated in (2), and sometimes n-gram probabilities estimated from large corpora [e.g., 9]. Thanks to recent advances in Natural Language Processing, there are pre-trained autoregressive large language models (LLMs) that can be used to calculate a word's conditional probability given some context. For example, [10] show that the surprisal value estimated from various GPT models provides a better fit for human reading time data than cloze and n-gram probabilities. [11] show that the surprisal value estimated from GPT-3 [12], RoBERTa [13], and ALBERT [14] can better account for the magnitude of N400 than human judgments.

(2) This morning she drank coffee with __.

However, can surprisal value estimated from LLMs capture all of human language processing? Here we examine an oftentimes overlooked factor in language processing, the *informativity bias*. Language is used as a communicative tool by which speakers share, among other things, newsworthy and informative messages, i.e., content that is otherwise unpredictable to the comprehender. Thus, comprehenders may have an informativity bias, where they prefer contents that are more informative or newsworthy. Highly predictable or boring information may incur even higher processing efforts, as shown in [15, 16]. Can LLMs trained only on next-word prediction over large corpora capture the informativity bias in human language processing? One hypothesis is that these LMMs can only reflect what's in the training data. In that case, it may favor more probable information, instead of more newsworthy information. However, it may still be possible that the training data contain less occurrences of boring information. For example, because "the house has a bedroom" provides so little information, it may not be produced at all. By contrast, "the house has an office" may be more likely to be produced. Another hypothesis is that these LLMs may have emergent functional linguistic competence [17]. However, this is not very likely considering recent evidence suggesting that although LLMs succeed at various formal linguistic tasks (e.g., syntactic and semantic understanding), they fail miserably in linguistic tasks that require world knowledge and communicative skills such as theory of mind reasoning, commonsense reasoning, and pragmatic inferences [18]. The bias towards more informative information essentially does not belong to formal linguistic competence, but to functional linguistic competence. We will discuss this further in later sections.

We will answer our research question by performing experiments on different versions of GPT-2 and -3 [12, 19], with varying numbers of parameters and different sizes of training data, as shown in Table 1. GPT-2 and -3, augmented with state-of-the-art Transformer architecture [20], have outperformed previous language models on many NLP tasks [21, 22], can capture various syntactic and semantic knowledge [23–25], and may better model human language cognition [11, 26]. We will reanalyze the behavioral experiments with surprisal value as predictors derived from the conditional probability estimated by these language models and compare the reanalysis with the original analysis. Section 2 describes the two behavioral experiments in detail. Simulation results from the language models are presented in Sect. 3. Section 4 concludes.

**Table 1.** Models used and their parameters of size of the training corpora

| Model | Parameters | Corpus Size |
|---|---|---|
| GPT-2 Base | 117M | 8B |
| GPT-2 Large | 774M | 8B |
| GPT3-Davinci | 175B | 300B |

## 2 Psycholinguistic Background

We will test LLMs using adapted stimuli from two behavioral experiments on language comprehension: Experiment 1 from [15] Rohde et al. (2021)] in which participants showed a preference for higher informativity when they were cued to expect informative content, and Experiment 1 from [16; He & Kaiser (2023)], which showed that a 'just right' amount of information was preferred over higher as well as lower levels of informativity. Both experiments used the self-paced reading methodology, and the dependent measure is reading time (RT). Longer RTs are usually linked with more processing efforts.

**Exp1, Rohde et al. (2021).** This reading experiment investigated how the way a protagonist is categorized can affect the processing cost of typical and atypical actions with a 2 x 2 design, as illustrated in (3). The protagonist has either a boring persona that always acts in a predictable way or a surprising persona that never acts the way people normally expect. In addition, the protagonist uses either a typical or an atypical instrument to perform an action.

(3) a. [boring protagonist / action-typical instrument] My cousin Mary is a *boring* person who always does things the way you'd expect. For instance, in order to dig a hole, she was using a *shovel* yesterday in the afternoon.
b. [surprising protagonist / action-typical instrument] My cousin Mary is a *surprising* person who never does things the way you'd expect. For instance, in order to dig a hole, she was using a *shovel* yesterday in the afternoon.
c. [boring protagonist / action-atypical instrument] My cousin Mary is a *boring* person who always does things the way you'd expect. For instance, in order to chop some carrots, she was using a *shovel* yesterday in the afternoon.
d. [surprising protagonist / action-atypical instrument] My cousin Mary is a *surprising* person who never does things the way you'd expect. For instance, in order to chop some carrots, she was using a *shovel* yesterday in the afternoon.

There are two reasons why the informativity bias of the comprehenders might emerge from these materials. First, they use the persona of the protagonist as contextual cue to gauge how newsworthy the messages might be. If it is a surprising subject, they expect to get surprising information in the following context. Second, they are given a context of an action and guided to make predictions about an action-typical or -atypical instrument, in accordance with the level of informativity they expect.

RT data showed a significant interaction effect, whereby the surprising persona altered the processing difficulty that is often associated with the action-atypical condition. The informativity bias towards newsworthy messages in the surprising persona conditions facilitated what usually induces more processing efforts in language processing.

However, it's unclear whether the design in [15] does not reflect a genuine expectation for newsworthy information, since participants were prompted with a surprising protagonist to expect unusual information. [16] tested whether the informativity bias can arise in a neutral context.

**Exp2, He & Kaiser (2023).** The article investigated how the likelihood of a *part-whole* relationship [27] modulates its processing effort in a neutral context. The experiment materials are two-sentence sequences, as in (4). The first sentence mentions a location (i.e., *whole*), and the second makes a statement about that location containing something (i.e., *part*) at varying likelihood levels.

(4) a. [high likelihood] Megan ordered food in a *cafeteria.* The cafeteria has a <u>counter</u>, Megan said.
b. [medium likelihood] Megan ordered food in a *cafeteria.* The cafeteria has a <u>clock</u>, Megan said.
c. [low likelihood] Megan ordered food in a *cafeteria.* The cafeteria has a <u>bookshelf</u>, Megan said.

If language processing is indeed modulated by informativity bias, even when the context is neutral, high likelihood will be dispreferred because it adds little new information to our common ground knowledge (i.e., world knowledge in their case). The reading time data confirmed the effect of informativity bias on the processing of high likelihood items. In addition to that, the medium likelihood condition was found to be the easiest to process, whereas high and low likelihood conditions were not significantly different from each other. These results indicate that, for humans, informativity bias leans towards more newsworthy information (the medium likelihood condition is easier to process than the low likelihood condition), but it also has a boundary (the high likelihood condition is not easier than the medium likelihood condition).

## 3 Simulation Results

### 3.1 Exp1, Rohde et al. (2021)

In replicating this experiment, we ask whether the state-of-art LLMs adapt language processing to the contextualized expectation for informativity. If so, we expect to see, first, in the boring persona, no significant difference in the surprisal values between action-typical and -Atypical instruments, and second, in the surprising persona, higher surprisal values of action-typical instruments than those of action-typical instruments. We semi-automatically generated 5,200 items in the form of (3).

Surprisal data at the critical word region (i.e., the instrument) were analyzed with *lmer* [28]. We included protagonist persona and action-typicality as the predictors, and random effects of items. The results were plotted using the ggplot2 package in R.

**GPT-2 and GPT-2L.** Results from these two LLMs are similar to each other (see Fig. 1). The action-atypical instruments were found to constantly induce significantly higher surprisal than the action-typical ones ($p$'s $< 0.001$), regardless of the specific protagonist persona. It is indicated that these two models showed little sensitivity to informativity bias.

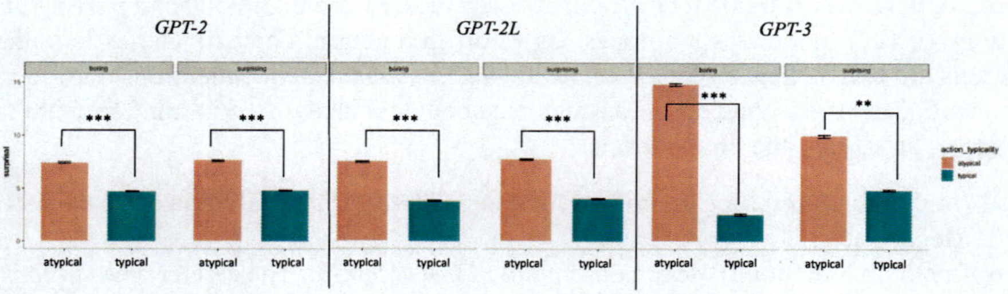


**Fig. 1.** Surprisal values of the critical word region in LLMs

**GPT-3.** The action-atypical instruments were still found to constantly induce significantly higher surprisal than the action-typical ones ($p$'s $< 0.01$), regardless of the specific protagonist persona (see Fig. 1). Nonetheless, results from this model also showed interesting signs that are different from the previous models. First, the surprisal value of the action-atypical condition is somewhat higher. Second, the surprisal value of the action-atypical condition increased by a greater degree in the boring persona than in the surprising persona. Third, the surprisal value of the action-typical condition was also affected by the protagonist persona, higher with a surprising persona than with a boring persona. These signs presented a reasonable contextual effect where atypicality became more surprising when one expects to receive predictable information and typicality became less predictable when one expects surprising news. However, we did not see an effect of informativity bias that is strong enough to alter the surprisal value of the action-atypical condition for the surprising persona.

### 3.2 Exp2, He & Kaiser (2023)

In replicating this other behavioral experiment, we wonder whether the state-of-art LLMs expect linguistic information in a 'just right' way. If so, we expect the surprisal values in the high and the low likelihood conditions to be higher than the medium likelihood level. We semi-automatically generated 1,800 items in the form of (4).

Surprisal data at the critical word region (i.e., *part*) were analyzed with *lmer*. We included likelihood as a three-level categorical predictor, and random effects of items. We contrast coded the categorical predictor in two ways that are most relevant to our research questions. The first contrast was coded with (high -0.25, low -0.25, and medium 0.5) and compared the medium group with the other two groups. The second contrast was coded with (high 0.5, low -0.5, medium 0) and compared the two poles of likelihood. We mentioned earlier that the medium is predicted to yield the lowest surprisal value, but we do not make specific predictions about the relationship between the other two levels of likelihood. The results were also plotted using the ggplot2 package in R.

**GPT-2 and GPT-2L.** Results from these two LLMs, which are similar to each other (see Fig. 2), did not show meaningful difference between the medium likelihood level and the other two levels ($p_{gpt\text{-}2} = 0.116, p_{gpt\text{-}2L} = 0.102$). Instead, it showed a significant negative correlation between likelihood and surprisal ($p$'s $< 0.001$). It indicates that these two models consistently favor higher-likelihood events in language processing and failed to capture the informativity bias or its limit.

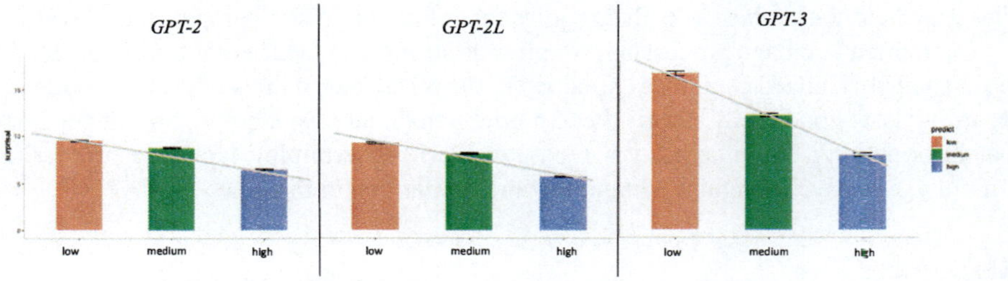


**Fig. 2.** Surprisal values of the critical word region in LLMs

**GPT-3.** The results still presented a significant negative correlation between likelihood and surprisal ($p < 0.001$), which in fact is more visually evident now. Likewise, we did not detect any trace of informativity bias, let alone the sensitivity to its boundary, in this language model.

### 3.3 Model Comparison

In order to quantitatively assess which LLM offers the best fit for human behavioral data, we additionally fit three separate models with the RT data from Exp1 in He & Kaiser (2023) as dependent variables and the surprisal value obtained from the three LLMs as predictor variables. The model with GPT-2 and GPT-2L surprisal values as predictors have very similar AICs (GPT-2: 6955.3; GPT-2 L: 6955.4). The model with GPT-3 surprisal as predictors have slight lower AICs (6953.5). GPT-3 may offer a better fit but it is statistically not significant.

## 4 Discussion

The answer to our research question concerning whether LLMs can capture the informativity bias in human language processing is clearly no. Across two experiments and three state-of-the-art LLMs, no human-like pattern is detected. For the replication of Exp1 of [15], LLMs consistently favor action-typical instruments over action-atypical ones, regardless of the specific protagonist persona. For the replication of Exp1 of [16], LLMs consistently favor more probable information. However, for both replications, GPT-3 show slightly better performance, despite not being able to capture the informativity bias. We discuss the implications of the current study on NLP and human linguistic cognition.

For NLP, we argue that even for the state-of-the-art unsupervised LLMs trained on billions of tokens, there is no evidence for emergent functional linguistic competence, corroborating [18]. It seems that LLMs can only reflect what's in the training data and thus favor more probable information. The lack of informativity bias may hinder the performance of LLMs in downstream NLP tasks, especially ones where newsworthy or interesting information is expected, such as story generation and creative writing. We argue that one way to solve this problem is to construct training corpora containing carefully selected data, with the help of both engineers and linguists with domain knowledge. This may help with other tasks that require functional linguistic competence as well.

For human language processing, we provide stronger evidence that the effects found in [15] and [16] indeed cannot be explained by surprisal, despite lots of existing evidence regarding how good LLM surprisal can model human language processing. It suggests that informativity bias may play a separate role in determining language processing difficulty, and psycholinguists should pay more attention to this bias.

## References

1. Hale, J.T.: A probabilistic Earley parser as a psycholinguistic model. In: Proceedings of the second meeting of the north American chapter of the association for computational linguistics and language technologies, pp. 1–8 (2001)
2. Levy, R.: Expectation-based syntactic comprehension. Cognition **106**(3), 1126–1177 (2008)
3. Jäger, L., Chen, Z., Li, Q., Lin, C.-J.C., Vasishth, S.: The subject-relative advantage in Chinese: Evidence for expectation-based processing. J. Mem. Lang. **79–80**, 97–120 (2015)
4. Frank, S.L., Otten, L.J., Galli, G., Vigliocco, G.: Word surprisal predicts N400 amplitude during reading. Proceedings of the 51st Annual Meeting of the Association for Computational Linguistics, pp. 878–883 (2013)
5. Aurnhammer, C., Delogu, F., Schulz, M., Brouwer, H., Crocker, M.W.: Retrieval (N400) and integration (P600) in expectation-based comprehension. PLoS ONE **16**(9), e0257430 (2021)
6. Delogu, F., Crocker, M.W., Drenhaus, H.: Teasing apart coercion and surprisal: Evidence from eye-movements and Erps. Cognition **161**, 46–59 (2017)
7. Brothers, T., Kuperberg, G.R.: Word predictability effects are linear, not logarithmic: Implications for probabilistic models of sentence comprehension. J. Mem. Lang. **116**, 104174 (2021)
8. Taylor, W.L.: Cloze procedure: A new tool for measuring readability. J. Q. **30**(4), 415–433 (1953)
9. Smith, N.J., Levy, R.: The effect of word predictability on reading time is logarithmic. Cognition **128**(3), 302–319 (2013)
10. Shain, C., Meister, C., Pimentel, T., Cotterell, R., Levy, R.: Large-scale evidence for logarithmic effects of word predictability on reading time. PsyArXiv preprint (2022)
11. Michaelov, J.A., Coulson, S., Bergen, B.K.: So Cloze yet so Far: N400 Amplitude is Better Predicted by Distributional Information than Human Predictability Judgments. arXiv:2109.01226 [Cs, Math] (2021)

12. Brown, T., et al.: Language models are few-shot learners. In: Larochelle, H., Ranzato, M., Hadsell, R., Balcan, M.F., Lin, H. (eds.) Advances in Neural Information Processing Systems, vol. 33, pp. 1877–1901. Curran Associates, Inc. (2020)
13. Liu, Y., et al.: RoBERTa: A robustly optimized BERT pretraining approach. arXiv abs/1907.11692 (2019)
14. Lan, Z., Chen, M., Goodman, S., Gimpel, K., Sharma, P., Soricut, R.: ALBERT: A lite BERTt for self-supervised learning of language representations. In: Proceedings of ICLR (2019)
15. Rohde, H., Futrell, R., Lucas, C.G.: What's new? A comprehension bias in favor of informativity. Cognition **209**, 104491 (2021)
16. He, M., Kaiser, E.: Not too surprising, not too boring? Goldilocks effects of likelihood in online processing. Poster at the 36th Annual Conference on Human Sentence Processing (2023)
17. Wei, J., et al.: Emergent abilities of large language models. Transactions on Machine Learning Research (2022)
18. Mahowald, K., Ivanova, A.A., Blank, I.A., Kanwisher, N., Tenenbaum, J.B., Fedorenko, E.: Dissociating language and thought in large language models: A cognitive perspective. arXiv preprint arXiv:2301.06627 (2023)
19. Radford, A., Narasimhan, K., Salimans, T., Sutskever, I., et al.: Improving language understanding by generative pre-training. OpenAI (2018)
20. Vaswani, A., et al.: Attention is all you need. In: Proceedings of the 31st Conference on Neural Information Processing Systems, pp. 6000–6010 (2017)
21. Chaitra, H., Shrikumar, P.: Unsupervised Paraphrase Generation using Pre-trained Language Models. arXiv preprint arXiv:2006.05477 (2020)
22. Zheng, X., Zhang, C., Woodland, P. C.: Adapting GPT, GPT-2 and BERT language models for speech recognition. ASRU, pp. 1–13 (2021)
23. Warstadt, A., et al.: BLiMP: The benchmark of linguistic minimal pairs for English. TACL, pp. 1–13 (2020)
24. Wilcox, E., Gauthier, J., Hu, J., Qian, P., Levy, R.P.: Learning syntactic structures from string input. In: Lappin, S. (ed.), Algebraic Systems and the Representation of Linguistic Knowledge, 2(5), pp. 1–13 (2022)
25. Hao, H.: Evaluating Transformers' Sensitivity to Syntactic Embedding Depth. Poster at the 36th Annual Conference on Human Sentence Processing (HSP 2023), Pittsburg, pp. 1–13
26. Merkx, D., Frank, S.L.: Human Sentence Processing: Recurrence or Attention? In: Proceedings of the 2021 Workshop on Cognitive Modeling and Computational Linguistics, Association for Computational Linguistics, Mexico City, Mexico, pp. 1–11 (2021)
27. Bhakthavatsalam, S., Richardson, K., Tandon, N., Clark, P.: Do dogs have whiskers? a new knowledge base of haspart relations. arXiv preprint arXiv:2006.07510, pp. 1–13 (2020)
28. Bates, D., Mächler, M., Bolker, B., Walker, S.: Fitting linear mixed-effects models using lme4. J. Stat. Softw. **67**(1), 1–48 (2015)

# A Typometrical Study of Greenberg's Linguistic Universal 1

Antoni Brosa-Rodríguez(✉) and M. Dolores Jiménez-López

GRLMC-Research Group in Mathematical Linguistics,
Universitat Rovira i Virgili, Tarragona, Spain
antoni.brosa@urv.cat, mariadolores.jimenez@urv.cat

**Abstract.** An interdisciplinary study that combines linguistic typology, corpus linguistics, and computational linguistics to quantitatively review Greenberg's Universal 1 is presented. By proposing new metrics and formalizations, we aim to provide a more accurate representation of language usage to solve the problem of basic word order in natural languages. A typological characterization of 143 languages is proposed. We perform a typometrical study using Universal Dependencies in order to quantitative evaluate new metrics. Our results show universal trends and can serve to categorize languages and to test universals with languages that have never been quantitatively checked.

**Keywords:** Linguistic Typology · Universal Dependencies · Typometrics · Greenberg's universals · Basic Word Order

## 1 Introduction

Language universals have been the subject of intense controversy throughout the history of linguistics. Their nature and existence have been questioned, and its analysis has been approached from many different perspectives. Although linguists have always been interested in discovering characteristics shared by languages, it was not until Greenberg's contribution that the research topic gained popularity and depth. A decade later, despite the interest aroused by Greenberg's findings, the impossibility of improving on these results caused the study of universals to lose interest and usefulness. This object of study became relevant again a decade later, thanks to the innovations in sampling proposed by linguistic typology. In recent years, the boost in Natural Language Processing has rescued linguistic universals from oblivion.

Today's technology and the computational power of many of the tools at our disposal allow us to use a huge set of uniformly annotated data for various languages. This allows us to discover new typological universals and of course offers the possibility to test those that were hypothesized in the past.

In this paper, we combine linguistic typology, corpus linguistics, and computational linguistics to quantitatively review the Universal 1, based on word order and proposed by Greenberg [9]. By proposing new metrics and formalizations,


R. Mehmood et al. (Eds.): DCAI 2023, LNNS 741, pp. 186–196, 2023.
https://doi.org/10.1007/978-3-031-38318-2_19

we aim to provide a more accurate representation of language usage, in order to solve the problem regarding basic word order. Our main objectives are to offer multipolar formalizations of basic word order to apply in a computational linguistic experiment; to compare results with other proposals in computational linguistics and canonical data of grammars; and to present Greenberg's Universal 1 from quantitative and real usage data and to offer it in different ways.

Joseph Greenberg [9] revolutioned linguistic typology by prioritizing word order over morphology. His proposal is based on the formulation of 45 different linguistic universals, from a sample of 30 languages.

The first of these universals is one of the most studied in linguistic typology by grammarians [12] [4, 18], who have tried to approach the phenomenon through slight modifications. Greenberg's Universal 1 literally states:

> Universal 1: In declarative sentences with nominal subject and object, the dominant order is almost always one in which the subject precedes the object. [9].

Beyond the formulated universal, what is truly complex in typology is the determination of what "dominant word order" means.

To make a faithful approach to Greenberg's universal, linguists must take several factors into account:

- Only declarative sentences should be analyzed, excluding other modalities like interrogatives or exclamatives.
- Greenberg focuses on two syntactic elements, the subject and object, and requires them to be nominal and to appear both.
- Certain orders dominate others, and the dominant one always has the S before the O. This leaves three possible orders (SVO, SOV, VSO) and three unlikely ones (VOS, OVS, OSV), with the addition of the NDO tag[1] .
- Greenberg notes that this universal is statistical and some languages may not follow it.

The paper is organized as follows. Firstly, we present a new formalization of what Greenberg understood as dominant basic word order. On this basis, we propose the typological characterization of 143 languages of the world, based on Universal Dependencies. Some of them have been analyzed for the first time in this paper. In other languages we have changed the linguistic type traditionally assigned to them. Secondly, we check in a quantitative way, what is the fulfillment of such universal if we rely on the multiple real texts of these languages. Finally, we evaluate new metrics that emerge from a quantitative study in a typometrics way.

[1] S = Subject; V = Verb O = Object, NDO = Non-Dominant Order.

## 2 State of the Art

Many linguists have attempted to determine the dominant basic word order in various languages worldwide since Greenberg's publication. The World Atlas of Linguistic Structures [5], particularly Chap. 81, summarizes this effort using grammar data, despite its limitations. The findings reveal that 83% of languages follow the universal order of SOV, SVO, or VSO, while 17% do not.

Computational linguistics has attempted to quantify the phenomenon of basic word order by analyzing real text frequencies. Liu's study [15] is a pioneering effort, but it has three main limitations. Firstly, the study main goal is based on an alternative to the object-subject distinction (head initiality vs. head finality) making it difficult to distinguish subjects and objects. Secondly, the researcher's classification of subjects and objects follows Dryer's style, which separates the analysis of SV-VS and OV-VO opposition, so it doesn't fit Greenberg's proposal. Thirdly, the study is limited to only 20 languages.

Other studies, such as Levshina's [13], focus on linguistic entropy in Universal Dependencies rather than strictly Greenberg's basic word order.

Recent proposals, such as those by Choi et al. [1,2], have a similar philosophy to our work. There are also more accurate proposals, such as Gerdes et al. [7], but they do not focus on Greenberg's Universals on basic word order, but in universals such as the position of adjectives modifying nouns.

## 3 Methodology

### 3.1 Dataset

In our approach, the language information is extracted from Universal Dependencies [17]. The main reason for this is the wide representation of languages in this source and, in addition, the desire to propose universal labels to facilitate comparison in linguistic typology and a better extrapolation of results.

The sample of languages we analyzed is a convenience sample. That is, given the existing difficulties and limitations in the availability of languages (and the lack of representation of some of them in a quantitative perspective) we decided to analyze all the available ones. The ultimate goal is that, once the number of languages available in UD grows sufficiently, the results will already be 100% reliable, based on the proposal of Guzmán and Becker [16], where it is mathematically demonstrated how no typological balancing is necessary if the sample is large enough.

The typology of the corpora in Universal Dependencies is varied, as is well known. This means that, once again, there may be some bias if a corpus is specific to a typology (legal, literary, etc.). In the case of languages that enjoy different types of corpora, we have chosen to offer a merging and weighting of the results, which allows us to neutralize this effect. A good summary of the limitations of a typological approach from computational and corpus linguistics can be found in [14].

Due to technical limitations, given the large volume of data managed in these corpora and in order to homogenize the results between larger corpora with millions of sentences and corpora with a few hundred, we work with a maximum input of 1000 sentences in each treebank. We work with 242 different treebanks, which means that the proposed analysis is based on slightly less than 242,000 sentences. We have selected the first 1000 sentences from each treebank for the analysis, to facilitate the falsifiability of our data and making it more easily retrievable. The tests we have performed comparing this selection of 1000 sentences with the whole corpus have not revealed any significant difference in the results.

Although 242 treebanks are analyzed, the languages taken into account for this analysis, prior to filtering, are 143 different, and can be consulted in the corresponding version of Universal Dependencies used, which is 2.11 [17].

### 3.2 Annotation Scheme and Tool

One of the main problems with the annotation scheme of Universal Dependencies are inconsistencies from a strictly syntactic point of view, such as the distinction between clausal and nominal subjects or the fact that the adposition is dependent on "mark" dependency. Most of these problems have been pointed out by linguists such as [3] or [6]. In order to solve those problems, we work with the conversion to the SUD (Surface Universal Dependencies) version, which has been devised with the main objective of working on syntax and typology in a more coherent and respectful way with the different phenomena. SUD are based on a strictly syntactic criterion for assigning dependency relations, while UD usually assigns content word as heads in many cases. This makes the comparison with Greenberg much more direct and less biased, since these universals have been formulated by fundamentally syntactic criteria. A deeper comparison of the differences and what is involved can be found in [6].

In order to work with the data available in Universal Dependencies, the platform itself recommends five different tools: SETS Treebank search, PML Tree Query, Kontext, Grew-Match and INESS.

After a detailed analysis of all of them that cannot be reproduced here due to lack of space, we have decided to work with Grew-Match [11]. It is defined as a Graph Rewriting tool dedicated to applications in Natural Language Processing, that basically allows to query any information annotated in UD's treebanks. In summary, we would like to highlight the multiple visualization and rewriting possibilities it allows, as well as the possibility of checking much more advanced metrics, such as the distance between dependents or the possibility of creating clusters of occurrences.

### 3.3 Formalization

In order to be able to work computationally with the typological data, we must first formalize the different linguistic elements we wish to analyze. The choice of one formalization or another can significantly change the result obtained.

Starting from Grew's own syntax, we propose the formalization to capture the order equivalent to SVO in Greenberg, for example, in formalization (1):

$$\boxed{\begin{array}{l} \%SVO \\ pattern\{V[upos = VERB]; V - [subj]- > S; S[upos = PROPN|NOUN]; V - [comp : obj]- > O; O[upos = PROPN|NOUN]; V >> S; S << O; V << O\} \\ without\{V - [punct]- > P; P[lemma = \text{“?”}|\text{“!”}]\} \end{array}} \quad (1)$$

The above formalization captures the following:

- In the first line, always after the symbol %, it is placed the info or heading of the formalization, useful for identifying it.
- We ask for a *pattern*, that will show the information inside the { and }.
- ; is used to separate the different elements.
- An element named $V$ is activated. It corresponds to a word which its universal part of speech is a verb [$upos = VERB$].
- The way of marking the whole dependency is HEAD ($V$), then −, plus [, plus "dependency relation tag" ($subj$), plus ], plus − >, ending with the DEPENENT ($S$). That is, we have an arrow (a dependency), with a HEAD in first position, a box in the middle marking the kind of dependency, and a DEPENDENT in the last position.
- The element named $S$ is activated. It corresponds to a word which its universal part of speech is a noun or a proper noun [$upos = PROPN|NOUN$]. | is used to add different elements at the same level.
- Another dependency in the structure we want to find is asked, which has the same element $V$ as a HEAD, the type (marked inside the box) is $comp : obj$ in SUD terms (equivalent to direct objects), and an element $O$ as a DEPENDENT.
- An element named $O$ is activated. It corresponds to a word whose universal part-of-speech is a noun (common or proper) [$upos = PROPN|NOUN$].
- We ask the $V$ element to appear after (>>) the $S$ element, the $S$ element before (<<) the $O$ element, and the $V$ element before (<<) the $O$ element. Resulting to a SVO order.
- We query at the same time a *without*, which is equal to a negative pattern or filtering. That is, from all the occurrences we find, all the cases inside the { } are discarded. In this case we don't want to take into account any sentence which has also the dependency of a $V$ (verb) being the HEAD of a punctuation dependency *punct* of a $P$ element as DEPENDENT. Then, we asign the $P$ element we named as any element that agrees with the lemma ? or !.

Subsequently, we repeat this same formalization with the different orders: SOV, VSO, VOS, OVS, OSV. That is, we modify the order restrictions we mention by the use of << or >>.

It should be made explicit that the main difference of this formalization with the one proposed by [1,2] is that they decide to include in the part of speech proper of the subject or object the pronouns. We claim that this approach is not respectful of Greenberg's universal, which explicitly mentions "nominal" subjects and objects. We consider that the confusion may be due to the fact that in Universal Dependencies there is a differentiation between the labels *csubj* and *nsubj*, corresponding to clausal subject and nominal subject (where pronouns are included). However, we consider that Greenberg's "nominal" opposition is precisely in front of pronominal.

Since we want to see the incidence of the above-mentioned decisions, we propose formalizations (which we do not present here for lack of space) where this same universal is captured by changing parameters: 1) taking only pronouns into account; 2) taking into account nouns and pronouns; 3) taking into account any possible part of speech; 4) eliminating the restriction of sentence type, etc. In addition, we also propose formalizations in which subject and object dependencies are analyzed independently and not jointly.

All this will allow us to assess whether a slight modification of Greenberg's universal can lead us to more interesting results or, on the contrary, to confirm that Greenberg's proposal is the most useful for linguistic typology.

### 3.4 Calculation

A minimum filter of 25 occurrences is established, which allows us not to give a representative data entity to a data that can be widely modifiable with more occurrences because it is too low.

In order to translate the label "dominant" into a quantitative and objective criterion, we use the proposal contained in Gries [10], with extensive validation in linguistics, in which it is determined that an option is dominant if it is the first in a frequency ranking and, in addition, it is twice or more frequent than the element that appears in second place in that ranking.

We explore a new approach based on the typometrics philosophy. Typometrics, a subfield of quantitative typology, is an open field of the study of the distribution of languages in a distributional scatter diagram based on empirical measures on corpora [8]. Our typometric approach consists of a purely quantitative calculation, individually for each language, in which the result is satisfactory if there is always a higher weight of one variable over another, which is what can be easily and visually seen in the original triangles in Typometrics [8].

## 4 Results

We have obtained typological classifications in relation to the basic word order à la Greenberg that are different or in more languages than those found in previous studies such as [2,5].

Table 1 presents all the languages in which we have detected a different data from the other two sources cited. The capitalized names correspond to data extracted by multiple treebank weighting.

**Table 1.** Analyzed languages that are different or new in comparison with Choi's et al. [2] and Wals' Basic Word Order [5]

| Language | OURS | Choi's | WALS' | Language | OURS | Choi's | WALS' |
|---|---|---|---|---|---|---|---|
| afrikaans | SVO | SOV | no data | kiche | SVO | no data | no data |
| AKKADIAN | SOV | NDO | no data | KOMI ZYRIAN | SVO | no data | SVO |
| akuntsu | SOV | no data | no data | kurmanji | SOV | no data | no data |
| amharic | SOV | NDO | SOV | LATIN | NDO | NDO | no data |
| ANCIENT GREEK | NDO | NDO | no data | maltese | SVO | SVO | no data |
| ancient hebrew | VSO | no data | no data | marathi | SOV | no data | SOV |
| ARABIC | VSO | NDO | VSO | moksha | SVO | no data | no data |
| belarussian | SVO | SVO | NDO | north sami | SVO | SVO | no data |
| buryat | SOV | no data | SOV | OLD EAST SLAVIC | NDO | no data | no data |
| Classical Chinese | SVO | SVO | no data | old french | SVO | NDO | no data |
| DUTCH | SVO | NDO | NDO | SANSKRIT | SOV | SOV | no data |
| FAROESE | SVO | SVO | no data | slovak | SVO | SVO | no data |
| GALICIAN | SVO | SVO | no data | TAMIL | SOV | no data | SOV |
| GERMAN | SVO | NDO | NDO | turkish_german | SOV | NDO | no data |
| gheg | SVO | no data | no data | upper sorbian | SVO | no data | no data |
| gothic | SVO | SVO | no data | western armenian | NDO | no data | SOV |
| greek | SVO | SVO | NDO | western puebla nahuatl | SVO | no data | no data |
| hindi_english | SVO | NDO | no data | xibe | SOV | no data | no data |
| ICELANDIC | SVO | NDO | SVO | yakut | SOV | no data | SOV |
| karelian | SVO | no data | no data | yoruba | SVO | no data | SVO |
| kazakh | SOV | SOV | no data | yupik | NDO | no data | NDO |

Going back to the universal itself and its quest to capture that the S preceding the V is much more frequent than following it in the basic orders, we present a classification of occurrences in Typometrics [8] style in Fig. 1. However, we decided to present it in a plot that, to our knowledge, has not been performed so far, adding also a comparison with subjects without any POS restriction. In Fig. 1, we place in the x-Axis the percentage of occurrences of nouns being the S of a V (subjects after verb) in all possible languages and in the y-Axis the percentage of occurrences of any part of speech being the S of a V (subjects after verb) in all possible languages. Thus, if a language is in the 0 in both axis means that is strictly SV (in nouns and any POS), while if, hypothetically, we would find a language in the 100 in the y-Axis and 0 in the x-Axis, would mean that it always (100% of occurrences) puts the subject after the verb, while it never (0% of ocurrences) puts the nominal subject after the verb or, in other words, it always puts the nominal subject before the verb. Languages that share same family also share same symbol and colour.

Finally, the data corresponding to the confirmation of the universal are shown in Table 2. *U1ori* corresponds to the percentage of languages that confirm the universal proposed by Greenberg in his original terms and by using the concept of "dominance", as he states. *U1pn* refers to exactly the same but checking the

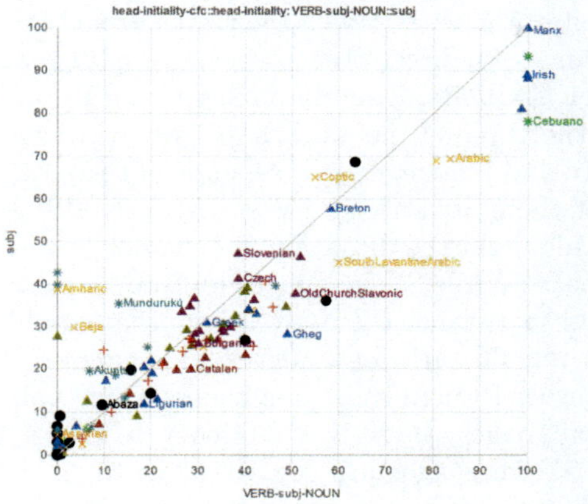


**Fig. 1.** Distribution of languages according to the position of subjects and verbs

occurrences of pronouns instead of nouns. *U1pnN* refers to exactly the same but checking the occurrences of pronouns and nouns at the same time. *U1any* refers to the same but taking into account any part of speech. The remaining columns, in which "q" is added refer respectively to the same values announced above, but without taking into account the concept of dominance, but simply looking at whether there is a higher proportion of the element highlighted by Greenberg.

**Table 2.** Percentages of certification of the U1 according to different parameters

| % | U1ori | U1pn | U1pnN | U1any | qU1 | qU1pn | qU1pnN | qU1any |
|---|---|---|---|---|---|---|---|---|
| True | 90.80 | 72.84 | 79.31 | 83.91 | 100.00 | 95.06 | 98.85 | 98.85 |
| Wrong | 9.20 | 27.16 | 20.69 | 16.09 | 0.00 | 4.94 | 1.15 | 1.15 |

## 5 Discussion

Our test has shown that Greenberg's Universal 1 it is largely fulfilled (more than 90%) in the bast majority of texts from multiple languages of the world. It is also interesting to note that his strict proposal is the one that produces the most satisfactory results from a categorical perspective. Considering Greenberg's Universal 1 in a typometrics way, all languages exhibit a basic word order greater in the number of subjects in front of the verb than behind it.

Thanks to the plot created, we can offer new typological information. The great majority of languages tend to be located in the left part (SV). It is also striking to see how there is a gradual presence of languages that is interrupted when it reaches 50%. This confirms the tendency of languages where the subject behaves better in front of the verb. When the restriction that sentences be enunciative or the appearance of a direct object is removed, languages tend to behave with relatively greater flexibility. Therefore, Greenberg's proposal, based on the basic word order of the three elements, has greater predictive power for other features. Moreover, we find languages at both ends of the plot, which means that we have attested that there are strictly SV and strictly VS languages.

It is important to highlight the deviation of behavior of VS languages depending on the part of speech. While SV languages tend to be close to the plot diagonal, VS languages tend to distance themselves from that equilibrium. Thus, what tends to be higher is the VS ratio in nouns, while there are considerably more instances of SV with other POS in those languages, which points in the direction that, even in those languages, there is a tendency to place the subject before the verb. Therefore, SV languages are consistent in their order either in their basic proposition or in behavior with any part of speech. In contrast, languages traditionally referred to as VS show greater variation and flexibility.

Our methodology has allowed to obtain new and more precise individual data regarding the dominant basic word order of each language. We now have this linguistic type information in languages not previously analyzed such as ancient Hebrew (VSO), Buryat (SOV), Gheg (SVO), Karelian (SVO), Old East Slavic (NDO) or Xibe (SOV). We can offer a closer approach to languages that have been traditionally considered as NDO. Although these languages can be considered NDO when general criteria can be met, if we deal with the concept of dominant basic word order we must be much stricter and more selective, which allows us to obtain more representative and explanatory data. This characterization allows, subsequently, to apply universals in a much more satisfactory way. We propose to try to use these new linguistic types in further analysis, something that we believe can help to perform better results in linguistic typology.

## 6 Conclusions

In this paper we have approached Greenberg's Universal 1 from a typometric perspective and by using Universal Dependencies. After our testing, we can determine that the proposed universal in Greenberg's terms is the most accurate and that all of its constraints make very useful sense. Any other choice would be less productive. A reformulation of the universals in a typometrics style allows for more flexibility and less reliance on part-of-speech, which often hinders work in typology. Our results are a good indicator of universal trends and can serve perfectly well to categorize languages and test universals with languages that have never been quantitatively checked.

The work that has been presented in this paper shows that it is possible to obtain a more productive quantitative representation when working with basic word order and with universals such as Universal 1. Our work implies a novel approach since it involves refined and new data which provide a new typological representation for some languages.

Since the exact data for a particular phenomenon may be slightly modified by changing the sampling, one of research lines for future work is to adapt this quantitative result to a categorical proposal based on a fuzzy logic model, which will allow us to present the data with a greater generality (with more confidence and representativeness) and at the same time with enough flexibility typical of this discipline. Our final goal is the use of the quantitative data obtained as a source for a trichotomous mathematical model of fuzzy logic in which we will be able to certify if a universal is fulfilled or not in a more representative way.

## References

1. Choi, H.S., Guillaume, B., Fort, K.: Corpus-based language universals analysis using universal dependencies. In: Proceedings of the Second Workshop on Quantitative Syntax (Quasy, SyntaxFest 2021), pp. 1–15 (2021). https://aclanthology.org/2021.quasy-1.3
2. Choi, H.S., Guillaume, B., Fort, K., Perrier, G.: Investigating dominant word order on universal dependencies with graph rewriting. In: Proceedings of the International Conference on Recent Advances in Natural Language Processing (RANLP 2021), pp. 281–290. Incoma Ltd (2021). https://doi.org/10.26615/978-954-452-072-4_033
3. Croft, W., Nordquist, D., Looney, K., Regan, M.: Linguistic typology meets universal dependencies. In: International Workshop on Treebanks and Linguistic Theories, pp. 63–75 (2017)
4. Dryer, M.S.: The Greenbergian word order correlations. Language **68**, 81–138 (1992)
5. Dryer, M.S., Haspelmath, M.: The world atlas of language structures online (2013). http://wals.info
6. Gerdes, K., Guillaume, B., Kahane, S., Perrier, G.: Starting a new treebank? GO SUD! theoretical and practical benefits of the surface-syntactic distributional approach. In: SyntaxFest Depling 2021 - 6th International Conference on Dependency Linguistics, pp. 35–46 (2022). https://hal.inria.fr/hal-03509136
7. Gerdes, K., Kahane, S., Chen, X.: Rediscovering Greenberg's word order universals in UD. In: Proceedings of the Third Workshop on Universal Dependencies (UDW, SyntaxFest 2019), pp. 124–131 (2019). https://doi.org/10.18653/v1/W19-8015
8. Gerdes, K., Kahane, S., Chen, X.: Typometrics: from implicational to quantitative universals in word order typology. Glossa **6** (2021). https://doi.org/10.5334/gjgl.764
9. Greenberg, J.H.: Universals of Language. The MIT Press, Cambridge (1963)
10. Gries, S.T.: Statistics for Linguistics with R: A Practical Introduction. Mouton De Gruyter (2013)
11. Guillaume, B.: Graph matching and graph rewriting: grew tools for corpus exploration, maintenance and conversion. In: Proceedings of the 16th Conference of the European Chapter of the Association for Computational Linguistics: System Demonstrations, pp. 1–9 (2021). https://hal.inria.fr/hal-03177701
12. Lehmann, W.P.: Syntactic typology: studies in the phenomenology of language. University of Texas Press (1978)
13. Levshina, N.: Token-based typology and word order entropy: a study based on universal dependencies. Linguist. Typology **23**, 533–572 (2019)
14. Levshina, N.: Corpus-based typology: applications, challenges and some solutions. Linguist. Typology **26**, 129–160 (2021). https://doi.org/10.1515/lingty-2020-0118

15. Liu, H.: Dependency direction as a means of word-order typology: a method based on dependency treebanks. Lingua **120**, 1567–1578 (2010)
16. Guzmán Naranjo, M., Becker, L.: Statistical bias control in typology. Linguist. Typology **26** (2021). https://doi.org/10.1515/lingty-2021-0002
17. Nivre, J., et al.: Universal dependencies (2023). https://universaldependencies.org/
18. Vennemann, T.: Categorial grammar and the order of meaningful elements, pp. 615–634. Anma Libri (1976)

# Session on New Perspectives and Solutions in Cultural Heritage (TECTONIC)

# Environmental Parameters that Negatively Impact Underwater Cultural Heritage

Marta Plaza-Hernández[1](✉), Mahmoud Abbasi[1], Vladimir Djapic[2], Javier Prieto-Tejedor[1], and Emilio Corchado-Rodríguez[1]

[1] BISITE Research Group, University of Salamanca, Edificio Multiusos I+D+I, Calle Espejo 2, 37007 Salamanca, Spain
{martaplaza,mahmoudabbasi,javierp}@usal.es
[2] H2O Robotics Ltd, Unska 3, 10 000 Zagreb, Croatia
vladimir@h2o-robotics.com

**Abstract.** Preventive conservation of Underwater Cultural Heritage requires complete knowledge of the environment in which an item/site is located, the materials from which it is made, and the degradation phenomena experienced in the surrounding environment. To understand the different decay phenomena, the first step is to identify the environmental stressors that participate in the deterioration processes. This paper conducts a literature review on the environmental parameters that play a key role in the degradation of Underwater Cultural Heritage. Sites/items suffer from natural degradation and change processes that are physicochemical, geological, and biological, being the leading agents that regulate and influence these processes water, sediment and living organisms. Understanding these agents and processes is needed to comprehend how heritage develops and degrades.

**Keywords:** Underwater Cultural Heritage · Degradation · Environmental parameters

## 1 Introduction

The protection of Underwater Cultural Heritage (UCH) is crucial to preserve society's history. However, conservation methods and processes, from the evaluation and analysis of the current state of the sites to restoration activities, still present multiple challenges [1], such as the elevated costs and the complexity of operating underwater, the lack of regulation, the inadequate protection of items and sites and the inability to use them for sustainable and responsible tourism development [2].

For preventive conservation, we need complete knowledge of the environment in which an item/site is located, the materials from which it is made, and the degradation phenomena experienced in the surrounding environment, which can be physicochemical (seawater), biological (living organisms), geological (type of substrates and sediments) or anthropogenic [3]. Therefore, environmental factors should be monitored and foreseen with enough time to respond, performing data analytics to detect patterns [4]. There is


R. Mehmood et al. (Eds.): DCAI 2023, LNNS 741, pp. 199–206, 2023.
https://doi.org/10.1007/978-3-031-38318-2_20

an urge to adopt innovative and sustainable solutions to preserve the heritage in situ and use the available information for decision-making. The Internet of Underwater Things (IoUT) could be one of the most cost-effective practices for evaluating the current state of sites, and facilitating the monitoring of environmental changes [5–7].

In previous work [8], the authors of this paper proposed an IoUT-based platform for the management of UCH that incorporates Artificial Intelligent (AI) techniques that support the monitoring of UCH [9]. The platform will combine information stored in databases with real-time data, working independently and in collaboration with other platforms and systems. The IoUT systems part of the proposed platform will be used for (i) monitoring the heritage's material conditions, detecting structural changes; (ii) monitoring the heritage's environment; and (iii) alerting of anomalous events. Among the main functionalities of the platform, we emphasise the ability to record environmental parameters and transmit the information wirelessly and encrypted. At the sensors level, the proposed architecture will be able to deal with high energy constraints and support fast-changing environments. It will provide elasticity at the storage level and facilitate access to data visualisation [10, 11]. The management of the different components of the platform will be performed by developing a layered architecture (Fig. 1).

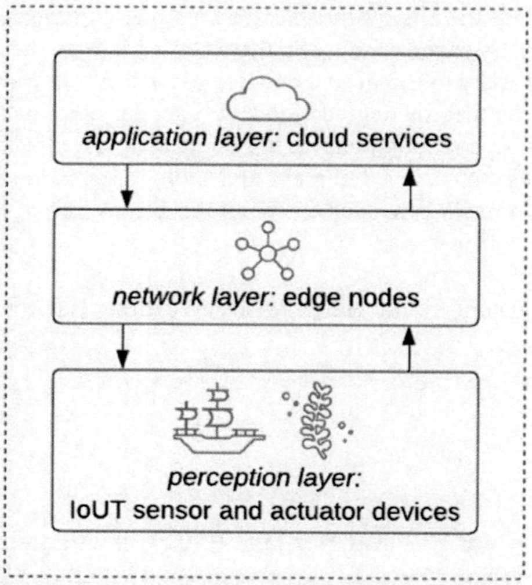


**Fig. 1.** Proposed architecture for the management of UCH.

To understand the various degradation phenomena that impact UCH, the first step is to identify the environmental factors that participate in the decay processes. This paper conducts a literature review on the environmental stressors that play a key role in the deterioration of UCH. This way, the platform's perception layer can be defined, and the necessary IoUT devices selected. The rest of the paper is organised as follows: Sect. 2 introduces the research methodology followed for this study, Sect. 3 presents the key findings, and Sect. 4 draws the main conclusions and describes the next research steps.

## 2 Research Methodology

The purpose of this review is to provide a summary of the available literature relevant to the topic. The first step of the search strategy was the definition of the research question:

*Which environmental parameters have been identified in the literature to play a key role in the deterioration of Underwater Cultural Heritage?*

Where the selected keywords are *underwater cultural heritage, environmental factors, environmental conditions, degradation, and deterioration.* Then, we defined the inclusion and exclusion criteria, presented in Table 1.

**Table 1.** Inclusion and exclusion criteria selected.

| Inclusion criteria (IC) | Exclusion criteria (EC) |
|---|---|
| IC1. Peer-reviewed book chapters and articles published in journals or conferences | EC1. Duplicated studies |
| IC2. Studies written in English | EC2. Non-open access studies |

The review was conducted using the Scopus search engine, narrowing the search to those studies that contain the keywords within the title, abstract and study's keywords. The search string built was:

TITLE-ABS-KEY ( "underwater cultural heritage") AND ( TITLE-ABS-KEY ( "environmental paramete*") OR TITLE-ABS-KEY ( "environmental facto*") OR TITLE-ABS-KEY ( "environmental conditio*")) AND ( TITLE-ABS-KEY ("degrad*") OR TITLE-ABS-KEY ( "deteriorat*")) AND ( LIMIT-TO ( DOCTYPE,"ar") OR LIMIT-TO ( DOCTYPE,"ch") OR LIMIT-TO ( DOCTYPE,"cp")) AND ( LIMIT-TO ( LANGUAGE,"English")).

The initial number of results was seven. After applying the inclusion and exclusion criteria, the final number of analysed studies was four (Fig. 2). Table 2 presents the findings.

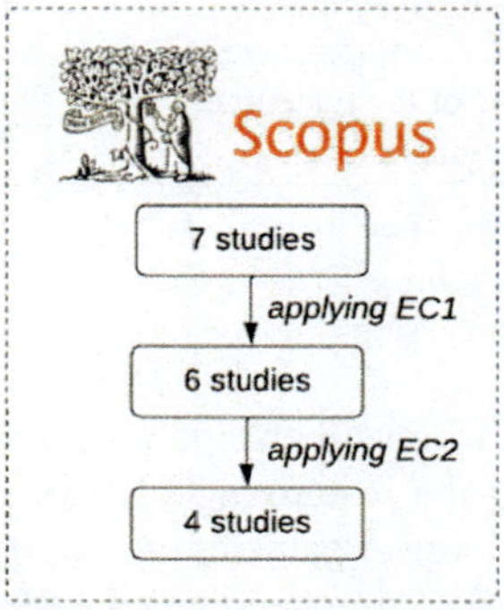


**Fig. 2.** Number of studies found in the Scopus search engine.

**Table 2.** Reference of studies found in Scopus after applying the search string.

| Authors | Title | Year | Source | Doc type |
|---|---|---|---|---|
| Randazzo L., Ricca M., Ruffolo S., Aquino M., Petriaggi B.D., Enei F., and La Russa M.F. [12] | An integrated analytical approach to define the compositional and textural features of mortars used in the underwater archaeological site of castrum novum (Santa marinella, rome, italy) | 2019 | Minerals | Article |
| Knight K.Y., Cousins T.A., and Parham D. [13] | A comparison of biodegradation caused by Teredinidae (Mollusca:Bivalvia), Limnoriidae (Crustacea:Isopoda), and C. terebans (Crustacea:Amphipoda) across 4 shipwreck sites in the English Channel | 2019 | Journal of Archaeological Science: Reports | Article |
| Bethencourt M., Fernández-Montblanc T., Izquierdo A., González-Duarte M.M., and Muñoz-Mas C. [14] | Study of the influence of physical, chemical and biological conditions that influence the deterioration and protection of Underwater Cultural Heritage | 2018 | Science of the Total Environment | Article |
| Fernández-Montblanc T., Bethencourt M., Izquierdo A., and González-Duarte M.M. [15] | Establishing the relationship between underwater cultural heritage deterioration and marine environmental factors. A comparative analysis of the bucentaure and fougueux sites | 2014 | Science, Technology and Cultural Heritage | Book chapter |

## 3 Findings

UCH experiences natural degradation and change processes that are physicochemical, geological, and biological, being the leading agents that regulate and influence these processes water, sediment and living organisms. A complete understanding of these agents and processes is required to comprehend how UCH develops and degrades since any shift in one element could impact the entire system.

## 3.1 Review of Selected Studies

In [12], authors conducted an archaeometric characterisation of mortar samples collected directly on-site for petrographic, mineralogical, and microchemical examinations. They identified the different biodegradation processes affecting the mortar samples by observing biological growth and colonisation. The samples showed heavy coverage from encrusting organisms like barnacles, tubeworms, bryozoans, molluscs, and coralline algae. Barnacles and tubeworms were the most abundant species. The biological growth and colonisation led to the deterioration of the mortar material through processes like scraping and boring.

The authors of [13] assessed the biodegradation rate provoked by wooden borers on four UCH locations in different marine environments. By placing wooden test panels in direct contact with the sites, they estimated the biodegradation rates at each location based on aspects like wood borer counts, the damage observed, and the material loss. The sediment type is considered a significant factor in determining the biodegradation rate. Different sediments likely impacted factors like oxygen, nutrients, and habitat for wood borers differently, influencing biodegradation. Environmental parameters such as temperature, salinity, and dissolved oxygen also affect the survival and growth of wood borers.

In [14] and [15], authors conducted non-destructive surveys to assess shipwreck cannons and anchors. They measured critical physical, chemical, and biological variables at each site to correlate the conservation status with the marine environment. They found that some artefacts were corroding faster due to high water movement, indicating water flow impacts deterioration. Sediment displacement and transport by waves also caused damage at the sites. In addition, the knowledge of the physical and hydro-chemical conditions is essential because they can contribute to developing sessile communities dominated by algae.

## 3.2 Other Potential Environmental Stressors

Many other environmental factors with the potential to impact UCH have not been covered in the studies compiled for this review. They should be paid attention to and monitored.

**Geological Parameters.** As identified by all the authors of the reviewed studies, sedimentation is one of the main stressors. Heavy sediment deposition buries artifacts, blocking access and oxygen supply. It also leads to anaerobic corrosion of metals and damage through abrasion. High turbidity from suspended sediments and particles should also be monitored. Geological properties of surrounding sediments, such as grain size, compactness, porosity, etc., impact deterioration through sediment instability, permeability, etc. loose, poorly sorted sediments with high porosity accelerate damage, while compact, fine-grained sediments stabilise the site and limit deterioration agents. Underwater tectonic activity like faulting or folding leads to damage through shifting, tilting, cracking, and dismantling of artefacts and their base support. Areas of active tectonics usually see rapid deterioration, while geologically stable areas best preserve cultural heritage remains. Finally, the nature of underwater subfloor geology on which UCH lies impacts damage through instability, permeability, compressibility, etc. Loose, compressible, permeable, or soluble rock types promote the sinking of objects. Dense with low permeability subfloors stabilise sites and support long-term preservation.

**Biological Parameters.** Only a Few Species Have Been Identified in the Literature Among the main biological communities that negatively affect UCH. Microorganisms (bacteria, fungi, algae, etc.) produce acids, enzymes, and other metabolic products that damage materials, accelerating decay. Macroalgae also damage surfaces through abrasion or penetration. Sponges erode materials through the secretion of acids and enzymes. They also facilitate the entry of other destructive organisms. Invertebrate organisms like barnacles and shipworms directly bore into materials. Their attachment strength also weakens structures. although fish do not directly damage artefacts, their activity may do. Scratching on surfaces by fins and spines erodes materials over time. Finally, some marine mammals, like seals, dig into sediments, causing damage.

**Physicochemical Parameters.** Temperature, salinity, pH, oxygen concentrations, and nutrient levels should be closely overseen. Fluctuations in temperature produce expansions and contractions that accelerate material weakening and cracking. Also, warm temperatures speed up chemical reactions and biological activity. High concentrations of salinity enable the corrosion of metals and the deterioration of other materials. Low-pH environments speed up the corrosion and dissolution of materials, while high-pH environments also damage materials through alkalosis. The lack of oxygen delays corrosion and deterioration, while high oxygen concentrations accelerate the breakdown of organic materials and the corrosion of metals. High levels of nutrients like nitrates and phosphates promote blooms of destructive microorganisms such as macroalgae, bacteria, sponges, etc. In addition to the impact that these factors have themselves, T, salinity, dissolved oxygen, light regime, chemical composition of the water column, nutrient availability, waves, and currents, etc., affect the growth, reproduction, and distribution of biological communities.

## 4 Conclusions and Future Work

The conservation of UCH helps preserve history and cultural memory, ensuring that communities' experiences, values, and identities are passed on to future generations. This paper compiled identified key environmental factors that play a primary role in the deterioration of UCH. A literature review was conducted in the Scopus search engine. Only four studies were analysed, complementing the findings with potential parameters that the authors of this paper consider should be paid attention to.

UCH experiences natural degradation and change processes that are physicochemical, geological, and biological, being the leading agents that regulate and influence these processes water, sediment and living organisms. A complete understanding of these agents and processes is required to comprehend how UCH develops and degrades, since any shift in one element could impact the entire system.

To expand the results, future work will include studies that contain the selected keywords not only in the title-abstract-authors' keywords; but also in the complete text. Additionally, anthropogenic stressors, focusing on climate change, will be considered.

Once this analysis is performed, the measuring devices (sensors) of the perception layer of the platform proposed in [8] will be selected.

**Acknowledgments.** This research has been supported by the project "Technological Consortium TO develop sustainability of underwater Cultural heritage (TECTONIC)", financed by the European Union (Horizon 2020 research and innovation programme under the Marie Skłodowska-Curie grant agreement No 873132).

## References


1. UNESCO: Convention on the Protection of the Underwater Cultural Heritage (2023). https://www.unesco.org/en/legal-affairs/convention-protection-underwater-cultural-heritage, last accessed: 11 April 2023
2. TECTONIC: The Underwater Cultural Heritage: an interdisciplinary challenge (2023). https://www.tectonicproject.eu/, last accessed: 4 April 2023
3. Memet, J.B.: Conservation of Underwater Cultural Heritage: characteristics and new technologies. Museum Int. **60**, 42–49 (2018)
4. Perles, A., et al.: An energy-efficient internet of things (IoT) architecture for preventive, conservation of cultural heritage. Futur. Gener. Comput. Syst. **81**, 566–581 (2018)
5. Qader Kheder, M., Aree Ali, M.: IoT-Based Vision Techniques in Autonomous Driving: A Review. ADCAIJ: Adv. Distri. Comp. Artifi. Intelli. J. **11**(3), 367–394 (2023)
6. Gupta, S., Ranga, V., Agrawal, P.: EpilNet: a novel approach to IoT based epileptic seizure prediction and diagnosis system using artificial intelligence. ADCAIJ: Adv. Distri. Comp. Artifi. Intelli. J. **10**(4), 435–452 (2022)
7. Astorga-González, E.M., Municio, E., Noriega-Alemán, M., Marquez-Barja, J.M.: Cultural Heritage and Internet of Things. 6th EAI International Conference on Smart Objects and Technologies for Social Good (2020)
8. Plaza-Hernández, M., Abbasi, M., Mezquita, Y.: An IoUT-Based Platform for Managing Underwater Cultural Heritage. In: Distributed Computing and Artificial Intelligence, Special Sessions, 19th International Conference. DCAI 2022. Lecture Notes in Networks and Systems, vol. 585. Springer, Cham (2023)
9. García-Retuerta, D., Casado-Vara, R., Martin-del Rey, A., De la Prieta, F., Prieto, J., Corchado, J.M.: Quaternion Neural Networks: State-of-the-Art and Research Challenges. In: Analide, C., Novais, P., Camacho, D., Yin, H. (eds.) IDEAL 2020. LNCS, vol. 12490, pp. 456–467. Springer, Cham (2020). https://doi.org/10.1007/978-3-030-62365-4_43
10. Márquez Sánchez, S.: Integral Support Predictive Platform for Industry 4.0. ADCAIJ. Adv. Distri. Comp. Artifi. Intelli. J. **9**(4), 71–82 (2020)
11. Rodríguez, S., Tapia, D.I., Sanz, E., Zato, C., de la Prieta, F., Gil, O.: Cloud Computing Integrated into Service-Oriented Multi-Agent Architecture. In: Ortiz, Á., Franco, R.D., Gasquet, P.G. (eds.) BASYS 2010. IAICT, vol. 322, pp. 251–259. Springer, Heidelberg (2010). https://doi.org/10.1007/978-3-642-14341-0_29
12. Randazzo, L., et al.: An integrated analytical approach to define the compositional and textural features of mortars used in the underwater archaeological site of castrum novum (Santa marinella, rome, italy). Minerals **9**(5), 268 (2019)
13. Knight, K.Y., Cousins, T.A., Parham, D.: A comparison of biodegradation caused by Teredinidae (Mollusca:Bivalvia), Limnoriidae (Crustacea:Isopoda), and C. terebans (Crustacea:Amphipoda) across 4 shipwreck sites in the English Channel. J. Archaeol. Sci. Rep. **23**, 854–867 (2019)

14. Bethencourt, M., Fernández-Montblanc, T., Izquierdo, A., González-Duarte, M.M., Muñoz-Mas, C.: Study of the influence of physical, chemical and biological conditions that influence the deterioration and protection of Underwater Cultural Heritage. Sci. Total Environ. **613–614**, 98–114 (2018)
15. Fernández-Montblanc T., Bethencourt M., Izquierdo A., González-Duarte M.M.: Establishing the relationship between underwater cultural heritage deterioration and marine environmental factors. A comparative analysis of the bucentaure and fougueux sites. Science, Technology and Cultural Heritage, edn M.A. Rogerio-Candelera. CRC Press/Balkema, The Netherlands (2014)

# A Preliminary Study of Unconventional Coatings for the Conservation of Underwater Cultural Heritage Within the TECTONIC Project

Antonio Donato[1](✉), Michal Novàk[2], Marie Novàkovà[2], Mauro Francesco La Russa[1], Silvestro Antonio Ruffolo[1], and Michela Ricca[1]

[1] Department of Biology, Ecology and Earth Sciences, University of Calabria, Via P. Bucci, 87036 Arcavacata di Rende (CS), Italy
antonio.donato93@unical.it

[2] SYNPO akciová společnost, S. K. Neumanna 1316, Pardubice 532 07, Czech Republic

**Abstract.** Underwater Cultural Heritage (UCH) refers to all the traces of human existence that are located underwater. These can include shipwrecks, submerged buildings, and other archaeological sites. The conservation of UCH can be challenging due to the unique environmental conditions, the difficulty of accessing and working in underwater environments, as well as the potential for damage from natural forces and human activity. The TECTONIC project aims to promote an intersectoral collaboration between academic and non-academic professionals, supporting the exchange of skill and expertise between them, aiming to implement, improve and assess innovative materials, techniques, tools, and methodologies for the Underwater Cultural Heritage. In this framework, the technology and scientific exchange between the University of Calabria (Italy) and Synpo akciova spolecnost (Czech Republic), is leading to the preparation of an underwater coating with antifouling properties suitable for the protection of underwater cultural heritage.

**Keywords:** Underwater · Cultural heritage · Coating · Epoxy Resin

## 1 Introduction

In recent decades, there has been a growing interest in studying degradation phenomena that occur at underwater archaeological sites. Various research efforts have focused on both the degradation of stone materials in underwater environments and on developing innovative approaches for their protection [1–5].

One of the most significant forms of damage to submerged archaeological artifacts is biological colonization (bio-fouling), which can develop differently depending on a variety of environmental conditions, as well as the textural and structural features of the materials [1, 4, 6, 7].

The latest guidelines from scientific and international cultural heritage protection organizations support the promotion, protection, and in-situ preservation of underwater archaeological and historical heritage, as outlined in the UNESCO Convention on the Protection of the Underwater Cultural Heritage from November 2nd, 2001.


R. Mehmood et al. (Eds.): DCAI 2023, LNNS 741, pp. 207–213, 2023.
https://doi.org/10.1007/978-3-031-38318-2_21

Epoxy resins find numerous applications due to their desirable properties. They are widely used as adhesives, coatings, and sealants in various industries, from the construction sector [8] to the electronics and crafts industry. Applications of resin for stabilization and restoration of Cultural Heritage can be found in literature [9, 10]. The present research, which is a part of the TECTONIC project (EU-H2020-MSCA-RISE-2019-873132-TECTONIC TEchnological Consortium TO develop sustaiNabIlity of underwater Cultural heritage), aims to develop an epoxy resin-based system with antifouling properties for the conservation and preservation of underwater stone materials [11].

The work is still in progress and therefore only the preliminary results based on the initial stages of laboratory experimentation will be discussed in the next paragraphs.

## 2 Material and Methods

Epoxy resin is a type of thermosetting polymer that is formed through the reaction of an epoxy resin and a curing agent. It is a versatile material known for its excellent adhesive properties, high strength, and chemical resistance [12, 13]. Typically, they are made by combining two components: an epoxy resin and a curing agent. The epoxy resin, also known as the "resin component," is a viscous liquid that contains epoxy groups. The curing agent, also called the "hardener," is a chemical compound that reacts with the epoxy resin, leading to cross-linking and the formation of a three-dimensional network. Once the epoxy resin and curing agent are mixed, a chemical reaction known as polymerization occurs. This curing process can be accelerated by the addition of heat or a catalyst. This reaction results in the formation of strong covalent bonds, creating a rigid and durable material. The cured epoxy resin has excellent mechanical properties, including high tensile and compressive strength, as well as good resistance to chemicals, moisture, and temperature extremes.

The first step of the research focused on a screening of the various epoxy resin-based systems produced by Synpo, trying to find out the most suitable and compatible formulation for underwater application. The products have been divided into series, according to the type of resin and the additives used. The colorimetric tests, the measurement of the contact angle, viscosity and underwater applicability were therefore evaluated, while the antifouling and ecotoxicity tests are still in progress. The products were applied on marble and limestone specimens sized 5 × 5 × 2 cm, in order to evaluate any different performance of the coatings on the lithotypes most commonly used in the field of cultural heritage.

Color variations induced by the treatment were evaluated by means of colorimetric tests [14], using a CM-2600d Konica Minolta spectrophotometer. Chromatic values are expressed in the CIE L*a*b* space, where L* is the lightness/darkness coordinate, a* the red/green coordinate (+a* indicating red and –a* green) and b* the yellow/blue coordinate (+b* indicating yellow and –b* blue).

Contact angle measurements [15] was carried out in order to determine the wettability of the treated surface, as this parameter may influence the adhesion of microorganisms in the submerged environment. The measurement consisted in the placement of a water drop of defined volume on the solid sample surface. Drop shape was recorded with a camera and automatically evaluated in terms of contact angle, which represents the angle between the substrate surface and the tangent from the edge to the contour of the drop.

Viscosity test were carried out by Brookfield DV-II+Pro RV viscometer, coupled with Vane V-74 spindle, in order to obtain information about the behavior of the compound under various shear rates.

Underwater applicability was evaluated using standardized two-component cartridges with a static mixer to obtain correctly mixed system under water-free conditions. Premixed coating was hereafter spread by various tools (brush, sponge, spatula) in an aquarium. It's important to note that for underwater tests, the original formulation was modified adding an UV-sensitive pigment, in order to check whether the product was well applied or not. The addition of this pigment did not alter any final thermos-mechanical property of the product.

In the next paragraphs, the discussion will be limited to the series of samples that proved to be suitable after the initial screening, that is Cx series, characterized by a modification of epoxy resin by thixotropic agent.

## 3 Results and Discussion

In order to narrow down the number of products to use in the research, a preliminary hydrophobicity test was carried out. Various types of epoxy resins and hardeners were placed in a container with water and mixed to observe their degree of immiscibility in water. The most immiscible samples, i.e., those that separated most rapidly from the water in a separate layer, keeping it clear, were subjected to further testing. This behavior ensures that mixed will not be dissolved to the environment when applied on the protected cultural heritage. Secondly, the selected epoxy resin and hardener were modified by thixotropic agents to improve rheological properties that suits underwater application. The aim of this modification was to find out the products with a pseudoplastic behavior. This behavior refers to a type of non-Newtonian fluid behavior where the viscosity of a material decreases as the shear rate or deformation rate increases. This means that the material becomes less resistant to flow as the applied force or stress increases. This character of resin ensures easy application on the sample and eliminates wash-out to the environment when the coating is applied. The C7 sample from Cx series modified with thixotropic agent, appeared to be the most suitable product (Fig. 1).

Hardener (B2) was chosen such that its viscosity curve was the closest matching one that of the selected resin: a modification of hardener by thixotropic agent (Fig. 2).

For what concerns colorimetric variation, we can notice a different behavior depending on the lithotype used: negligible variation for marble specimens and important color changes for limestone ones (Table 1).

Contact angle measurement showed an increase in hydrophobicity, which should decrease the adhesion capability of microorganisms in a marine environment; further confirmation or denial will come from bacterial tests (Table 2).

The underwater applicability tests focused on several aspects: possible dispersion of the product during application, ease of application depending on the tool used, and hardening of the product. The results confirmed what had previously emerged from the viscosity tests: C7 sample showed the best applicability in the underwater environment, using soft brushes and/or sponges (Fig. 3), even though issues with short pot life of the mixed product (resin + hardener) emerged.

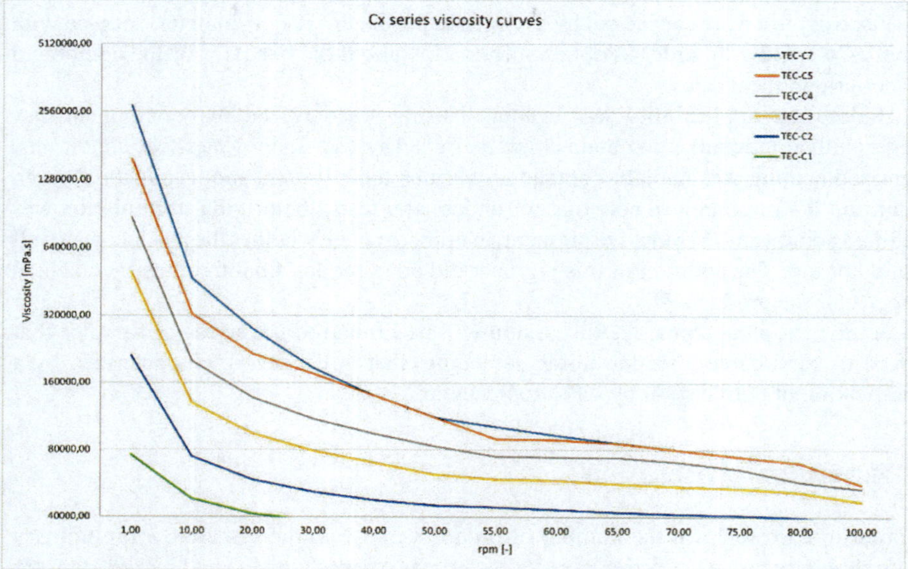


**Fig. 1.** Viscosity test results for Cx series.

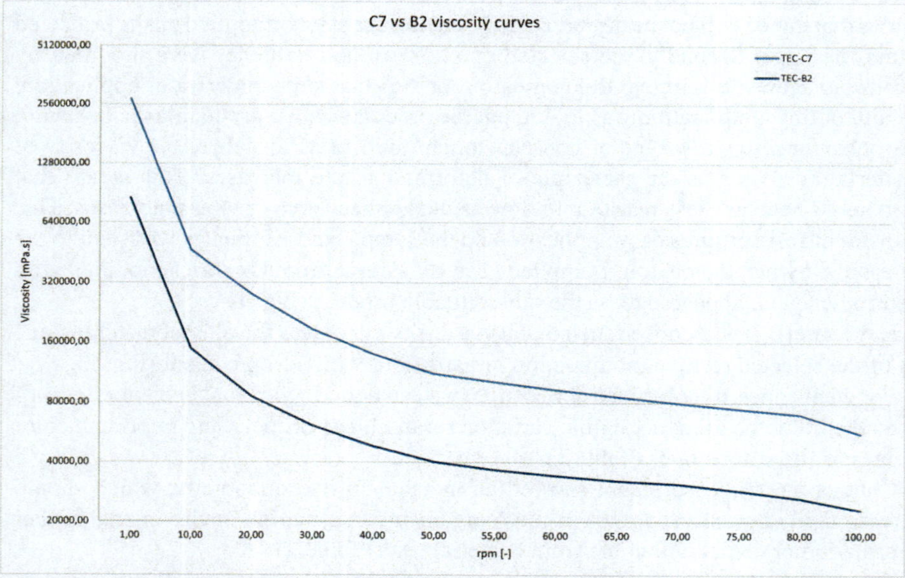


**Fig. 2.** C7 vs B2 viscosity curves.

**Table 1.** Colorimetric test results.

| Colorimetric Measurements | | | | | |
|---|---|---|---|---|---|
| Lithotype | Reference value | | After treatment | | ΔE |
| Marble | L* | 79.1 | L* | 77.4 | 1.7 |
| | a* | − 1.1 | a* | − 1.4 | |
| | b* | 1.0 | b* | 0.8 | |
| Limestone | L* | 82.7 | L* | 72.2 | 13.8 |
| | a* | 0.3 | a* | 1.2 | |
| | b* | 7.2 | b* | 16.4 | |

**Table 2.** Contact angle results.

| Contact angle Measurements | | |
|---|---|---|
| Lithotype | Reference value | After treatment |
| Marble | 50.86 | 69.3 |
| Limestone | 21.20 | 68.6 |

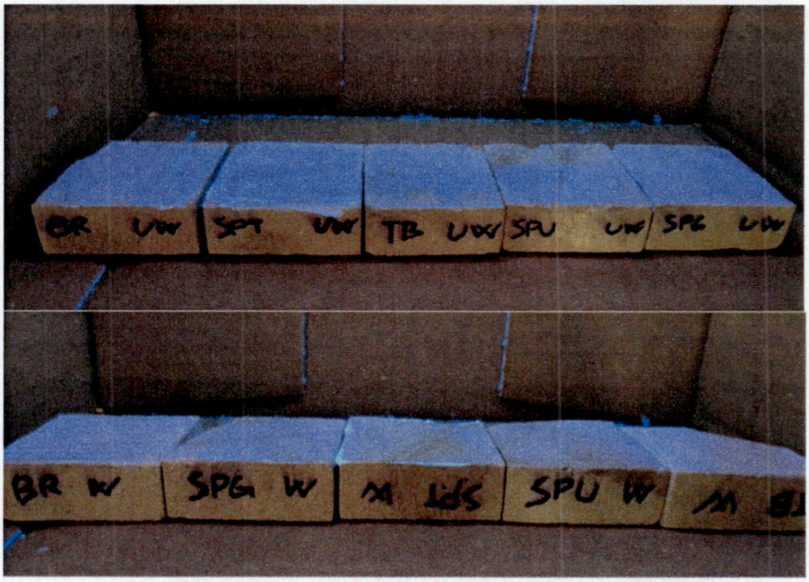


**Fig. 3.** Different application methods on limestones for C7 product. W = wet samples, on air application; UW = underwater application; BR = brush, SPT = spatula, TB = toothbrush, SPU = PVA sponge, SPG = sponge.

## 4 Conclusion

The research carried out in the frame of the TECTONIC project is still in progress, but the preliminary results are well promising. Colorimetric test showed negligible color variation on marble and significant color change on limestone that was caused by high gloss of the coating. The development of protective coating still continues and current studies are focused on the incorporation of a matting agent in order to reduce color variation on limestone samples. Also, further tests on wet samples will be performed to provide an evaluation representing marine environment. Contact angle measurements showed the increase in hydrophobicity of the treated surface, which will, hopefully, discourage the growth of microorganism on it. Adjustments need to be done on formulation to slow down the pot life of the product, without altering the current other properties. The next stages of laboratory experimentation will focus on antifouling tests on the already formulated products in order to guarantee their use for the protection of underwater objects and stone materials of cultural interest.

## References

1. Ricci, S., Davidde, B., Bartolini, M., Priori, G.F.: Bioerosion of lapideous artefacts found in the underwater archaeological site of Baia (Naples). Archaeol. Marit Mediterr. Int. J. Underw. Archaeol. **6**, 167–186 (2009)
2. Crisci, G.M., La Russa, M.F., Macchione, M., Malagodi, M., Palermo, A.M., Ruffolo, S.A.: Study of archaeological underwater finds: deterioration and conservation. Appl. Phys. A **100**, 855–863 (2010)
3. La Russa, M.F., et al.: A Multidisciplinary approach for the study of underwater artefacts: the case of Tritone Barbato marble statue (Grotta Azzurra, Island of Capri, Naples). Period Miner. **82**, 101–111 (2013)
4. La Russa, M.F., Ricca, M., Belfiore, C.M., Ruffolo, S.A., Álvarez Ballester De Buergo, M., Crisci, G.M.: The contribution of earth sciences to the preservation of underwater archaeological stone materials: an analytical approach. Int. J. Conserv. Sci. **6**, 335–348 (2015)
5. La Russa, M.F., Ruffolo, S.A., Ricca, M., Rovella, N., Comite, V., Barca, D.: Archaeometric approach for the study of mortars from the underwater archaeological site of Baia (Naples) Italy: preliminary results. Period Miner. **84**, 1–16 (2015)
6. Davidde, B., Ricci, S., Poggi, D., Bartolini, M.: Marine bioerosion of stone artefacts preserved in the museo archeologico dei Campi Flegrei in the castle of Baia (Naples). Archaeol. Marit Mediterr. Int. J. Underw. Archaeol. **7**, 75–115 (2009)
7. Relini, G.: Il Biofouling. Biol. Marit Mediterr. **10**, 285–326 (2003)
8. Dębska, B., Brigolini Silva, G.J.: Mechanical properties and microstructure of epoxy mortars made with polyethylene and poly(ethylene terephthalate) waste. Materials **14**(9), 2203 (2021). https://doi.org/10.3390/ma14092203
9. Stumes, P.: The application of epoxy resins for the restoration of historic structures. Bull. Assoc. Preserv. Technol. **3**(1), 59–63 (1971)
10. Selwitz, C.: The use of epoxy resins for the stabilization of deteriorated masonry. APT Bull. J. Preserv. Technol. **26**(4), 27–34 (1995). https://doi.org/10.2307/1504447
11. Ricca, M., et al.: A sustainable approach for the management and valorization of underwater cultural heritage: new perspectives from the TECTONIC project. Sustainability **12**, 5000 (2020). https://doi.org/10.3390/su12125000

12. Capricho, J.C., Fox, B., Hameed, N.: Multifunctionality in epoxy resins. Polym. Rev. **60**(1), 1–41 (2020). https://doi.org/10.1080/15583724.2019.1650063
13. Ruffolo, S.A., Ricca, M., Macchia, A., La Russa, M.F.: Antifouling coatings for underwater archaeological stone materials. Prog. Org. Coat. **104**, 64–71 (2017). https://doi.org/10.1016/j.porgcoat.2016.12.004
14. NorMaL 43/93: misure colorimetriche di superfici opache. Raccomandazioni NorMaL: alterazioni dei materiali lapidei e trattamenti conservativi: proposte per l'unificazione dei metodi sperimentali di studio e di controllo. CNR, Roma, Italy (1993)
15. NorMaL 33/89: misura dell'angolo di contatto. Raccomandazioni NorMaL: alterazioni dei materiali lapidei e trattamenti conservativi: proposte per l'unificazione dei metodi sperimentali di studio e di controllo. CNR, Roma, Italy (1989)

# Architecture of a Low-Cost Solution for ROVs to Improve Navigation and Data Collection

Umberto Severino[1](✉), Santina Fortuna[1], Antonio Lagudi[1], Fabio Bruno[1], Nikola Mišković[2], and Vladimir Djapic[3,4]

[1] Department of Mechanical, Energetic and Management Engineering, University of Calabria, 87036 Arcavacata di Rende, Italy
{umberto.severino,santina.fortuna,antonio.lagudi, fabio.bruno}@unical.it

[2] Faculty of Electrical Engineering and Computing, University of Zagreb, 10 000 Zagreb, Croatia
nikola.miskovic@fer.hr

[3] H2O Robotics Ltd, 10 000 Zagreb, Croatia
vladimir@h2o-robotics.com, vladimir.djapic@ivi.ac.rs

[4] Institute for Artificial Intelligence Research and Development, 21000 Novi Sad, Serbia

**Abstract.** The documentation, conservation, and preservation of underwater archaeological sites is conventionally performed by divers in shallow waters. However, when dealing with significant depths, the sites can be inaccessible for humans, presenting significant challenges and risks, making these activities hazardous. Recently, the use of robotic platform such as remotely operated vehicles (ROVs) allows to overcome the difficulty of such a harsh environment. Nevertheless, professional-grade ROVs employed for this purpose are expensive and specifically designed with integrated advanced sensor systems for data acquisition and navigation. In the context of the MSCA-RISE H2020 Technological Consortium TO develop sustainability of underwater Cultural heritage (TECTONIC) project, a solution consisting of an optical payload and an acoustic localization system for low-cost ROV, was defined in order to support ROV pilots during documentation and monitoring activities. The project aims to preserve underwater cultural heritage (UCH) encompassing the selection of a commercially available ROV and the definition of a customized hardware/software (HW/SW) architecture. After market research, a suitable commercial off-the-shelf (COTS) low-cost ROV was defined and used as the foundation for integration. Mechanical, electric, and data interface modules were designed to seamlessly integrate and manage the sensor suite. The results of the integration demonstrate the potential of the chosen algorithms and the sensor suite for improved navigation and data collection in challenging underwater environments, where high-grade and expensive sensors are not available, thereby making more accessible the field of underwater survey missions.

**Keywords:** Underwater Cultural Heritage (UCH) · Underwater 3D Reconstruction · Underwater Mapping · Underwater Localization · Underwater Survey


R. Mehmood et al. (Eds.): DCAI 2023, LNNS 741, pp. 214–223, 2023.
https://doi.org/10.1007/978-3-031-38318-2_22

# 1 Introduction

Preserving UCH sites, such as shipwrecks, sunken cities, and prehistoric artwork and treasures, is of utmost importance [1, 2]. Documentation, conservation, and preservation measures must be implemented, while also increasing public awareness and knowledge. The EU-funded TECTONIC project aims to foster interdisciplinary collaboration among academic and non-academic professionals in the UCH field. By leveraging their expertise, the project seeks to address the complex challenges [3] that still exist in this domain. The project will facilitate knowledge exchange and provide training for the utilization of advanced techniques, instruments, and methods. It aims to establish strong links between research, higher education, and business. In the TECTONIC project [4], was evaluated the feasibility of employing cost-effective robotic systems for exploring, documenting, and safeguarding UCH. The majority of robotic systems consist of ROVs equipped with sophisticated sensors, but their cost makes them inaccessible to a broader audience. At the same time, such sensors are necessary to prevent damage on the inspection site. Precise and controlled manoeuvring plays a critical role in specialized fields like cultural heritage preservation, where unintentional movements can cause damage to valuable artifacts [5]. Likewise, in photogrammetry mapping missions, a deficiency in environmental perception can result in incomplete area coverage and the production of low-fidelity 3D reconstructions. Another relevant issue is the accurate georeferencing of defined assets in such as environment. Underwater localization technologies exploit the physical properties of sound and signal processing algorithms to determine the position of a sound source concerning a set of beacons arranged in reference points [6]. By incorporating a stereo camera, employing computer vision algorithms for both localization and mapping, it becomes possible to enhance environmental perception [7, 8]. This technological approach offers several benefits, particularly in terms of real-time data processing [9]. This capability allows for immediate verification of collected data, ensuring that ongoing missions adhere to the required standards. The stereo camera configuration provides on-the-fly quantitative spatial information, thereby facilitating piloting and operational tasks.

Consequently, this paper presents the architectural design of a robotic platform specifically developed to integrate optical sensors and an acoustic localization system. The main objective of this platform is to support the ROV pilot in navigation tasks and data georeferencing by utilizing a computer vision based solution to get proprioceptive parameters [10] and an acoustic localization system. The proposed solution mainly includes valuable feedback on navigation velocity, the distance between the camera reference frame and the target, and the ground sample distance.

The structure of the paper is as follows: Sect. 2 presents a comprehensive overview of the system architecture, outlining the components and the integration of the proposed solution into a commercially available and cost affordable small ROV. Following this, Sect. 3 presents the results obtained from controlled environment experiments and simulations. Finally, the last section concludes the paper and discusses potential avenues for future testing and application.

## 2 System Architecture

The typical survey scenario is depicted in Fig. 1. The ROV moves on the seafloor while acquiring images of the relevant asset with both the cameras and being located by the means of the acoustic localization system. The cable connection between the ROV and the surface control unit facilitates offloading a portion of the computational load onto the surface control unit, which can manage camera control, acquisition parameters, and execute computer vision/artificial intelligence algorithms for mapping and, eventually, target classification in the underwater environment.

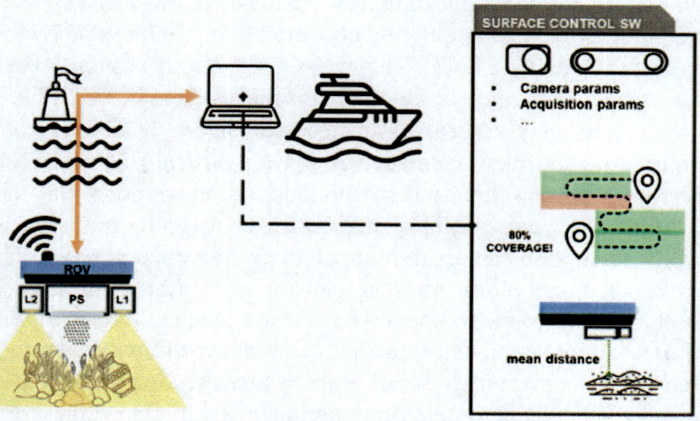


**Fig. 1.** Schematic diagram of the survey scenario.

An extensive market survey was conducted to identify and select the appropriate COTS platform. The choice was to develop the proposed solution in the environment of the BlueROV2 Rev3 [11] due to its affordability for small research groups and small size. Various architectural options were assessed based on factors such as cost, requirements, and application scenarios. The specification and the goal of the TECTONIC project leaded to the architecture of Fig. 2. The system consists of a mono-camera, a stereo-camera, a lighting system, a battery pack and an acoustic localization system.

Considering the above-described scenario, the mono-camera is employed for offline high-resolution 3D reconstruction, whereas the stereo-camera is responsible for real-time extraction of spatial information from the scene. An internal power supply is necessary since the ROV's tether does not provide power. The acoustic localization system allows georeferenced data and to supply additional data for navigation.

With the aim of remote controlling the camera, the GoPro Hero 9 Black [12] was chosen. This camera offers a wide field of view (FOV) and a resolution suitable for high-resolution 3D reconstruction. The GoPro API allows for sending setting commands through an easily implementable communication protocol.

A thorough evaluation of commercial stereo-cameras was conducted to identify the most suitable option for the project's objectives. The ZED2i [13] camera offers a wide field of view (FOV) and a robust design that aligns well with the compact dimensions of

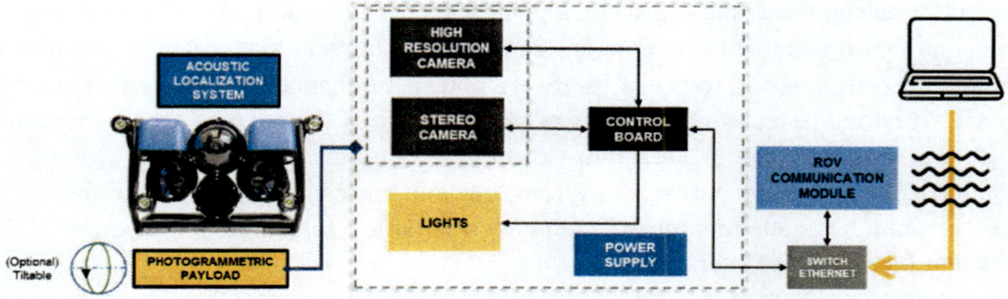


**Fig. 2.** Architecture of the sensor with connections. The photogrammetric payload is attached to the bottom of the ROV and an optional tiltable system can allows to set the proper acquisition angle.

the ROV. Additionally, it is equipped with an integrated IMU, and a dedicated processing unit tailored for artificial intelligence applications. Although the camera's low frame rate may be perceived as a limitation in some scenarios, it does not pose a drawback in our specific application. Moreover, it allows to resize the images to increase the frame rate if necessary.

Both the cameras are synchronized and managed by a custom software installed on the surface control unit.

Given the need to have an accurate position of the underwater vehicle, essential for operating in complex underwater environments, long-baseline (LBL) systems are the most suitable solution, demonstrating a theoretical greater accuracy compared to setups with a lower baseline (short- or ultra-short- baseline) [14, 15]. Figure 3 illustrates the architectural framework delineating the constituent elements of the system and the communication logic.

To address the challenges related to rapid deployment and recovery, as well as to avoid mooring on UCH sites, the proposed architecture incorporates a network of buoys equipped with GPS and beacons. This design allows for georeferencing the buoys solely based on GPS, eliminating the need for fixed point positioning.

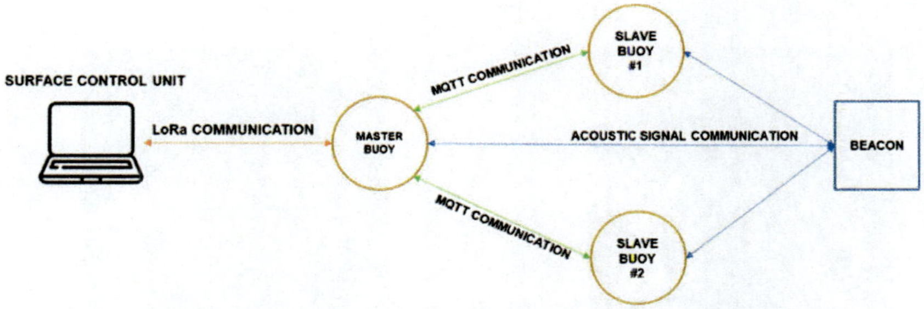


**Fig. 3.** LBL system architecture and communication connections.

The acoustic localization system comprises a group of three buoys (a master and two slaves), and an underwater acoustic beacon. On the surface, the communication between the master buoy and the surface control unit is through LoRa [16] technology. That technology is capable to cover a long range with relatively low energy consumption and being cost-effective in terms of hardware and maintenance. The slave buoys communicate with the master using a MQTT protocol [17], in a shorter range. In underwater, through the acoustic channel, each buoy knows the range from the beacon. The master buoy interrogates the slave buoys at different time and knowing the position of the buoys themselves and the acoustic range of each one is capable to calculate the position of the underwater beacon by trilateration.

The control software is integrated in the same console that manage the optical payload. Considering the general architecture, the system allows the user different options regards the data georeferencing:

- The images are collected by the photogrammetric system and georeferenced for the sake of the acoustic localization system.
- The images are collected and georeferenced by the fusion of both acoustic localization data, additional sensors and visual odometry.
- The images are collected and georeferenced by the deployment of a SLAM algorithm which provide a scaled and globally consistent reconstruction of the seafloor.

The open-source architecture of the low-cost ROV allows strong customizations of the environment. The software is developed in the framework of the ROV and integrates the mission planner, the sensors manager module, and the elaboration module.

## 3 Experimentation

The system is currently undergoing development, but preliminary experimentations and simulations were already conducted. Tests and simulations involve separately the optical payload and the acoustic localization system. The optical payload was tested using a functional prototype to assess the feasibility of using computer vision to get proprioceptive parameters to aid the ROV pilot during navigation (Fig. 4). Meanwhile, the trilateration algorithms of the acoustic localization system were tested in the MATLAB [18] environment, varying the system's characteristic parameters.

(a) (b)

**Fig. 4.** Functional prototype of the optical payload (a), installed on the bottom of a professional-grade ROV (b).

The aim of the test for the optical payload is to compare the navigation solution obtained from the computer vision algorithm of the optical payload, with the fusion of professional-grade sensors, such as the Nortek DVL [19] and an accurate IMU. In particular, the comparison concerns proprioceptive parameters such as position, velocity, and additionally the distance from the target. The experimentation was carried out in controlled environment at the University of Calabria. To replicate artificial seafloor features detectable by the computer vision algorithms, a high-resolution printed canvas was placed at the bottom of a pool of dimensions 6 × 4 × 4 m. While various SLAM algorithms can be utilized, the images were used to fed the ZED SLAM developed by Stereolabs in order to compute proprioceptive parameters of the ROV.

Prior to the experimentation, the proposed sensor was calibrated, acquiring images at different distance and angles, in the same underwater environment using a 22 × 22 checkerboard with each square side measuring 45 mm. The camera model parameters were computed scaling the 3D reconstructed scene with reference object and optimizing the calibration process in Agisoft Metashape [20]. Following the calibration dataset, two separate datasets were obtained by navigating the ROV along two trajectories and varying the height offsets by increments of 1 m, up to a maximum of 3 m above the pool bottom. The first dataset was acquired systematically as the ROV circumnavigated the pool perimeter, while the second dataset simulated the acquisition of transects comprising three parallel lines.

The results of the first test are shown in Fig. 5. Considering a NED reference frame, the figure depicts on the plane N-E the trajectories of the ROV navigating around the perimeter of the pool at a 1 m height from the bottom, computed by both the fusion of professional-grade sensors installed on the ROV and the ZED SLAM. On the right of the figure, the velocity computed by the sensor fusion, the SLAM, and the DVL are compared. The Root Mean Square Error (RMSE) between the SLAM and DVL velocity data is 0.095 m/s. A comparison between the distance from the bottom of the pool (in means of altitude) computed by the SLAM algorithm and the ones of the DVL sensor, along the followed path, is presented in Fig. 6. The different estimates between altitudes decreases after the closure of the loop of the SLAM algorithm.

The statistics in Table 1 were computed by considering the fitting plane of the points in the 3D reconstruction provided by the stereo-camera, which approximates the entire 3D reconstructed surface for each pair of stereo images.

The results indicate that the errors in mean distance and velocity computed by the SLAM algorithm at each epoch do not significantly affect the performance of a photogrammetric mission. In fact, considering the characteristics of the mono-camera sensor, the RMSE in velocity only causes a variation in the overlap of approximately 2% at a frame rate of 15 frames per second and an altitude of 1 m when advancing in the direction of the sensor's height.

Regards the acoustic localization system, the goal of the simulations is to assess the performance of state-of-the-art trilateration algorithms in order to define the one that guarantees the minimum error for a given setup. In addition to the classic spherical trilateration algorithm, two iterative algorithms were evaluated in terms of performances: Gauss-Newton and Levenberg-Marquardt. Following the guidelines and assumptions of the simulation methodology proposed in literature [21], the schematic model of the

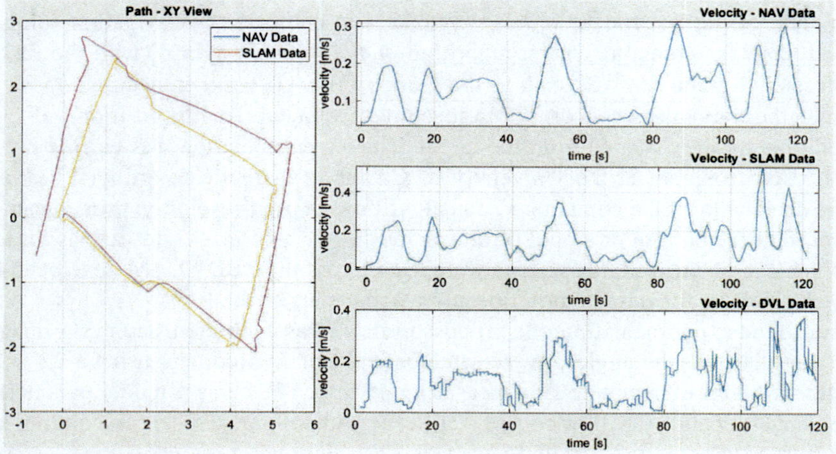


**Fig. 5.** Comparison between professional-grade sensor (blue path) and ZED SLAM (yellow path) navigation solution, along with velocity plots on the right (Color figure online).

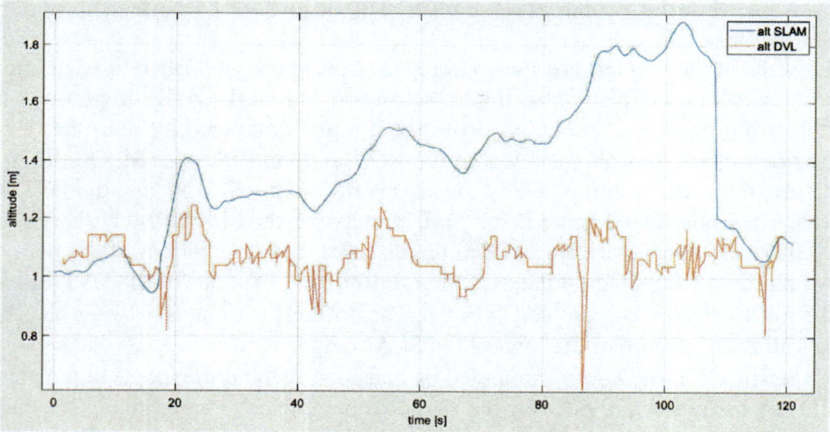


**Fig. 6.** Altitudes computed by SLAM (blue) and acquired by DVL (red). The error decreases when the SLAM algorithm is able to close the loop (Color figure online).

**Table 1.** Descriptive statistics of the ZED2i camera's distance estimation in relation to the 3D point cloud, such as mean distance, standard deviation, and percentage error.

| Nominal distance [m] | Mean distance [m] | STD deviation [m] | % Error [–] |
|---|---|---|---|
| 1 | 0.988 | 0.028 | 1.2 |
| 2 | 1.968 | 0.053 | 1.6 |

process, depicted in Fig. 7, was defined. Assuming the error of each sensor involved in the LBL system [21, 22], the ranges and the GPS position of the buoys were computed as input for the simulation. Then, the positioning of the acoustic unit is estimated and compared with the reference position using the different algorithms.

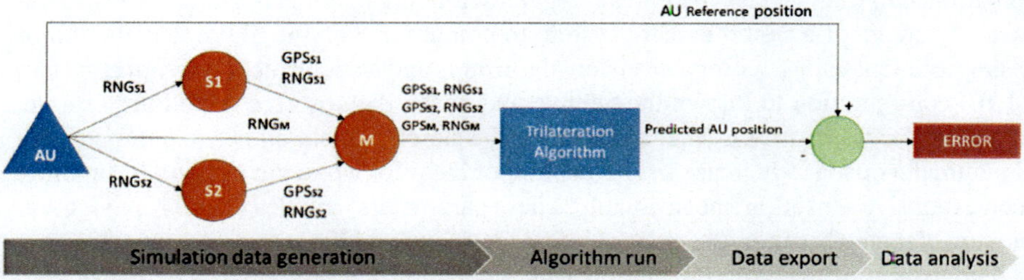


**Fig. 7.** Schematic model of the simulation process. AU – underwater beacon; M – Master buoy; S – Slave buoys.

It has been assumed that the acoustic unit moves according to a spiral trajectory at a given depth which remains within the perimeter identified by the three buoys. For the simulation it is considered a baseline of 100 m, a mean error on ranging of 1 $\pm$ 0.5 and a circular error probable on GSP of 3 m.

Figure 8a illustrates a sample of the reference spiral trajectory of the vehicle and the estimated trajectory. It can be observed that the estimated positions obtained from Gauss-Newton and Levenberg-Marquardt algorithms are very similar. After 400 simulation runs, the cumulative distribution function (CDF) of the absolute error was evaluated, Fig. 8b. The behavior of the functions confirms that the spherical trilateration algorithms outperform Levenberg-Marquardt and Gauss-Newton algorithms for the given system configuration. Specifically, for the simulation runs, the 95% error is estimated to be 9.31 m for Levenberg-Marquardt, 6.88 m for spherical trilateration, and 9.00 m for Gauss-Newton.

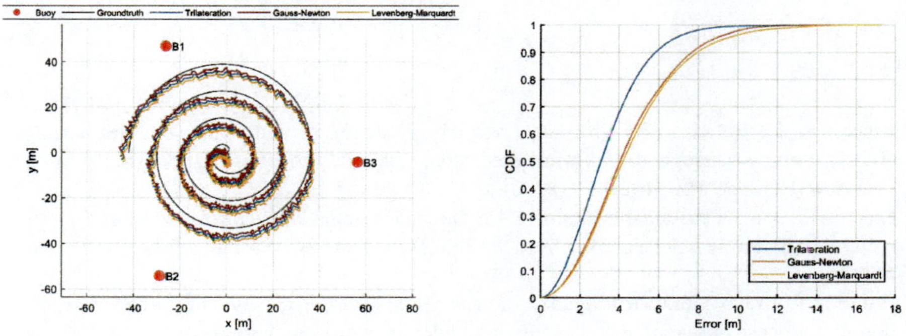


(a) (b)

**Fig. 8.** Results of the simulation carried out in MATLAB for the given system configuration.

## 4 Conclusions

The paper presents the architecture of a low-cost robotic platform system, proposed in the MSCA-RISE H2020 TECTONIC, for the documentation and mapping of UCH sites. In particular, the solution here presented aim to support ROV pilots in underwater documentation activities such as mapping and georeferencing of collected data. The solution takes advantage of a stereo-camera system to compute an online 3D reconstruction of the environment using a computer vision algorithms and an acoustic localization system in LBL configuration to known the vehicle and hence the acquired data. Furthermore, the data of the underwater scene, acquired by the stereo-camera allows to compute useful parameters used to increase the perception of the pilot while the mission is ongoing, when expensive sensors are not available. These parameters includes position, navigation velocity, distance between the camera reference frame and the target.

The results of the experimentation with a functional prototype can be considered satisfactory; comparing the SLAM results with the navigation solution provided by the sensor fusion, the RMSE on the velocity can cause a variation in the overlapping of two consecutive images of about 2% on a range of 70–80%. Regarding the acoustic localization system, the results concerning the performance of the three tested trilateration algorithms indicate that the spherical trilateration algorithm performs better than the iterative methods. However, further investigation is deemed necessary to assess the influence of the system's characteristic parameters and improve its performance. This could involve conducting a Design of Experiments (DOE) analysis to explore potential enhancements or further experimental tests on the field. Although development is ongoing and further testing is needed, the results are promising and demonstrate that the proposed solution can be used complementarily to expensive sensors with an acceptable degree of accuracy.

**Acknowledgments.** This research has been supported by the project "Technological Consortium TO develop sustainability of underwater Cultural heritage (TECTONIC)", financed by the European Union (Horizon 2020 research and innovation programme under the Marie Skłodowska-Curie grant agreement No 873132).

## References

1. Ricca, M., La Russa, M.F.: Challenges for the protection of underwater cultural heritage (UCH), from waterlogged and weathered stone materials to conservation strategies: an overview. Heritage **3**(2), 402–411 (2020)
2. Ricca, M., et al.: A sustainable approach for the management and valorization of underwater cultural heritage: new perspectives from the TECTONIC project. Sustainability **12**(12), 5000 (2020)
3. Gambin, T., Bravo-Morata Rodríguez, A., Sausmekat, M.: from discovery to public consumption: the process of mapping and evaluating underwater cultural heritage in Malta. Heritage **4**(4), 2732–2745 (2021)
4. TECTONIC Project Website: Available online: https://www.tectonicproject.eu/. Last accessed 1 May 2023

5. Bruno, F., Lagudi, A., Barbieri, L., Rizzo, D., Muzzupappa, M., De Napoli, L.: Augmented reality visualization of scene depth for aiding ROV pilots in underwater manipulation. Ocean Eng. **168**, 140–154 (2018)
6. Su, X., Ullah, I., Liu, X., Choi, D.: A review of underwater localization techniques, algorithms, and challenges. J. Sensors **2020**, 1–24 (2020)
7. Rossi, M., Dooly, G.: Real-time vision systems for underwater robotics. Sensors **18**(11), 3936 (2018)
8. Laranjeira, M., Arnaubec, A., Brignone, L., Dune, C., Opderbecke, J.: 3D perception and augmented reality developments in underwater robotics for ocean sciences. Curr. Robot. Rep. **1**, 123–130 (2020)
9. Tippetts, B., Lee, D.J., Lillywhite, K., Archibald, J.: Review of stereo vision algorithms and their suitability for resource-limited systems. J. Real-Time Image Proc. **11**(1), 5–25 (2013). https://doi.org/10.1007/s11554-012-0313-2
10. Conte, G., Zanoli, S.M., Scaradozzi, D.: Guide and control of an unmanned underwater vehicle by visual-feedback techniques. IFAC Proc. **35**(1), 67–92 (2002)
11. BlueROV2 product website: Available online: https://bluerobotics.com/store/rov/bluerov2/. Last accessed 1 May 2023
12. GoPro Hero 9 Black product page. https://gopro.com/it/it/shop/cameras/hero9-black/CHDHX-901-master.html. Last accessed 1 May 2023
13. Zed2i product page: https://www.stereolabs.com/zed-2i/. Last accessed 1 May 2023
14. Cario, G., Casavola, A., Gagliardi, G., Lupia, M., Severino, U.: Accurate localization in acoustic underwater localization systems. Sensors **21**(3), 762 (2021)
15. Bruno, F., et al.: Development and integration of digital technologies addressed to raise awareness and access to European underwater cultural heritage. An overview of the H2020 i-MARECULTURE project. In: Oceans 2017-Aberdeen, pp. 1–10 (2017)
16. LoRa website. Available online: https://lora-alliance.org/about-lorawan/, last accessed 2023/05/01.
17. MQTT website: Available online: https://mqtt.org/. Last accessed 1 May 2023
18. MATLAB website: Available online: https://it.mathworks.com/products/matlab.html. Last accessed 1 May 2023.
19. Nortek DVL 1000 product website. Available online: https://www.nortekgroup.com/products/dvl-1000-300m. Last accessed 1 May 2023
20. Agisoft website: Available online: https://www.agisoft.com/. Last accessed 1 May 2023
21. Aparicio, J., Jiménez, A., Álvarez, F.J., Ruiz, D., De Marziani, C., Ureña, J.: Characterization of an underwater positioning system based on GPS surface nodes and encoded acoustic signals. IEEE Trans. Instrum. Meas. **65**(8), 1773–1784 (2016)
22. Turetta, A., Casalino, G., Simetti, E., Sperindè, A., Torelli, S.: Analysis of the accuracy of a LBL-based underwater localization procedure, pp. 1–7. In: 2014 Oceans-St. John's, IEEE (2014)

# Student Training on IoT Marine Surveying: A Hands-On Perspective by Means of Remotely Operated Underwater Vehicles

Triantafyllia Anagnostaki, Sofia Sarakinioti, Dimitrios Tziourtzioumis, and Theodoros Kosmanis(✉)

International Hellenic University, Alexander University Campus, 57400 Sindos, Greece
{dtziour,kosmanis}@ihu.gr

**Abstract.** The detailed steps and procedures taken during a hands-on course of a smart marine and maritime surveying postgraduate program are described in this work. The purely practical and experimental course combined with student internship, forced the students to engage themselves to real condition procedures and measurements, focusing in the development of a sensor cluster for the assessment of water quality. Starting from the setup of the sensors and their integration on a Remotely Operated Underwater Vehicle, the course also incorporates the calibration of sensors, their testing in safe and clear water and finally their use in order to assess the environmental conditions of a fish farm. All measurements, performed in two coastal areas of Greece and France, were be uploaded live to the cloud. Although the structure was not combined with Artificial Intelligence during the course, it is designed and developed in order to incorporate bidirectional communication with the Remotely Operated Underwater Vehicle in the near future.

**Keywords:** Marine Surveying · Maritime Surveying · Training · ROUV · Environmental Measurements

## 1 Introduction

Internet of Things (IoT) and Artificial Intelligence (AI) are technologies that literally rule the technical everyday life for the past decade. There is no technological sector that is not connected even in a small portion with IoT or AI [1]. Similarly, in the marine and maritime sector these technological changes are also with a strong tendency to replace or improve traditional equipment, operations or procedures with new ones based on IoT and AI [2–6].

Supported by European Union's Erasmus+ Programme: Knowledge Alliances call under grant agreement 612198-EPP-1-2019-1-ES-EPPKA2-KA.


R. Mehmood et al. (Eds.): DCAI 2023, LNNS 741, pp. 224–231, 2023.
https://doi.org/10.1007/978-3-031-38318-2_23

Towards this direction, a piloting postgraduate course has been implemented during the past three years, offering the participants specialized training on modern subjects and techniques based on IoT and AI for performing surveys and monitoring in maritime and marine applications [7]. The piloting course with the acronym SMARTSEA, produced out of its title "Surveying and MARiTime internet of thingS EducAtion", entails a significant laboratory character. Based on the large scale experiments, the program provides hands-on experience that combines IoT and AI with marine and maritime measurement procedures.

This paper presents in details the large scale marine experiments taken place in the frame of SMARTSEA course, as the most challenging activity of the overall educational scheme that was launched after two years of preparatory work [8,9]. The experiments are based on a sensoring cluster appropriately developed to measure water quality parameters either as standalone on a testing and calibration phase or adjusted to a Remotely Operated Underwater Vehicle (ROUV), thus providing additional flexibility. It incorporates two typical temperature and pressure sensors and a series of environmental ones, namely Oxygen Reduction Potential (ORP), pH, dissolved oxygen (DO) and conductivity sensors. The calibration procedure as well as the testing and validation phase performed in the coastal areas of Thessaloniki, Greece and la Seyne-sur-Mer, France analyzed in the following sections.

## 2 Measuring Equipment

A cluster of sensors was the basic equipment used by the students in order to perform the on site measurements. The sensor cluster could be implemented either as standalone either attached on a ROUV for more flexibility in underwater measurements. The second part of the equipment was, as mentioned, an ROUV especially acquired for the SMARTSEA educational course [9] (Fig. 1).

**Fig. 1.** pH and ORP sensors connected to the ROUV.

### 2.1 Sensor Cluster

The sensor cluster was composed of six (6) sensors, measuring the Temperature, Pressure, Oxygen Reduction Potential (ORP), pH, dissolved oxygen (DO) and conductivity of the water. Table 1 presents sensors' specifications whereas the communication protocol and port used for their connection with the on board computing platform are presented in [9].

**Table 1.** Specifications of the sensor cluster items.

| Sensor Type | Range | Accuracy |
|---|---|---|
| Temperature [°C] | $-45 : +125$ | $\pm 2\%$ |
| Pressure [bar] | $0 : 30$ | $\pm 0.2$ |
| ORP [mV] | $\pm 2000$ | $\pm 1$ |
| pH | $0 : 14$ | $\pm 0.001$ |
| Conductivity [mS/cm] | $0.01 : 1000$ | $\pm 2\%$ |
| Dissolved Oxygen [mg/lt] | $0 : 100$ | $\pm 0.05$ |

According to the teaching schedule of the SMARTSEA program, the expected values of the sensors according to the water environment were described and analyzed to the students. Thus, measurements' evaluation was possible. The expected values per sensor are noted below.

**Pressure Sensor.** The pressure sensor is used to indicate the exact depth of the sensor cluster at each measurement. Specifically, one should estimate 1 bar of pressure per 1m of depth.

**Temperature.** In seawater, temperature can be affected by environmental conditions. Factors affecting temperature are sunlight, solar radiation, heat transfer due to convergence of atmospheric flows, and turbidity. Water temperature is a parameter that decreases gradually with increasing depth.

**ORP.** In healthy water bodies, ORP levels should be between 300 and 500 mV. ORP is dependent on the amount of dissolved oxygen in the water and the amount of other elements that act like oxygen, such as sulfur (S) and selenium (Se). Low ORP can lead to low dissolved oxygen and increased toxicity of certain metals and contaminants, resulting in many dead putrefactive substances in the water that cannot be purified or decomposed. Such an aquatic environment description does not match a healthy environment for fish and bugs.

**pH.** pH is a measure of the acidity or alkalinity of a liquid solution representing the concentration of hydrogen ions (H+) and hydroxide ions (OH−) on a scale of 0–14. Solutions with a pH below 7 are considered acidic and solutions with a pH above 7 are considered alkaline. The average pH of the ocean is about 8.1, meaning basic or alkaline. However, human activity has fundamentally changed the chemical composition of the ocean leading to a drop in seawater pH, a phenomenon called ocean acidification that threatens marine ecosystems.

**Dissolved Oxygen.** Dissolved oxygen (DO) refers to the amount of unbound free oxygen present in the water. It constitutes a parameter that evaluates water quality based on the impact on organisms inhabiting the water area. Dissolved oxygen levels can be high or low. Both can harm aquatic life and affect water quality. The dissolved oxygen concentration in seawater is typically 7 to 8 milligrams per liter (mg/lt). Dissolved Oxygen is affected by phenomena like diffusion, aeration, photosynthesis, respiration, and decomposition, while it varies significantly with temperature, pressure, and salinity. The solubility of dissolved oxygen decreases with increasing temperature, meaning that larger depths dissolved oxygen is expected to be lower.

**Conductivity.** The conductivity of a solution, that is concentration of ions in water is a measure of its ability to conduct electricity, sound, and heat. The conductivity of seawater depends on temperature and salinity and is independent of pressure (depth). Salts such as sodium chloride (NaCl) dissociate in water to form cations and anions that migrate in the presence of an electric field to generate an electric current. As a result, conductivity increases with the concentration of dissolved salt, i.e. salinity. In addition, the conductivity increases with increasing temperature due to the increased ion mobility. Sea water has a typical conductivity between 30 mS/cm and 60 mS/cm.

### 2.2 Remotely Operated Underwater Vehicle

The sensor cluster was hosted by an appropriately developed ROUV. The vehicle used for the measurements was based on the Blue ROV2 model of Blue Robotics [10], the chassis of which was modified in order to host the extra environmental sensors. Similarly, a graphical user interface (GUI) was developed in order to acquire sensor data and upload them on the cloud [9]. Besides the sensors, it is worth mentioning that the ROUV operation is powered by a 14.8 V, 15.6 Ah Li-ion battery providing an average of 2 h autonomy under constant thruster use.

## 3 Measurements

The equipment described in the previous section was utilized by the students in two cases. As it was important for the sensors to be tested in a real water environment, the two sets of measurements took place in coastal areas, one clear of any aquatic entrepreneurial activity in Thessaloniki, Northern Greece and one in the premises of a fish farm in La Seyne-sur-Mer, Southern France (Fig. 2).

Initially, and prior to any measurements at sea, the sensors were calibrated. Afterwards, a test measurement procedure was followed in order to stress the equipment under the real conditions. The water proof of the ROUV chambers was checked as well as the ability of the sensors to provide rational, in the expected range measurements. It is noted that all procedures were performed by the students as part of their training, under appropriate supervision. The educational added value lies not only in the preparation phase of the equipment or in the measurements themselves but also in the post-measurement procedure, when the outcomes had to be assessed. Data were acquired via the secure shell (ssh) protocol and uploaded to the cloud enabling any possible bidirectional reaction, e.g. ROUV depth stabilization or specific movement.

### 3.1 Sensors' Calibration

Essential to improve sensor performance, calibration procedure was preceded any measurement performed. The procedure took place in two parts. First, the temperature and pressure sensors, built in the ROUV, were calibrated through the ROUV control software. The four environmental sensors were calibrated via a specific procedure defined by the manufacturer using appropriate chemical solutions. The sensors were put in sample liquids and their control software was adjusted to read each sample as an appropriate value. At the end, all sensors were able to measure according to their respective range.

Although trivial for any measuring equipment, the calibration procedure had to be performed by the students themselves in order to realize its significance.

### 3.2 Measurements in Greece

The first set of measurements took place in the coastal area of Thessaloniki prefecture in Northern Greece on June 1st 2022. The area characterized by the absence of any aquatic entrepreneurial activity that would cause the water parameters to deviate from their typical values, was ideal to test the ability of the environmental sensors to measure correctly. Additionally, the ROUV chambers accommodating the battery and the electronic components was tested for being leakproof. Thus, a series of tests from an appropriate checklist has been performed prior to any ROUV dive. Measurements from the sensors were taken along various paths and depths up to 2 m, as the waters were shallow. Such a path is depicted in Fig. 3 whereas in Table 2 the values acquired by all sensors in the three measurement spots 3, 4 and 5 at a constant depth of 1.5 m are presented. The spots are characterized as points G3, G4 and G5, with direct correspondence to points 3, 4 and 5 of Fig. 3. The results were typical of environmentally healthy sea water, according to the description of Sect. 2, with sensor values being relatively constant.

**Fig. 2.** Map with the locations of test measurements in Greece and France.

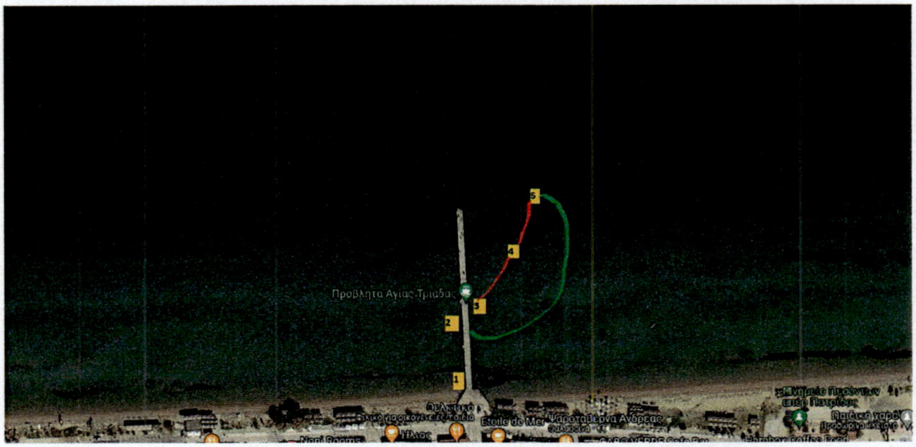


**Fig. 3.** Measurement points in a coastal area in Thessaloniki, Northern Greece.

**Table 2.** Specifications of the sensor cluster items.

| Sensor Type | G3 | G4 | G5 | FW | FN | FS | FE |
|---|---|---|---|---|---|---|---|
| Conductivity [mS/cm] | 48.10 | 48.20 | 48.40 | 56.998 | 56.923 | 56.978 | 56.974 |
| ORP [mV] | 115.1 | 115.0 | 115.1 | – | – | – | – |
| pH | 8.75 | 8.77 | 8.73 | 8.20 | 8.20 | 8.17 | 8.19 |
| Dissolved Oxygen [mg/lt] | 6.36 | 7.31 | 7.76 | 7.89 | 7.92 | 7.90 | 7.91 |
| Temperature [°C] | 25.8 | 25.2 | 24.8 | 24.9 | 24.7 | 24.7 | 24.8 |

**Fig. 4.** Measurement points around a fish farm in La Seyne-sur-Mer, Southern France.

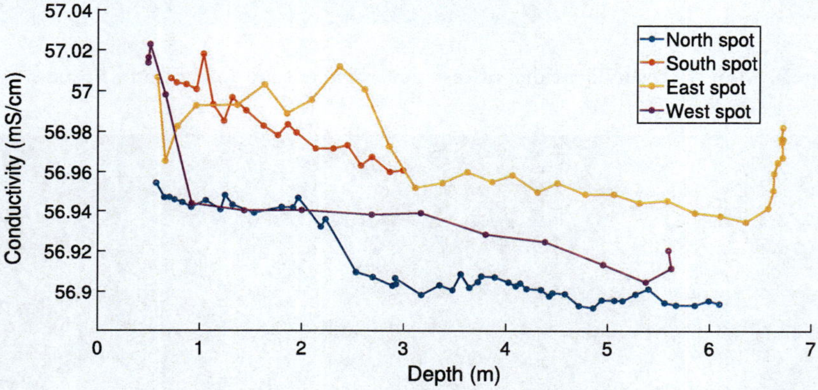


**Fig. 5.** Conductivity graph versus depth from the four measuring spots around a fish farm in La Seyne-sur-Mer, France.

### 3.3 Measurements in France

The second set of measurements took place around a fish farm in La Seyne-sur-Mer, Les Sablettes, France on June 29th 2022 in the frame of the industrial internship of the students. The sensor cluster was already calibrated and tested and could be used for the assessment of the water quality around the fish farm (Fig. 4). Indicatively, data were acquired at four (4) measurement spots around a fish cage and one in the middle of it. Table 2 presents the average values of the data acquired from each sensor at the specific spots around the fish cage (FE, FN, FW, FS correspond to E, N, W, S respectively). Averaging was performed on all values acquired by each sensor from each spot during a dive from the sea surface until a maximum depth of about 7 m. The variation of values measured per sensor along this distance was insignificant as depicted in Fig. 5. Similar results with those taken from the first measurement set in Thessaloniki district were observed, as expected since the water quality was equally good. Apparently, any deviation would be an indication of abnormal water conditions. The findings were similar as those of the first set in Greece. It is noted that due to a sudden, irreversible sensor fault, no ORP data were possible to be acquired.

### 3.4 Conclusions

As the need to increase the use of smart measuring equipment and approach in maritime and marine sector increases, the procedures of a purely laboratorial course in the frame of SMARTSEA postgraduate program was presented in this paper. The development and calibration of a sensor cluster was initially described. The measurements performed in two coastal areas of different characteristics were presented emphasizing mostly on the safety procedures and the acquisition of reliable sensor values. The overall structure provides increased flexibility as connected in a wireless network it can provide remote access to underwater premises by means of the IoT technology, fully aligned with the current trends of marine and maritime sector.

## References

1. OECD: Artificial Intelligence in Society. OECD Publishing, Paris (2019)
2. Cancela, J.-J., González, X.-P., Vilanova, M., Mirás-Avalos, J.-M.: Water management using drones and satellites in agriculture. Water **11**(5), 874 (2019). https://doi.org/10.3390/w11050874
3. Chen, C.-H., Wu, Y.-C., Zhang, J.-X., Chen, Y.-H.: IoT-based fish farm water quality monitoring system. Sensors **22**(17), 6700 (2022)
4. Lafont, M., Dupont, S., Cousin, P., Vallauri, A., Dupont, C.: Back to the future: IoT to improve aquaculture : real-time monitoring and algorithmic prediction of water parameters for aquaculture needs. In: 2019 Global IoT Summit (GIoTS), Aarhus, Denmark, pp. 1–6 (2019). https://doi.org/10.1109/GIOTS.2019.8766436
5. Adhipramana, M., Mardiati, R., Mulyana, E.: Remotely operated vehicle (ROV) robot for monitoring quality of water based on IoT. In: 2020 6th International Conference on Wireless and Telematics (ICWT), Yogyakarta, Indonesia, pp. 1–7 (2020). https://doi.org/10.1109/ICWT50448.2020.9243614
6. Kabanov, A., Kramar, V.: Marine internet of things platforms for interoperability of marine robotic agents: an overview of concepts and architectures. J. Mar. Sci. Eng. **10**(9), 1279 (2022)
7. SMARTSEA MSc homepage. http://www.smart-sea.eu. Accessed 12 Mar 2023
8. Kravari, K., Tziourtzioumis, D., Kosmanis, T.: AIRUV: a remotely operated underwater vehicle with artificial intelligence perspectives. In: Rodríguez González, S., et al. (eds.) DCAI 2020. AISC, vol. 1242, pp. 116–125. Springer, Cham (2021). https://doi.org/10.1007/978-3-030-53829-3_11
9. Charitonidis, A., Tziourtzioumis, D., Kravari, K., Kosmanis, T.: Marine surveying education by implementing artificial intelligence on remotely operated underwater vehicle sensors. In: González, S., et al. (eds.) DCAI 2021. LNNS, vol. 332, pp. 110–119. Springer, Cham (2022). https://doi.org/10.1007/978-3-030-86887-1_10
10. Blue Robotics Inc. homepage. http://www.bluerobotics.com. Accessed 25 May 2021

# Special Session on Power System Optimization and Machine Learning Models (PSO-ML)

# Analyzing Electric Vehicle Charging Behaviour Using Advanced Clustering Tools

Carlos Ernani da Veiga[1](✉), Carlos Ramos[2], Juan Manuel Corchado[3], Piara Fernandes[1], and João Soares[2]

[1] DAE - Academic Department of Electrotechnics, IFSC - Federal Institute of Santa Catarina, Florianopolis, Brazil
{cveiga,piara}@ifsc.edu.br

[2] GECAD - Research Group On Intelligent Engineering and Computing for Advanced Innovation and Development, LASI - Intelligent Systems Associate Laboratory, Polytechnic of Port, Porto, Portugal
{csr,jan}@isep.ipp.pt

[3] BISITE Research Group, University of Salamanca, Edificio I+D+I, 37008 Salamanca, Spain
jm@corchado.net

**Abstract.** Electric vehicle (EV) owners often have electric vehicle charging stations (EVCS) in their homes and leave the EV connected when it remains in the garage. This procedure does not guarantee that EV charging is adequate, and monitoring this period is necessary to identify anomalies. A study was developed where the EV charging curves was analyzed by clustering with K-means. The advantage presented in this work is that the necessary data for the charging analysis come from the EVCS power supply circuit. This proposal allows monitoring and rapidly classifying the data according to the form of the EV charging curves, with no need for the supervision of the input data. The unsupervised machine learning method can classify any situation by collecting data on the electrical power supplied to the EVCS. In this study, the classification was applied with learning performed in the history, identifying the clusters representing abnormal situations to identify anomalies in future charging curves. The results of the proposed approach presented around 27.8% as an anomaly in the EV charging, allowing to identify the moment in which it occurs and to make the necessary adjustments to avoid them.

**Keywords:** Charge battery EV · Classifier Curve Charge · Standard views curves · K-means

## 1 Introduction

In the analysis of the behavior of the charge in an EV has been observed standard characteristics in the charge current. This can be observed in the study of A. Rashidi et al. [1] where the optimized battery curve was indicated, and in the study of R. German et al. [2] where the charging of a set of batteries was exemplified. Moreover, in their article, V. B. Vu et al. [3] discussed the use of current and voltage constants in battery charging, which results in standard charge curves. An example of an EV charging curve


R. Mehmood et al. (Eds.): DCAI 2023, LNNS 741, pp. 235–244, 2023.
https://doi.org/10.1007/978-3-031-38318-2_24

is shown in the blue curve of Fig. 1 (**B**), which considers a complete battery charging process. However, the analysis of this behavior can be broadened by considering different possibilities, such as that the battery is not always fully charged, that the battery was not fully discharged before charging or that it was disconnected before being fully charged. This analysis can be performed using the curves of the charge current of the battery, and the information can be provided by either the EVCS or by collecting the current (or power) supplied to the EVCS.

In a situation where the number of charges and EVs is low, it is feasible to individually analyze the charging graphs generated by the data captured from a EVCS feed. However, in situations where there are many EVs and charging records, the analysis is extremely costly. Our study is based on applying an analysis of the generated curves through machine learning (ML), which would classify the curves according to predetermined criteria and, thus, perform the analysis automatically, allowing for a rapid response to a large volume of data, as addressed by A. K Jain [4]. Other battery applications such as the paper by N. Shateri et al. [5] which proposes a classification technique to estimate the State of Charge of Li-S batteries. The paper by H. Beltran, E. Sansano and M. Pecht [6] analyzes several machine learning algorithms to monitor a battery system through the partial charge and discharge curves. In the paper by K. Chen et al. [7] a hybrid model with a combination of linear discriminant analysis (LDA) and elastic net to detect battery failures through charging curves. Still on the classification of charging and discharging batteries, have the paper by S. Gorrachategui et al. [8] which addresses the clustering of battery charge and discharge waveforms in a photovoltaic system. However, the classification of curves by clustering is also used in other applications, such as the clustering of charging curves made by S. Balasubramanian and P. Balachandra [9].

In the proposed study, the current of the EVCS power circuit was analyzed and classified it into clusters to identify any anomalies in the battery charging process. The implementation of the procedure will be through the detection of the period of the charging curves, transforming the curves into standard periods with the extraction of the period time and, finally, performing the classification. The main contributions of this work are the following:

- Analysis of residential EV charging demand curves, only with information from the EVCS power supply.
- Identify when an abnormal EV charging occurs.
- Visualization of the type of EV charging curve.
- Aggregation of historical charging data of the EVs.

In addition to the Introduction Section, the paper has five other sections, as described below: Section 2 describes the collection and processing of data, which has allowed to correct any information-related problems, as well as the detection of the times to which the EV charging curves corresponded. Section 3 identifies the means of standardizing the number of records for each curve, which are to be used in the K-means module; this makes it necessary to reduce some charging curves or expand others. Section 4, describes the classification procedure performed in the study, justifying the defined choices. Then, Sect. 5 describes the results obtained from the investigation that has been carried out using the K-means module from the Scikit-Learn [10] library. Finally, in Sect. 6, conclusions are drawn from the conducted study regarding the classification of charging curves in EV.

## 2 Data Processing

For the analysis of charging curves by ML, it has been necessary to standardize the data to be used as in the analysis. In the case study conducted in this paper, the information has been obtained from the Pecan Street Dataport [11], which provides data on power consumption, registered per minute, at different sockets to which a EVCS was connected.

It should be noted that the raw data collected from the Pecan Street Dataport source, from January 2018 to December 2019, may have information gaps, such as null data or missing records in some periods. So, it has been necessary to follow a series of steps prior to using the clustering module, such as collection and correction; processes which are described below. The evaluated dataset refers to 12 homes located in the states of Texas (7) and New York (5), which have EVs, being analyzed the charging of 12 electric vehicles.

### 2.1 Collection and Correction of Data for Analysis

The data recorded by the data source provided information on one EVs each home at 1-min interval, and in the data obtained for the EV, there were approximately 10% of records with null values. To repair data with null values, the criterion of using the closest value was adopted, prioritizing the preceding record, as the value of the evaluated field remained constant for long periods and null values had sparse behavior, with correlation of the data in the sequence. According to [12], Linear Interpolation, Cubic Spline Interpolation, Lagrange Polynomial Interpolation or Newton's Polynomial Interpolation could be used but cites the Nearest Interpolation that proved to be adequate for the situation, in addition to requiring less processing charging. The result of the repair can be seen in Fig. 1 (**A**), where most of the null fields changed to 0.002 kW, that is, close to zero. With this result, the repair of data was considered satisfactory, allowing for the use of these data with an adequate level of quality for the study analyses.

### 2.2 Detection of EV Charging Times

The variation in current has been used to identify the start and the end of charging or, in this case, the power supplied to the EVCS. The method that has been used to detect is an adaptation of the Roberts operator [13] used for edge detection, where the detection of the edge was the start and the end of the EV charging curves. The result can be seen in Fig. 1 (**B**) where the calculations indicating the percentage of variation are highlighted in red.

Some observations, that are necessary for the correct detection of the curve, are described below: the end of the curve only exists after the detection of the beginning of the curve; the start of the curve always precedes the end; the definition of the edge detection threshold must be in accordance with the characteristics of the charging; and, an interval should be defined at the end of the curve to observe variations.

In this analysis, the value of 0.3 kW was used as a threshold, by analyzing the characteristics of the sample data and according to the indication of stoppage in the charging [2] when the current is lower than a certain limit. With the edge detection procedure, 3964 charging curves were detected with charging times ranging from 3 min to 60 h, and the average was 1 h and 28 min. After defining the periods for each curve to be registered in the database, it was possible to carry out the next step of preparing the data to be analyzed by the clustering module.

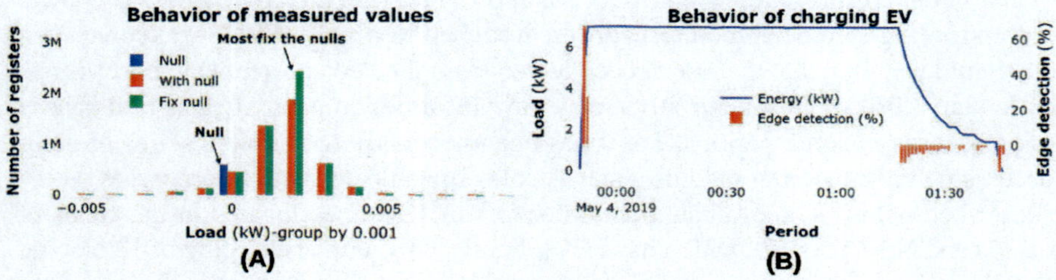


**Fig. 1.** Data behavior in the correction of null values (A) and edge detection (B).

# 3 Standardization of the Number of Records for Each Curve

The ML method needs to be fed with all sets of input features have the same dimension, therefore, to use the data from the charging curves, it is necessary to convert the information to standard dimensions. The EV charging curves contain information regarding the energy delivered to the EVCS as a function of the time that this process lasts. Information about charging times will be evaluated separately from the classification of curves.

The next step was to standardize the number of records for each curve, as this was necessary to feed the clustering module. The adjustment of the number of records can be done in several ways, as described below.

## 3.1 Techniques for Reducing the Number of Records

For the reduction, each technique has its own advantages and disadvantages. In the next subsection, each technique is described and assessed, with the aim of identifying the most suitable one for situations where the number of records in the EV charging curve is reduced.

Two methods use existing point associations as defined below: The first is the average of the points in each step used to identify the points to be represented in the sample, a simple average of the values contained in each time step is made; The second method is the point closest to the center of the step, consists of evaluating the original points to the left and right of the central point of each step in the analyzed data, selecting the value that corresponds to the closest point and, in the case of a tie, the point to the left is selected. The start and end points of the sample correspond to the original start and end points, respectively.

Two other methods represent the curves by wave composition at different frequencies, as described below; In Discrete Fourier Transform (DFT), the graphic representation is decomposed by a series of sine waves with different frequencies and amplitudes, which together reconstruct the represented wave. The description and application in Python is in Python Programming and Numerical Methods [12]; The other way is the Discrete Wavelet Transform (DWT) [14], described as an efficient application for temporal series in [15] and for applications in the financial market in [16], separates the original data into two groups, one representing the high frequencies and the other representing the low frequencies.

Another five methods for reducing samples from a curve are presented by S. Steinarsson [17] and J. Xu et al. [18], which are the following: The Largest Triangle One Bucket algorithm; The Largest Triangle Three Bucket; The Mode Median Bucket algorithm; The Min Std Error algorithm; and the Longest Line Bucket (LLB) algorithm.

### 3.2 Techniques for Enlarging the Number of Records

In this Section, methods that calculate the new values to complement the sample are analyzed, considering the following methods. In Partial lines algorithm, the angular and linear coefficients of the lines between the points of the original data are calculated, with these coefficients, the points are calculated for each step, resulting in the sample points. The Discrete Fourier Transform (DFT) use to build a data sample had already shown inefficiency in the data studied, for the case of reduction discussed in Section Discrete Fourier Transform and, in data expansion, it was less efficient. For Nearest point procedure, the choice of the closest point to the defined step was adopted. The algorithm defines the points for each step by selecting the closest point to the analyzed step, resulting in an expansion of the existing data.

## 4 Methodology

To classify the EV charging curve data, data processing was first performed as described in Sect. 2 and shown in the flowchart in the Fig. 2. Once the data had been corrected and the periods of each curve were known, the algorithm was implemented to build the sample of each curve. For the study, 60 steps were adopted as standard for the sample of graphs, resulting in 62 records for each graph, due to the addition of the start and end points. With this standardization, the existing data were divided between 45.5% of the data with larger records that were then reduced, 53.7% of the curves whose records had to be expanded and less than 1% whose records did not require any changes.

The LLB method was adopted for use in reducing curve records and for data expansion, two analyses were performed, one using the Partial lines method and the other using the Nearest Point. In the case of the Partial lines method, it provided a result that was close to the representations of the graphs showing the original data.

The data needed to feed the clustering method were organized in a matrix with 62 columns and 3964 lines, represented the curves detected in the data. The information on

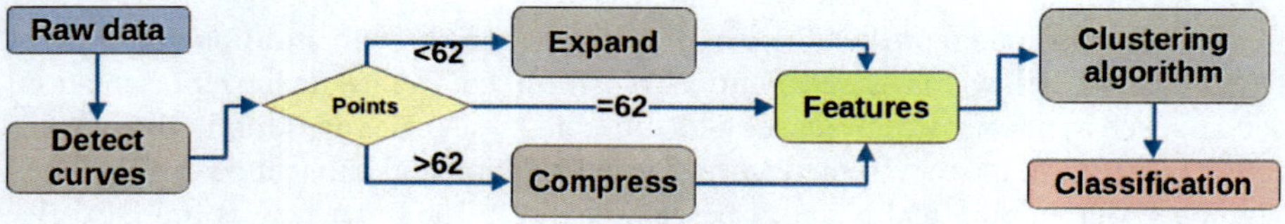


**Fig. 2.** Flowchart of the classification process.

the period during which the charging occurred was recorded but not used in the clustering method.

All algorithms were built in Python language, with PostgreSQL DB and libraries like Pandas for data analysis. Including the algorithms for edge detection with the Roberts operator [13], the algorithms tested for data reduction such as the algorithm for calculating the DFT from the book Python programming and numerical methods [12], the algorithm developed to determine the developed DWT, as described in efficient time series matching by wavelets [15], as well as the algorithms presented by S. Steinarsson on Downsampling Time Series for Visual Representation [17] and the algorithm that determines the shortest path implemented by Max Burstein [19]. The algorithms to determine the near point and to calculate the points using partial lines were based on techniques presented in the book Python programming and numerical methods [12].

## 5 Implementation

The results of three Scikit-Learn libraries were analyzed, balanced iterative reducing and clustering using hierarchies (BIRCH), Gaussian Mixture and K-Means, as they allow using learning to classify new data. The initial analysis was to identify the ideal number of clusters to automatically classify the data, with the curve that relates the number of clusters [20] with the calculated distortion value for each set as a way of choosing the number of clusters. The result obtained in this analysis presents the best result in between 4 and 7 clusters, represent an ideal value in the relationship between distortion and number of clusters to classify the data.

After analyzing several values for the number of clusters, where it was possible to verify the main characteristics of the existing curves in the data, however, it was not a reasonable value for the classification of the curve data. After the tests, the best results were identified with 7 random centroids, so this will be the choice. The difference between classification with random centroids and classification with initially selected centroids shows the number of curves classified in each type of cluster with similar values.

In this study, there was no significant difference in the use of expanding the sample records by the method of the closest point or by partial straight lines, when applied to centroids that had initially been selected.

Another analysis of the results refers to the period of the curves for each type of cluster. In Table 1, the statistics of the data referring to the period are presented. In the analysis, there were type 0, 3 and 5 clusters displayed characteristics that were not consistent with EV charging, the 25% and 75% percentiles represent a very small-time

**Table 1.** Analysis of the period of each curve in relation to the classification clusters.

| Type | count | mean | std | min | 25% | 50% | 75% | Max |
|---|---|---|---|---|---|---|---|---|
| 0 | 331.0 | 80.63 | 403.20 | 4.0 | 10.00 | 14.0 | 17.00 | 4224 |
| 1 | 670.0 | 69.50 | 46.31 | 8.0 | 39.00 | 53.0 | 86.00 | 295 |
| 2 | 457.0 | 57.24 | 52.48 | 7.0 | 34.00 | 50.0 | 64.00 | 457 |
| 3 | 162.0 | 11.39 | 41.59 | 4.0 | 4.00 | 4.0 | 4.00 | 330 |
| 4 | 774.0 | 171.90 | 141.43 | 24.0 | 76.25 | 149.0 | 211.75 | 815 |
| 5 | 608.0 | 10.92 | 2.42 | 6.0 | 9.00 | 11.0 | 12.00 | 22 |
| 6 | 962.0 | 119.38 | 72.49 | 12.0 | 65.00 | 100.0 | 176.00 | 342 |

range. Therefore, the classification of these curves could be used to indicate the curves that should be considered anomaly in the functioning of the EVCS.

Normalizing the amplitude of the curves by the maximum value of each curve results in a better classification, as seen in Fig. 3 (**A**), where the result with 7 initially selected centroids was like the result with 7 random centroids.

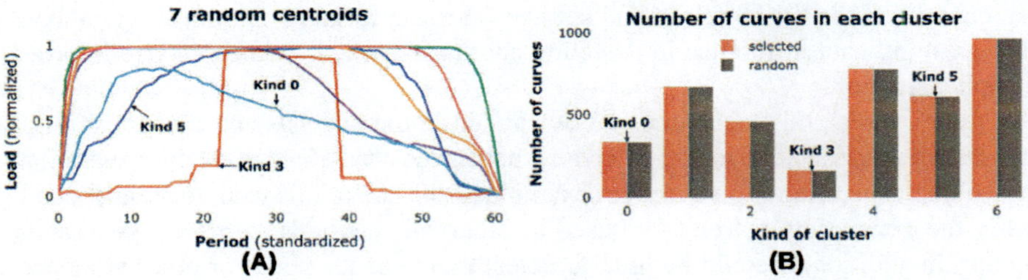


**Fig. 3.** Random centroids (A) and initially selected and random centroids (B).

The classification carried out with the selected centroids has resulted in a very similar number of selected curves in each cluster, as can be seen in the graph in Fig. 3 (**B**).

In a general analysis, the normalization of curve values resulted in a more adequate classification and has allowed for the identification of curves that present problems, to be investigated with more certainty.

In the examined EV charging history, there was a 27.8 percent incidence of anomalous charging curves, which are classified as clusters 0, 3, and 5. When analyzing a sample of these charging curves, it was discovered that some instances involved the use of the EVCS feeder for another purpose, rapid connections without establishing the charging process, interrupted loads, and other anomalies. With the standardization of the number of points for each curve and the information on the total period of the curve, it was possible to compactly store the occurrences for later analysis or control of EVCS behavior.

The ability to analyze the behavior of the charging curves directly from the EVCS feeder enables the energy management of an EV to be independent of the adopted technology and eliminates the need to extract data from the EVCS.

## 6 Conclusion

Currently, the use of EVs is on the rise, with complete replacement targets adopted in the production of internal combustion vehicles. This new scenario impacts the electricity distribution system, making it necessary to monitor the electrical system and verify the status of each element, mainly because they are interconnected. However, big data are generated by the elements involved in the electrical system, requiring automatic tools for this monitoring.

ML tools, such as the K-means algorithm, an unsupervised method, allow efficient detection of patterns in large data volumes. In the presence of big data, this process would be unfeasible using manual data classification. Another positive side of the method used in this study is that there is no need to label data to make the classification, unlike supervised ML methods, where labeled data are needed for both learning and testing. In the classification resulting from the study, which involved a standardization of records and a normalization of values, it was observed that three curves could be separated according to two main characteristics: those that showed an abnormal situation of the EV charging, represented by three clusters in our study; and the situations that were consistent with EV charging.

The result of the classification of the K-means and the BIRCH methods showed that the groups had very close means and the identification of anomalies is equivalent. Still, when comparing the K-means method and the Gaussian Mixture classification method, there is a significant difference in detecting anomalies, which is necessary to improve the implementation.

In future works, this technique can be applied in the analysis and classification of numerous situations with cyclical behavior, such as the case of photovoltaic generation. Photovoltaic generation is an excellent example of the use of this technique, as it could classify the generation curves to be used in other ML methods to predict generation behavior. In addition, it could be used to detect maintenance needs or other variables. The technique will be used to classify large volumes of unlabeled information, allowing it to be used later for prediction purposes or behavior analysis.

**Acknowledgment.** This work has received funding from National Funds through FCT -- Portugal and CAPES -- Brazil, under the project 2019.00141.CBM Desenvolvimento de Técnicas de Inteligência Artificial para a Otimização de Sistemas de Distribuição de Energia Elétrica. It was also supported in part by the Coordination for the Improvement of Higher Education Personnel (CAPES) -- Finance Code 001, and by the São Paulo Research Foundation (FAPESP), under grants 2015/21972-6 and 2018/20355-1.

The authors acknowledge the work facilities and equipment provided by GECAD research center (UIDB/00760/2020) to the project team and also acknowledge the support provided by the Thematic Network 723RT0150 "Red para la integración a gran escala de energías renovables en sistemas eléctricos (RIBIERSE-CYTED)" financed by the call for Thematic Networks of the CYTED (Ibero-American Program of Science and Technology for Development) for 2022. João Soares is also supported by National funds from FCT CEECIND/00420/2022.

## References


1. Rashidi, A., Namazi, M.M., Karshenas, H.R., et al.: An optimized simple current sharing function of SRM with intrgrated battery charger for EV drive. In: 2016 IEEE Transportation Electrification Conference and Expo, Asia-Pacific (ITEC Asia-Pacific), pp 1–5. IEEE (2016)
2. German, R., Delarue, P., Bouscayrol, A.: Battery pack self-heating during the charging process. In: Proceedings IEEE International Conference Industrial Technology 2018-February, pp. 2049–2054 (2018). https://doi.org/10.1109/ICIT.2018.8352504
3. Vu, V.B., Tran, D.H., Choi, W.: Implementation of the constant current and constant voltage charge of inductive power transfer systems with the double-sided lcc compensation topology for electric vehicle battery charge applications. IEEE Trans. Power Electron **33**, 7398–7410 (2018). https://doi.org/10.1109/TPEL.2017.2766605
4. Jain, A.K.: Data clustering: 50 years beyond K-means. Pattern Recognit. Lett. **31**, 651–666 (2010). https://doi.org/10.1016/j.patrec.2009.09.011
5. Shateri, N., Shi, Z., Auger, D.J., Fotouhi, A.: Lithium-sulfur cell state of charge estimation using a classification technique. IEEE Trans. Veh. Technol. **70**, 212–224 (2020)
6. Beltran, H., Sansano, E., Pecht, M.: Machine learning techniques suitability to estimate the retained capacity in lithium-ion batteries from partial charge/discharge curves. J. Energy Storage **59**, 106346 (2023). https://doi.org/10.1016/j.est.2022.106346
7. Chen, K., Zheng, F., Jiang, J., et al.: Practical failure recognition model of lithium-ion batteries based on partial charging process. Energy **138**, 1199–1208 (2017)
8. Sanz-Gorrachategui, I., Pastor-Flores, P., Guillén-Asensio, A., et al.: Unsupervised clustering of battery waveforms in off-grid PV installations. In: 2020 Fifteenth International Conference on Ecological Vehicles and Renewable Energies (EVER), pp. 1–6 (2020)
9. Balasubramanian, S., Balachandra, P.: Characterising electricity demand through load curve clustering: a case of Karnataka electricity system in India. Comput. Chem. Eng. **150**, 107316 (2021)
10. Pedregosa, F., Michel, V., Grisel, O., et al.: Scikit-learn: Machine learning in python. Gaël Varoquaux Bertrand Thirion Vincent Dubourg Alexandre Passos PEDREGOSA, VAROQUAUX, GRAMFORT ET AL. Matthieu Perrot. J. Mach. Learn. Res. **12**, 2825–2830 (2011)
11. Certification L (2020) Pecan Street Dataport. In: Pecan Str. Inc. https://www.pecanstreet.org/dataport/. Accessed 15 Nov 2021
12. Kong, Q., Siauw, T., Bayen, A.M.: Python programming and numerical methods : a guide for engineers and scientists, 1st Editio. © Academic Press 2020 (2020)
13. Davies, E.R.: Computer and Machine Vision, 4th edn. Academic Press (2012)
14. Lee, G., et al.: PyWavelets - Wavelet Transforms in Python — PyWavelets Documentation. https://pywavelets.readthedocs.io/en/latest/index.html. (2006). Accessed 17 Nov 2021
15. Chan, K.P., Fu, A.W.-C.: Efficient time series matching by wavelets. In: Proceedings International Conference Data Engineering, pp. 126–133 (1999). https://doi.org/10.1109/icde.1999.754915
16. Nobre, J., Neves, R.F.: Combining principal component analysis, discrete wavelet transform and XGBoost to trade in the financial markets. Expert. Syst. Appl. **125**, 181–194 (2019). https://doi.org/10.1016/j.eswa.2019.01.083
17. Steinarsson, S.: Downsampling Time Series for Visual Representation. University of Iceland (2013)
18. Xu, J., Qiu, Y., Zhang, H., et al.: Large-scale time series data down-sampling based on map-reduce programming mode. In: Proceedings 2017 IEEE 2nd Advanced Information Technology Electronic Automation Control Conference IAEAC 2017, pp. 409–413 (2017). https://doi.org/10.1109/IAEAC.2017.8054047

19. Burstein, M.: Introduction to graph theory: finding the shortest path - max Burstein's Blog. In: Max Burstein. https://www.maxburstein.com/blog/introduction-to-graph-theory-finding-shortest-path/. (2013). Accessed 22 Nov 2021
20. Raschka, S.: Python machine learning: unlock deeper insights into machine learning with this vital guide to cutting-edge predictive analytics (2015)

# Dynamic Parameterization of Metaheuristics Using a Multi-agent System for the Optimization of Electricity Market Participation

João Carvalho[1], Tiago Pinto[1,2](✉), Juan M. Home-Ortiz[3], Brigida Teixeira[1], Zita Vale[1], and Ruben Romero[3]

[1] School of Engineering, Polytechnic of Porto, Porto, Portugal
tiagopinto@utad.pt
[2] Department of Engineering, University of Trás-os-Montes e Alto Douro and INESC-TEC, UTAD's Pole, Vila Real, Portugal
[3] Department of Electrical Engineering, São Paulo State University, Ilha Solteira, SP, Brazil

**Abstract.** Metaheuristic optimization algorithms are increasingly used to reach near-optimal solutions for complex and large-scale problems that cannot be solved in due time by exact methods. Metaheuristics' performance is, however, deeply dependent on their effective configuration and fine-tuning to align the algorithm's search process with the specific characteristics of the problem that is being solved. Although the literature already offers some solutions for automatic algorithm configuration, these are usually either algorithm-specific or problem-specific, thus lacking the capability of being used for diverse metaheuristic models or diverse optimization problems. This work proposes a new approach for the automatic optimization of metaheuristic algorithms' parameters based on a multi-agent system approach. The proposed model includes an automated fine-tuning process, which is used to optimize a given function in an algorithm- and problem-agnostic manner. Results show that the proposed model is able to achieve better optimization results than standard metaheuristic algorithms, with a negligible increase in the required execution time.

**Keywords:** automatic algorithm configuration · dynamic parameterization · metaheuristic optimization · multi-agent systems · particle swarm optimization

## 1 Introduction

Most real-world optimization problems have a large solution space. Due to resource and time constraints, it is often not possible to apply an exact method to solve such problems [1]. For this reason, metaheuristic optimization algorithms have earned an increase in solving complex combinatorial optimization problems.

However, despite being problem-agnostic techniques, metaheuristics' optimization results are highly impacted by the defined parameters [2]. The best parameterizations are highly impacted by the metaheuristic version and by the addressed objective function.


R. Mehmood et al. (Eds.): DCAI 2023, LNNS 741, pp. 245–255, 2023.
https://doi.org/10.1007/978-3-031-38318-2_25

For this reason, with each new development, it is necessary to optimize the metaheuristic parameters practically from scratch. Thus, and given the increasing complexity of metaheuristics and the problems to which they are normally applied, there has been a growing interest in the problem of optimal configuration of these algorithms. Some notable examples are [3], which presents a gradient-based parameter adaptation method for metaheuristic's parameter tuning and control; the development of fuzzy systems for online parameter adaptation [4]; and [5], which presents a self-parameter adaption mechanism that can be used with the Genetic Algorithm (GA). However, existing models are usually developed for specific metaheuristics or for concrete optimization problems, which brings up the need to develop generalized solutions that can be used for different algorithms and problems.

In this work, a new approach for the automatic optimization of metaheuristic algorithms' parameters is presented. This approach is based on a dynamic strategy in which the parameterization changes during the optimization. This solution divides the optimization process into three stages, forcing, in the first stage, a high level of exploration of the search space, followed by an intermediate exploration, and in the last stage, fostering local search focused on the points of greatest potential. Furthermore, in order to allow an efficient and effective solution, a multi-agent system is used to optimize a given function following the automated fine-tuning process. Consequently, the process of integrating a new metaheuristic or a new objective function is facilitated.

After this introductory section, Sect. 2 discusses the most relevant related work. Section 3 describes the proposed solution, and Sect. 4 presents the achieved results. Finally, Sect. 5 finalizes the paper with the most relevant conclusions.

## 2 Related Work

Influenced by the informatization and the amount of available data, multiple metaheuristics have been proposed during the past years. Generally, metaheuristics are grouped and classified by some characteristics, e.g., nature-inspired versus non-nature inspired or population-based versus single-point search [1]. The number of algorithm parameters has a direct effect on the complexity and, consequently, on the time/iterations required to optimize the metaheuristic parameters. Among these characteristics, the fine-tuning of population-based models is usually more complex because it implies the parameterization of a whole family of solutions [1, 6].

*Dokeroglua et al.* classified the metaheuristics created for the first time before the year 2000 as "classical" metaheuristics and the others as "new generation". Based on this assumption, a state of art review was done, and the most implemented classical metaheuristics are the GA, Particle Swarm Optimization (PSO), Ant Colony Optimization, and Genetic Programming (GP). On the other hand, the most implemented new generation metaheuristics are Artificial Bee Colony (ABC), Bacterial Foraging (BF), Bat Algorithm (BA), and Biogeography-based Optimization (BFO) [7].

Another reason for the success of metaheuristics is their versatility and application to all real-world combinatorial optimization problems. So, these algorithms are being implemented for many different contexts, such as financial, medical, chemical, electrical/energy, and social.

Corazza et al. [8] implemented a hybrid PSO-based algorithm for costly portfolio selection problems, comparing an exact method against a classic PSO and a PSO with dynamic parameterization. In the last approach, a huge reduction in computational time was obtained.

Paul et al. [9] implemented a new feature selection strategy, GA based on Random Forest (GARF) in order to predict esophageal cancer. In this paper, when compared with the existing algorithms in the literature, the outcome was improved by more than 8% for the predictive study and more than 11% for the prognostic study. These improvements were obtained by the optimization of the selected features and by the tuning of the metaheuristic parameters.

El-Gendy et al. [10] implemented a hybrid of a GA and a PSO for tuning a Proposal-Integral-Derivative controller parameters used in a chemical process. The proposed solution gets better results when compared with the existing solutions in the literature. All the parameters of GA and PSO are configured without turning the metaheuristic parameters, i.e., the used parameterization is defined based on the existing papers in the literature.

In order to satisfy the user demands and make the best use of the available power on a community parker charging station, Álvarez et al. [11] proposed an ABC algorithm with local search. The implemented solution has been compared with the state of the art, and a similar result was obtained. However, the computation time was reduced significantly by allowing the utilization of real environments for online scheduling. The time was optimized by some manual updates on the metaheuristic parameters.

In Ozbay and Alatas [12], two novel optimization approaches for fake news detection were implemented. These approaches are implemented following the Grey Wolf Optimization (GWO) and the Salp Swarm Optimization principles (SSO). These two metaheuristics are considered very promising; however, to improve the algorithm's performance, the researchers suggest a dynamic parameterization implementation and a refactor of the metrics used in the model construction.

In summary, metaheuristics are very dynamic approaches, and for this reason, they have been used in several different contexts in order to solve combinatorial optimization problems. However, a problem identified in the existing algorithms in the literature is the metaheuristics parameterization. The parameterization significantly influences the algorithm's performance and the computation time required to archive a good result. In the papers referred on this section, the problems were already partially addressed with some manual testing for each implementation or by problem specific dynamic online parameterization. In this regard, the code implemented cannot be reused in other contexts, and the research must start optimizing from scratch. A possible way to overcome the limitation is to implement metaheuristic models with dynamic online parameterization.

In the past years, several online parameter adaption algorithms have been developed. Generally, the metaheuristics' efficiency and effectiveness have increased significantly with these algorithms. However, the existing online parameter adaption methods were optimized to the problem in the study (overspecialized), implying the inclusion and configuration of new critical parameters [13]. Consequently, the need to develop generalized parameter-tuning models is increasing significantly.

# 3 Proposed Model

## 3.1 Agents

This work proposes a metaheuristic agnostic multi-agent system developed in Python that optimizes different objective functions. The algorithm reads the configurations from an externalized configurations file and executes the optimization using an external service. The system is developed using the Spade library and is composed of three agents: the Timer agent, a Coordinator agent, and an Executer agent. The communication between these agents is direct, without a facilitator, and the implemented communication strategy is the transmission of messages between agents. From the coordination point of view, the model follows the master-slave architecture. The timer agent requests services from the Coordinator, and the Coordinator requests services from the Executer.

The Timer agent is a cognitive agent capable of initializing the Coordinator, reading the queue, starting the optimization, managing the execution time, and requesting a faster solution if the time to run is exceeded. To do that, the agent is composed of three behaviors. A one-shot behavior is responsible for starting the Coordinator, a period behavior is responsible for periodically checking the training queue, and a cyclic behavior is responsible for processing the received messages.

The Coordinator is a cognitive agent capable of initializing the Executer, calculating the best parameterizations for each execution, requesting an optimization, storing and managing the best responses and sending the optimization result to the Timer agent. To do that, two behaviors were implemented. A one-shot behavior is responsible for starting the Executer, and a cyclic behavior is responsible for processing the received messages.

The Executer agent is a reactive agent capable of receiving a request to execute the necessary optimization, processing the optimization response, and sending back the answer. This agent is composed of only one cyclic behavior, which is executed every time the agent receives a request.

## 3.2 Multi-agent System

These agents form a multi-agent system that enables the execution of the full flow of an optimization request. The Timer agent periodically checks the optimization queue until a new optimization request is registered. Once a request arrives, the input file is processed, and an asynchrony message is sent to the Coordinator. In the meantime, the Timer starts an iterative process of checking the elapsed time since the optimization started. If the time to run is close to being exceeded, a new message is sent requesting a valid result. In the end, this agent stores the result in the configured output directory.

Regarding the Coordinator agent, after receiving a message to start, it reads the metaheuristic parameterization and sends a message to the Executer indicating the required data to run the algorithm, parameterization, the number of trials, the function identifier, the metaheuristic identifier, the execution mode, and the initial solutions. Afterward, the agent waits for the executer response. When the response is received, the agent calculates the next iteration initial solutions, reads the parameterization of the next iteration, and sends another message to the Executer agent so that it may start the new execution using the new parameter set. Generally, the loop ends after the end of the high exploitation execution. However, if the agent receives a timeout notification message, it retrieves the current best response and stops the execution after the next iteration.

The Executer agent reacts based on the received messages. When it receives a message, it performs the optimization using the metaheuristic and functions repositories, and it sends the response back. During the optimization, a configured number of trials received on the request is executed. These are executed using asynchronous threads. Each thread executes only one trial, and the parallel processing includes the optimization and the processing of the optimization results. In the end, the results are aggregated, and the response is sent back to the Coordinator agent.

### 3.3 Multi-agent System Communication

The developed multi-agent system uses FIPA ACL messages to communicate. The body of all the messages is a Python dictionary converted into a valid json structure. The agents communicate using a XMPP server. In this project, the used server is an instance of the ejabberd installed locally [14]. The chosen XMPP server is irrelevant and can be replaced by any other similar application. The supported messages are:

- **Start optimization** from **Timer** to **Coordinator**. This has the **inform** performative code indicating to the Coordinator that an optimization must be supported.
- **Notify timeout** from **Timer** to **Coordinator**. This has the **request** performative code meaning the timeout is being exceeded, and a faster response is needed.
- **Run optimization** from **Coordinator** to **Executer**. This has the **request** performative code, requesting for an optimization execution.
- **Optimization result** from **Coordinator** to **Timer**. This has the **inform** performative code and it informs of the optimization result.
- **Execution result** from **Executer** to **Coordinator**. This has the **inform** performative code, and it contains the optimization execution result.

### 3.4 Metaheuristics Response Processing

In order to improve the optimization results, the three best solutions are selected. This is important since, for each optimization, the system executes several trials. So, if a specific trial does not retrieve good results, it can be completely excluded and not processed on the following iterations. After being selected, the best solutions of all the trials are analyzed, and the most appropriate are used as the initial solutions of the next iteration.

Figure 1 shows a diagram explaining the first step of the mentioned process. The example is purely exemplificative, and it is not a result of an optimization.

The best responses are processed in three steps, as shown in Fig. 1. Initially, the best 100 solutions obtained are selected, sorted by fitness values, and the best one is selected. The sort can be descending or ascending, depending on the optimization mode: maximization or minimization. In the second step, the variation of each objective function result is calculated between the selected best solution and the others. Then, the solutions are sorted by the calculated value and the solution with the highest variation is chosen. This is performed so that the three selected solutions, to be used as initial solutions for the next execution, are sufficiently spread along the solution space and are not focused on the same search point. In the third step, the processing is similar; however, the variation is

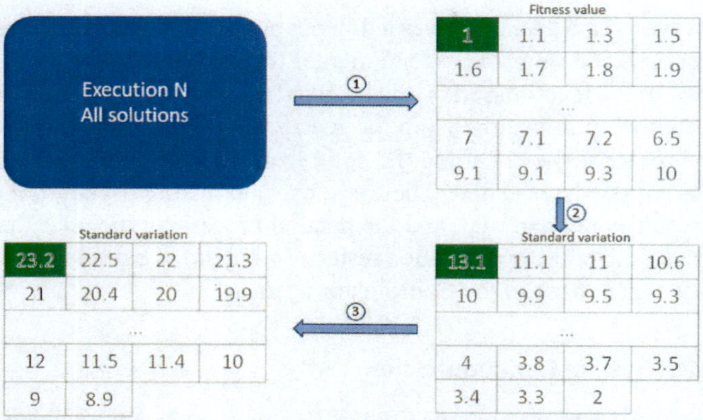


**Fig. 1.** Example of the process of selecting the best solutions after a minimization

calculated using the two already selected solutions. In this way, a third, distinct solution is identified.

These selected solutions are combined with the solutions of the other trials and enable starting the process that filters the most appropriate solutions. This process is conditioned by the current step of the optimization process. Figure 2 explains how the initial solutions are selected for the intermediate configuration when the number of trials configured is 4.

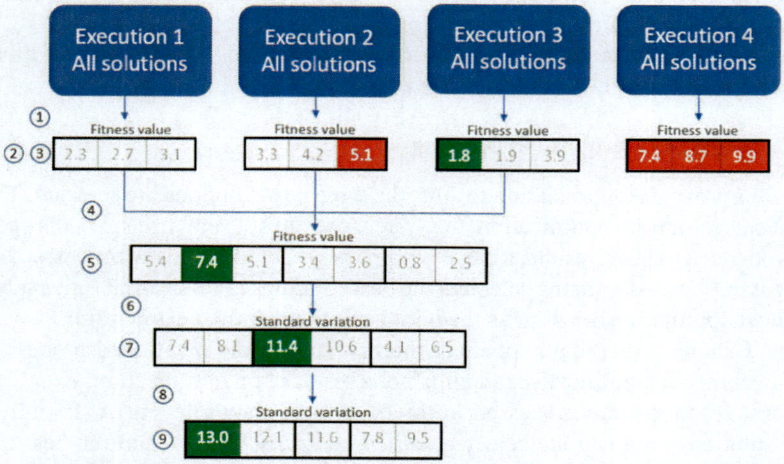


**Fig. 2.** Example of the process of select the initial solution in a minimization

The process, as depicted in Fig. 2, is executed by the following steps:

1. Process the three best solutions for each trial. The process described in Fig. 1

2. Exclude the worst 1/3 solutions. The highest values for minimizations and the smallest values for maximizations.
3. Sort by fitness value and select the smaller value in minimizations and the highest value in maximizations.
4. Calculate the variation between the selected solution and the others.
5. Select the solution with the highest variation. It means that, theoretically, it is, among the best solutions, the solution farthest from the already selected.
6. Calculate the variation between the two selected solutions and the others.
7. Select the solution with the highest variation.
8. Calculate the variation between the three selected solutions and the others.
9. Select the solution with the highest variation.

Following this approach, it is possible to ensure that the best solution is always considered in the next iteration. Besides that, if an optimization retrieves bad results, it can be excluded, and it is not taken into consideration in the second step. At the same time, by selecting using variation instead of fitness value, the chances of selecting two similar initial solutions are reduced significantly.

For the high exploration parameterization, the initial solution is not defined and for the high exploitation, the process is similar to the intermediate parameterization. However, taking into consideration that the focus of the high exploitation optimization is the local search, instead of selecting the initial solutions by the variation, the list is sorted by fitness value, and the best solutions are selected.

# 4 Experimental Findings

## 4.1 Metaheuristic

This work uses the PSO metaheuristic as a basis since it is one of the most widely implemented population-based metaheuristics with several parameters to configure. Besides, the existing online parameter adaptation methods applied to PSO demonstrate a good improvement in the metaheuristic efficiency. The PSO algorithm is a population-based stochastic optimization technique proposed by Kennedy and Eberhart in 1995 [15]. It is inspired by the social behavior of swarming animals like insects, herds, birds, and fishes. The swarms work together to establish cooperation, changing their behavior based on previous experience [16].

In order to get valid results in a reasonable time, the PSO parameters must be optimized. The parameter list may depend on the PSO version, but generally, it includes the number of particles, acceleration constant, inertia weight, and maximum limited velocity [16, 17].

In the past years, several PSO libraries, like PySwarms [18], Purpose Optimization Library (GPOL) [19], and Pyticle Swarm [20], have been published. In this project, the Pyticle Swarm is used as a baseline since it is an open-source library aiming at a fully flexible version of PSO, offering the possibility to configure all parameters and apply the PSO to any optimization problem. Using Pyticle Swarm, it is possible to optimize very distinct types of problems. To do so, the library allows the configuration of several parameters, such as the initial and final weigh inertia, the learning factors, the total amount of particles, and the total number of iterations.

## 4.2 Execution Time Analysis

As described in Sect. 3, the proposed model is a multi-agent that optimizes a configured number of trials in parallel. This developed model incorporates additional processing in the basis optimization, namely communication between agents, thread management and handling optimization responses. The additional processing execution time in real-time environments cannot be significant. Table 1 presents a performance analysis, in milliseconds, of the proposed model by number of optimization trials. Table 2 summarizes the most relevant times of the mentioned analysis in percentage. The processor used in these tests is the AMD Ryzen 7 4700 (8 threads).

**Table 1.** Optimization module execution time (milliseconds) analysis.

| Trial Number | 1 | 2 | 3 | 4 | 5 | 6 | 7 | 8 | 9 | 10 |
|---|---|---|---|---|---|---|---|---|---|---|
| Optimization execution time | 1377 | 1367 | 1402 | 1434 | 1467 | 1565 | 1622 | 1643 | 1581 | 1630 |
| Process response | 227 | 234 | 222 | 230 | 236 | 242 | 256 | 262 | 256 | 261 |
| Parallel proc. Time | 1606 | 1636 | 1681 | 1725 | 1806 | 1915 | 1978 | 3898 | 3894 | 4095 |
| Thread management | **2** | **35** | **57** | **61** | **103** | **108** | **100** | **1993** | **2057** | **2204** |
| Total Time | 1672 | 1688 | 1728 | 1788 | 1869 | 1978 | 2057 | 3945 | 3958 | 4156 |
| Algorithm additional time | **66** | **52** | **47** | **63** | **63** | **63** | **79** | **47** | **64** | **61** |
| Difference to 1 trial time | **0** | **16** | **56** | **116** | **197** | **308** | **385** | **2373** | **2386** | **2484** |

**Table 2.** Optimization module time analysis summary

| Trial Number | 1 | 2 | 3 | 4 | 5 | 6 | 7 | 8 | 9 | 10 |
|---|---|---|---|---|---|---|---|---|---|---|
| Algorithm additional time (%) | 3.9 | 3.1 | 2.7 | 3.5 | 3.4 | 3.2 | 3.8 | 1.2 | 1.6 | 1.5 |
| Threading management time (%) | 0.1 | 2.0 | 3.3 | 3.4 | 5.5 | 5.5 | 4.9 | 50.5 | 52.0 | 53.1 |
| Process response (%) | 13.6 | 13.9 | 12.9 | 12.8 | 12.6 | 12.2 | 12.4 | 6.6 | 6.5 | 6.2 |
| Optimization time (%) | 82.4 | 81.0 | 81.1 | 80.3 | 78.5 | 79.1 | 78.9 | 41.6 | 39.9 | 39.2 |

As one can see from Table 1 and Table 2, the additional algorithm time (agents communication, reading, storing files, etc.) is not impacted by the number of trials, and it is not significant. In percentage, the highest value is less than 4%, and in absolute numbers, it is stable, always below 80 ms.

Thread management time is the time it takes to initialize the optimizing threads. This time is highly impacted by the number of trials configured. For one trial, the number is close to 0; the time goes up slightly until reaching eight trials, which has a sharp rise representing 50% of algorithm's execution time. It can be explained by the characteristics of the used processor that supports only eight threads running simultaneously. So, for this processor, the maximum number of trials that should be used is 7. However, this analysis is only valid for this processor, performing a similar analysis using the proposed model in different processors.

The process response time is not impacted by the number of trials, and it is always below 250 ms. The optimization time, time spent running the optimization algorithm, excluding when the number of trials is higher than 7, represents about 80% of the total time. So, using the proposed algorithm, it is possible to automatize the process of running a metaheuristic. This algorithm includes additional processing that, for a reasonable number of trials (below the number of threads supported by the processor) does not significantly impact the optimization time.

### 4.3 Proposed Solution Applied to a Real-World Problem

In order to validate the suitability of the proposed model in dealing with real-world problems, a case study considering an existing problem in the power and energy systems' domain is presented. The portfolio optimization problem aims at optimizing the investment of a supported player in different electricity market opportunities. Several market opportunities are considered, associated with an expected transaction price per negotiation period. Accordingly, the optimization model aims at finding the optimal combination of trading volume (sale and purchase) that should be performed according to the expected prices and specific market constraints. Full details of this problem can be consulted in [21].

This model was used to validate the proposed solution. Two training executions with 2197 combinations are executed, one for the traditional mode (simply running the PSO algorithm with the standard parameterization) and the other for the dynamic mode, using the proposed model. The fine-tuned parameters are evaluated, and the comparison between the results obtained by number of trials after 1000 repetitions are presented in Fig. 3.

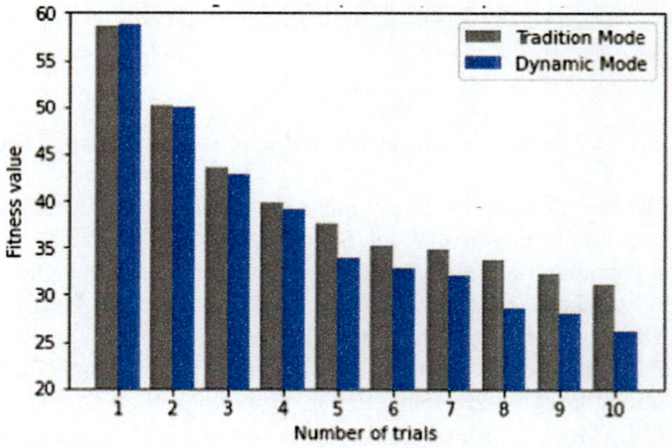


**Fig. 3.** Impact of the trials number on the optimization results using a real word problem

As one can see from Fig. 3, for a low number of trials, the results are very similar. However, by increasing the number of trials, the results using the dynamic mode improve considerably compared with the same test using the traditional mode. It occurs because with the proposed solution, using several trials, the bad results regarding parameter combinations obtained on the first iterations are not considered in the following executions, thus improving the optimization process and, consequently, the achieved results.

## 5 Conclusions

This work proposed an automatic parameter configuration solution based on a multi-agent system to deal with the impact of metaheuristic parameterization on the optimization results in solving complex real-world problems. The proposed model optimizes a function using optimized parameters. The solution introduces extra processing that increases optimization execution time. However, most of the time is spent on optimizing, and this approach considerably improves the obtained results, as shown through its application to the portfolio optimization problem. Furthermore, for each execution, the maximum execution time is defined, and if the time is exceeded, the system stops the execution and retrieves the current best solution. In summary, using the proposed model, the fine-tuning process is automatized, and consequently, the process of supporting different functions or metaheuristics is facilitated.

**Acknowledgements.** The present work funds from FCT Portuguese Foundation for Science and Technology. This work has received funding from FEDER Funds through COMPETE program and from National Funds through (FCT) under the project PRECISE (PTDC/EEI-EEE/6277/2020). This work is also supported by National Funds through FCT - Portugal and CAPES - Brazil, under project 2019.00141.CBM Desenvolvimento de Técnicas de Inteligência Artificial para a Otimização de Sistemas de Distribuição de Energia Elétrica. Brígida Teixeira is supported by FCT with Ph.D. Grant 2020.08174.BD. The authors acknowledge the work facilities and equipment provided by the GECAD research center (UIDB/00760/2020 and UIDP/00760/2020) to the project team.

## References

1. Gogna, A., Tayal, A.: Metaheuristics: review and application. J. Exp. Theor. Artif. **25**(4), 503–526 (2013)
2. Peres, F., Castelli, M.: Combinatorial optimization problems and metaheuristics: review, challenges, design, and development. Appl. Sci. **11**(14), 6449 (2021)
3. Tatsis, V.A., Parsopoulos, K.E.: Reinforced online parameter adaptation method for population-based metaheuristics. In: 2020 IEEE Symposium Series on Computational Intelligence (SSCI), Canberra, ACT, Australia (2020)
4. Melin, P., et al.: Optimal design of fuzzy classification systems using PSO with dynamic parameter adaptation through fuzzy logic. Expert Syst. Appl. **40**, 8 (2013)
5. Moctezuma, C.J.M., Mora, J., Mendoza, M.G.: A self-adaptive mechanism using weibull probability distribution to improve metaheuristic algorithms to solve combinatorial optimization problems in dynamic environments. Math. Biosci. Eng. **17**(2), 975–997 (2019)

6. Boussaïd, I., Lepagnot, J., Siarry, P.: A survey on optimization metaheuristics Inf. Sci. **237**, 82–117 (2013)
7. Dokeroglu, T., Sevinc, E., Kucukyilmaz, T., Cosar, A.: A survey on new generation metaheuristic algorithms. Comput. Ind. Eng. **137**, 106040 (2019)
8. Corazza, M., di Tollo, G., Fasano, G., Pesenti, R.: A novel hybrid PSO-based metaheuristic for costly portfolio selection problems. Ann. Oper. Res. **304**(1–2), 109–137 (2021)
9. Paul, D., Su, R., Romain, M., Sébastien, V., Pierre, V., Isabelle, G.: Feature selection for outcome prediction in oesophageal cancer using genetic algorithm and random forest classifier. Pattern Recogn. **60**, 42–49 (2017)
10. El-Gendy, E.M., Saafan, M.M., Elksas, M.S., Saraya, S.F., Areed, F.F.G.: Applying hybrid genetic–PSO technique for tuning an adaptive PID controller used in a chemical process. Soft. Comput. **24**(5), 3455–3474 (2019). https://doi.org/10.1007/s00500-019-04106-z
11. Álvarez, J.G., Gonzalez, M.A., Vela, C., Varela, R.: Electric vehicle charging scheduling by an enhanced artificial bee colony algorithm. Energies **11**, 10 (2018)
12. Ozbay, F.A., Alatas, B.: A novel approach for detection of fake news on social media using metaheuristic optimization algorithms. Elektronika ir Elektrotechnika **25**(4), 62–67 (2019)
13. Tatsis, V.A., Parsopoulos, K.E.: Dynamic parameter adaptation in metaheuristics using gradient approximation and line search. Appl. Soft Comput. **74**, 368–384 (2019)
14. Ejabberd, "Ejabberd," https://docs.ejabberd.im/get-started/. Acedido em 2022
15. Eberhart, R., Kennedy, J.: Particle swarm optimization. In: MHS'95 6th International Symposium on Micro Machine and Human Science, Nagoya, Japan (1996)
16. Wang, D., Tan, D., Liu, L.: Particle swarm optimization algorithm: an overview. Soft Comput. **22**, 387–408 (2018)
17. Jain, N.K., Nangia, U., Jain, J.: A review of particle swarm optimization. J. Inst. Eng.: Ser. B **99**, 407–411 (2018)
18. Miranda, L.J.: PySwarms: a research toolkit for particle swarm optimization in Python. J. Open Source Softw. **3**(21), 433 (2018)
19. Bakurov, I., Buzzelli, M., Castelli, M., Vanneschi, L., Schettini, R.: General purpose optimization library (GPOL): a flexible and efficient multi-purpose optimization library in Python. Appl. Sci. **11**(11), 4774 (2021)
20. Veiga, B., Faia, R., Pinto, T., Vale, Z.: https://pypi.org/project/Pyticle-Swarm/. Accessed 14 Jan 2022. https://pypi.org/project/Pyticle-Swarm/#description. Acedido em 22 Jan 2022
21. Pinto, T., et al.: Adaptive portfolio optimization for multiple electricity markets participation. IEEE Trans. Neural Netw. Learn. Syst. **27**(8), 1720–1733 (2015)

# Sizing of Urban Power Systems Based on Renewable Sources

Dani Vidal[1], Tiago Pinto[1,2(✉)], and José Baptista[1,2]

[1] Department of Engineering, University of Trás-os-Montes and Alto Douro, 5000-801 Vila Real, Portugal
al64795@alunos.utad.pt

[2] INESC-TEC, UTAD's Pole, 5000-801 Vila Real, Portugal
{tiagopinto,baptista}@utad.pt

**Abstract.** In recent years, sustainable power supply has become a necessary asset for the daily survival and development of populations. The incentive to the use of renewable energies has been increasing worldwide. Solar energy, in particular, is widespreading fast in countries whose location allows to obtain excellent radiation conditions. In this work, autonomous photovoltaic (PV) systems are studied, having as main aim its application in the supply of urban loads. For this purpose, a PV system is designed to supply the decorative lighting of a monument. Particular emphasis is given to studying the behavior of the energy storage system. The achieved results demonstrate that the use of this type of systems is a very efficient solution for the municipalities to supply several urban loads such as fountains, traffic lights, decorative lighting, among others.

**Keywords:** Solar energy · Stand alone photovoltaic systems · Energy storage systems

## 1 Introduction

Energy has always played a key role in history. Nowadays more than 75% of the energy produced comes from fossil fuels that are increasingly scarce, therefore becoming increasingly more expensive [5]. At the same time, energy is a key resource that is crucial with regards to sustainable development, as it is responsible for a big portion of the total amount of pollution produced by human activities. To reduce or eliminate these effects without causing damage to day-to-day activities, the best option is to change the paradigm by promoting renewable energy sources. Given the excellent conditions of solar radiation that the European southern countries have, the use of solar energy for use in the most diverse

This research presents the results of a training course framed in the project: Energy Transition Entrepreneurs in Action (ETEIA), funded by the European Institute of Innovation and Technology (EIT) under the HEI Initiative: Innovation Capacity Building for Higher Education. HEI Grant Agreement N 10036.


R. Mehmood et al. (Eds.): DCAI 2023, LNNS 741, pp. 256–266, 2023.
https://doi.org/10.1007/978-3-031-38318-2_26

purposes is increasing significantly. Currently, photovoltaic systems (PV) are being used in self-consumption systems but also to supply stand alone systems, of which urban equipment is an example. Several research works involving the design of PV systems have emerged, sparking the interest of many researchers from all over the world. Several studies, namely in stand alone systems can be found in the literature. As notable examples, El Shenawy et al. [5], Zubi et al. [16], and Akram et al. [1] present cases of stand alone residential systems. In these works, PV systems are sized in order to produce the necessary power demand to support lighting and basic household equipment's such as a washing machine, refrigerator, and stove.

According to Wassie e Adaramola [15], solar electricity systems can represent an economical alternative to conventional high-cost generator systems, especially in developing countries where the majority of the population lives in rural and isolated areas. In the studies carried out by Kuandinya et al. [9] Chakrabarty and Islam [3], a technical feasibility analysis was conducted on PV systems both on-grid and off-grid depending on their utilization.

Regarding stand alone urban photovoltaic systems, the published studies are still very scarce, mainly focused on urban grid-connected systems, as can be observed in Junior et al. [8]. Grid-connected PV systems are not essential when we need to supply loads with higher power demand than what is produced, either due to the lack of conditions for the application of a larger PV system or because the location is not the most advantageous in terms of solar radiation incidence.

This paper addresses the design of an stand alone urban system based on a photovoltaic solar energy with an energy storage system (ESS). A study is also conducted on the behavior of ESS in stand alone systems, which is a subject that has been studied by several authors who address the aspect related to electric vehicles, presenting models and solutions regarding battery charging and discharging [4] and [11].

The case study refers to a PV system to lighting a monument in the town of Vila Real, with the main objectives of studying low-power photovoltaic systems, designing and sizing photovoltaic systems for supplying energy to decorative urban lighting, especially a monument, and analyzing relevant parameters related to the battery behaviour.

After this introductory section, the document is structured as follows: Sect. 2 briefly discusses the topic of PV systems, Sect. 3 described the components of a stand alone PV system, Sect. 4 presents the case study, and finally in Sect. 5 the most relevant conclusions from this work are drawn.

## 2 Photovoltaic System

The PV Effect consists of converting solar light (radiation) into electrical energy through photovoltaic cells. These cells consist of a larger diode, i.e., a substrate of semiconductor material where a permanent internal electric field is created. The semiconductor material is composed of atoms that, when hit by radiation, release electrons that can be conducted by the internal electric field of the contacts, producing current through the cell. The material of choice for photovoltaic cells is crystalline silicon, however, for there to be electrical current, this cannot be pure, and a potential difference between two zones of the cell is necessary. The reasons for choosing this material are associated with its special characteristics, such as its abundance on the planet, its non-toxicity, and the fact that it is the raw material of the electronics industry. A representation of the PV cell can be described through the equivalent circuit presented in Fig. 1.

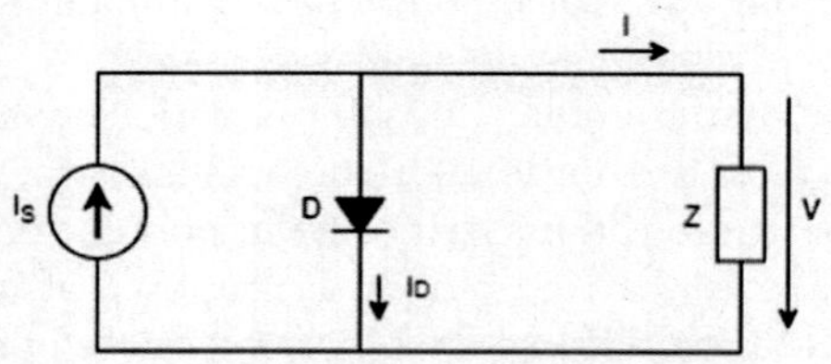


**Fig. 1.** Equivalent circuit of photovoltaic cell

In order to study the behavior of a PV system, it was necessary to develop mathematical models for the constituent components of the system [7]. In the scheme presented earlier, $I_S$ represents the electric current generated by solar radiation and the diode represents the exchange of electrons at a p-n junction, through which the internal current $I_D$ flows, which depends on the voltage at the PV cell terminals and it's possible calculated using (1):

$$I_D = I_0 * (e^{\frac{V}{mVT}} - 1) \tag{1}$$

where:

$I_0$ is the maximum reverse saturation current of the diode;
$V$ is the voltage across the cell terminals;
$m$ is the ideality factor of the diode;
$V_T$ is a thermal potential;
$T$ is the cell temperature in K;

The current $I$ passing through the load is obtained from Eq. (2):

$$I = I_S - I_D = I_S - I_0 * (e^{\frac{V}{mVT}} - 1) \tag{2}$$

The performance of a photovoltaic module is extremely important for that losses that occur in the system can be controlled, making it possible to maximize its efficiency. Considering that the main function of a photovoltaic module is the conversion of solar energy into electrical energy, the verification of factors that negatively influence this conversion is fundamental [14]. According to Kurbonov [10] and Lopes [12], radiation and temperature are the main factors that influence the performance of a PV cell.

The power output from PV sources is dependent on the incident solar radiation and temperature at a specific location and time interval, as well as the specific characteristics of the PV panel being utilized. The maximum power produced by the PV can be calculated from the given Eq. (3).

$$P_{max} = V_{max} \cdot I_{max} \tag{3}$$

where the maximum voltage is obtained by Eq. (4).

$$V_{max} = m \cdot V_T \cdot ln\left(\frac{I_{SC}^r - I_{max}^r \cdot \frac{G}{G^r}}{I_0}\right) \tag{4}$$

and the max current is given by Eq. (5).

$$I_{max} = \frac{G}{G^r} \cdot I_{max}^r \tag{5}$$

Besides the radiation and temperature, other parameters that influence module production are the ideality factor ($m$), thermal potential ($V_T$), short-circuit current ($I_{SC}^r$), maximum reference current ($I_{max}^r$), maximum reverse saturation current ($I_0$) and incident reference radiation ($G^r$) at Standard Test Condition ($1000\,\mathrm{W/m^2}$).

## 3 Stand Alone Systems and Their Constituents

One of the main components of stand alone PV systems is the battery charge controller, which measures the voltage and protects it against overloads. Their main functions are to ensure the correct charging of the battery, block reverse current between it and the panel, and prevent deep discharges that occur in cases of lead-acid batteries. In this way, the charge controller allows the battery life to be extended and ensure that the energy produced by solar panels is stored efficiently. In a stand alone system, the use of ESS is essential if a load is used during the night. A correct choice of the battery is extremely important to allow the autonomy of the system, because your capacity will determine the time that the system can operate during a night time period or even during periods of low or no radiation. In the studied case, no inverter was needed, because the lighting of the monument is done through DC current lights. Figure 2 shows a scheme of the PV system implemented and sized in this work.

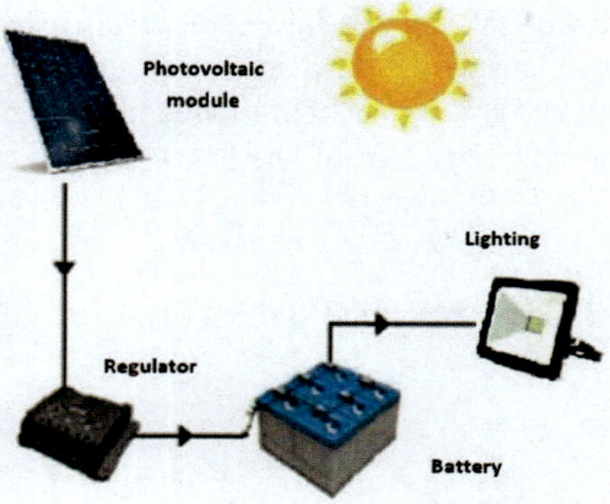


**Fig. 2.** Schematic diagram of isolated photovoltaic system created

## 4 Case Study

This section presents the case study, as well all the necessary calculations for the design of an isolated PV system that will provide lighting for a monument located in the city of Vila Real with coordinates (41°17′42.0″N 7°44′49.2″W). The data used in this study were taken from the PVGIS tool [6] and are presented in Figs. 3 and 4. These were obtained considering an optimal angle of 32°.

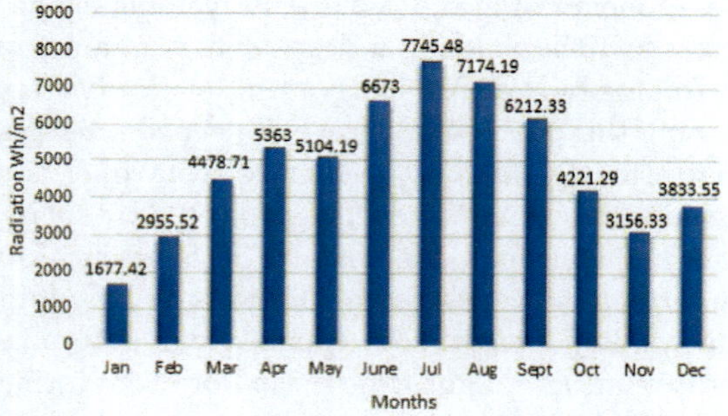


**Fig. 3.** Average radiation at the optimum angle throughout the year

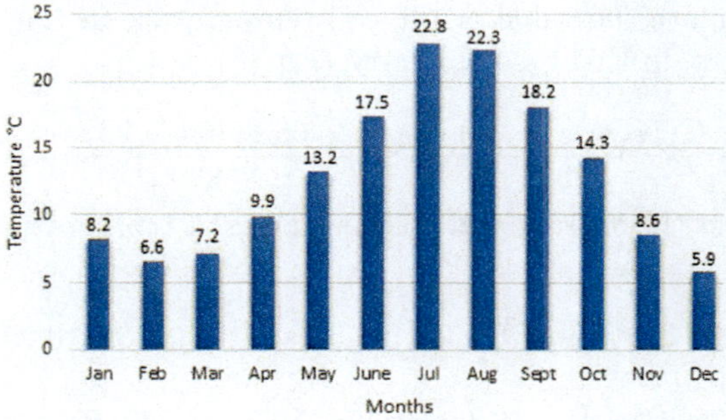


**Fig. 4.** Average temperature throughout the year

The photovoltaic module chosen for the design of this system was the Panasonic 340, whose characteristics are listed in Table 1.

**Table 1.** Panasonic N340 Module Specifications

| Model VBHN340SJ53 | Values |
|---|---|
| Maximum power (Pmax) | 340 W |
| Maximum voltage (Vmax) | 59.7 V |
| Maximum current (Imax) | 5.7 A |
| Short circuit current (Isc) | 6.13 A |
| Open circuit voltage (Voc) | 71.3 V |
| Nominal Operating Cell Temperature (NOCT) | 44 °C |
| Current temperature coefficient ($\alpha$) | 3.37 mA/°C |
| Voltage temperature coefficient ($\beta$) | −0.235 V/°C |

### 4.1 PV System Sizing

This system was sized based on an energy consumption of 3 kWh/day (in this case at night) and to meet these daily energy needs, it is necessary to calculate the number of PV panels, given by Eq. (6) [2]:

$$N_m = \frac{E_{load}}{E_{daily} * NDCL} \tag{6}$$

To be able to perform this calculation, it is first necessary to calculate other important parameters:

1st step: Calculate the average hourly solar radiation value (7) [2]:

$$G(h) = \frac{G_{day}}{24} = \frac{1677.42}{24} = 69.89\,\mathrm{W/m^2/h} \tag{7}$$

2nd step: Calculate the value of the cell temperature in the month with the lowest temperatures, in this case, January (Eq. (8)):

$$\theta_c = \theta_a + \frac{G(NOCT - 20)}{800} = 8.2 + \frac{69.89(44 - 20)}{800} = 10.29\,^{\circ}\mathrm{C} \tag{8}$$

3rd step: Correct the current and voltage values at temperature using Eqs. (9) and (10) [2]:

$$I_{max}^{T}(\theta_c) = I_{max}(25\,^{\circ}\mathrm{C}) * [1 + \alpha(\theta_c - 25)] = 5.7 * [1 + 0.00337(10.29 - 25)] = 5.41\,\mathrm{A} \tag{9}$$

$$V_{max}^{T}(\theta_c) = V_{max}(25\,^{\circ}\mathrm{C}) + [1 + \beta(\theta_c - 25)] = 59.7 - 0.235(10.29 - 25)] = 63.16\,\mathrm{V} \tag{10}$$

4th step: Calculate the daily hours with radiation at 1 kW/m$^2$ (11) [2]:

$$HPD = \frac{G}{1000} = \frac{1677.42}{1000} = 1.67\,\mathrm{h} \tag{11}$$

5th step: Calculate the energy produced per day per module considering the worst month of radiation (Eq. (12)) [2]:

$$E_{daily} = HPD * I_{max}^{T} * V_{max}^{T} = 1.67 * 5.41 * 63.16 = 570.6\,\mathrm{Wh} \tag{12}$$

Finally, calculate the number of modules required using Eq. (6):

$$N_m = \frac{3 * 10^3}{570.6 * 1} = 5 \tag{13}$$

Thus, at least 5 modules will be required to meet the daily consumption of 3 kWh per night.

The appropriate choice of battery is an essential task to ensure the proper operation of the system and provide the battery with a longer lifespan. After consulting several manufacturers, the chosen battery was the Exide Technologies Classic Solar lead-acid battery with a maximum capacity of 245Ah and a depth of discharge (DOD) of 70%. In order to store energy for nighttime lighting, it is necessary to calculate the battery capacity using Eq. (14) [2], considering one day of autonomy:

$$C = \frac{\frac{E_{LOAD}}{V_{bat}} * NDA}{NDCL * DOD * \eta_{bat}} = \frac{\frac{3000}{24} * 1}{1 * 0.7 * 0.85} = 210.084\,\mathrm{Ah} \tag{14}$$

Next, the number of batteries is determined using Eq. (15) [2]:

$$N_{bat} = \frac{C}{C_{market}} = \frac{210.084}{245} = 0.86 \tag{15}$$

It is thus determined that one 245Ah battery is necessary. For the electrical installation of a PV system, the used cables should comply with the standards, and therefore, the sizing of the wiring should be carried out following the technical rules of low voltage electrical installations. Therefore, in order to perform a correct sizing of the wiring, the following calculations are made:

1st step: Calculate the installed power in the lighting systems (16) [2]:

$$P = \frac{E}{t} = \frac{3000}{12} = 250\,\mathrm{W} \tag{16}$$

with an installed voltage of

$$V_{bat} = 24\,\mathrm{V} \tag{17}$$

the necessary current $I_B$ (rated current) to supply the system is calculated from Eq. (18) [2].

$$I_B = \frac{P}{V} = \frac{250}{24} = 10.4\,\mathrm{A} \tag{18}$$

2nd step: Verify the heating conditions (Eqs. (19) and (20)) [2]: Using a 16 A circuit breaker and using table 52C1 of RTIEBT [13], with the reference method B, copper conductors, and a section of 10 mm$^2$, the value of max current capacity $I_Z = 57$ A is obtained.

$$I_B < I_N < I_Z = 10,4\,\mathrm{A} < 16\,\mathrm{A} < 57\,\mathrm{A} \tag{19}$$

$$I_2 \leq 1.45 * I_Z = 1.3 * I_N \leq 1.45 * 57 = 20.8 \leq 82.65\,\mathrm{A} \tag{20}$$

After verifying these two conditions, the resistance is calculated using (21) [2]:

$$R = 2\rho * \frac{l_{TOTAL}}{S} = 2 * 0.0225 * \frac{20}{10} = 0.09\,\Omega \tag{21}$$

Finally, the maximum voltage drop is calculated using Eq. (22) in photovoltaic systems [2],

$$\Delta V = R * I_B = 0.09 * 10.4 = 0.936\,\mathrm{V} \tag{22}$$

ensuring that the voltage drop condition is less than 1 V.

For sizing the controller, it is necessary to start by knowing the maximum current that can be obtained from PV panels, which is considered as the minimum current ($I(REG)$) for the controller's choice and is calculated from (23) [2].

$$I_{REG} \geq 1.3 * I_{max}(25\,°\mathrm{C}) * \frac{N_M}{N_M inseries} \Leftrightarrow I_{REG} \geq 1.3 * 5.7 * \frac{5}{1} \geq 37.05\,\mathrm{A} \tag{23}$$

In this way, an OutBack MX60 controller was used, which has a maximum input current ($I_{REG,NOM}$) of 60 A.

To calculate the number of useful regulators, we need to divide the minimum regulator current by the rated current of the charge controller, requiring in this case only one 60 A regulator.

### 4.2 Sizing the Storage System

To make this study more complete, a study was carried out on the battery used in sizing this photovoltaic system. The battery state of charge varies according to the current ($I_{bat}$) requested from it in a given time interval. In order to obtain the state of charge of a battery, Eq. (24) [11] is used:

$$SoC = 1 - \frac{1}{3600} * \int_{t_1}^{t_2} \frac{I_{bat}^k}{C_p} dt \tag{24}$$

where $k$ is the Peukert constant and $C_p$ is the Peukert capacity.

In order to obtain the state of charge in hours, it is necessary to replace 3600 s seconds with 1 h. $SoC$ values are represented between 0 and 1, with 1 being the maximum energy of the battery. By substituting the values of the case under study in the equation and with the aid of Excel, we obtain the graph of Fig. 5.

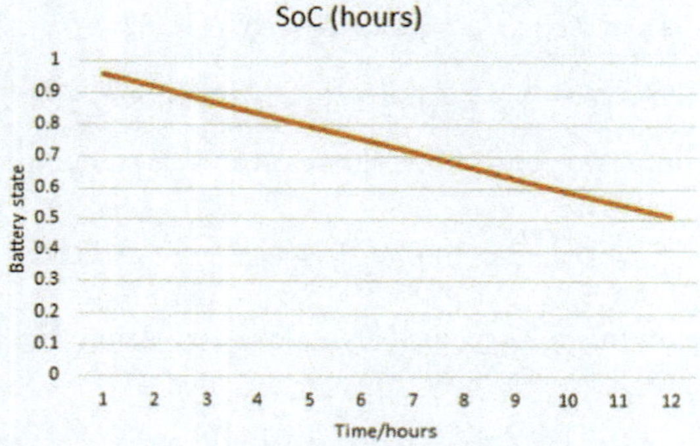


**Fig. 5.** State of charge over 12 h

The depth of discharge of a battery is the ratio between the removed charge and the Peukert capacity and can be obtained through Eq. (25) [11]:

$$DoD = \frac{CR_{(n)}}{C_p} \tag{25}$$

The calculation of the removed charge is done through Eq. (26) [11]:

$$CR_{(n+1)} = CR_{(n)} + \frac{\delta t * I^k}{1} \tag{26}$$

To calculate the Peukert capacity, it is necessary to multiply the hour rate by the division of the battery capacity by the hour rate, and this division is raised to the Peukert constant, which results in (27) [11]:

$$C_p = R * (\frac{C}{R})^k \Leftrightarrow C_p = 20 * (\frac{245}{20})^{1.2} = 404.38\,\text{Ah} \tag{27}$$

The voltage across the battery terminals can be determined through (28) [11]

$$V_{bat} = E - R_{int} * I \tag{28}$$

However, it was essential to calculate previously the open-circuit voltage with Eq. (29) and the battery output current through Eq. (30) [11].

$$E = n * (2.15 - DoD * (2.15 - 2)) \tag{29}$$

where n represents the number of battery cells, as in a lead-acid battery the voltage is 2 V per cell, so 12 cells are needed to reach the 24 V.

$$I = \frac{E - \sqrt{E^2 - 4RP}}{2R} \tag{30}$$

where the installed power was previously calculated and the internal resistance ($R_{int}$) of the battery has a value of 0.00105 Ω.

Finally, with the help of Excel, it is possible to obtain the battery output voltage through Eq. (31) [11], which is represented by Fig. 6.

$$V_{out} = E - I * R_{int} \tag{31}$$

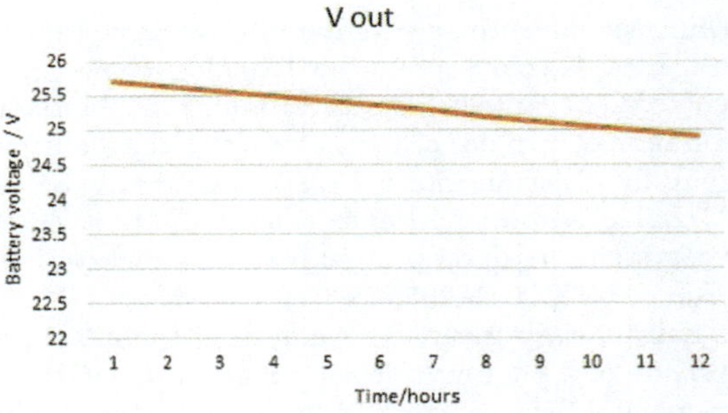


**Fig. 6.** Voltage over 12 h

## 5 Conclusions

Fossil fuels such as coal, oil, and gas are sources of energy that may soon be depleted. Although they are relatively cheap sources of energy, they have a significant impact in the environment. This work addresses one of the types of alternative energies, namely PV solar energy, focusing on the design of isolated PV systems. To do so, the components and calculations for each used component are described. With the completion of this work, it was concluded that the design of PV systems is an extremely important and relatively complex stage, as there are several factors to consider when installing a photovoltaic system. During this work, it has been possible to verify that the battery chosen for the system meets its requirements, as during the 12 h of operation during the monument's lighting period, the depth of discharge does not exceed 70%, and the battery output voltage does not drop below 24 V.

## References


1. Akram, F., Asghar, F., Majeed, M.A., Amjad, W., Manzoor, M.O., Munir, A.: Techno-economic optimization analysis of stand-alone renewable energy system for remote areas. Sustain. Energy Technol. Assess. **38** (2020)
2. Baptista, J., Morais, R.: Fundamentos de Energia Solar Fotovoltaica. Vila Real, Série Didática, Ciências Aplicadas 383. Edição do Núcleo Editorial e Gráfico dos SDB da UTAD (2010)
3. Chakrabarty, S., Islam, T.: Financial viability and eco-efficiency of the solar home systems (SHS) in Bangladesh. Energy **36**(8), 4821–4827 (2011)
4. Corrêa, F.C., et al.: Desenvolvimento e análise de estratégias de gerenciamento de potência em veículo elétrico híbrido de configuração paralela (2013)
5. El Shenawy, E., Hegazy, A., Abdellatef, M.: Design and optimization of stand-alone PV system for Egyptian rural communities. Int. J. Appl. Eng. Res. **12**(20), 10433–10446 (2017)
6. European Commission Joint Research Centre: JRC Photovoltaic Geographical Information System (PVGIS) (SD-B). https://re.jrc.ec.europa.eu/pvg_tools/en/#MR
7. Freitas, S.S.A.: Dimensionamento de sistemas fotovoltaicos. Ph.D. thesis, Instituto Politécnico de Bragança, Escola Superior de Tecnologia e de Gestão (2008)
8. Junior, J.U., Tiepolo, G.M., Junior, E.F.C., Tonin, F.S., Mariano, J.D.: Geração distribuída fotovoltaica: O caso dos sistemas fotovoltaicos da utfpr em curitiba. In: Congresso Brasileiro de Planejamento Energético. Gramado-RS (2016)
9. Kaundinya, D.P., Balachandra, P., Ravindranath, N.H.: Grid-connected versus stand-alone energy systems for decentralized power—a review of literature. Renew. Sustain. Energy Rev. **13**(8), 2041–2050 (2009)
10. Kurbonov, Y.M., Saitov, E., Botirov, B.: Analysis of the influence of temperature on the operating mode of a photovoltaic solar station. In: IOP Conference Series: Earth and Environmental Science, vol. 614, p. 012034. IOP Publishing (2020)
11. Larminie, J., Lowry, J.: Electric Vehicle Technology Explained. Wiley, Hoboken (2012)
12. Lopes, R.J.C.: Efeito do sombreamento nos painéis fotovoltaicos. Ph.D. thesis, Instituto Superior de Engenharia de Lisboa (2013)
13. Ministério da Economia e da Inovação: Rtiebt - regras técnicas das instalações eléctricas de baixa tensão. https://dre.pt/application/conteudo/303262
14. Mohamed, S.A., Abd El Sattar, M.: A comparative study of P&O and INC maximum power point tracking techniques for grid-connected PV systems. SN Appl. Sci. **1**(2), 174 (2019)
15. Wassie, Y.T., Adaramola, M.S.: Socio-economic and environmental impacts of rural electrification with solar photovoltaic systems: evidence from southern Ethiopia. Energy Sustain. Dev. **60**, 52–66 (2021)
16. Zubi, G., Dufo-López, R., Pasaoglu, G., Pardo, N.: Techno-economic assessment of an off-grid PV system for developing regions to provide electricity for basic domestic needs: a 2020–2040 scenario. Appl. Energy **176**, 309–319 (2016)

# Study of Forecasting Methods' Impact in Wholesale Electricity Market Participation

Brígida Teixeira[1](✉), Ricardo Faia[1], Tiago Pinto[1,2], and Zita Vale[1]

[1] GECAD, LASI, Polytechnic of Porto (ISEP/IPP), Porto, Portugal
{bct,rfmfa,tcp,zav}@isep.ipp.pt

[2] University of Trás-os-Montes and Alto Douro, Vila Real, Portugal

**Abstract.** Renewable energy sources have transformed the electricity market, allowing virtual power players or aggregators to participate and benefit from selling surplus energy. However, meeting demand and ensuring energy production stability can be challenging due to the intermittent nature of renewable sources. Accurate forecasting of energy consumption, generation, and electricity prices is critical to address these issues. Moreover, the selection of the best algorithm for forecasting is usually based on the predictions' accuracy, neglecting other factors such as the impact of errors on the real context. This paper presents a study on the economic risk of price forecasting errors on the electricity market's trading. For this, a simulation model is proposed to analyze the deviations between actual and predicted prices and how these deviations can affect trading in the electricity market, where the main purpose is to maximize profit, depending on whether the player is buying or selling electricity. The economic risk analysis and the predictions scores are used to improve the forecasts, using an approach based on reinforcement learning to evaluating and selecting which models demonstrated better performance in past transactions. The study involved simulating an aggregator's transactions in the Iberian electricity market for two consecutive days in October 2021. Real data from the market operator between 2020 and 2021 and seven forecasting models were used. The findings showed that errors have a significant impact on profit. Including the economic impact analysis and score evaluation of forecasting methods to determine which method can offer better results has proven to be a viable approach.

**Keywords:** Accuracy analysis · Forecasting methods · Wholesale electricity market

The present work funds from FCT Portuguese Foundation for Science and Technology. Brígida Teixeira is supported by FCT with Ph.D. Grant 2020.08174.BD. This work has received funding from FEDER Funds through COMPETE program and from National Funds through (FCT) under the project PRECISE (PTDC/EEI-EEE/6277/2020). The authors acknowledge the work facilities and equipment provided by the GECAD research center (UIDB/00760/2020 and UIDP/00760/2020) to the project team.


R. Mehmood et al. (Eds.): DCAI 2023, LNNS 741, pp. 267–276, 2023.
https://doi.org/10.1007/978-3-031-38318-2_27

## 1 Introduction

The increased use of renewable energy sources (RES) has brought about significant transformations in the electricity market (EM), granting more influence on smaller entities such as prosumers, virtual power players, and aggregators [1]. These entities have been able to leverage the cost-saving benefits associated with generating their own renewable electricity and subsequently selling any surplus energy through local or wholesale markets. Consequently, these actors have profited from this agreement, with the producers earning extra revenue and the consumers being able to access electricity at lower costs. Nonetheless, the intermittency of RES poses a significant challenge in maintaining production stability and satisfying demand [2]. Consequently, it is imperative to carefully plan and assess the sustainable integration of RES into the market.

To address challenges associated with energy resources managing (ERM), an increasing use of artificial intelligence (AI) techniques has emerged to monitor and control energy usage in real-time. These techniques have been applied to forecasting models based on different algorithms, such as artificial neural networks, support vector machines, decision trees, nearest neighbors, and hybrid models [3]. These models analyze multiple sources of heterogeneous data, detect complex daily life patterns, and use them to predict the future demand and generation of the players [4]. The use of AI techniques in ERM allows for improved analysis of data and more accurate predictions, leading to better decisions and ultimately, increased profitability [5].

One of the limitations of selecting the optimal algorithm for energy consumption forecasting is that the focus is often placed solely on accuracy analysis, while critical factors, such as the impact of errors in EM transactions, are overlooked. This approach can have significant limitations in certain contexts, especially for players participating in the wholesale EM, who handle substantial amounts of energy [6, 7]. In this context, players require forecast models with a high degree of accuracy to ensure profitability. However, overall model accuracy may not be sufficient to ensure the best results, particularly if the highest errors occur during specific periods, resulting in more significant impacts. For instance, if players purchase electricity during periods of high error, the impact on profitability can be significant. Similarly, if actual demand exceeds expectations, costs will increase as players are forced to purchase additional electricity to avoid a blackout and potential penalties. Therefore, accurate forecasting is crucial for players to make informed decisions and achieve optimal economic outcomes. In this regard, careful consideration of the impact of market errors on profitability is crucial when selecting the best algorithm for energy forecasting.

This paper aims to analyze the impact of errors in electricity consumption, generation, and price forecasting on the day-ahead (DA) and intraday (ID) sessions in EMs. Additionally, the study aims to assess the economic risk associated with these deviations and propose a methodology to leverage this information to enhance future forecasts. To optimize the performance of forecasting models during EM participation, the paper proposes a methodology that employs continuous simulation model updates by using an approach based on reinforcement learning (RLA). This process involves training several forecasting machine learning (ML) models with different hyperparameters that can adapt to the context and minimize the impact of errors, namely Multi-Layer Perceptron (MLP), Support Vector Machines (SVM), Random Forest (RF), Nearest Neighbors

(KNN), Adaboost (ADA), Gradient Boosting (GB), and Extreme Gradient Boosting (XGB), to achieve the desired outcome.

To investigate the effectiveness of the proposed methodology, a case study using actual data from an aggregator is presented in this paper. The study incorporates information on energy consumption and generation, as well as prices from the Mercado Ibérico de Electricidade (MIBEL) [8] EM for both DA and ID market sessions. The case study findings demonstrate the superior performance of the proposed approach, achieving higher levels of accuracy and better economic impact compared to traditional forecasting methods. The case study highlights the potential consequences of errors in electricity prices forecasting for market participants in the wholesale EM, emphasizing the need for improved forecasting methods.

The proposed methodology is designed to enhance the accuracy and reliability of forecasting models by continuously analyzing the impact of forecast errors. By leveraging this approach, market participants can make better-informed decisions that optimize their economic outcomes while minimizing the risks associated with forecast errors. This methodology enables market participants to gain deeper insights into the effects of forecast errors on market outcomes, which can help them make better-informed decisions and ultimately achieve better economic outcomes.

The paper is structured as follows: Sect. 2 outlines the methodology, Sect. 3 defines the case study, Sect. 4 presents the study results, and Sect. 5 concludes and discusses future research directions.

## 2 Methodolody

In this section, the methodology employed to address the challenges of forecasting in the context of wholesale EMs is presented.

The proposed model aims to predict energy consumption, generation, and market prices utilizing smart energy management systems to enhance ERM in several contexts. Smart meters installed in the grid collect data, which is subsequently preprocessed to extract new time-based features such as weekends, days of the week, and others. The preprocessed data is then fed into several ML forecasting models during the training phase to determine which ones yield the best results. The selected models are used to predict energy consumption, generation, and market prices for each market bidding session. Predictions are updated prior to the bidding sessions to ensure profitable operations while maintaining the stability of production and meeting demand. Figure 1 illustrates the proposed model.

The methodology is divided into two distinct phases. In the first phase, the aim is to select the best forecasting models to determine the market session in which the supported player will participate based on its predicted demand. The second phase involves simulating the player's participation in the EM and analyzing the resulting economic impact. The following subsections provide a detailed description of these phases.

### 2.1 Selection of Best Forecasting Models

During the first phase of the proposed model, the selection of the best forecasting models is conducted to generate accurate predictions for energy consumption, generation, and

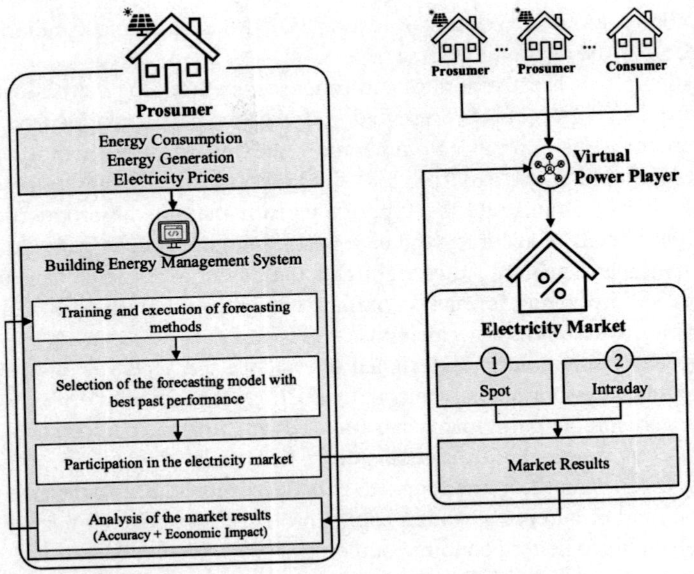


**Fig. 1.** Overview of the proposed model

EM prices. Market price data is obtained from the MIBEL's market operator website [9]. In this phase, several ML forecasting models are employed, including MLP, SVM, RF, KNN, ADA, GB, and XGB. The use of multiple ML models provides greater flexibility to adapt to different contexts.

To prepare the data for the training of the ML models, a preprocessing phase was performed. This phase involved the extraction of new time-based features such as the day of the week and whether it was a weekend, followed by the application of feature transformation techniques to create new features that could capture trends and patterns in the data that were not apparent from the original features. Outlier detection techniques were applied to remove any anomalous data points that could skew the training results. The data was then normalized and used to train the forecasting models.

The evaluation of each ML model's performance is conducted by training and testing them using the coefficient of determination ($R^2$), expressed in Eq. 1, where $n$ is the number of observations, $P_i$ is the $i$-th predicted value of electricity, and $A_i$ is the $i$-th actual value.

$$R^2 = \left( \frac{\sum\left[\left(A_i - \left(\frac{1}{n}\sum_{i=1}^{n} A_i\right)\right) * \left(P_i - \left(\frac{1}{n}\sum_{i=1}^{n} P_i\right)\right)\right]}{\sqrt{\left[\left(A_i - \left(\frac{1}{n}\sum_{i=1}^{n} A_i\right)\right)^2 * \left(P_i - \left(\frac{1}{n}\sum_{i=1}^{n} P_i\right)\right)^2\right]}} \right)^2 \quad (1)$$

To select the best models for generating accurate forecasts, an Automated ML technique known as grid search is used. Once the best models are identified, they are used to generate updated forecasts at the beginning of each market session. These sessions include the DA and the six ID bidding sessions. The model was designed to ensure that the most accurate forecasts were used to inform market participation decisions, helping to guarantee the stability of energy production while meeting demand and ensuring good profitability for players in the EM.

## 2.2 Simulation of Electricity Market

The second phase of the methodology involves simulating a player's participation in the EM with two main objectives: meeting their own demand and maximizing profits by buying energy when prices are low and selling energy when prices are high. To participate in the market, a player can bid in both the DA and ID sessions. The DA market allows energy to be traded for the following day, while ID sessions enable players to adjust their bids based on changes in expected demand and supply. A player has seven opportunities to negotiate energy. The player's success in the market depends on the accuracy of the forecasts. Therefore, it is essential to evaluate the impact of forecast errors on the player's profits and adjust the forecasting models accordingly.

During each period, the player must decide which market session to participate in, based on their current position and the forecasts for each session. For instance, if the player is selling energy and the forecasts predict that prices will be higher in the ID session than in the DA market, the player may choose to participate in the ID session. Conversely, if the player is buying energy and the forecasts predict that prices will be lower in the DA market, the player may choose to participate in the DA market. The goal is to maximize profits by participating in the market session with the most promising results, based on the player's current position and the forecasts. Therefore, accurate forecasting is crucial to the player's success in the EM.

After each market session, the market results are released and made available. In this way, players can evaluate the accuracy and performance of the forecasts and refine their models before the start of the next session using RLA-based technique. To evaluate the performance of the forecasting model, two factors must be considered: the accuracy of the price predictions and the economic impact of forecast errors. The accuracy of the forecasting models in predictions is measured using the root mean square error (RMSE), expressed in Eq. (2). A lower RMSE score indicates better accuracy and match of predictions with actual values.

$$RMSE = \sqrt{\frac{\sum_{i=1}^{n} (P_i - A_i)^2}{n}} \tag{2}$$

The RMSE is a useful metric because it is easy to interpret and can be used to compare different models or forecasting methods. The score of the model is calculated based on total RMSE between the predicted and actual values, as shown in Eq. 3.

$$S = \sum\nolimits_{i=1}^{n} RMSE(P_i, A_i) \tag{3}$$

Additionally, the economic risk assessment of the predicted values is calculated. The economic risk refers to the financial implications that result from deviations between actual energy demand and what was predicted. These deviations can have significant consequences for players in the wholesale EM who participate in DA and ID sessions, as errors in forecasting can lead to missed opportunities for profitable trades or additional costs to meet unexpected demand. Therefore, the proposed approach aims to minimize these errors and their economic impact with continuous simulation model updates to optimize energy consumption and increase profitability. The economic impact is one of the components used to calculate the overall performance of the proposed approach, as described by the Eq. 4.

In addition to evaluating the accuracy of the predicted values through the RMSE score, an economic risk assessment is calculated to determine the financial implications of deviations between the predicted and actual energy demand. These deviations can result in significant consequences for players in the wholesale EM who participate in DA and ID sessions, as it can lead to missed opportunities for profitable trades or additional costs to meet unexpected demand. The economic impact is mathematically represented by Eq. 4:

$$EI = \sum\nolimits_{i=1}^{n} (P_{ap_i} - P_{pp_i}) \tag{4}$$

where $P_{api}$ represents the predicted transacted energy with the application of the real prices and $P_{ppi}$ represents the predicted transacted energy with the application of the predicted prices. The sum is taken over all the instances ($i$) in the results dataset.

Once the score and economic impact have been calculated, the proposed model's performance can be assessed. The performance of the model is measured using a composite metric that takes into account both the accuracy of the forecasts and their economic impact. The formula used to calculate the performance is given in Eq. (5), where $P$ represents the performance, $EI$ represents the economic impact, $S$ represents the accuracy score of the forecasts, and $\beta$ is a weighting factor that determines the influence of the economic impact on the performance. This composite metric allows for a more comprehensive evaluation of the model's effectiveness in predicting energy demand and generation in the wholesale EM, as it considers both the accuracy of the forecasts and their financial impact. The performance is used to evaluate the effectiveness of the proposed approach compared to traditional forecasting methods.

$$P = \beta * EI + (1 - \beta) * S \tag{5}$$

The performance of the forecasting model is ultimately determined by applying a RLA approach that incorporates past performance into the calculation. This approach is represented by the Eq. (6), where $P_f$ denotes the final performance score, $P$ is the current performance score calculated in (5), $P_p$ is the past performance score, and $\alpha$ is a weighting factor that dictates the influence of past performance on the current performance. By employing RLA principles, the model adapts to changes in the market, resulting in improved performance and more accurate predictions.

$$P_f = \alpha * P + (1 - \alpha) * P_p \tag{6}$$

## 3 Case Study

In this section, the proposed model is used to conduct a real case study aimed at exploring the influence of forecasting errors on the bids in the wholesale EM. The study aims to determine if this information can be leveraged to make a more informed selection of forecasting models, leading to increased participation profits. To this end, the study simulates the participation of an aggregator in the EM over a continuous period of 48 h spanning 11$^{th}$ and 12$^{th}$ October 2021. Tests were conducted on machine with Intel Core i7-9700K CPU (3.60 GHz Octa Core) and 16 GB RAM.

The first step of the case study involves performing electricity consumption, generation, and price forecasts using different forecasting ML models, including MLP, SVM, RF, ADA, GB, and XGB. The electricity price predictions were generated by training the selected models using data collected from the MIBEL market operator [9], during the years 2021 and 2022. For consumption and generation predictions, and data from an office building were used for the same period. The models were trained by fine-tuning their hyperparameters to enhance accuracy and prevent overfitting, using time series cross-validation with 3 splits. The hyperparameters utilized for fine-tuning each model are presented in Table 1. The top five models with the highest scores for each forecasting target, e.g., consumption, generation, and session prices, were selected based on the $R^2$ score. The RLA-based approach aims to assess if any other of these five forecasting models can provide better results at the end of the session.

**Table 1.** Forecasting models and hyperparameters

| # | Models | Hyperparameters |
|---|---|---|
| 1 | SVM | {'kernel': ['linear'], 'C': [0.1, 1, 10, 100]}, {'kernel': ['rbf'], 'C': [0.1, 1, 10, 100], 'gamma': [0.1, 0.01, 0.001]} |
| 2 | MLP | {'hidden_layer_sizes': [(50, 50, 50)], 'activation': ['tanh', 'relu'], 'solver': ['adam'], 'alpha': [0.0001, 0.05], 'learning_rate': ['constant', 'adaptive'], 'max_iter': [10000]} |
| 3 | RF | {'n_estimators': [10, 100, 1000], 'max_depth': [None, 10, 50], 'max_features': ['sqrt', 'log2']} |
| 4 | KNN | {'n_neighbors': [5, 10, 20], 'weights': ['uniform', 'distance'], 'algorithm': ['ball_tree', 'kd_tree', 'brute'], 'leaf_size': [10, 50]} |
| 5 | ADA | {'n_estimators': [50, 100, 200], 'learning_rate': [0.01, 0.1, 1]} |
| 6 | GB | {'n_estimators': [50, 100, 200], 'learning_rate': [0.01, 0.1, 1], 'max_depth': [3, 5, 10], 'max_features': ['sqrt', 'log2']} |
| 7 | XGB | {'n_estimators': [100, 500, 1000], 'learning_rate': [0.01, 0.1, 1], 'max_depth': [3, 5, 10]} |

After the forecasting models selection, the next step consists in proceeding with the wholesale EM simulation. The aggregator participates in both the DA and ID sessions by placing bids for buy or sell energy. The information regarding the different EM sessions is described in [9]. To participate effectively in the market, the aggregator needs to have an accurate estimate of the expected demand, which is determined by computing the difference between the predicted consumption and generation values, and it is used to determine the optimal bidding strategy for the aggregator. To make well-informed bidding decisions, the aggregator must constantly update its demand and price forecasts based on the latest information available. This necessitates redoing the forecasts at the beginning of each session to reflect any changes in market conditions.

The case study evaluated the economic impact of learning algorithms on electricity price forecasting in the wholesale EM through two tests. The first test used the best forecasting model for each session without reevaluating its performance. In contrast, the second test reevaluated and selected one of the other top 5 models based on their historic performance to determine if it had better performance. The simulation results from both tests were compared to assess the proposed approach's effectiveness. The study's analysis of forecasting errors on the aggregator's performance provides valuable insights into the effectiveness of various forecasting methods and the optimal bidding strategies for different market conditions. The weights used for performance calculation are $\alpha = 0.35$ and $\beta = 0.60$, for giving more consideration to economic impact, and emphasis on past performance, respectively.

## 4 Results and Discussion

Table 2 presents the bidding strategy of the aggregator for the first day, based on three scenarios: Ideal, where no errors are considered; Forecasted, which uses the same forecasting method for each session; and Forecasted with RLA, which adjusts the selection of the forecasting method based on performance.

**Table 2.** Bidding strategy for day 1, for DA and ID sessions

| | 11th October | | | | | | | | | | | | | | | | | | | | | | | |
|---|---|---|---|---|---|---|---|---|---|---|---|---|---|---|---|---|---|---|---|---|---|---|---|---|
| | 1 | 2 | 3 | 4 | 5 | 6 | 7 | 8 | 9 | 10 | 11 | 12 | 13 | 14 | 15 | 16 | 17 | 18 | 19 | 20 | 21 | 22 | 23 | 24 |
| Forecasted | ID2 | ID1 | ID3 | ID1 | ID3 | ID3 | ID3 | ID5 | ID3 | ID3 | ID4 | DA | ID6 | ID6 | ID4 | ID4 | ID2 | ID6 | ID5 | ID4 | ID2 | ID3 | ID1 | ID3 |
| Forecasted w/ RLA | ID2 | ID1 | ID3 | ID1 | ID3 | ID4 | ID4 | ID5 | ID3 | ID4 | ID4 | DA | ID6 | ID6 | ID4 | ID4 | ID1 | ID6 | ID2 | ID3 | ID2 | ID6 | ID3 | ID4 |
| Ideal | ID3 | ID1 | ID3 | ID3 | ID4 | ID4 | ID4 | ID5 | ID4 | ID4 | ID4 | ID1 | ID6 | ID6 | ID6 | ID6 | ID1 | ID6 | ID5 | ID5 | ID5 | ID6 | ID6 | ID4 |

Table 3 presents the economic impact of the aggregator's bidding strategies in the wholesale EM, with and without the use of learning algorithms. In addition, Table 4 provides a summary of the profits for each day and session. The results presented in both tables demonstrate the effectiveness of using learning algorithms to improve the economic performance of the aggregator's bidding strategies in the wholesale EM. Across all market sessions and days, the economic impact with learning was lower than without learning for the ID sessions and the DA session.

Table 3 demonstrates that the use of RLA-based selection of forecasting models had a positive impact on the aggregator's economic performance, resulting in lower economic impact values in most cases. The economic impact with learning was on average 5.69% lower than without learning. In the DA session on Day 1, the economic impact with and without learning was the same. However, in the DA session on Day 2, the economic impact without learning was 7.76% higher compared to the approach with learning. Similarly, in the ID sessions, the economic impact values were mostly lower with learning compared to without learning, with an average difference of 13.64%. ID 3 showed the most significant improvement with learning, with a 40.88% economic impact value. The XGB provided the best results with RLA, with hyperparameters set to 'learning_rate' of 0.01, 'max_depth' of 10, and 'n_estimators' of 1000. Conversely, the best forecasting model without learning was also XGB, but with different hyperparameters, including a learning rate of 1, a maximum depth of 3, and 100 estimators. ID sessions 1 and 2 showed the worst performance, possibly due to higher market volatility and unpredictability during those sessions, as well as a larger number of periods.

**Table 3.** Economic impact results

| Market Session | Day 1 | | Day 2 | |
|---|---|---|---|---|
| | Economic Impact (EUR) | Economic Impact w/ RLA (EUR) | Economic Impact (EUR) | Economic Impact w/ RLA (EUR) |
| DA | 2.00 | 2.00 | 165.34 | 155.18 |
| ID 1 | 129.52 | 690.51 | 145.06 | 189.39 |
| ID 2 | 1103.50 | 1146.69 | 1252.30 | 1730.48 |
| ID 3 | 705.63 | 406.64 | 1818.93 | 1075.23 |
| ID 4 | 1216.52 | 986.32 | 784.16 | 137.03 |
| ID 5 | 748.73 | 294.24 | 207.29 | 825.73 |
| ID 6 | 1248.85 | 1276.97 | 582.20 | 53.98 |

**Table 4.** Summary of profit calculated from economic impact in EUR.

| Market Session | Profit (EUR) | Profit w/ RLA (EUR) | Total |
|---|---|---|---|
| DA | 0.00 | 10.15 | **10.15** |
| ID 1 | −560.99 | −44.33 | **−605.32** |
| ID 2 | −43.19 | −478.19 | **−521.38** |
| ID 3 | 298.99 | 743.71 | **1042.70** |
| ID 4 | 230.21 | 647.13 | **877.34** |
| ID 5 | 454.49 | −618.49 | **−164.00** |
| ID 6 | −28.12 | 528.23 | **500.10** |
| **Total** | **351.38** | **788.21** | **1139.59** |

Table 4 shows the profits calculated from the economic impact for DA and ID sessions. The table reveals that the use of learning enhances the economic performance of the aggregator's bidding strategies, leading to higher profits in most cases. In ID3 it is possible to observe a significant increase in profits of 743.71 EUR with use of learning, compared to only 298.99 EUR without learning. Despite some negative profits observed in ID1 and ID5 without learning, the overall total profit was higher with the use of learning, resulting in a profit of 788.21 EUR. With the continuous learning to select the best forecasting method, it was possible to save 1139.59 EUR.

In general, the results demonstrate that using learning algorithms to improve demand and price forecasts and to optimize bidding strategies can have a significant impact on the economic performance of the aggregator. By continually improving the accuracy of forecasting and bidding strategies, aggregators can achieve higher profits and contribute to more efficient and sustainable use of energy resources.

## 5 Conclusions

The paper proposes an approach to study the impact of forecasting errors on energy consumption in wholesale EMs. It uses a continuous simulation model updates to maximize the forecasting performance, considering the economic impact of the errors. The proposed model was evaluated using real-world data and outperformed traditional methods. The approach provides insights into the impact of forecasting errors on profitability and offers a viable solution for players to optimize energy consumption and increase profitability. Further research could explore the effectiveness of other forecasting algorithms and techniques and their impact on profitability.

## References

1. European Commission. Directive (EU) 2019/944 of the European Parliament and of the Council of 5 June 2019 on common rules for the internal market for electricity 2019
2. Impram, S., Varbak Nese, S., Oral, B.: Challenges of renewable energy penetration on power system flexibility: a survey. Energ. Strat. Rev. **31**, 100539 (2020). https://doi.org/10.1016/J.ESR.2020.100539
3. Amasyali, K., El-Gohary, N.M.: A review of data-driven building energy consumption prediction studies. Renew. Sustain. Energy Rev. **81**, 1192–1205 (2018). https://doi.org/10.1016/J.RSER.2017.04.095
4. Halkos, G.E., Gkampoura, E.C.: Reviewing usage, potentials, and limitations of renewable energy sources. Energies **13**, 2906 (2020). https://doi.org/10.3390/EN13112906
5. Ali, S.S., Choi, B.J.: State-of-the-art artificial intelligence techniques for distributed smart grids: a review. Electronics **9**, 1030 (2020). https://doi.org/10.3390/ELECTRONICS9061030
6. Scharff, R., Amelin, M.: Trading behaviour on the continuous intraday market Elbas. Energy Policy **88**, 544–557 (2016). https://doi.org/10.1016/J.ENPOL.2015.10.045
7. Silva, A.R., Pousinho, H.M.I., Estanqueiro, A.: A multistage stochastic approach for the optimal bidding of variable renewable energy in the day-ahead, intraday and balancing markets. Energy **258**, 124856 (2022). https://doi.org/10.1016/J.ENERGY.2022.124856
8. Mibel – Mercado Ibérico de Electricidade 2023. https://www.mibel.com/en/home_en/. Accessed 24 Apr 2023
9. OMIE 2023. https://www.omie.es/en/mercado-de-electricidad. Accessed 13 Apr 2023

# A Variable Neighborhood Descent Algorithm for the Service Restoration Problem of Radial Distribution Systems

Gabriel F. Puerta[1,2(✉)], Leonardo H. Macedo[2,3], João Soares[1,4], and Rubén Romero[2]

[1] GECAD – Knowledge Engineering and Decision-Support Research Centre of the School of Engineering – Polytechnic Institute of Porto (ISEP/IPP), Porto, Portugal
gabriel.puerta@unesp.br, jan@isep.ipp.pt

[2] LaPSEE – Power Systems Planning Laboratory, Department of Electrical Engineering, São Paulo State University, Ilha Solteira, Brazil
leohfmp@ieee.org, ruben.romero@unesp.br

[3] Department of Engineering, São Paulo State University, Rosana, Brazil

[4] Innovation and Development, LASI - Intelligent Systems Associate LAboratory, Polytechnic of Porto, Porto, Portugal

**Abstract.** This paper presents a variable neighborhood descent (VND) algorithm for the service restoration problem (SRP) in distribution systems. The restoration problem appears when a permanent fault occurs in the system. The fault must be localized and isolated from the rest of the system. As a consequence of the isolation, the downstream area is de-energized and requires restoration. The VND algorithm features its ability to solve the SRP for radial distribution systems and introduces a strategy that helps to deal with the system's radiality. Tests are performed using a 53-node test system. The results show that the proposed algorithm can efficiently solve the SRP in a distribution system, and as a deterministic algorithm, the results can be used to help non-deterministic algorithms.

**Keywords:** Electrical distribution systems · Metaheuristic · Service restoration · Variable neighborhood search

This work has received funding from National Funds through FCT – Portugal and CAPES – Brazil, under the project 2019.00141.CBM Desenvolvimento de Técnicas de Inteligência Artificial para a Otimização de Sistemas de Distribuição de Energia Elétrica. It was also supported in part by the Coordination for the Improvement of Higher Education Personnel (CAPES) – Finance Code 001, and by the São Paulo Research Foundation (FAPESP), under grants 2015/21972-6 and 2018/20355-1. The authors acknowledge the work facilities and equipment provided by GECAD research center (UIDB/00760/2020) to the project team and also acknowledge the support provided by the Thematic Network 723RT0150 "Red para la integración a gran escala de energías renovables en sistemas eléctricos (RIBIERSE-CYTED)" financed by the call for Thematic Networks of the CYTED (Ibero-American Program of Science and Technology for Development) for 2022. João Soares is also supported by National funds from FCT CEECIND/00420/2022.


R. Mehmood et al. (Eds.): DCAI 2023, LNNS 741, pp. 277–287, 2023.
https://doi.org/10.1007/978-3-031-38318-2_28

# 1 Introduction

The electrical distribution system (EDS) establishes a direct relationship between the consumers and the utilities. Usually, a distribution network operates with a radial topology for a low-cost and more simple design of the protection coordination scheme. This type of topology also facilitates the isolation of permanent faults in the systems due to the easy isolation and detection of the fault [1].

Power outages in an EDS can occur for various reasons, such as equipment failures, natural disasters, or human errors. Thus, after the occurrence of a permanent outage in a distribution network, the area affected by the fault is isolated from the rest of the system. Due to the isolation, the downstream areas are disconnected from the system and need to have their supply service restored. The service restoration problem (SRP) refers to the process of elaborating a restoration plan that re-establishes the supply of out-of-service areas and also minimizes the required amount of switching operations [2].

The topology change of the EDS is characterized by switching operations that can restore the faulted area by reconnecting it with its feeder or even transferring to another feeder. This process must consider the system's physical and operational constraints. The SRP must satisfy the constraints of both Kirchhoff's laws, current capacity on branches, minimum and maximum voltage magnitude boundary on nodes, the power capacity of substations, and the radiality conditions [3].

The SRP is a nonlinear and multi-objective combinatorial problem and has been extensively studied in the literature. Initially, the problem was resolved using a set of heuristic strategies and specialized systems that relies on the operator's experience and knowledge of the system [4,5]. Posteriorly, hybrid algorithms that combine heuristic rules with mathematical optimization were proposed [6]. A heuristic algorithm that also prioritizes critical loads and a heuristic algorithm that design the system using graph theory have also been proposed to solve the problem [7,8].

In addition to heuristic algorithms, metaheuristic-based algorithms are also popular in the specialized literature: genetic and evolutionary algorithms [9–11], particle swarm optimization [12], and hybrid algorithms [13] have been used to solve the SRP in distribution systems.

Mathematical models are also very popular in specialized literature. In this approach category, it is possible to model the SRP as a mixed-integer linear programming (MILP) model [3,14] and as a mixed-integer second-order cone programming (MISOCP) model [1,15,16].

In [12], a hybrid algorithm divided into three stages is proposed for the SRP. The first stage is based on a binary PSO algorithm. The second stage is a user-centered stage that defines the initial solution in the third stage, characterized by a local search based on the user's preferences. Reference [3] solves the SRP in conjunction with the switch sequence problem for unbalanced three-phase EDS. At first, a mixed-integer nonlinear programming model is proposed. Then, the model is subject to a sequence of linearizations which transforms the model into an MILP model. An MISOCP model is proposed in [16] to solve the SRP in large radial EDS. The main feature of the work is its ability to deal with a reduced number of binary variables, which reduces the search space and, consequently, the necessary time to find the optimal solution. In [14] an MILP model is proposed for the SRP. This model considers voltage control device as on-load tap changes, voltage regulators and capacitor banks, and distributed generation (DG), including its ability of island operation. Finally, an evolutionary multi-objective algorithm that combines the use of a sub-population table is proposed in [11] to solve the SRP in distribution systems with the presence of DG and islanded operation.

This work uses VND's deterministic characteristic to solve the SRP for radial EDS. The novel contributions of this paper include: 1) a VND algorithm specialized in the SRP; 2) the strategy of working with a fictitious substation; and 3) as a deterministic algorithm, it can be used to verify other algorithms' convergence.

The paper is organized as follows: Sect. 2 discusses the problem's formulation. The proposed method is presented in Sect. 3. Section 4 discusses the simulation results. Finally, the paper's conclusion and description of future work are given in Sect. 5.

## 2 Problem Formulation

The service restoration problem is a critical issue in a radial distribution system. Its primary goal is to restore as many loads as possible while minimally changing the network topology. In this section, the SRP is modeled as an MISOCP; the model was proposed in [16], the objective function is defined by (1) and the set of constraints by (2)–(27).

$$\text{minimize } v = \sum_{i \in \Omega_b} \alpha_i P_i^D y_i + \sum_{i,j \in \Omega_{o-}} \beta_{i,j} x_{i,j} + \sum_{i,j \in \Omega_{o+}} \mu_{i,j}(1 - x_{i,j}) \quad (1)$$

$$\sum_{j,i \in \Omega_l} P_{j,i} - \sum_{i,j \in \Omega_l} (P_{i,j} + R_{i,j} I_{i,j}^{sqr}) + P_i^G = P_i^D (1 - y_i) \quad \forall i \in \Omega_b \quad (2)$$

$$\sum_{j,i \in \Omega_l} Q_{j,i} - \sum_{i,j \in \Omega_l} (Q_{i,j} + X_{i,j} I_{i,j}^{sqr}) + Q_i^G = Q_i^D (1 - y_i) \quad \forall i \in \Omega_b \quad (3)$$

$$\sum_{j,i \in \Omega_l \cup \Omega_h} H_{j,i} - \sum_{j,i \in \Omega_l \cup \Omega_h} H_{i,j} + H_i^G = y_i \quad \forall i \in \Omega_b \quad (4)$$

$$P_i^G = 0 \quad \forall i \in (\Omega_b - \Omega_s) \cup S^f \quad (5)$$

$$Q_i^G = 0 \quad \forall i \in (\Omega_b - \Omega_s) \cup S^f \quad (6)$$

$$H_i^G = 0 \quad \forall i \in \Omega_b, i \neq S^f \quad (7)$$

$$\sum_{i,j \in \Omega_o \cup \Omega_h} x_{i,j} = |\Omega_b| - |\Omega_s| - |\Omega_c| \quad (8)$$

$$V_i^{sqr} - V_j^{sqr} = 2(R_{i,j} P_{i,j} + X_{i,j} Q_{i,j}) + Z_{i,j}^2 I_{i,j}^{sqr} + b_{i,j} \quad \forall i,j \in \Omega_l \quad (9)$$

$$|\Omega_b| \leq (\overline{V}^2 - \underline{V}^2)y_i \qquad \forall i,j \in \Omega_c \qquad (10)$$

$$|\Omega_b| \leq (\overline{V}^2 - \underline{V}^2)(1 - x_{i,j} + y_i + y_j) \qquad \forall i,j \in \Omega_o \qquad (11)$$

$$V_j^{sqr} I_{i,j}^{sqr} \geq P_{i,j}^2 + Q_{i,j}^2 \qquad \forall i,j \in \Omega_l \qquad (12)$$

$$(P_i^G)^2 + (Q_i^G)^2 \leq (\overline{S}_i^G)^2 \qquad \forall i \in \Omega_s, i \neq S^f \qquad (13)$$

$$\underline{V}^2(1 - y_{i,j}) \leq V_i^{sqr} \leq \underline{V}^2(1 - y_{i,j}) \qquad \forall i \in \Omega_b \qquad (14)$$

$$0 \leq I_{i,j}^{sqr} \leq \overline{I}_{i,j}^2(1 - y_{i,j}) \qquad \forall i,j \in \Omega_c \qquad (15)$$

$$0 \leq I_{i,j}^{sqr} \leq \overline{I}_{i,j}^2 x_{i,j} \qquad \forall i,j \in \Omega_o \qquad (16)$$

$$|H_{i,j}| \leq M y_i \qquad \forall i,j \in \Omega_c \qquad (17)$$

$$|H_{i,j}| \leq M x_{i,j} \qquad \forall i,j \in \Omega_o \cup \Omega_h \qquad (18)$$

$$|P_{i,j}| \leq \overline{VI}_{i,j}(1 - y_i) \qquad \forall i,j \in \Omega_c \qquad (19)$$

$$|Q_{i,j}| \leq \overline{VI}_{i,j}(1 - y_i) \qquad \forall i,j \in \Omega_c \qquad (20)$$

$$|P_{i,j}| \leq \overline{VI}_{i,j} x_{i,j} \qquad \forall i,j \in \Omega_o \qquad (21)$$

$$|Q_{i,j}| \leq \overline{VI}_{i,j} x_{i,j} \qquad \forall i,j \in \Omega_o \qquad (22)$$

$$|y_i - y_j| \leq (1 - x_{i,j}) \qquad \forall i,j \in \Omega_o \cup \Omega_h \qquad (23)$$

$$y_i = y_j \qquad \forall i,j \in \Omega_c \qquad (24)$$

$$y_{S^f} = 1 \qquad (25)$$

$$x_{i,j} \in \{0,1\} \qquad \forall i,j \in \Omega_o \cup \Omega_h \qquad (26)$$

$$y_{i,j} \in \{0,1\} \qquad \forall i,j \in \Omega_b \qquad (27)$$

The objective function (1) is a weighted sum that aims to minimize the unsupplied loads and decrease the number of switching operations required to restore the system. Kirchhoff's current law is imposed by constraints (2)–(4). The active, reactive, and artificial power generations are limited by (5)–(7). The necessary conditions to maintain the system radiality are imposed by (2)–(8). Kirchhoff's voltage law is imposed by constraint (9) and (12) when the branches are energized. Constraints (10) and (11) supports the constraint (12) when the branches are de-energized. The operating limits of the radial distribution system are imposed by (13)–(16). In addition, the artificial power flow is limited by constraints (17) and (18). The main purpose of constraints (19)–(24) is to decrease the search space and, consequently, to help the model to find the optimal solution faster. Constraint (25) imposes that the fictitious substation must be connected to the de-energized area. Finally, constraints (26) and (27) define the binary variables $x_{i,j}$ and $y_i$, respectively.

## 3 Variable Neighborhood Search

The variable neighborhood search (VNS) algorithm is a metaheuristic optimization algorithm that has been strongly applied to a extensive range of combinatorial optimization problems, introduced by Mladenović and Hansen in 1997

[17]. This metaheuristic is known for its various features like universality, applicability, precision, and efficacy. Moreover, it is a very flexible metaheuristic that can be applied to solve different optimization problems. In this section, we propose a VNS algorithm based on a variation called variable neighborhood descent (VND) algorithm to solve the radial distribution SRP.

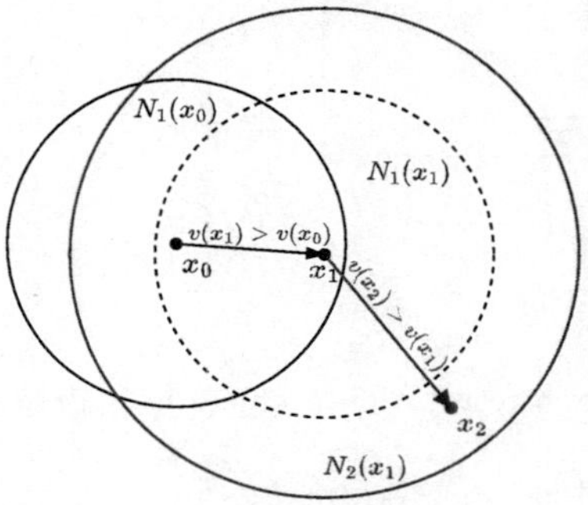


**Fig. 1.** Illustration of the VND algorithm search procedure

### 3.1 Variable Neighborhood Descent

The VND algorithm is a local search metaheuristic used for optimization problems and is a variation of the VNS algorithm. Its main feature is the deterministic exploration of different neighborhoods in an iterative way. The algorithm starts with a random initial solution; then, each iteration evaluates the current solution and moves to the best solution in a defined neighborhood.

The space search exploration of the VND algorithm is shown in Fig. 1. In the illustrated case, the search starts centralized at the initial solution $x_0$. The first neighborhood of $x_0$ is evaluated, and its best solution is $x_1$. Thus, the search is centralized in $x_1$, and this is the new current solution. At the first neighborhood of the new current solution, the algorithm could not find a neighboring solution that is better than the current one. In this case, the search is centralized again at the current solution $x_1$, and its second neighborhood is visited. This process is repeated until the search reaches the last neighborhood, and there is no better solution than the current solution.

### 3.2 Radiality Control

The metaheuristic-based algorithm's ability to generate only radial solutions for the SRP is indispensable for its efficacy. In this work, the radiality is controlled by the initial solution, the neighborhood structures, and the use of a fictitious substation. The last one is one of the main features of this work.

The main objective of the fictitious substation is to maintain the out-of-service area connected with the rest of the system. Thus, the de-energized loads are connected to the fictitious substation through fictitious branches. There is one fictitious branch between every load in the system and the fictitious substation.

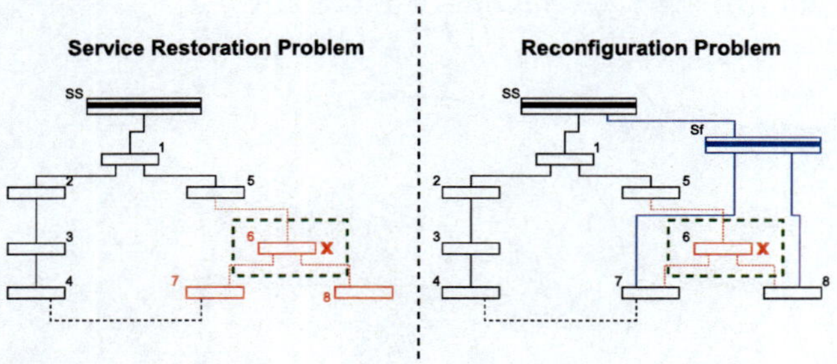


**Fig. 2.** Transformation of a restoration problem into a reconfiguration problem

Initially, every load island located in the out-of-service area is connected to the fictitious substation by closing one fictitious branch per load island. Thus, the SRP can be dealt as a reconfiguration problem where the objective is to minimize the active power flow at the branch that connects the real substation with the fictitious one. This process is illustrated in Fig. 2.

### 3.3 The VND Algorithm Proposed

This work proposes a VND algorithm specialized for the SRP in radial distribution systems. The initial solution is the system topology after the fault identification and isolation. Then, the faulted area is connected to a fictitious substation node supplied by a real substation node. Using a fictitious substation helps the algorithm control the system's radiality in the faulted area. Besides the substation, every node in the system has a fictitious branch that connects the node with the fictitious substation.

The operation of a pair of switches defines the neighborhood structure used in this work. In the first neighborhood ($N_1(x)$), only one pair of switches is operated; in the second neighborhood ($N_2(x)$), two pair of switches is operated; and in the last neighborhood ($N_3(x)$), three pairs of switches is operated. The flowchart of the proposed VND algorithm is presented in Fig. 3. Moreover, the proposed VND algorithm has three neighborhood levels and its stopping criterion is when is not possible to find a neighboring solution that is better than the incumbent at the third neighborhood level ($N_3(x)$). When the algorithm finds a feasible solution, i.e., a radial solution with switching operations of real branches, the fictitious part of the system is disconnected and the search process continues with the objective of reducing the switching operations.

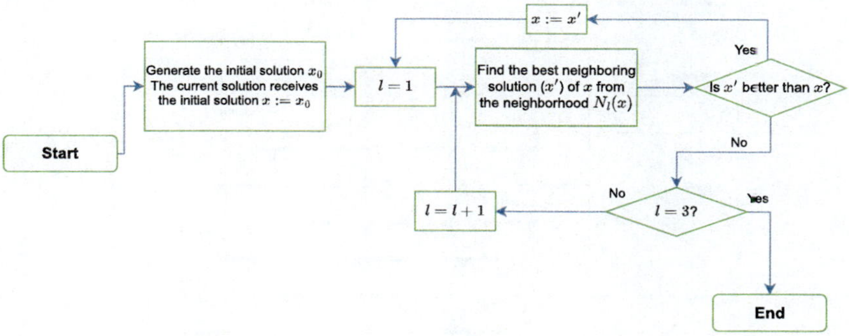


**Fig. 3.** Flowchart of the proposed VND algorithm

# 4 Tests and Results

This section applies the proposed VND algorithm in a 53-node test system adapted from [15]. Single faults were simulated at critical nodes. Simulations were performed on a computer with a 3.30 GHz Intel® Core ™ i7-8700 processor and 32 GB of RAM using MATLAB [18].

## 4.1 53-Node Test System

This test system has 53 nodes, within 50 load nodes, three substations, and 61 branches. In this system, the nominal voltage is equal to 13.8 kV, and the lower and upper voltage limits are, respectively, equal to 0.95 and 1.0 p.u.. The total load of the system is 45,668.70 kW of active power and 22,118.24 kVAR of reactive power. More details about the system are available in [15].

The initial topology of the system is shown in Fig. 4, and single faults were simulated at every node of the system. When a permanent fault occurs in a distribution system, it needs to be isolated, and, as a consequence of the fault isolation, the area post-fault remains de-energized until the restoration plan is executed. In this study case, only the scenarios where the faulted node is at the exit of the substations, consequently the scenarios with a great amount of de-energized active load, will be analyzed. Information about the critical fault scenarios is shown in Table 1. The results for the scenarios are shown in Table 2.

For the scenario with a fault at node 1, 6,791.40 kW of active load is de-energized after the fault isolation. The restoration plan obtained by the VND algorithm suggests the operation of three branches (17-9, 18-17, and 104-22), in which the first is opened and the last two are closed. This solution presents three switching operations. However, the neighborhood structure of the VND algorithm is based on the operation of pairs of switches, which indicates that a fictitious branch was operated together with a real branch, totaling four switching operations. In this case, the total restoration was not possible; the restoration plan can only restore 99.32% of the de-energized load (without considering the

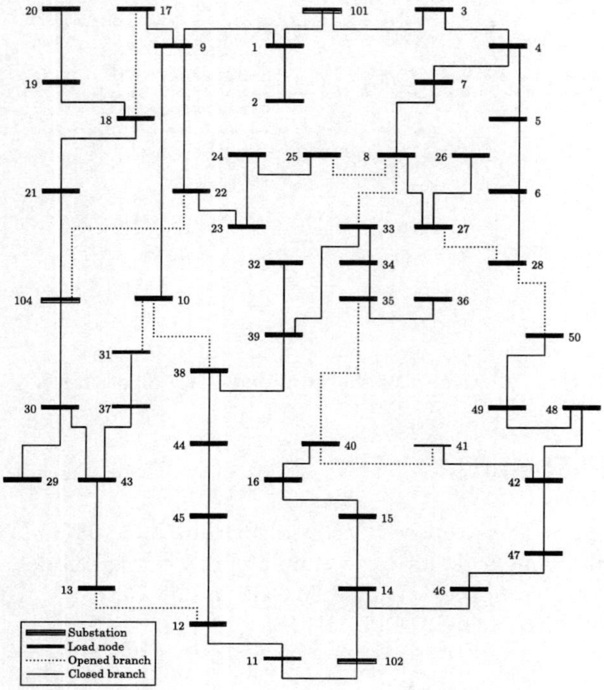


**Fig. 4.** Initial topology of the 53-node system

**Table 1.** 53-nodes single fault scenarios information

| Faulted nodes | De-energized nodes | Active load disconnected (kW) |
|---|---|---|
| 1 | 2, 9, 10, 17, 22, 23, 24, 25 | 6,791.40 |
| 3 | 4, 5, 6, 7, 8, 26, 27, 28 | 7,415.10 |
| 11 | 12, 32, 33, 34, 35, 36, 38, 39, 44, 45 | 9,078.30 |
| 14 | 15, 16, 40, 41, 42, 46, 47, 48, 49, 50 | 8,108.10 |

faulted area), which means that an amount of 1,039.50 kW of the active load remains de-energized.

In the case of a fault at node 3, an amount of 7,415.10 kW load is disconnected. In order to restore the system, the VND algorithm proposes the opening of branches 5-4, 27-8, 26-27, 28-6, and 34-33 and the closing of branches 28-27, 8-33, 35-40, and 28-50. Thus, the restoration plan can restore 57,94% of the de-energized load, totaling an energy-not-supplied equal to 3,118.50 kW.

**Table 2.** Results for critical faults at the 53-node system

| Faulted Nodes | Active load restored (%) | Active load not restored (kW) | Opened branches | Closed branches | Time (s) |
|---|---|---|---|---|---|
| 1 | 99.32% | 1,039.50 | 17-9 | 18-17, 104-22 | 30 |
| 3 | 57.94% | 3,118.50 | 5-4, 27-8, 26-27, 28-6, 34-33 | 28-27, 8-33, 35-40, 28-50 | 56 |
| 11 | 100.00% | 0.00 | 45-12, 39-38, 34-33 | 13-12, 8-33, 35-40, 10-18 | 105 |
| 14 | 40.17% | 4,435.20 | 36-35, 16-40, 42-41, 47-42 | 40-41, 35-40 | 45 |

For the fault at node 11, after the isolation, 9,078.30 kW of active load is de-energized. The restoration plan obtained by the VND algorithm suggests opening branches 45-12, 39-38, and 34-33 and closing branches 13-12, 8-33, 35-40, and 10-18. In this case, the restoration plan was able to restore 100% of the de-energized load.

Finally, in the case of a fault at node 14, 8,108.10 kW of active load is disconnected. In this case, the VND algorithm proposes the operation of six branches: 36-35, 16-40, 42-41, 47-42, 40-41, and 35-40, in which the first four are opened, and the last two are closed. Thus, the restoration plan can restore 40,17% of the de-energized load, totaling an energy-not-supplied value of 4,435.20 kW.

The results shown in Table 2 can be compared to the results obtained by [15]. It can be verified that all of the VND algorithm's solutions are globally optimal.

## 5 Conclusion

In this work, we propose a variable neighborhood descent (VND) algorithm to solve the service restoration problem (SRP) in radial distribution systems. The algorithm features the strategy of using a fictitious substation for problem resolution. This strategy is responsible for enabling the algorithm to deal with the SRP as a reconfiguration problem, which causes the complexity of dealing with the problem to decrease. Metaheuristic algorithms are known for its stochastic characteristic, in contrast, the VND algorithm is a technique purely deterministic, that will always find the local optimal of a given neighborhood. The algorithm was tested using a 53-node system which is a typical test system. The solutions obtained by the algorithm were the same solutions obtained from a mixed-integer second-order cone programming model, proving the algorithm's efficiency and validating both metaheuristic and mathematical modelling approaches. Future work will solve the SRP for more complex distribution systems with distributed generation.

## References

1. Vargas, R., Macedo, L.H., Home-Ortiz, J.M., Romero, R.: Optimal restoration of distribution systems considering temporary closed-loop operation. IEEE Syst. J. **15**, 5483–5494 (2021)
2. Zidan, A., et al.: Fault detection, isolation, and service restoration in distribution systems: state-of-the-art and future trends. IEEE Trans. Smart Grid **8**(5), 2170–2185 (2016)
3. López, J.C., Franco, J.F., Rider, M.J., Romero, R.: Optimal restoration/maintenance switching sequence of unbalanced three-phase distribution systems. IEEE Trans. Smart Grid **9**(6), 6058–6068 (2017)
4. Morelato, A., Monticelli, A.: Heuristic search approach to distribution system restoration. IEEE Trans. Power Deliv. **4**(4), 2235–2241 (1989)
5. Lee, S., Kim, K., Kim, H., Lee, J., Nam, K.: Expert system-aided service restoration in distribution automation. In: Proceedings of the 1992 IEEE International Conference on Systems, Man, and Cybernetics, pp. 157–161. IEEE, Chicago (1992)
6. Ciric, R.M., Popovic, D.S.: Multi-objective distribution network restoration using heuristic approach and mix integer programming method. Int. J. Electr. Power Energy Syst. **22**(7), 497–505 (2000)
7. Kleinberg, M.R., Miu, K., Chiang, H.D.: Improving service restoration of power distribution systems through load curtailment of in-service customers. IEEE Trans. Power Syst. **26**(3), 1110–1117 (2010)
8. Dimitrijevic, S., Rajakovic, N.: Service restoration of distribution networks considering switching operation costs and actual status of the switching equipment. IEEE Trans. Smart Grid **6**(3), 1227–1232 (2015)
9. Luan, W., Irving, M.R., Daniel, J.S.: Genetic algorithm for supply restoration and optimal load shedding in power system distribution networks. IEE Proc.-Gener. Transm. Distrib. **149**(2), 145–151 (2002)
10. Marques, L.T., Delbem, A.C.B., London, J.B.A.: Service restoration with prioritization of customers and switches and determination of switching sequence. IEEE Trans. Smart Grid **9**(3), 2359–2370 (2017)
11. Faria, W.R., et al.: Service restoration in modern distribution systems addressing grid-connected and islanded operations. Electr. Power Syst. Res. **196**, 107238 (2021)
12. Wang, S., Chiang, H.D.: Multi-objective service restoration of distribution systems using user-centered methodology. Int. J. Electr. Power Energy Syst. **80**, 140–149 (2016)
13. Carvalho, P., Ferreira, L., Barruncho, L.: Optimization approach to dynamic restoration of distribution systems. Int. J. Electr. Power Energy Syst. **29**(3), 222–229 (2007)
14. Macedo, L.H., Muñoz-Delgado, G., Contreras, J., Romero, R.: Optimal service restoration in active distribution networks considering microgrid formation and voltage control devices. IEEE Trans. Ind. Appl. **57**(6), 5758–5771 (2021)
15. Romero, R., Franco, J.F., Leão, F.B., Rider, M.J., de Souza, E.S.: A new mathematical model for the restoration problem in balanced radial distribution systems. IEEE Trans. Power Syst. **31**(2), 1259–1268 (2016)
16. Souza, E.S., Puerta, G.F., Romero, R.: An ac mathematical model for solving complex restoration problems in radial distribution systems in a treatable runtime. IEEE Access **8**, 228303–228314 (2020)

17. Mladenović, N., Hansen, P.: Variable neighborhood search. Comput. Oper. Res. **24**(11), 1097–1100 (1997)
18. MATLAB: 9.13.0.2105380 (R2022b) Update 2. The MathWorks Inc., Natick, Massachusetts (2022)

# Special Session on Smart Factories of the Future (SmartFoF)

# Sensitivity Analysis of Surrogate Modeling for Manufacturing in Digital Twins

Alejandro Gonzalez Casal[1] and Abraham Prieto García[2(✉)]

[1] Universidade da Coruña, A Coruña, Spain
[2] Universidade da Coruña, Campus Industrial de Ferrol, CITENI, A Coruña, Spain
abraham.prieto@udc.es

**Abstract.** In this study, we address the limitations of digital twins, including significant computational time and the complexity of real-world processes, by utilizing surrogate models (SMs) to partially or entirely represent digital twins. We investigate the performance of surrogate models in a manufacturing system scenario through sensitivity analysis and adaptive sampling strategies with incremental learning. Our experimental setup involves a synthetic digital twin and a neural network-based surrogate model. The study explores the impact of various parameters such as dataset size, number of epochs, and batch size on the neural network's performance. We also analyze the effectiveness of different adaptive sampling strategies in the incremental learning process.

**Keywords:** Digital twins · Surrogate modeling · Neural networks · Manufacturing industry · Adaptive sampling · Smart factories · Predictive modeling

## 1 Introduction

In recent years, digital twins have gained significant attention for their ability to enhance the performance of processes they represent, predict future states, prevent or anticipate potential failures, and enable remote system monitoring [1]. A digital twin is a virtual representation or digital counterpart of a physical object, system, or process, created by capturing real-time data and using advanced simulation techniques to mirror the behavior and performance of the original. Industries such as construction, manufacturing, propulsion systems, energy, and healthcare have increasingly adopted digital twins for these purposes. However, current digital twins face limitations stemming from two primary issues: the significant computational time required for simulations [2], and the complex nature of real-world processes, which often involve discrete events, stochastic elements, and highly variable behaviors [3]. These challenges become particularly critical for applications requiring rapid responses (control) or the evaluation of numerous scenarios (optimization) [4].


R. Mehmood et al. (Eds.): DCAI 2023, LNNS 741, pp. 291–300, 2023.
https://doi.org/10.1007/978-3-031-38318-2_29

In this scientific article, we propose the use of surrogate models (SM) to partially or entirely represent digital twins, thereby addressing these limitations. Surrogate models, also known as metamodels or emulators, are computationally efficient approximations of simulation model behavior , enabling tasks such as optimization, design space exploration, and sensitivity analysis that would otherwise be impossible due to the need for thousands or even millions of simulation evaluations.

## 2 State of the Art

Digital Twins are virtual replicas of physical systems, enabling real-time monitoring, prediction, and optimization through data-driven models [5]. Their importance in various applications, such as aerospace, manufacturing, and energy management, highlights their significance in modern industry. However, as Digital Twins become increasingly complex and detailed, the computational cost associated with these models can become a significant challenge [6].

To address this issue, researchers have been exploring various methods to reduce the computational burden while maintaining the accuracy and fidelity of the digital twin representation. Some of these approaches include dimensionality reduction [7], model order reduction [8], and parallel computing [9]. One promising approach is the use of surrogate models, which are approximate representations of complex, computationally expensive models. Surrogate models, also known as metamodels or response surface models, are designed to deliver quick, accurate predictions while significantly reducing the computational cost. There are several types of surrogate models, including polynomial regression, Kriging, radial basis functions, and artificial neural networks [10]. These models have been widely applied to various fields, such as optimization, design space exploration, and uncertainty quantification [10].

In recent years, researchers have focused on adaptive sampling strategies to further improve the efficiency and accuracy of surrogate models [11]. Adaptive sampling involves selecting training data points based on specific criteria, such as the highest expected improvement or the largest prediction uncertainty. This approach aims to minimize the number of data points required to achieve a desired level of model accuracy. Some popular adaptive sampling techniques include Efficient Global Optimization (EGO), Gaussian Process Regression (GPR), and Monte Carlo methods [11].

## 3 Experimental Setup

In this section, we outline the experimental setup used to investigate the performance of surrogate models in the context of a digital twin representation of a manufacturing system.

### 3.1 Synthetic Digital Twin

In this study, we aim to construct a digital twin test-bed of a manufacturing system, encompassing $N \in \mathbb{N}$ production lines and $M \in \mathbb{N}$ designated products, represented by the set $P = \{A, B, C, D, E\}$. Each of these products possesses a unique quantity ($q_i \in \mathbb{N}$) and priority level ($p_i \in \mathbb{N}$), both of which are contingent on the input parameters specified for each scenario. Furthermore, the master data attributes a specific number of operators ($o_i \in \mathbb{N}$) and a designated unit production time ($t_i \in \mathbb{R}^+$) to each product $i \in P$.

The manufacturing system under investigation is characterized by machines with the capability to produce only certain products, as indicated by a binary matrix $C$ of dimensions $N \times M$, where $c_{n,m} \in 0, 1$ denotes whether the machine in production line $n$ can produce product $m$ or not, with $n \in 1, 2, \ldots, N$ and $m \in 1, 2, \ldots, M$. A priority function $f(p_i)$ governs the production sequence, such that $f(p_i) < f(p_j)$ if product $i$ has a higher priority than product $j$ in a given scenario.

Building upon the digital twin framework described previously, we generate a complete dataset comprising 20,000 data points. In a real-world digital twin or manufacturing process, data generation would typically occur online and adaptively. However, to expedite the experimental configuration process and facilitate the study of various modeling algorithms, we employ a pre-generated dataset and conduct our analysis on it.

Each element of the dataset includes the quantity of products from the $M \in \mathbb{N}$ categories, denoted by $q_i \in \mathbb{N}$ for each product $i \in P$, as well as the priority level ($p_i \in \mathbb{N}$) for each product. The output of each data point is the time required to produce each of the products, represented by $t_i \in \mathbb{R}^+$ for each product $i \in P$.

To represent each data point in the dataset, we can define a vector that encapsulates both the input and output information. Let $D$ be the dataset, containing 20,000 data points, and let $d_k \in \mathbb{R}^{2M+1}$ represent the $k$-th data point, with $k \in \{1, 2, \ldots, 20,000\}$. Each vector $d_k$ can be described as follows:

$$d_k = [q_1, q_2, \ldots, q_M, p_1, p_2, \ldots, p_M, T_1, T_2, \ldots, T_k]$$

### 3.2 Surrogate Model

We aim to determine the production end time for each product individually, based on the priorities ($p_i \in \mathbb{N}$) and quantities ($q_i \in \mathbb{N}$) of the products for each scenario.

We use a neural network as the operator to model the digital twin, with the output being a size 5 vector containing the estimated production end times ($T_k \in \mathbb{R}^+$) for each product. The input consists of information on the quantities and priorities of each product, represented as a $5 \times 5$ matrix, where each column corresponds to a specific product. The column contains four zeros and a value, with the non-zero row indicating the product's priority order.

$$\begin{bmatrix} B & C & D & E & A \end{bmatrix} \begin{bmatrix} q_2 & q_3 & q_4 & q_5 & q_1 \end{bmatrix} \Longrightarrow \underbrace{\begin{bmatrix} 0 & q_1 & 0 & 0 & 0 \\ 0 & 0 & q_2 & 0 & 0 \\ 0 & 0 & 0 & q_3 & 0 \\ 0 & 0 & 0 & 0 & q_4 \\ q_5 & 0 & 0 & 0 & 0 \end{bmatrix}}_{\text{5x5 Input Matrix}} \Longrightarrow \underbrace{\text{MLP (3 x 625), ReLU}}_{\text{Neural Network Architecture}} \Longrightarrow \underbrace{\begin{bmatrix} T_1 \\ T_2 \\ T_3 \\ T_4 \\ T_5 \end{bmatrix}}_{\text{Output Vector}} \qquad (1)$$

For the neural network architecture, we employ a multilayer perceptron with three hidden layers (625 neurons each) and ReLU activation functions. We use the Keras library for efficient neural network training and implementation. The ANN is trained using the Adam optimizer with a default learning rate of 0.001. The batch size strategy will be one of the parameters for the sensitivity analysis conducted later in the study.

## 4 Methodology

The methodology involves training and testing various aspects of the neural network to model the digital twin, with a focus on sensitivity analysis, adaptive sampling strategies, and incremental learning. The training process includes standard training error analysis, validation error based on 20% of the available data, and a full validation error obtained by testing the model on the entire 20,000 data points generated initially from the digital twin.

### 4.1 Sensitivity Analysis

To increase the representativeness of these experiments, we executed 10 replications of each configuration presented.

1. **Dataset Size:** Dataset size is a critical aspect as it affects the relevance of the generated data for training and may take long to be produced by the model. Large datasets can provide more accurate models but they may take longer to have them generated, while smaller datasets may result in overfitting or underfitting.
2. **Number of Epochs:** This parameter is especially important when training time is significant relative to data generation time. It may be preferable not to use an excessive number of epochs and instead incorporate new data as it becomes available. The number of epochs also influences the model's generalization capability and convergence rate.
3. **Batch Size:** The batch size is an internal parameter of the trainig algorithm that has a significant impact on the learning outcome. Some combinations of batch size (specially with low sizes) may lead to suboptimal results or instability during training. We have included a procedure to fix inestability in conflicting batch sizes and studied the effect of this parameter on the learning process.

### 4.2 Adaptive Sampling Strategies

The adaptive sampling strategies use an incremental training process with data packets of size 25. A pool of candidate samples is generated for training, and the top 25 candidates that best meet the adaptive sampling criteria are selected. The parameter K, representing the size of the pool, is varied from K=1 (random selection) to K=100 (strong pressure) in the experiments. Uniform sampling, a data range of 50–500, proportional batch size, and five epochs per sample are used. Ten replications were performed for each strategy and k value.

The following selection criteria have been studied:

1. **Input Space Distance-based Selection:** In order to thoroughly explore the input space, candidates exhibiting the largest distance from previously employed training samples are chosen. Various distance metrics are taken into account, such as a) the Euclidean distance, grounded in the neural network input, b) distances derived from the digital twin input (priority order and quantities) using the Kendall Tau distance to evaluate the sequence differences.
2. **Real Error-based Selection:** Using the real error (evaluation of each of the candidates from the pool), we select candidates where the current model produces a higher or lower estimation error of the output. In a real-world scenario, this would be achieved using an error estimator, such as a Gaussian Process Regression (GPR) model or a parallel neural network trained to estimate the error alongside the output prediction model. This strategy enables the model to focus on regions of the input space where its performance is currently either comparatively low or comparatively high.
3. **Stability-based Selection:** This criterion aims to assess the model's performance without relying on ground truth or parallel error models. Instead, it analyzes characteristics of the model response in specific regions. The model is evaluated with slight changes in product quantities, and the variability in the response is employed to assess the suitability of the model in that region.

## 5 Results

This section presents the results obtained from our experiments focusing on sensitivity analysis and adaptive sampling strategies with incremental learning. The subsections are organized according to the methodology.

### 5.1 Sensitivity Analysis Results

We now present the findings of our sensitivity analysis, exploring the effects of dataset size, and batch size for different numbers of epochs.

- We implemented two batch size approaches: a fixed 50 samples (Fig. 1a) and a proportional batch size equal to one-tenth of the dataset size (Fig. 1b) to maintain a constant number of learning steps per epoch. Results showed

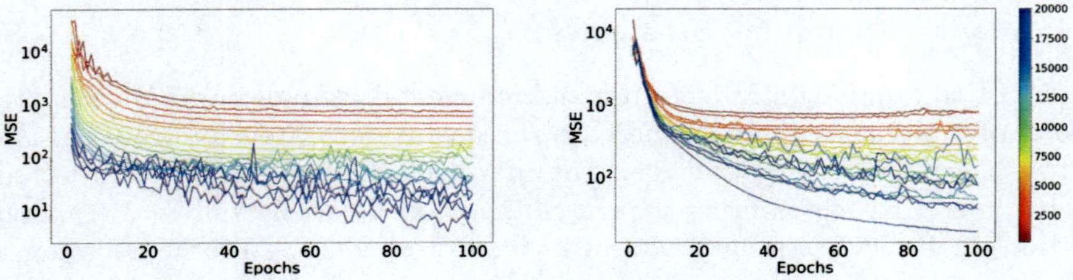


(a) Constant Batch (50) (b) Proportional Batch (n/10)

**Fig. 1.** Data-set Size and Epochs

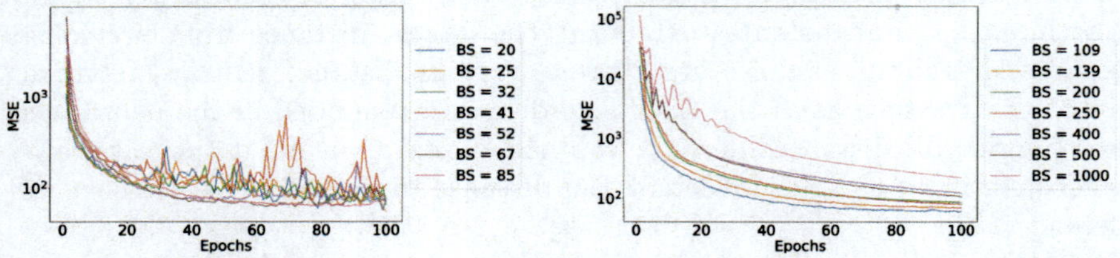


(a) Batch 20-85 (b) Batch 109-1000

**Fig. 2.** Batch Size and Epochs - Constant Data-set

the substantial impact of sample number on model performance, with MSE reducible by up to 100 times through dataset size increase. The study underscores the limitations of the Keras library in handling specific batch sizes and the necessity of proper batch size selection for effective training.
The experiment showed instabilities in both approaches. The constant ones were likely due to excessive learning steps per epoch as dataset size increased, while the proportional approach showed inconsistent instabilities across all dataset sizes, suggesting batch size as the cause. To examine this parameter, additional analyses were conducted with a constant number of samples.

– Batch Size: The study aimed to evaluate batch size's impact on learning outcomes and training stability. We used a fixed dataset of 2000 samples and experimented with various batch sizes from the proportional batch experiment (Fig. 2). Results show that decreasing batch size reduces saturation error; however, instabilities arise when batch size falls below 100 samples, affecting the final model performance. Instabilities increase with batch size reduction, and specific batch sizes exhibit higher fluctuations, seemingly triggering or amplifying these fluctuations.

We determined dataset and batch sizes, and their relationship, as crucial parameters through several independent experiments. We conducted a dual sweep with dataset sizes from 200 to 20,000 samples and batch sizes from 1/160 to the full dataset size, limiting to 1,600 training steps for fair comparison. Heat maps of performance for both parameters were generated (Fig. 3). The analysis

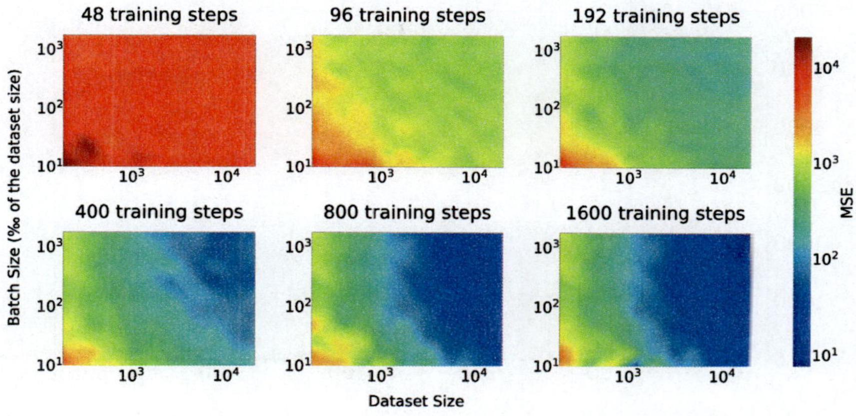


**Fig. 3.** Data-set and Batch Size Proportion - Evolution through the training steps

revealed low sensitivity to both parameters during initial steps, with benefits observed from using larger datasets and batch size proportions.

### 5.2 Adaptive Sampling Strategies Results

First, we present a preliminary analysis of the response of the incremental learning process with random sampling, then we study the different selection strategies.

- Fig. 4 illustrates the incremental learning approach with random sampling. Three error types are observed as learning progresses : training error, validation error (20%), and complete validation error. Initially, the model quickly learns and reduces validation error until saturation. With new data, the training error increases while the model adjusts, leading to lower validation error. Complete validation, indicated by green peaks, is performed before adding new datasets. It is observed that standard validation error accurately estimates the model's global performance.
- Input Space Distance-based Selection: For both distance approaches, no significant improvement in the final training performance or speed was observed, indicating that selecting random samples provides enough diversity to learn the data without additional benefits in selecting the most distant samples. (See Fig. 5.)
- Real Error-based Selection: In this experiment, a slight improvement was observed in the first epochs of the training. However, as more samples were added, the learning speed of the experiments with $k > 1$ started to decrease until it was surpassed by the random approach ($k = 1$). The higher the k, the greater the improvement in the first selections, but the faster it slows down, and the worse its final performance becomes. (See Fig. 6a.)

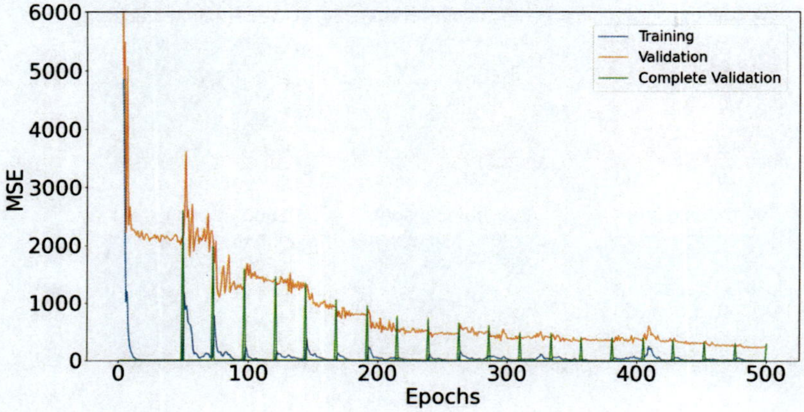


**Fig. 4.** Training Development for Adaptative Sampling

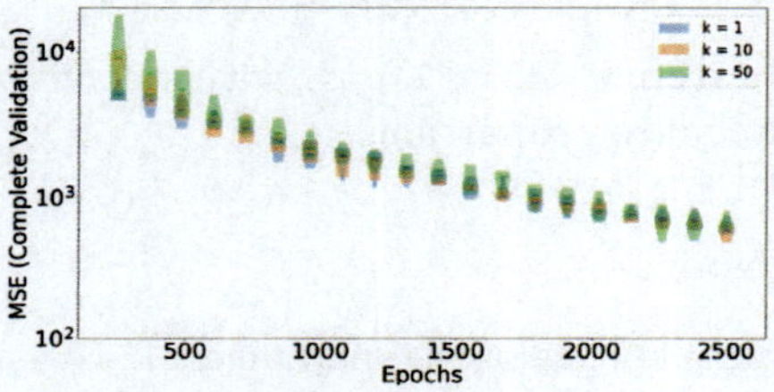


(a) ANN Input Distance Selection

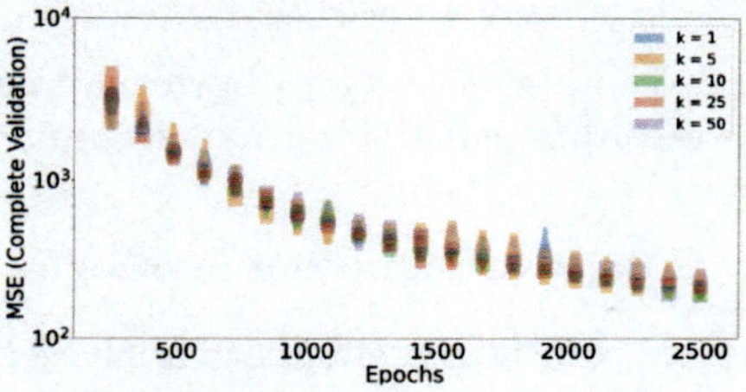


(b) DT Input Distance Selection

**Fig. 5.** Adaptive Sampling Strategies (I)

- Stability-based Selection: We studied only a few values of k (2, 3, and 5) due to the higher time required to select samples using this approach. The results showed that, although there were no significant differences for the initial steps, the random selection approach performed better as the training progressed. (See Fig. 6b.)

## 6 Conclusion

Our study shows that surrogate models can improve digital twin performance in manufacturing systems. Sensitivity analysis highlights the importance of dataset size, number of epochs, and batch size for the neural network's performance. Adaptive sampling strategies, however, didn't show major improvements compared to random sampling.

Proper dataset and batch sizes are crucial for effective training. Random sampling provides adequate data diversity for incremental learning. Adjusting

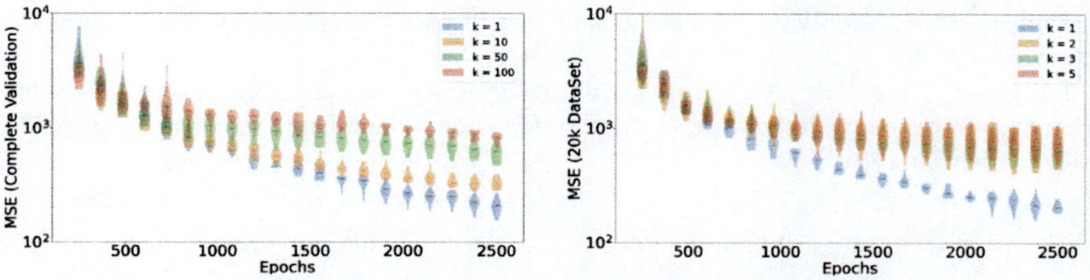


(a) Max Error Selection Strategy (b) Stability-based Selection Strategy

**Fig. 6.** Adaptive Sampling Strategies (II)

batch size proportion can significantly reduce MSE, while insufficient dataset size can cause a substantial increase.

Surrogate models offer better computational efficiency than digital twins, providing predictions up to 30,000 times faster on average. Adaptive sampling has benefits with limited data points but underperforms compared to random selection for larger steps. Future research should focus on combined approaches and new strategies to reduce required samples.

Further studies may explore applying these methods to complex systems, improving learning efficiency, and examining alternative machine learning algorithms or hybrid approaches for digital twin modeling.

## References

1. Tao, F., Sui, F., Liu, A., Qi, Q., Zhang, M., Nee, A.Y.C.: Digital twin-driven product design, manufacturing and service with big data. Int. J. Adv. Manuf. Technol. **94**(9), 3563–3576 (2018)
2. Negri, E., Fumagalli, L., Macchi, M.: A review of the roles of digital twin in CPS-based production systems. Procedia Manuf. **11**, 939–948 (2017)
3. Leng, J., et al.: Digital twin-driven rapid reconfiguration of the automated manufacturing system via an open architecture model. Robot. Comput.-Integr. Manuf. **63**, 10 (2020)
4. Kritzinger, W., Karner, M., Traar, G., Henjes, J., Sihn, W.: Digital twin in manufacturing: a categorical literature review and classification. IFAC-PapersOnLine **51**(11), 1016–1022 (2018)
5. Grieves, M.: Digital twin: manufacturing excellence through virtual factory replication. White Paper **1**(2014), 1–7 (2014)
6. Tao, F., Zhang, H., Liu, A., Nee, A.Y.: Digital twin in industry: state-of-the-art. IEEE Trans. Ind. Inf. **15**(4), 2405–2415 (2018)
7. Chakraborti, A., Heininen, A., Koskinen, K.T., Lämsä, V.: Digital twin: multi-dimensional model reduction method for performance optimization of the virtual entity. Procedia CIRP **93**, 240–245 (2020)
8. Benner, P., Gugercin, S., Willcox, K.: A survey of projection-based model reduction methods for parametric dynamical systems. SIAM Rev. **57**(4), 483–531 (2015)
9. Uhlemann, T.H.-J., Lehmann, C., Steinhilper, R., Verl, A.: The digital twin: demonstrating the potential of real time data acquisition in production systems. Procedia Manuf. **9**, 113–120 (2017)

10. Enders, M.R., Hoßbach, N.: Dimensions of digital twin applications-a literature review (2019)
11. Dias, L., Bhosekar, A., Ierapetritou, M.: Adaptive sampling approaches for surrogate-based optimization. In: Computer Aided Chemical Engineering, pp. 377–384. Elsevier (2019)

# Context-Aware Applications in Industry 4.0: A Systematic Literature Review

Pedro Monteiro[1], Claudio Lima[1], Tiago Pinto[1](✉), Paulo Nogueira[2], Arsénio Reis[1], and Vitor Filipe[1]

[1] Universidade de Trás-os-Montes e Alto Douro, Vila Real, Portugal
al69605@alunos.utad.pt, {claudiol,tiagopinto,ars,vfilipe}@utad.pt
[2] Continental Advanced Antenna, Vila Real, Portugal
paulo.nogueira@continental-corporation.com

**Abstract.** Industry 4.0 was publicly introduced in Germany in 2011 and is known as the fourth industrial revolution, whose goal is to improve manufacturing processes and increase the competitiveness of the manufacturing industry. Industry 4.0 uses technological concepts such as Cyber-Physical Systems, Internet of Things and Cloud Computing to create services, reduce costs and increase productivity in industry. This paper aims to explore the use of context-aware applications in Industry 4.0 in order to assist workers in decision making and thus improve the performance of factory production lines. This literature review is part of the project "Continental AA's Factory of the Future" (Continental FoF) and will integrate a context-aware system in Industry 4.0 of the mentioned company, which is a manufacturer of radio frequency devices for the automotive industry. This systematic literature review identifies, from the researched solutions, the concept of context and context-awareness, the main technologies used in context-aware systems, how context management is performed, as well as the most used integration and communication protocols.

**Keywords:** Context-aware applications · Context-Awareness · Industry 4.0 · Internet of Things · Literature review

## 1 Introduction

Modern industrial production requires the coordinated collaboration of people, machines and equipment. Production information systems play a key role in optimizing the entire manufacturing operation, as they are responsible for important

This work was supported by the RD Project "Continental Factory of Future (CONTINENTAL FoF)/POCI-01-0247-FEDER-047512", financed by the European Regional Development Fund (ERDF), through the Program "Programa Operacional Competitividade e Internacionalização (POCI)/PORTUGAL 2020", under the management of AICEP Portugal Global - Trade & Investment Agency. This work also funded by the R&D Project "A-MoVeR–Agenda Mobilizadora para o Desenvolvimento de Produtos e Sistemas Inteligentes de Mobilidade Verde", financed by POCI European Structural and Investment Funds (FEEI).


R. Mehmood et al. (Eds.): DCAI 2023, LNNS 741, pp. 301–311, 2023.
https://doi.org/10.1007/978-3-031-38318-2_30

tasks such as information processing, monitoring and control of the production process, among others [1]. Companies face enormous challenges related to the management and complexity of the production system, and in order to perform well, they need to be consistent in their manufacturing processes.

In 2011, it became publicly known in Germany, the term "Industrie 4.0" which refers to the fourth industrial revolution. This revolution aims to strengthen the competitiveness of the manufacturing industry by targeting improvements in industrial processes [2]. The core idea of Industry 4.0 is centered on technology and especially in the production area, but it is possible to apply this concept to any type of company. Industry 4.0 enables new technological ideas, such as Cyber-Physical Systems (CPS), Internet of Things and Cloud Computing, to develop new services, decrease expenses and increase efficiency [3].

Industry 4.0 is based on communication between people, machines, and resources to reach common goals, thus creating "Smart Factories". These factories use context awareness to help people and machines perform their tasks. This is accomplished through systems that operate in the background, performing their activities based on information received from both the physical and virtual worlds [2]. The goal of context-aware systems in I4.0 is to assist workers in decision making by providing relevant information that can improve their performance in their activities and, consequently, improve the factory performance.

One of the technological ideas of Industry 4.0 is the Internet of Things (IoT). Many organizations are beginning to recognize and value the possibilities and benefits of IoT to enhance their operational activities. The IoT is a system that connects everyday and industrial objects, making them intelligent and able to interact with humans. In industry these technologies can be used to collect and analyze information to improve operations and services provided [4].

This article presents a literature review that addresses the concept of context-awareness, identifies the main technologies used in context-aware systems, explores how the context management is performed and also surveys the most used integration and communication protocols. This literature review integrates the project "Continental AA's Factory of the Future" (Continental FoF) and will compose a context-aware system in I4.0 in the aforementioned company.

The rest of the paper is divided into the following sections. Section 2 presents the main aspects regarding context-awareness and IoT devices in Industry 4.0. Section 3 presents the steps used in the systematic literature review. Section 4 presents the results obtained from the data reading and extraction. Section 5 presents the concluding remarks.

## 2 Background

Context-aware computing was first addressed by Schilit and Theimer in 1994, and since then there have been numerous attempts to define this concept [5]. Context is any information that can be used to characterize the situation of an entity (person, place or object) relevant to its interaction with a context-aware application. These applications can provide support in three types of situations:

display of information and services to a user, automatic execution of a service for a user and association of context to information to facilitate its retrieval [5].

Context-aware applications react specifically to location, time, or other attributes and adapt their behavior without explicit user intervention, increasing usability and output efficiency [6]. These context-aware applications can increase the visibility and performance of operations in a manufacturing environment. In industry their goal is to enable factory and office workers to make decisions based on an understanding of the system, derived from the real-time factory context [7]. In addition, the response provided varies according to the user's task.

Currently, in the industry, there is a considerable amount of sensors integrated into elements of the production environment (such as parts, machines, and tools), which are processed and used to generate production context data, and context-aware applications play a significant role in analyzing and interpreting this data[1].

The implementation of the architecture of context-aware systems or applications can vary depending on various requirements and conditions, such as the number of users, the storage and processing capacity of the devices, and the expandability of the system [6]. The choice of approach will depend on these specific factors.

## 3 Methodology

This section presents the review methodology that was used to conduct the literature survey. From the several methods currently used for conducting systematic reviews, the one chosen in this work was the PRISMA [8] method.

### 3.1 Research Strategy

In order to define a goal for the analysis and synthesis of the literature review, as well as the search for relevant articles on the research topic, the following research questions were defined:

- RQ1: What is the concept of context and context-awareness that the authors address in their studies?
- RQ2: What are the main technologies used in I4.0 context-aware systems?
- RQ3: What are the main architecture models used for context management in the analyzed solutions?
- RQ4: How are the most commonly used integration and communication protocols for exchanging information between system layers/applications or external applications?

### 3.2 Systematic Review Protocol

To search the research articles related to context awareness, context detection architectures and context-aware system architectures the information search was conducted in Google Scholar.

Subsequently, the first inclusion criterion (IC1) was defined, which is composed of the search according to the following "search string":

- ("context-aware systems") AND ("industry 4.0") AND (("Architectures") OR ("frameworks")))

Next, the exclusion criteria were defined:

- EC1: the article was not published between 2017 and 2022.
- EC2: the articles found do not include patents or citations.
- EC3: the articles, after reading the title and abstract, are not related to the subject of this review.
- EC4: After reading, it was found that the articles do not provide good practical considerations on the subject.

The review protocol and the implementation of the Inclusion and Exclusion criteria are presented in Fig. 1.

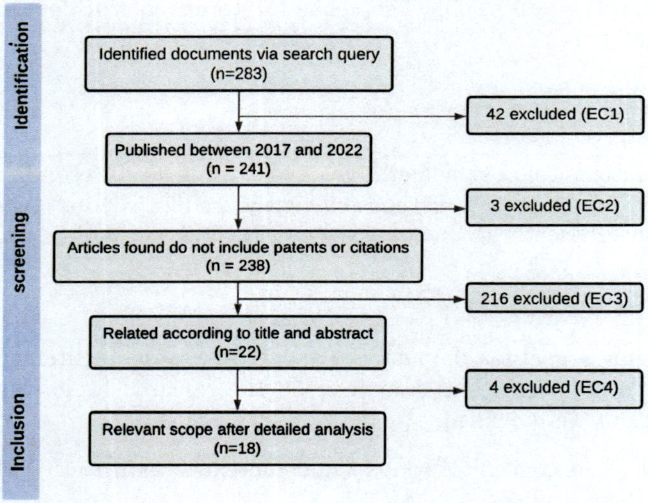


**Fig. 1.** Review methodology

## 4 Results and Discussion

After applying the review methodology, 18 articles were obtained. These articles are listed in Table 1, presenting their respective titles and authors. The articles resulting from the review will be analyzed and grouped into four main categories, which correspond to the research questions:

**Table 1.** Papers analysed

| # | Title | Authors | Year |
|---|---|---|---|
| P1 [9] | Development of a Context-aware framework for the Integration of Internet of Things and Cloud Computing for Remote Monitoring Services | Ali Al-Shdifat et al | 2018 |
| P2 [10] | An industrial Internet of things based platform for and context-aware information services in manufacturing | Kosmas Alexopoulos et al | 2018 |
| P3 [11] | Context Aware Site-visit Scheduling for Asset Management in Oil and Gas Sector | Ravi Shankar et al | 2019 |
| P4 [12] | Cloud-Based Architectures for Environment Monitoring | Vasile M. Tovarnitchi | 2017 |
| P5 [13] | Context-Aware End-User Development Review | Victor Ponce et al | 2022 |
| P6 [14] | Real-Time Context-Aware Microservice Architecture for Predictive Analytics and Smart Decision-Making | Guadalupe Ortiz et al | 2019 |
| P7 [15] | Context-aware scheduling in Fog computing: A survey, taxonomy, challenges and future directions | Mir Salim Ul Islam et al | 2021 |
| P8 [16] | Middleware technologies for cloud of things: a survey | Amirhossein Farahzadi et al | 2019 |
| P9 [17] | Software Architecture Solutions for the Internet of Things: A Taxonomy of Existing Solutions and Vision for the Emerging Research | Aakash Ahmad et al | 2019 |
| P10 [18] | Internet of Things for enabling smart environments: A technology-centric perspective | Carles Gomez et al | 2019 |
| P11 [19] | eHealth: A Survey of Architectures, Developments in mHealth, Security Concerns and Solutions | Caroline Alenoghena et al | 2020 |
| P12 [20] | IoT data processing pipeline in FoF perspective | Pedro Corista | 2017 |
| P13 [21] | Modeling and executing IoT-enhanced business processes through BPMN and microservices | Pedro Valderas et al | 2022 |
| P14 [22] | Uma arquitetura orientada a serviços de nuvem para apoiar testes experimentais de sistemas ubíquos | Heder Soares et al | 2021 |
| P15 [23] | Geospatial business intelligence and cloud services for context aware digital twins development | Mezzour Ghita et al | 2020 |
| P16 [24] | Artificial Intelligence for Cloud-Assisted Smart Factory | Jiafu Wan et al | 2020 |
| P17 [25] | Smart Lab: an application of Industry 4.0 design for principles to calibration laboratories | Gean Geronymo | 2021 |
| P18 [26] | An Approach for Context-Aware Cyber-Physical Automation Systems | Nada Sahlab et al | 2021 |

- Concept of Context and Context-Awareness
- Context Management Models
- Technologies used in context-aware systems in I4.0
- Integration and communication protocols between layers/applications within context-aware systems or external applications

### 4.1 Resulting Articles

Articles [9–11] and [25] propose information systems that use context awareness integrated with IoT. [9] also integrates cloud computing into the system and uses context awareness to improve the health status of factory equipment thanks to effective fault prediction. In [10] the system aims to provide decision support, using context awareness, for operators and supervisors. [11] uses microservices and context awareness to assist in maintenance and management in the oil and gas industry to decrease operation costs. [25] presents a Smart Lab following Industry 4.0 guidelines to develop a low-cost environmental surveillance system.

Articles [13–15] and [26] also present context-aware systems. Authors of [13] present an analysis of existing "End-User Development" approaches to building context-aware applications. The [14] implements a microservice-based architecture that makes use of context-awareness to provide users with personalized recommendations based on their contexts. The [15] provides a comprehensive review of the literature regarding programming these systems in "Fog computing", addressing their advantages and disadvantages. [26] presents a flexible and reusable approach including heterogeneous data acquisition, dynamic modeling and automated context reasoning approaches as requirements.

In addition, [12] addresses environmental monitoring systems, highlights the most important features of such a system and the suggested technologies and guidelines. [16] addresses the characteristics of intermediary devices, as well as different architecture styles and service domains. It also makes a study of middleware technologies for the Cloud of Things. [17] reviews challenges, solutions and future efforts needed to implement software architectures for IoT-based systems. In [18] the authors analyze the specific requirements of smart environments, including smart factories, and how they relate to currently available IoT communication solutions. In [19] the authors explore the latest innovations in eHealth architectures, focusing on their respective benefits and drawbacks, while providing an overview of their implementation and use. In [20] a set of existing theories and models for achieving continuous and homogeneous data exchange and contextual interpretation is evaluated, with the goal of creating a framework that uses these theories to formalize the situational context. [21] proposes a microservices architecture to run business process model and notation models and facilitate their integration with the physical world. The work in [22] presents a service architecture model that employs the concept of "Sensors as a Service" for sharing and managing IoT resources in the cloud. This architecture enables the creation of an intelligent environment that enables real-time communication between multiple mobile devices. [23] presents a framework for designing digital twins. Finally [24] presents how cloud computing and artificial intelligence can be used to improve the performance of smart factories.

### 4.2 Concepts of Context and Context Awareness

Context consists of any information that can be used to characterize the situation of an entity that is relevant to the interaction between the user and the application [13,15]. The relevance of context information depends on the user's task [9]. The authors of [26] emphasize that only information that can contribute meaningful knowledge to the system should be considered as context, and that context helps in interpreting or understanding events or situations. Alternatively, [21] states that context is composed of relevant data from the physical world that is used to perform a task of detecting or acting upon it. This may contradict the definition presented by [5], because if context is any information that can be used to characterize the situation of an entity this information can in a sense also come from the virtual world. According to [26], in the computer science domain, context has been commonly defined as implicitly available information that reflects the situation of an entity during its interaction with a system.

The term Context awareness was first introduced in 1994 by B.Schild, and subsequently used by several researchers [11,12]. This term can be defined as the property of providing appropriate services to a user by analyzing their context [10,15]. [9] highlights that context awareness is increasingly needed in IoT solutions to aid adaptation to dynamic changes in the environment and decision making. Context awareness is considered a key feature in ubiquitous and pervasive computing systems that provide intelligent services to users [15]. In the manufacturing domain, context awareness can be used to increase visibility into operations and performance in a manufacturing environment [10]. In [24] context awareness services are defined as typical applications of artificial intelligence in smart factories and [26] describes context awareness as the ability to model and incorporate contextual information to adapt and optimize a system behavior.

### 4.3 Context Management Models

After analyzing the various articles, it was found that several authors present a mechanism capable of detecting, acquiring, handling and sending the contexts/data, which could be called a "context engine". Although some articles have named it this way, in other works it received another denomination or was not named, since the applications that perform these tasks were not addressed within a specific system. In general this "context engine" or the architectures of the systems studied are responsible for the following tasks:

- Context Acquisition: This is the process where data from IoT sensors/devices is received ("Context Acquisition" [9], "Acquisition Layer" [20], "Context Acquisition" [15] and "IoT Integration Platform" [26]).
- Context Builder: This is where data is processed and refined, resulting in information relevant to the system. This information is usually stored in its own database, and this part may also be responsible for updating or deleting old data ("Context modeling" [9], "Context Knowledge Base" [11], "Sensor Data Harmonization Layer" [20], "Context DataBase" [14], "Context Repository" [15] and "Context Middleware" [26]).
- Context Reasoning: Consisting of the logic functions, where high-level context information is obtained from data/context processing ("MAPE Layer" [11] and "Context Reasoning" [9,15]).

– Context Dissemination: Process of distributing the contextual information to interested parties ("Context REST/Broker Service" [14], "services layer" [10], "Context Analyzer" [15] and "Services" [26]).

It was also found that this "context engine", in many articles, has its own database to insert, modify and analyze the data pertinent to the functions it will perform, without interfering with other databases in the overall system. Other architectures reviewed do not divide these tasks into applications/components, but consider them as a single layer responsible for them, or for similar tasks (such as "Middleware" [15], "Cloud Layer" [22,24], "Cognition Layer" [23]). This type of layer is usually referred to as an intermediate layer that connects the devices layer with the users layer, and is always or predominantly described this way.

### 4.4 Technologies Used in Context-Aware Systems in I4.0

The most used or recommended technologies for the implementation of these systems were cloud computing ([9,10,16,19,22–24]), presenting numerous advantages, such as cost reduction, scalability, data analysis, storage, performance, reliability, transparency, accessibility, flexibility, extensibility, and maintenance. Micro-services ([11,12,14,20,21,25]) are also recommended, due to ease of implementation, scalability, extensibility, resilience, technological independence, flexibility, evolution, and maintenance. In addition, Middleware ([26]) is another technology that has advantages, such as flexibility, extensibility, and scalability. Finally, Fog Computing ([15,18]) is recommended due to latency reduction and transparency.

### 4.5 Integration and Communication Protocols

There are several ways to perform the integration of a system. This is a fundamental part of the architecture and allows different systems to communicate and cooperate effectively. The choice of the type of communication depends on the specific requirements of each system and can be used for various situations, including data acquisition, information exchange between layers/applications, communication with IoT devices and even communication with external applications. A system can use several communication protocols or different ways to perform integration. Table 2 presents the protocols used by the different works.

In [10], the protocols used for data acquisition are HTTP/REST and MQTT. MQTT is seen as ideal for limited networks due to its lightness. In [11], the MQTT protocol is used to send temperature readings from the gateway to the broker. [12] uses CoAP and MQTT for communication in low power systems. APIs are also used to integrate diverse systems using REST. RESTful and publish/subscribe connectors are used by [14] for communications among different system components. [20] uses Fiware Orion for communication between system layers. [21] uses HTTP/REST and event-based messaging for its microservices architecture. [22] mentions that integration with cloud tools and applications can be done through Publish/Subscriber communication or REST API. In [25], the

**Table 2.** Integration and communication protocols

| # | HTTP | REST | MQTT | COAP | RESTful | Fiware Orion | TCP/IP | BLE |
|---|---|---|---|---|---|---|---|---|
| [10] | ✓ | ✓ | ✓ | - | - | - | - | - |
| [11] | - | - | ✓ | - | - | - | - | - |
| [12] | - | ✓ | ✓ | ✓ | - | - | - | - |
| [14] | - | - | - | - | ✓ | - | - | - |
| [20] | - | - | - | - | - | ✓ | - | - |
| [21] | ✓ | ✓ | - | - | - | - | - | - |
| [22] | - | ✓ | - | - | - | - | - | - |
| [25] | - | ✓ | - | - | - | - | - | - |
| [26] | - | - | ✓ | - | - | - | ✓ | ✓ |

integration of system participants is made through an API following REST principles. Finally, [26] highlights that heterogeneous data acquisition is essential for realizing context-aware automation systems, and that different communication protocols, such as MQTT, TCP/IP and BLE, can be used for this purpose.

## 5 Concluding Remarks

I4.0 is a technological approach that aims to improve production efficiency and increase industry competitiveness. Context-aware applications/systems applied to I4.0 have assisted workers in decision making, thus increasing factories performance while helping companies to be consistent in their manufacturing processes.

This work reviews the concept of context-awareness present in the literature, as well as the main technologies used in context-aware systems. These goals are achieved by reaching answers to several research questions.

It was observed that context refers to any relevant information capable of characterizing an entity's situation, and it is used to interpret events or situations. Context awareness is a property that allows to provide adequate services by analyzing the user's context, becoming a key feature in ubiquitous and pervasive computing systems. The development of context-aware systems usually comprises a context engine that is typically responsible for four tasks: acquisition, construction, reasoning, and context disclosure. Moreover, in many works it has been found that this context engine has its own database to insert, modify and analyze the pertinent data. It has also been verified that the most relevant technologies for system implementation are cloud computing, microservices, middleware and Fog Computing. System integration is a fundamental part of the architecture and there are several ways to accomplish it, including the use of different communication protocols and APIs. The choice of the appropriate method depends on the specific requirements of each system.

Future research aims to develop an architecture for a context-aware engine, grounded in I4.0 and the requirements and specifications of our project. Furthermore, it is intended to establish the communication and integration protocols used between the components of this engine, as well as to define the implementation technology.

**Acknowledgements.** This work was supported by the RD Project "Continental Factory of Future (CONTINENTAL FoF)/POCI-01-0247-FEDER-047512", financed by the European Regional Development Fund (ERDF), through the Program "Programa Operacional Competitividade e Internacionalização (POCI)/PORTUGAL 2020", under the management of AICEP Portugal Global – Trade & Investment Agency.

## References

1. Alexopoulos, K., Makris, S., Xanthakis, V., Sipsas, K., Chryssolouris, G.: A concept for context-aware computing in manufacturing: the white goods case. Int. J. Comput. Integr. Manuf. **29**, 839–849 (2016)
2. Hermann, M., Pentek, T., Otto, B.: Design principles for industrie 4.0 scenarios. In: 2016 49th Hawaii International Conference On System Sciences (HICSS) (2016)
3. Petrasch, R., Hentschke, R.: Process modeling for industry 4.0 applications: towards an industry 4.0 process modeling language and method. In: 2016 13th International Joint Conference On Computer Science and Software Engineering (JCSSE), pp. 1–5 (2016)
4. Breivold, H., Sandström, K.: Internet of things for industrial automation – challenges and technical solutions. In: 2015 IEEE International Conference On Data Science And Data Intensive Systems, pp. 532–539 (2015)
5. Dey, A.: Understanding and using context. Pers. Ubiquit. Comput. **5**, 4–7 (2001)
6. Baldauf, M., Dustdar, S., Rosenberg, F.: A survey on context-aware systems. Int. J. Ad Hoc Ubiquit. Comput. **2**, 263–277 (2007)
7. Alexopoulos, K., Makris, S., Xanthakis, V., Sipsas, K., Liapis, A., Chryssolouris, G.: Towards a role-centric and context-aware information distribution system for manufacturing. Procedia CIRP **25**, 377–384 (2014)
8. Liberati, A., et al.: The PRISMA statement for reporting systematic reviews and meta-analyses of studies that evaluate health care interventions: explanation and elaboration. J. Clin. Epidemiol. **62**, e1–e34 (2009)
9. Al-Shdifat, A., Emmanouilidis, C.: Development of a context-aware framework for the integration of internet of things and cloud computing for remote monitoring services. Procedia Manuf. **16**, 31–38 (2018)
10. Alexopoulos, K., Sipsas, K., Xanthakis, E., Makris, S., Mourtzis, D.: An industrial Internet of things based platform for context-aware information services in manufacturing. Int. J. Comput. Integr. Manuf. **31**, 1111–1123 (2018)
11. Shankar, R., Sasirekha, G.V.K., Ramanathan, C., Bapat, J.: Context aware site-visit scheduling for asset management in oil & gas sector. In: 2022 IEEE 5th International Conference On Industrial Cyber-Physical Systems (ICPS), pp. 01–06 (2022)
12. Tovarnitchi, V.: Cloud-based architectures for environment monitoring. In: 2017 21st International Conference On Control Systems And Computer Science (CSCS), pp. 708–714 (2017)
13. Ponce, V., Abdulrazak, B.: Context-aware end-user development review. Appl. Sci. **12**, 479 (2022)
14. Ortiz, G., Caravaca, J.A., García-de-Prado, A., Boubeta-Puig, J.: Real-time context-aware microservice architecture for predictive analytics and smart decision-making. IEEE Access **7**, 183177–183194 (2019)

15. Islam, M., Kumar, A., Hu, Y.: Context-aware scheduling in fog computing: a survey, taxonomy, challenges and future directions. J. Netw. Comput. Appl. **180**, 103008 (2021)
16. Farahzadi, A., Shams, P., Rezazadeh, J., Farahbakhsh, R.: Middleware technologies for cloud of things: a survey. Digit. Commun. Netw. **4**, 176–188 (2018)
17. Ahmad, A., Abdulaziz, S., Alanazi, A., Alshammari, M., Alhumaid, M.: Software architecture solutions for the internet of things: a taxonomy of existing solutions and vision for the emerging research. Int. J. Adv. Comput. Sci. Appl. **10** (2019)
18. Gomez, C., Chessa, S., Fleury, A., Roussos, G., Preuveneers, D.: Internet of things for enabling smart environments: a technology-centric perspective. JAISE - J. Ambient Intell. Smart Environ. **11**, 23–43 (2019)
19. Alenoghena, C., et al.: eHealth: a survey of architectures, developments in mHealth, security concerns and solutions. Int. J. Environ. Res. Pub. Health **19**(20), 13071 (2022)
20. Corista, P.: IoT data processing pipeline in FoF perspective (2017)
21. Valderas, P., Torres, V., Serral, E.: Modelling and executing IoT-enhanced business processes through BPMN and microservices. J. Syst. Softw. **184**, 111139 (2022)
22. Soares, H.: Uma arquitetura orientada a serviços de nuvem para apoiar testes experimentais de sistemas ubíquos (2021)
23. Ghita, M., Siham, B., Hicham, M., Abdelhafid, A., Laurent, D.: Geospatial business intelligence and cloud services for context aware digital twins development. In: 2020 IEEE International Conference Of Moroccan Geomatics, pp. 1–6 (2020)
24. Wan, J., Yang, J., Wang, Z., Hua, Q.: Artificial intelligence for cloud-assisted smart factory. IEEE Access **6**, 55419–55430 (2018)
25. Geronymo, G.: Smart lab: an application of industry 4.0 design principles to calibration laboratories. J. Phys.: Conf. Ser. **1826** (2021)
26. Sahlab, N., Jazdi, N., Weyrich, M.: An approach for context-aware cyber-physical automation systems. IFAC-PapersOnLine **54**, 171–176 (2021)

# A Computer Vision Approach for Level Measurement of Refilling Stations in Industrial Scenarios

José Ribeiro[1](✉), Rui Pinheiro[1,3], Paulo Nogueira[4], Arsénio Reis[1,2], and Vitor Filipe[1,2]

[1] School of Science and Technology, Universidade de Trás-os-Montes e Alto Douro (UTAD), 5000-801 Vila Real, Portugal
{joseribeiro,arsenio.reis,vfilipe}@utad.pt
[2] INESC TEC—INESC Tecnologia e Ciência, 4200-465 Porto, Portugal
[3] Neoception, Incubadora de Empresas da Universidade de Trás-os-Montes e Alto Douro (UTAD), Edifício P2, Gabinete 16, 5000-081 Vila Real, Portugal
rpinheiro@neoception.pt
[4] Continental Advanced Antenna, Vila Real, Portugal
paulo.nogueira@continental-corporation.com

**Abstract.** In industrial environments, the measurement and monitoring of filling levels (FL) in refilling stations (RS) are critical for quality control processes. Traditional methods used for this purpose, such as manual inspection and sensor-based techniques, have proven to be costly and time-consuming. As an alternative, this paper proposes a novel approach that leverages computer vision (CV) and advanced image processing techniques. This approach provides a more efficient and accurate method for monitoring filling levels in refilling stations, thereby reducing operational costs. The system operates through a comprehensive five-stage pipeline, including pre-processing, perspective transformation, thresholding and edge detection, post-processing and filling level calculation. The performance evaluation of this approach demonstrated promising results in accurately determining filling levels in most scenarios. However, we also identified challenges such as overlapping columns and occlusions in the camera's field of view that require further improvements. By addressing these challenges, our research aims to develop a streamlined and automated method for filling level measurement in refilling stations, thereby enhancing productivity in industrial environments. Ultimately, this proposed approach holds potential to significantly improve the efficiency of refilling stations across multiple sectors.

**Keywords:** Industry 4.0 · Refilling Stations · Assembly Lines · Computer Vision · Measurement

## 1 Introduction

In today's fast-paced industrial landscape, efficient management and control of production processes are crucial. A critical aspect is continuously monitoring available stocks and stringent quality control measures to ensure a seamless production flow. Accurate


R. Mehmood et al. (Eds.): DCAI 2023, LNNS 741, pp. 312–321, 2023.
https://doi.org/10.1007/978-3-031-38318-2_31

measurement of stock levels is a vital component in inventory management and assembly line (AL) optimization [1].

ALs are a popular manufacturing process in which a product is assembled from components or sub-assemblies at numerous stations as it proceeds through the line, resulting in a high degree of standardization and specialization. This leads to increased efficiency and lower production costs [2]. Companies are continuously seeking ways to integrate new technologies into their operations to improve efficiency and responsiveness [3]. Refilling stations (RS) are one of the areas where new technologies can be applied, as manual labor is often required to monitor and replenish materials, which can lead to potential production delays. One problem that affects RSs is the challenge of accurately measuring filling levels (FL), meaning the levels of raw materials used for the AL. This can cause a delay in resupplying the RS due to a lack of monitoring, resulting in setbacks in the AL itself.

A RS, represented in Fig. 1, is a crucial point on the production line where materials or supplies are topped up to maintain the production process. It is usually located near the point of use on the AL and may be equipped with storage areas, handling equipment, and automation technology to keep workers informed when materials need to be refilled.

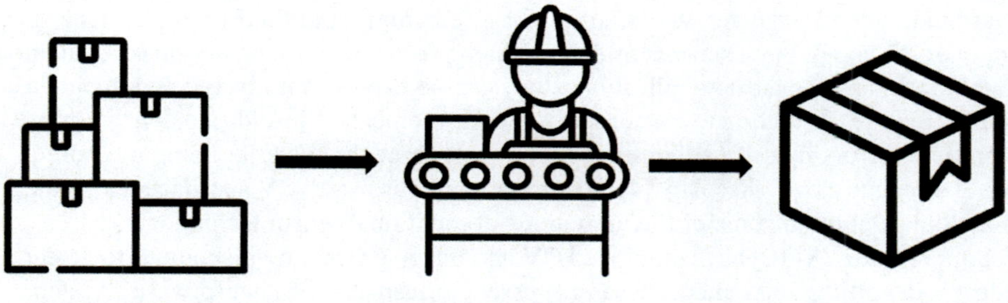


**Fig. 1.** Generic manufacturing process from a RS to an AL and as a finalized product.

Traditionally, RS measurements have relied on manual inspection or basic sensor technologies, which can be expensive but provide good precision. However, these methods have some drawbacks. For instance, they may not be suitable for harsh industrial environments and often require a significant amount of manual labor to set them up, which can lead to human errors, inaccurate measurements, and delays in the production process. Additionally, these sensor technologies may be prone to errors due to variations in the properties of the material being measured, such as conductivity or viscosity [4].

This paper proposes a Computer Vision (CV) approach that uses advanced image processing techniques and deep learning algorithms to measure and monitor RSs in real-time, improving the efficiency and accuracy of quality control processes and reducing operational costs associated with manual inspection and maintenance. Real-time assessment of stock levels is crucial for efficient production management, enabling informed decisions, optimized resource allocation, and achieving manufacturing goals [5].

The rest of this paper is organized as follows: Sect. 2 reviews the related work on traditional FL measurement methods and the use of CV in industrial settings. Section 3 outlines the proposed methodology for FL measurement, detailing a comprehensive six-stage pipeline that includes image pre-processing, perspective transformation, edge detection, post-processing, and FL calculation. Section 4 presents the results of the proposed methodology's implementation, providing a discussion on the success rate, and addressing the challenges and limitations encountered. Finally, Sect. 5 concludes the paper with key takeaways, potential improvements, and directions for future work on this topic.

## 2 Related Work

In recent years, various studies have proposed different techniques for detecting and measuring the FLs of containers in diverse industries, such as food, cosmetics, healthcare, chemicals, and waste management. These techniques aim to ensure product quality, maintain filling accuracy, enable remote monitoring of manufacturing lines, and optimize waste management processes.

Traditionally, FL measurements have relied on manual inspection or basic sensor technologies [6]. Sensor technologies are commonly used in various industrial environments to measure FLs. These sensors work based on physical principles such as ultrasound, electromagnetic waves, and light to determine the fluid level in a tank or container. Although these sensors often have high precision and accuracy, making them reliable for many industrial applications, they can be expensive to install and maintain and may require calibration to ensure accurate measurements [7]. Additionally, they may be prone to errors due to variations in the properties of the material being measured, such as conductivity or viscosity [4]. As proposed in this paper, CV approaches provide a potential solution to some of the limitations of traditional sensor technologies.

Some studies [8–10] have employed CV and image processing techniques to detect fill levels in bottling lines and transparent vessels. Anush et al. [8] put forward a system that utilizes digital image processing and machine learning for detecting bottle fill levels. In a different approach, Pithadiya et al. [9] examined various edge detection methods for inspecting overfilled and underfilled liquid levels in machine vision systems. Meanwhile, Felipe et al. [10] introduced a vision-based technique that leverages CV methods to detect liquid levels in amber glass bottles.

Two studies [11, 12] concentrate on fill-level assessment in industrial waste management environments. One study [11] explores the possibility of using common monocular cameras for fill-level assessment in industrial waste management, comparing the proposed solution to simpler methods to evaluate the effectiveness of the emptying process. Another paper [12] describes a CV system that employs deep learning and convolutional neural networks to estimate fill levels in industrial waste containers of metal scrap.

## 3 Methodology

In this paper, we introduce a methodology to extract FLs from images of RSs using a five-stage pipeline, as depicted in Fig. 2. Each stage—*pre-processing*, *perspective transformation*, *thresholding and edge detection*, *post-processing*, and *FL calculation*—is described in detail below.

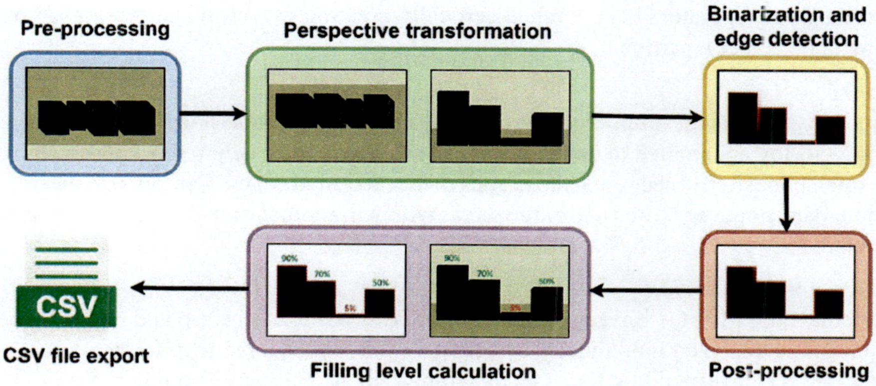


**Fig. 2.** Representation of the pipeline for filling level measurement

***Pre-processing***
To maintain consistency and enhance the quality of input data, the captured image undergoes a series of pre-processing steps. First, brightness and contrast are adjusted using a function that applies alpha and beta factors for contrast and brightness control, respectively. This enhances the visibility of FLs, facilitating subsequent processing.

***Perspective Transformation***
The top-down view of the image is transformed into a frontal view using perspective transformation, allowing for more accurate FL measurements. By defining reference points corresponding to the compartments' corners, we employ our custom function for four-point transformation (Fig. 3) to calculate the transformation matrix and warp the image. If necessary, the image is then rotated using a dedicated function to align compartments horizontally.

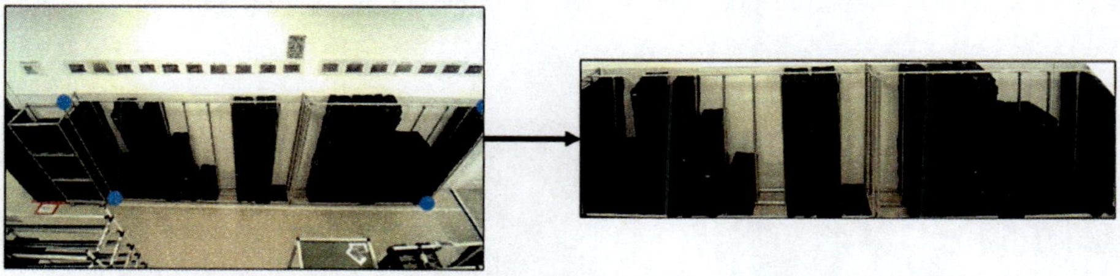

**Fig. 3.** Example of the four-point transformation function application, which calculates the transformation matrix using a method for obtaining perspective transformation and applies it to the image with a method for warping perspective. (Image previously rotated for better presentation)

***Binarization and Edge Detection***
After perspective transformation, binarization, and edge detection techniques are applied to differentiate filled and empty areas within RS columns. Utilizing Otsu's method [13], we automatically compute an optimal threshold value to binarize the transformed image.

Then, the Sobel operator [14] is applied vertically, creating an edge map that emphasizes FLs within each compartment.

***Post-processing***
To enhance the detected edges, post-processing operations such as morphological opening and closing are applied to the edge map. These procedures help reduce noise, eliminate small undesired objects, and close gaps in the identified edges. Contours delineating the boundaries between filled and empty areas within the compartments are subsequently extracted.

***Filling Level Calculation***
To calculate the FLs, the image is separated into two portions: the top and bottom. The top portion is analyzed individually by cropping the refined edge map to the region of interest (ROI) corresponding to a specific compartment. Filling percentages are calculated by comparing the mode of the contour's vertical coordinates to the total height of the compartment. The resulting FLs are visualized by overlaying percentages and compartment numbers on the original image.

In the bottom portion, an empty filling station detector is used. As the image perspective is unwarped in the bottom-front part, the presence of contours is detected. If contours are present, it indicates that the compartment is filled. If there are no contours, it indicates that the station is empty.

## 4 Results and Discussion

In this section, we test the application proposed in the methodology and evaluate its performance. We present successful and unsuccessful results to assess the effectiveness and limitations of the proposed approach.

To evaluate this methodology, images were obtained using a fixed camera mounted on the ceiling of the industrial facility Continental Advanced Antenna in Vila Real, Portugal (as depicted in Fig. 4). The use of a stationary camera guarantees consistent image capture from a stable perspective, which is essential for precise analysis and interpretation.

**Fig. 4.** Installed sensor and example of a captured image.

Throughout the image acquisition process, a pre-trained deep learning model for object detection, known as YOLOv4 [15], is employed to identify individuals. In this instance, the YOLOv4-Tiny version is utilized, which offers quicker inference times on edge devices. This approach enables the model to refrain from saving or utilizing the detected images for inference, thereby safeguarding privacy and preventing potential image compromises due to the presence of passersby.

Owing to the lack of labeled data, we categorized the images into five levels to assess the model's performance. Level 1 corresponds to 0–19% filling, level 2 ranges from 20–39%, level 3 spans 40–59%, level 4 covers 60–79%, and level 5 includes 80–100%. Here, 0% represents an empty FL, while 100% signifies a full FL.

To evaluate the predicted (P) FL generated by the algorithm, we compare it with the actual (A) level visually inferred from the image. Additionally, since the image's perspective is warped, we allow for a tolerance of 10%. This enables us to accept FLs that are close to two different levels.

From the image capture process, around 1000 images were captured. However, since those images are not labeled, and labeling them would be a laborious task, we created a subset of 80 random images. We applied the presented evaluation process to these images and achieved a correct FL measurement of 94.4%, which demonstrates the effectiveness and reliability of the proposed approach in handling real-world scenarios.

In Fig. 5, we can see an image where the algorithm successfully calculates all levels. However, it is not uncommon for the algorithm to detect areas as part of the RS even though there isn't one, but these areas can be easily discarded. For instance, there is a small area detected between columns #10 and #11 that could have raised the FLs to 5. However, since it is minimal, it is disregarded.

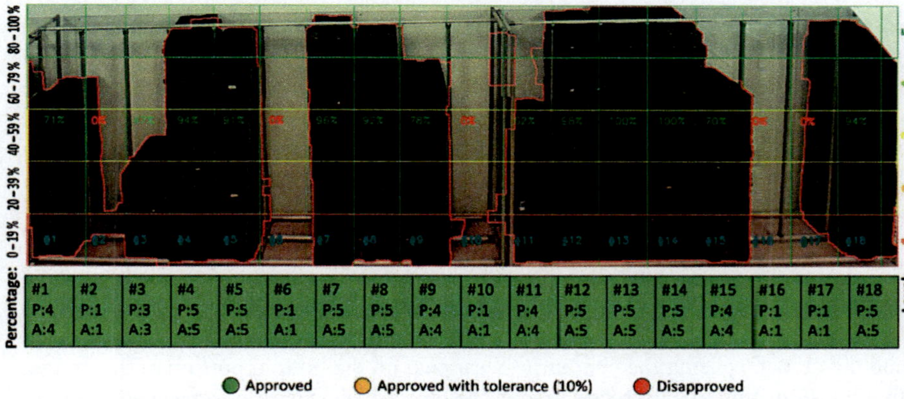


**Fig. 5.** An example of a successfully processed image with FLs accurately calculated by the algorithm.

In the image depicted in Fig. 6, all FLs are accurately calculated except for column #17. Upon manual inspection, it is evident that column #17 is in the 4th level (bordering on the 5th), but it is calculated at 93%, which cannot be approved even with the allowed tolerance. This discrepancy is likely due to the adjacent column overlapping its area.

While algorithms have been applied to address this issue, the image's perspective prevents a completely flawless solution. The overlapping of neighboring columns is visible in the first and last columns but not in the middle ones due to the image's perspective.

A potential solution to this problem is to divide the image into smaller segments and perform the perspective warp on each segment. However, this approach is heavily dependent on the camera's perspective. Therefore, implementing additional cameras at wider angles could also help resolve this issue.

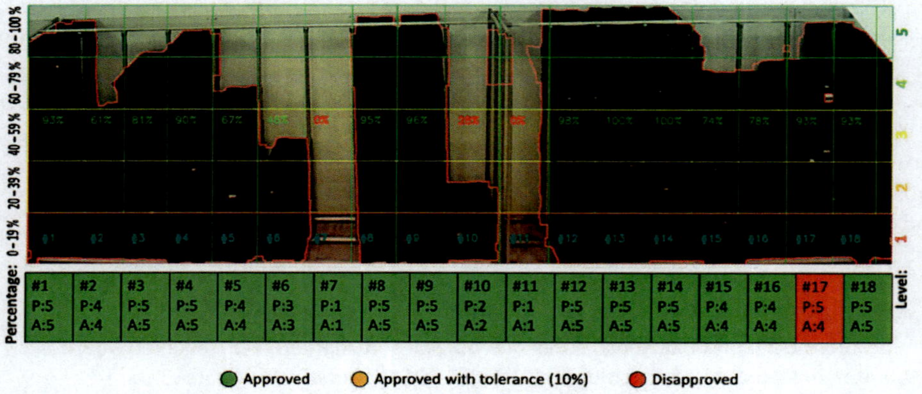


**Fig. 6.** An example of a partially successful processed image with FLs calculated by the algorithm.

In the subsequent image (Fig. 7), we observe the first example of the first column approved with tolerance. The predicted value lies between levels 2 and 3, making it a challenging task even for a human evaluator. Moreover, in columns #15 and #17, there are issues with occlusion from neighboring columns. In column #15, shadows are detected as part of the RS, possibly due to the threshold that binarizes the image. In column #17, the components in column #16 are arranged in a manner that creates a bridge from column #17 to column #18, misleading the algorithm into believing that the top of the column is higher than it truly is.

The last image (Fig. 8) presents the worst-case scenario: when an occlusion blocks the camera's view of the RS. As observed in columns #7 and #8, they are detected as empty, despite being well-filled in reality. This is due to bad contours being created around the object, causing the contour to disappear on the bottom portion of the image, leading the algorithm to incorrectly predict the levels. This underlines the need for more robust algorithms to identify and discard such erroneous frames, as it could undermine future applications built on this foundation.

Moreover, in column #1, the topmost part of the column is not detected, leading to a marginally lower estimated FL. However, given the 10% tolerance, this outcome is considered acceptable. In contrast, in the arrows on the left and right, the RSs are detected as empty using the bottom portion of the image, which shows the absence of contours. The algorithm can correctly identify the absence of filling due to the presence of an object causing bad contours, which underscores the importance of using multiple approaches to mitigate the effect of occlusions on the system's accuracy.

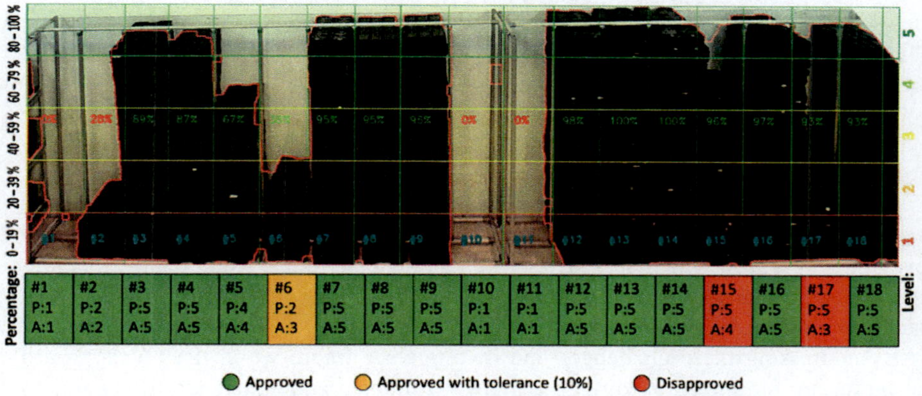


**Fig. 7.** An image illustrating some challenges faced by the algorithm.

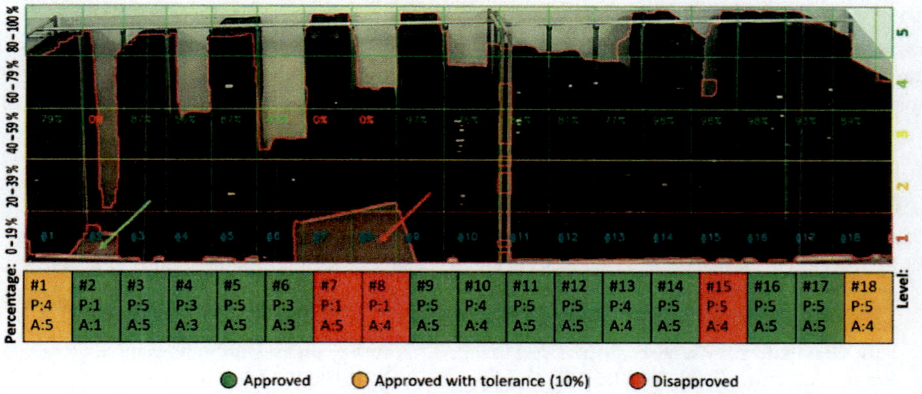


**Fig. 8.** An image demonstrating the worst-case scenario with obstructions blocking the camera's view of the RS.

In summary, in most cases, the proposed CV approach demonstrates promising results in accurately measuring FLs. In addition, the method successfully handles various situations, such as ambiguous FLs and minor occlusions. However, some challenges remain, particularly with overlapping columns and obstructions blocking the camera's view. These issues highlight the need for further improvements to the algorithm and potential changes in the camera setup or image processing techniques.

## 5 Conclusion and Future Work

In this paper, we presented a CV approach for real-time measurement and monitoring of RSs using advanced image processing techniques. This method offers an efficient and accurate alternative to traditional manual inspection and sensor-based approaches while reducing operational costs.

Our results demonstrate promising performance in determining FLs; however, challenges such as overlapping columns and camera obstructions remain. Addressing these issues is crucial for future research and development. Additionally, it is important to note that the results presented in this paper were obtained using a small subdataset. Evaluating the proposed approach on a larger and more diverse dataset would be necessary to further validate its effectiveness and generalizability.

Potential enhancements and research directions include refining the algorithm to better handle overlaps and occlusions, improving image processing techniques, and developing approaches to cope with varying lighting conditions. Additionally, integrating forecasting and alerting mechanisms for timely refueling could improve supply chain management and resource allocation.

By addressing these challenges and investigating future work, the CV approach can be further optimized, leading to a more streamlined and automated filling station measurement method that adapts to the dynamic nature of industrial environments.

**Acknowledgments.** This work was supported by the RD Project "Continental Factory of Future (CONTINENTAL FoF)/POCI-01-0247-FEDER-047512", financed by the European Regional Development Fund (ERDF), through the Program "Programa Operacional Competitividade e Internacionalizacao (POCI)/PORTUGAL 2020", under the management of AICEP Portugal Global – Trade Investment Agency.

## References

1. Rezg, N., Xie, X., Mati, Y., Rezgy, N., Xiey, X., Matiy, Y.: Joint optimization of preventive maintenance and inventory control in a production line using simulation. Int. J. Prod. Res. **42**, 2029–2046 (2004)
2. Saif, U., Guan, Z., Wang, B., Mirza, J., Huang, S.: A survey on assembly lines and its types. Front. Mech. Eng. **9**(2), 95–105 (2014). https://doi.org/10.1007/s11465-014-0302-1
3. Pérez-Lara, M., Saucedo-Martínez, J.A., Marmolejo-Saucedo, J.A., Salais-Fierro, T.E., Vasant, P.: Vertical and horizontal integration systems in Industry 4.0. Wirel. Netw. **26**(7), 4767–4775 (2018). https://doi.org/10.1007/s11276-018-1873-2
4. Dybko, A.: Errors in chemical sensor measurements. Sensors **1**, 29–37 (2001)
5. Rácz-Szabó, A., Ruppert, T., Bántay, L., Löcklin, A., Jakab, L., Abonyi, J.: Real-time locating system in production management. Sensors **20**, 6766 (2020)
6. Yang, J., et al.: Implementation of ultrasonic sensor as a chemical percol fluid level control based on Atmega 16. IOP Conf. Ser. Mater. Sci. Eng. **1098**, 042046 (2021)
7. Martins, A.B., Farinha, J.T., Cardoso, A.M.: Calibration and certification of industrial sensors – a global review. WSEAS Trans. Syst. Control **15**, 394–416 (2020)
8. Anush, C., Yashwanth, K., Shashank, S., Venkat Reddy, M., Kumar, A.: Bottle line detection using digital image processing with machine learning. J. Phys. Conf. Ser. **1998**, 012033 (2021)
9. Pithadiya, K.J., Nagar, V.V., Modi, C., Chauhan, J.D.: Selecting the most favourable edge detection technique for liquid level inspection in bottles (2011)
10. Felipe, M.A.A., Olegario, T.V., Bugtai, N.T., Baldovino, R.G.: Vision-based liquid level detection in amber glass bottles using OpenCV. In: 2019 7th International Conference on Robot Intelligence Technology and Applications (RiTA 2019), pp. 148–152 (2019)
11. Jungåker, J.: Monocular depth estimation for level assessment in an industrial waste management environment. Master Science Thesis, KTH Royal Institute of Technology (2021)

12. Alexopoulos, K., et al.: Deep learning for estimating the fill-level of industrial waste containers of metal scrap: a case study of a copper tube plant. Appl. Sci. **13**, 2575 (2023)
13. Otsu, N.: A threshold selection method from gray-level histograms. IEEE Trans. Syst. Man Cybern. **9**, 62–66 (1979)
14. Kanopoulos, N., Vasanthavada, N., Baker, R.L.: Design of an image edge detection filter using the Sobel operator. IEEE J. Solid-State Circuits **23**, 358–367 (1988)
15. Bochkovskiy, A., Wang, C.-Y., Liao, H.-Y.M.: YOLOv4: optimal speed and accuracy of object detection (2020)

# Design of Context-Aware Information Systems in Manufacturing Industries: Overview and Challenges

Arlindo Santos[1], Claudio Lima[2], Arsénio Reis[2], Tiago Pinto[2](✉), Paulo Nogueira[3], and João Barroso[2]

[1] Polytechnic Institute of Bragança, Campus de Santa Apolónia, 5300-253 Bragança, Portugal
acsantos@ipb.pt
[2] University of Trás-os-Montes and Alto Douro, Quinta de Prados, 5000-801 Vila Real, Portugal
{claudiol,ars,tiagopinto,jbarroso}@utad.pt
[3] Continental Advanced Antenna, Zona Industrial de Constantim, 5000-082 Vila Real, Portugal
paulo.nogueira@continental-corporation.com

**Abstract.** In the last 30 years, several academic and commercial projects have explored the context-awareness concept in multiple domains. Ubiquitous computing and ambient intelligence are features associated with the 4th generation industry empowering space to interact and respond appropriately according to context. In the scope of Industry 4.0, context-aware systems aim to improve productivity in smart factories and offer assistance to workers through services, applications, and devices, delivering functionalities and contextualised content. This article, through descriptive research, discusses the concepts related to context, presents and analyses projects related to ubiquitous computing and associated with Industry 4.0, and discusses the main challenges in systems and applications development to support intelligent environments for increased productivity, supporting informed decision-making in the factories of the future. The study results indicate that many research questions regarding the analysed projects remain the same, leading the research in the context-aware systems area to minimise issues related to context-aware features, improving the incorporation of Industry 4.0 paradigm concepts.

**Keywords:** Context-aware · context-aware systems · manufacturing industries · ubiquitous computing

## 1 Introduction

In 1970 the mainframe was a resource shared by several people, but in 1980 this concept changed, and the personal computer has become part of our lives. In the 1990 decade, computational advances, such as increased processing power and storage, size reduction, interconnectivity, increasing battery capacity, wireless networking, and software technologies, allowed the computer to become an integrated part of everyday life. It led to an era where technology is present and seamlessly integrated into performing the most diverse tasks. Weiser [1] called this new paradigm ubiquitous computing, and he believed that "*the most profound technologies are those that disappear. They weave themselves into the fabric of everyday life until they are in*".


R. Mehmood et al. (Eds.): DCAI 2023, LNNS 741, pp. 322–331, 2023.
https://doi.org/10.1007/978-3-031-38318-2_32

Later, Lyytinen and Yoo [2] state that devices can obtain contextual information from the environment and use it for one purpose. In this sense, the environment becomes smart because it acquires the ability to detect other devices. Thus, the ubiquitous computing paradigm presents interesting concepts from which the industrial community and visions of the factory of the future (FoF) can benefit [3]. As a consequence of technological advances in the area of ubiquitous computing, semantic and soft computing, Thaduri et al. [4] report an increasing demand in the research of context-aware applications, such as medical applications, smart homes, smart cities, e-commerce, web services, and industrial applications. Within industrial environments, spaces incorporate the context-aware concept to assist people and machines in performing their tasks, and Weiser's approach is visible in the factory environment [5].

Considering the increasing presence of context-aware systems in the industry, this paper reviews the context concept by exploring projects related to ubiquitous computing and associated with Industry 4.0 (i4.0). Consequently, we identified the main challenges of systems and applications development to support intelligent environments for increased productivity and highlighted promising paths for future research and development.

## 2 Background

The term context was first used by Schilit et al. [6], where context-aware systems adapt according to the location of use, the collection of nearby people, hosts, and accessible devices. as well as changes to such things over time. These authors define three aspects of context as essential: where you are, who you are with, and what resources are available in the surroundings. According to Dey [8], context is any information that can be used to characterize the situation of an entity (person, place, or object) and that is considered relevant to the interaction. To differentiate and capture the various types of context information, Dey et al. [9] suggests finding answers to five basic questions: who? related to the user identification; where? related to the user location; when? related to the temporal context; what? related to the identification of the user's activities, and why? related to the information that provokes specific user actions.

Other definitions emphasize contextual information used to provide relevant services, which depend on user characteristics and tasks performed [7]. The diversity of definitions is so complex that researchers Bazire and Brézillon [10] examined 150 definitions of the word "context" from various research areas, concluding their analysis by stating that creating a unified definition of context is an arduous and probably impossible endeavour. The context definition depends mainly on the application domain, and in general, context is a set of constraints that influences behaviour relative to a given task.

Nowadays, the availability of sensors, actuators and the extensive use of intelligent devices allow the development of systems that can "feel" the environment and recognize the specific context. This study discusses context concerns in the industrial environment, where context-aware systems development has gained popularity, as they can provide contextual assistance and automate processes. However, according to Rosenberger et al. [11], despite the advantages of context-aware systems, they still play a minor role in industrial applications. Moreover, compared to the personal domain, their use is incipient in the industrial environment [12] due to the absence of methods and techniques to support context documentation, analysis, and optimization, hindering the more intensive development of context-aware systems [13]. This way, it presents a relevant overview and challenges of context-aware systems development in manufacturing industries.

## 3 Methodology

In this study, we used descriptive research methodology to identify facts and phenomena. In addressing issues in a permanent change process, descriptive research searches obtain and analyse data to present an accurate description of the studied phenomenon through a critical examination of the information [14]. In this study, the addressed phenomenon is the development of context-aware systems for manufacturing industries.

The first three selected projects are the most representative of the ubiquitous computing field that embeds Weiser's vision, followed by four projects demonstrating the application of context awareness in the manufacturing industries where Weiser's vision is present. Finally, the selected projects are comprehensively described in Sect. 4, and analysed in Sect. 5, making it possible to reach the research conclusions in Sect. 6.

## 4 Related Projects

We conduct descriptive research of systems associated with the ubiquitous computing concept and oriented to the industry domain, considering that industrial spaces incorporate properties of ubiquitous computing and act as smart spaces that aggregate smart objects [3]. These projects provide relevant architecture insights incorporating the context-aware concept and facilitating rapid application development.

### 4.1 Context ToolKit

Dey et al. [15] presented the Context Toolkit, which provides tools that facilitate the development of context-aware applications and contribute to standardising how sensors and actuators access context data. These encapsulate standard methods where applications can access and share data transparently. In this approach, context is information that applications use and is part of the space. This strategy pretended to meet some requirements for handling context and contextual information [8, 15] to facilitate application development.

The architecture introduces the widget component, which executes the mediation between the user and the environment and provides a standard interface for applications that use the context independently of the sensors used to capture the information. Aggregators are responsible for gathering the context information associated with a real-world entity, such as a person or a place, and interpreters are responsible for transforming one form of context into another. The services embedded in widgets can execute determined actions according to the interest and needs of a given application, i.e., when a specific combination of conditions occurs, any predefined action must be taken at the application's request. Finally, the discoverers are responsible for keeping track of all the components' information and what each one can accomplish.

### 4.2 Aura

Garlan et al. [16] present a framework for supporting user mobility in ubiquitous computing environments and facilitating the task of developing context-oriented applications. The base of this system relies on the concept that the most precious resource in a computing system is no longer the processor, memory, disk, or network, but human attention, i.e., the ability of a user to dedicate himself to primary tasks, is a priority.

The framework monitors the actors in the context and acts to be proactive and anticipate needs, adapting to the resources available and the requests made. In this way, it provides opportunities for adaptation to changing resources, user mobility, changes in user needs, and system failures [17]. The proposal supports applications that benefit from the co-knowledge of available computing resources.

In the lower layers of the architecture are the components that monitor the network conditions or determine the user's location. Finally, above the operating system layer are the components which allow access to remote files considering the network state and the monitoring and adaptation of resources and applications: the Spectra, which, based on context, determines the best mechanism for the execution of remote procedure calls and, at the highest level of the architecture, the Prism that supports task execution, is context-sensitive, allows adaptation to the environment, and detects the user's intentions, assisting before any need is declared.

### 4.3 Gaia

Román and Campbel [18] proposed a distributed software infrastructure for ubiquitous applications development and publishing. The project abstracts access to physical resources and several concepts related to an operating system, such as events, file systems, processes, and others. The result culminated in a middleware and framework development, aiming to support creating user-centric, resource-aware, multi-device, and context-sensitive applications for active spaces, performing resource and service management and offering services related to context, events, and localisation.

The architecture has three main layers: kernel, application framework, and applications. The kernel provides essential services such as events, the presence of entities, context notification, discovery, and others. The Application Framework provides mechanisms to build, run and adapt the Applications executed over the third layer. The great advantage is the ease of generating scripts to create distributed applications by instantiating components present in the space while allowing rapid application development.

### 4.4 CIConAInd

Thaduri et al. [4] propose a meta-database generated by contextual information and use a rule-based approach, thus providing decision-making using various soft computing algorithms. In the industrial application setting, context often considers users, machines, the environment, and their interactions that affect performance.

The users have interests, qualifications, education, and others. In contrast, the environment has variables such as weather information, location, and temperature, among others, and the machine is an abstraction of devices, resources, and events. Consequently, implementing context-aware systems depends on requirements, availability, cost, and business objectives. Therefore, the proposed framework includes module integration, like data acquisition, pre-processing, network, data storage, and user interface layer.

## 4.5 Three-Tier Architecture

Alexopoulos et al. [19] proposed a system designed according to the requirements of the shop floor, thus offering a context-aware solution to support operators and activities. The architecture has three levels: Edge tier, Platform tier, and Enterprise tier. In the Edge tier, data is obtained from the devices in the physical environment (assets) and pre-processed. Then, audio and visual alerts can be created in the Integration layer according to pre-configuration rules. Subsequently, the low-level sent information to the Platform tier, and here contextual information is extracted (context engine) and available to services (Services layer) and applications embedded in the Enterprise tier.

The shop floor actors validated the proposed approach and allowed application development independent of heterogeneous data sources (assets), which can be easily reconfigured without additional development effort.

## 4.6 Context-Aware Programming (CAP)

Gaur et al. [20] propose a declarative approach for application development called Context-Aware Programming (CAP), allowing developers to write autonomous code blocks promoting abstraction, modularity, and mobility. As a result, low-level issues are discarded, ignoring heterogeneity issues, and code reuse and availability are promoted on different devices and execution contexts.

The proposal provides a component called Context Manager that keeps a complete record of each context and the applications associated with each context. Moreover, for an application to be associated with a context, it must already have been created using the Applications Manager component. The Resource Administrator provides automated CoAP operations (such as pull, push, get, and observe) to maintain the state of available devices and the resources provided by those devices.

## 4.7 Digital Factory Assistant

Belkadi et al. [3] designed a contextual knowledge-based system that supports the factory actors with the correct information at the right time and in the appropriate format according to their work context and level of expertise for machine configuration and manufacturing operations. Also, they present a matrix of comparison of this system with five other systems based on each system's ability to understand the situation, assist the user in real-time, integrate the ability to provide collaboration, support human-machine interaction, access context, provide interoperability of legacy tools, and mapping to different application domains.

The authors proposed a context model with organisational, operational, and user-centric dimensions, as well as the relationship between them for intensive knowledge reuse. The model aims to help shop floor workers in their daily activities and improve the efficiency and effectiveness of the operational process in assisting decision-making.

## 5 Analysis and Discussion

Although society has permanent contact with solutions that take advantage of different context variables, the definition of the concept is still superficial and with some limitations [13], and the development of applications that use simple context data, with a minimum of variables and without any complexity, is "trivial". The issue arises when the objective is to develop computational systems with the skill to comprehend contexts beyond the scope of simplicity and encompassing a high degree of complexity. In industrial settings, particularly in the manufacturing industry, computational systems traditionally incorporate cloud and edge computing technologies. More recently, the discourse has arisen regarding the application of fog computing in such contexts [21]. At the extreme, we can design multiple layers of context storage and computation for subsequent decision-making (supervised, semi-supervised, or unsupervised). Under such circumstances, complexity escalates, and responses to the 5W inquiries posed by Dey [9] do not remain facile.

The discussion initiated by Dey et al. [15] concerning minimizing development is related to the development process of applications that use contextual information. These authors identified a set of system requirements to facilitate the developer's task, shifting to the system the responsibility of managing the contextual life cycle: acquisition, interpretation, storage, dissemination, as well as resource discovery in the environment. Moreover, this approach is followed in some previously presented works and in literature review articles [22, 23], where the focus is on creating an architecture that hides the complexity of the context and provides access interfaces for applications. According to Gaur et al. [20], the facilitation of contextual understanding necessitates providing suitable tools for application development and concealing knowledge about specific resources.

Baldauf et al. [23] consider that the design of context-aware systems should include two principles: the architecture and the context model. However, in the industry domain, software system architectures are available specifically targeting the application environment in I4.0, where every physical object will always be connected with its digital twin [19]. Also, Alexopoulos et al. [19] argue that IIoT (Industrial Internet of Things) architectures oriented to industrial and context-aware environments should privilege a layered approach to simplify the development of a complex system, isolating functionalities, as well as an event-driven approach and the use of ontologies or semantics in the context specification in a dynamic environment.

Another challenge that adds complexity is related to the interactions of the environment's actors, be they people, applications, machines, robots, etc. Instead of the traditional model, where the computer is the centre of the interactions, the new scenario allows users to interact with the system at any time, in any space, and from any device. Kymäläinen et al. [24] contend that forthcoming control operatives must oversee intelligent systems via novel modes of interaction and that tasks will be distributed flexibly between human workers and systems. And the interaction between humans and machines is still underestimated by developers, which makes it challenging to design a system that can interact with multiple users and support their daily activities [3]. More, in the smart factory context that aims to build the "Factory of Things" concept, the factory incorporates smart objects (physical and/or virtual) that interact together to exchange information [3] in the framework of the Cyber-Physical System (CPS) concept.

The Aura project [16] privileges the focus on the user and the system's ability to prevent this actor from being disturbed in his activities. For proactivity to be effective, a system must track the user's intent to determine which system actions will help [25]. This is, i.e., the concern of World Wide Web search engines such as Google Search that "guesses" the user's intent when they type a term into the search bar. Extending to an industrial environment populated by other actors, each with intelligence capabilities, the challenges increase. Conversely, the ability to forecast actors' intentions within a given environment engenders several significant research inquiries. According to Gaur et al. [20], the context-aware concept can anticipate the needs of a user/environment and act accordingly. Therefore, providing users with the appropriate tools to write applications context-aware is essential. The main goal of such a programming approach should be to let the user express the desired goals without requiring knowledge about specific resources. More, we only visualize the concept of usability addressed by Alexopoulos et al. [19] intakes at the project evaluation stage.

We are recovering the concept of Active Spaces [26] as a collection of dynamic components installed and distributed to meet the requirements of a user or a group of users. Dynamism is arguably the most critical aspect of an application, necessitating a flexible, component-based application architecture capable of modifying its composition during runtime. This requirement presupposes that the mobility of entities is a constant requirement [16], and the challenge focuses on capturing the attention of the interaction actors.

The advent of the Internet and mobile technologies has undoubtedly transformed how users acquire and engage with information. Nevertheless, it is crucial to recognize that users transition between physical or virtual spaces, just as sensors and actuators may move within space, potentially persevering or relinquishing connection to the same environment. Besides these, we have the mobility of applications and activities between spaces, devices, and users. Garlan et al. [16] refer that mobility is a constant, and the challenge is capturing the interaction actors' attention.

With the merging of wireless communication and device functionality, devices can be employed in highly mobile situations. Thus, entities must be able to operate in different and dynamic communication characteristics, possibly changing service access providers. Furthermore, the incorporation of new entities may lead to a surge in the volume of data that necessitates effective data management. Considering that context is dynamic and multi-variant, working with the complexity, heterogeneity, and data management of context-aware systems requires the adoption of sophisticated hardware and software computing, and with the emergence of the soft computing concept, in contrast to traditional computing, may be an opportunity [4]. Still, context-aware information is deemed even more pertinent in industrial settings characterised by dynamic and semi-structured conditions, where job responsibilities are subject to change, and workers are mobile [19].

Applying Weiser's vision [1] to secure increased availability and proactivity is a concept that could enhance the capacity to provide real-time assistance to users. Belkadi et al. [3] contend that, although ubiquitous computing is not novel, it presents an intriguing scientific challenge that industrial engineering research communities have yet to explore comprehensively. Incorporating such concepts may offer significant advantages to these communities and facilitate the realisation of visions about the factory of the future.

## 6 Conclusions

Currently, many commercial companies offer assistance to the user by delivering context-aware content and functionality through services, applications, and devices. Since the start of the 21st century, several academic projects have demonstrated the efficacy of implementing context awareness, resulting in the term "context" proliferating over the years. However, the generalised and unique definition of the context concept does not exist and differs from domain to domain. As a result, projects embodying this concept within industrial spaces are reduced, but the encouragement in the development of applications is an option to meet the challenges of the market.

In this sense, this paper has analysed and presented a set of discussion topics related to the meaning of the word context and its challenges in industrial systems, as well as enumerating academic projects, and from this leverage a set of challenges for the development of context-aware systems and applications to support decision making in manufacturing industries. The advances in digital technologies and systems have opened new opportunities in the soft computing field, where industry practices have incorporated no-code and low-code development strategies. With the ability to design systems capable of facilitating real-time decision-making and reducing the application development process, effort and time, we can surmount new challenges in the soft computing domain.

While the early analysed projects did not address the concepts of cloud computing, fog computing, and edge computing, in recent projects, these concepts are part of the discussion in the smart factories design solution. In the context of Industry 4.0 and the production domain, it is crucial to acknowledge that the underlying assumptions differ from those of other knowledge domains. The final analysis indicates that many issues addressed by projects in the first two decades of the 21st century remain relevant in the current literature, such as mobility, heterogeneity, interoperability, resource discovery, and system integration. Therefore, future system development and improvements should minimise issues related to context-sensitive features, improving the speed of incorporating I4.0 paradigm concepts into these systems.

**Acknowledgements.** This work was supported by the R&D Project "Continental Factory of Future, (CONTINENTAL FoF) / POCI-01-0247-FEDER-047512", financed by the European Regional Development Fund(ERDF), through the Program "Programa Operacional Competitividade e Internacionalização (POCI) / PORTUGAL 2020", under the management of aicep Portugal Global – Trade & Investment Agency.

## References

1. Weiser, M.: The Computer for the 21st Century. Sci. Am. **265**, 94–104 (1991)

2. Lyytinen, K., Yoo, Y.: Issues and Challenges in Ubiquitous Computing. Commun. ACM. **45**, 62–65 (2002)
3. Belkadi, F., Dhuieb, M.A., Aguado, J.V., Laroche, F., Bernard, A., Chinesta, F.: Intelligent assistant system as a context-aware decision-making support for the workers of the future. Comput. Ind. Eng. **139**, 105732 (2020)
4. Thaduri, A., Kumar, U., Verma, A.K., Ajit, Verma, K., Verma, A.K.: Computational intelligence framework for context-aware decision making. Int. J. Syst. Assur. Eng. Manag. **8**, 2146–2157 (2017)
5. Lucke, D., Constantinescu, C., Westkämper, E.: Smart Factory - A Step towards the Next Generation of Manufacturing. Manuf. Syst. Technol. New Front. 115–118 (2008)
6. Schilit, B., Adams, N., Want, R.: Context-aware computing applications. In: Mobile Computing Systems and Applications - Workshop Proceedings. pp. 85–90. IEEE (1995)
7. Abowd, G.D., Dey, A.K., Brown, P.J., Davies, N., Smith, M., Steggles, P.: Towards a better understanding of context and context-awareness. Lect. Notes Comput. Sci. (including Subser. Lect. Notes Artif. Intell. Lect. Notes Bioinformatics) **1707**, 304–307 (1999)
8. Dey, A.: Providing Architectural Support for Building Context-Aware Applications, https://www.cc.gatech.edu/fce/ctk/pubs/proposal.pdf (2000)
9. Dey, A.: Understanding and Using Context. Pers. Ubiquitous Comput. **51**(5), 4–7 (2001)
10. Bazire, M., Brézillon, P.: Understanding context before using it. Lect. Notes Comput. Sci. (including Subser. Lect. Notes Artif. Intell. Lect. Notes Bioinformatics). 3554 LNAI, 29–40 (2005)
11. Rosenberger, P., Grafinger, M., Gerhard, D., Hennig, M., Dumss, S.: The CASAD matrix method: introduction of a technique for the documentation, analysis, and optimization of context-aware systems. Procedia CIRP. **91**, 427–432 (2020)
12. Perera, C., Zaslavsky, A., Christen, P., Georgakopoulos, D.: Context aware computing for the internet of things: A survey. IEEE Commun. Surv. Tutorials. **16**, 414–454 (2014)
13. Alegre, U., Augusto, J.C., Clark, T.: Engineering context-aware systems and applications: A survey. J. Syst. Softw. **117**, 55–83 (2016)
14. Cordova, F.P., Silveira, D.T.: Métodos de Pesquisa: A Pesquisa Científica. In: Métodos de Pesquisa. pp. 33–44. Editora da UFRGS (2009)
15. Dey, A., Abowd, G., Salber, D.: A conceptual framework and a toolkit for supporting the rapid prototyping of context-aware applications. Human-Computer Interact. **16**, 97–166 (2001)
16. Garlan, D., Siewiorek, D.P., Smailagic, A., Steenkiste, P.: Project Aura: Toward distraction-free pervasive computing. IEEE Pervasive Comput. **1**, 22–31 (2002)
17. Soldatos, J., Pandis, I., Stamatis, K., Polymenakos, L., Crowley, J.L.: Agent based middleware infrastructure for autonomous context-aware ubiquitous computing services. Comput. Commun. **30**, 577–591 (2007)
18. Román, M., Campbell, R.H.: A middleware-based application framework for active space applications. Lect. Notes Comput. Sci. (including Subser. Lect. Notes Artif. Intell. Lect. Notes Bioinformatics) **2672**, 433–454 (2003)
19. Alexopoulos, K., Sipsas, K., Xanthakis, E., Makris, S., Mourtzis, D.: An industrial Internet of things based platform for context-aware information services in manufacturing. **31**, 1111–1123 (2018)
20. Gaur, S., Almeida, L., Tovar, E., Reddy, R.: CAP: Context-Aware Programming for Cyber Physical Systems. IEEE Int. Conf. Emerg. Technol. Fact. Autom. ETFA. 2019-Septe, 1009–1016 (2019)
21. Matt, C.: Fog Computing: Complementing Cloud Computing to Facilitate Industry 4.0. Bus. Inf. Syst. Eng. **60**, 351–355 (2018)
22. Hong, J. yi, Suh, E. ho, Kim, S.J.: Context-aware systems: A literature review and classification. Expert Syst. Appl. **36**, 8509–8522 (2009)

23. Baldauf, M., Dustdar, S., Rosenberg, F.: A survey on context-aware systems. Int. J. Ad Hoc Ubiquitous Comput. **2**, 263–277 (2007)
24. Kymäläinen, T., et al.: A creative prototype illustrating the ambient user experience of an intelligent future factory. J. Ambient Intell. Smart Environ. **9**, 41–57 (2017)
25. Satyanarayanan, M.: Pervasive computing: Vision and challenges. IEEE Pers. Commun. **8**, 10–17 (2001)
26. Román, M., Hess, C., Cerqueira, R., Ranganathan, A., Campbell, R.H., Nahrstedt, K.: A middleware infrastructure for active spaces. IEEE Pervasive Comput. **1**, 74–83 (2002)

# Wearable Devices in Industry 4.0: A Systematic Literature Review

Hugo Anes[1], Tiago Pinto[1(✉)], Cláudio Lima[1], Paulo Nogueira[2], and Arsénio Reis[1]

[1] Universidade de Trás-os-Montes e Alto Douro Vila Real, Vila Real, Portugal
al68571@alunos.utad.pt, {tiagopinto,claudio,ars}@utad.pt
[2] Continental Advanced Antenna, Vila Real, Portugal
paulo.nogueira@continental-corporation.com

**Abstract.** Over the years, industrial evolution has proved to be a complex process, since there are several aspects that need to be considered to achieve highly functional processes and differentiated quality products. To date, four industrial revolutions have been implemented. Thus, the paradigm of Industry 4.0 (I4.0) was born, a concept that aims to improve the efficiency, productivity, automation, and safety of industrial processes, but which also considers the operator's relevance and centrality in these processes. Besides these four revolutions one more concept is emerging, called Industry 5.0 (I5.0). In recent years, and with the advance of scientific research, the implementation of wearables has proven to be the ideal solution to move towards the digitisation of Industrial sector. In this sense, the aim of this work is to provide a systematic review on the currently available knowledge about wearable technology and its applicability within I4.0. Through these technologies, both processes and operators can be monitored in real time, actively contributing to the identification of limitations and to the implementation of improvements. On the other hand, studies on the acceptance of these devices have shown a certain apprehension by users regarding the security and privacy of collected data. Therefore, studies should be conducted to analyse in depth these limitations, to raise users' confidence and contribute, in a broader perspective, to the success of industrial processes.

**Keywords:** Wearable · Wearable Technology · Wearable Devices · Industry 4.0 · Internet of wearable Things

---

This work was supported by the RD Project "Continental Factory of Future (CONTINENTAL FoF)/POCI-01-0247-FEDER-047512", financed by the European Regional Development Fund (ERDF), through the Program "Programa Operacional Competitividade e Internacionalização (POCI)/PORTUGAL 2020", under the management of AICEP Portugal Global - Trade & Investment Agency. This work also funded by the R&D Project "A-MoVeR-Agenda Mobilizadora para o Desenvolvimento de Produtos e Sistemas Inteligentes de Mobilidade Verde", financed by POCI European Structural and Investment Funds (FEEI).


R. Mehmood et al. (Eds.): DCAI 2023, LNNS 741, pp. 332–341, 2023.
https://doi.org/10.1007/978-3-031-38318-2_33

## 1 Introduction

Over time, the industry has been facing and overcoming several challenges related not only to the production processes themselves, but also to their management and that of the work teams. In fact, the industrial context has several aspects that can prove problematic and represent significant losses, such as: (i) malfunctions and technical failures of equipment; (ii) loss of manpower as a result of the occurrence of work accidents; (iii) loss of material (either in the form of raw material or final product) as a result of the reduced functionality of the production processes; (iv) among others [1,2]. Thus, the Fourth Industrial Revolution (or Industry 4.0) emerged, a paradigm focused on the functionalisation of industrial processes - optimisation, monitoring and automation - through the implementation of the technological advances achieved to date [2].

It is in this perspective that wearables can help overcome many of these challenges: as easy-to-use, low-cost devices with a wide range of applications, their implementation in the industry allows real-time monitoring of the progress of production processes, but also of the operator's status. In this way, the use of wearables contributes to the increase of process efficiency, also ensuring the well-being of the work teams and the maintenance of the necessary safety conditions [3]. Despite being widely used in the most diverse sectors (health, sports, education, among others) [4,5], the application of wearables in industry is still uncommon [1]. However, with the continuous technological advances, the opportunity has arisen to develop strategies that promote operator centricity but supports their cooperation with the equipment and the technology [5,6].

The aim of this work is to explore the application of wearables in the most varied aspects of the industrial process, providing the necessary information for the development of new technologies directed to this sector. Therefore, in Sect. 2, the concepts of wearable and I4.0 are explored. In Sect. 3 the applied review methodology is presented. Section 4 presents and discusses the results. Finally, in Sect. 5 the final considerations and conclusions are presented.

## 2 Background

Over the years, the industry sector has been progressively revolutionising as a result of the evolution of the external environment that surrounds it. Thus, the three major industrial revolutions have occurred because of the mechanisation, modernisation and digitisation of the industrial sector [7]. Recent technological advances have given rise to a new paradigm called the Internet of Things (IoT), which seeks to integrate users and devices that communicate with each other [8,9]. In this sense, wearables have emerged as electronic devices that integrate the IoT and can be applied to several areas, including industry [10]. The concept of I4.0, initially proposed by the German government [7], refers to the digitisation and interconnection of all stages of industrial production. I4.0 aspires to establish intelligent production, responding to changes in manufacturing processes in real time [11,12].

The implementation of the Industrial Internet of Things (IIoT) is intended to: increase the efficiency and productivity of manufacturing processes; facilitate and streamline the decision-making process; increase the efficiency of resource use and worker safety and improve the management of work teams in this environment, increasing the financial profitability of the processes [2,5,7,13]. These objectives can be achieved by using wearables, which, in a simplified way, consist of mobile and particularly discrete electronic devices that can be incorporated into the most varied daily tasks of humans and into the processes of a company. Moreover, these are easily customisable according to their intended use, allowing not only active data collection in real time, but also assisting the user in performing their tasks by increasing their skills and capabilities. All together, these characteristics make wearables great tools for monitoring industrial processes at different levels [10].

Although industrial production is evolving towards replacing human labour in the future in heavy and repetitive jobs [6], the truth is that manual labour remains indispensable to the success of this industries, not only because some processes are still performed by human hand, but also because the presence of human in the industry provides another level of monitoring, predicting the occurrence of malfunctions and therefore avoiding production stoppages that, in turn, compromise the financial and resource profitability of companies. In general, the IIoT has focused primarily on ensuring the viability and correct operation of equipment. However, more recent studies have pointed to the applicability of these technologies in monitoring and evaluating the productivity of work teams [2], contributing to the improvement of the worker-machine-process dynamics.

## 3 Methodology

For this systematic review, the Preferred Reporting Items for Systematic Reviews and Meta-Analyses (PRISMA) technique has been used. With this review we aim to: (i) Compile the existing information to date on the implementation of wearables in the industry; (ii) Identify the existing challenges in this topic; and finally, (iii) Provide an overview on the use of these devices in the industrial context [14].

### 3.1 Research Strategy

As a research strategy, a set of terms that guided the process were defined in the preparatory phase. To answer the review's research questions the concepts of productivity were identified and then the application of wearable devices in the context of I4.0. This systematic literature review was conducted using the Google Scholar search engine, delimiting the databases IEEE Xplore, ACM Digital Library, Springer, Elsevier Science Direct, Wiley Online Library, and Taylor and Francis. After formulation, tests and adjustments, Inclusion Criterion CI1 was reached, composed of the following search query: *("wearable devices" OR "wearable technology" OR "wearable technologies" OR "Internet of wearable" OR "Smart wearables") AND ("Industry 4.0" OR "industrie 4.0").*

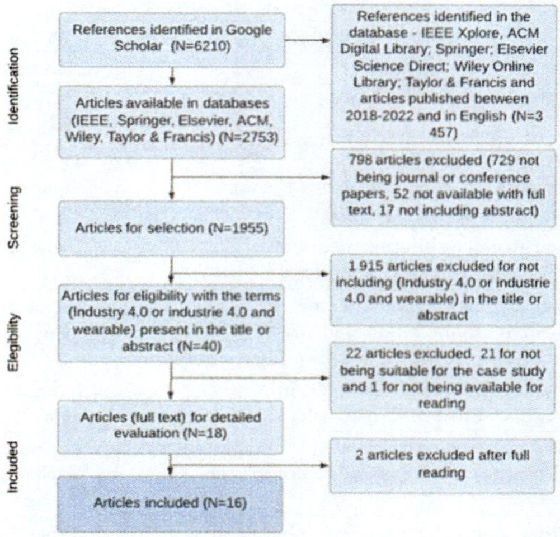


**Fig. 1.** Review methodology

### 3.2 Review Protocol

From the definition of the research query, the literature review protocol was guided by the formulation and execution of the following steps: Research Questions The following research questions were established to guide this work:

- RQ1: What are the main features of wearable devices in I4.0?
- RQ2: What are the main benefits and challenges of wearables in I4.0?
- RQ3: What applications currently exist for wearables in the industrial sector?

After formulating and refining the search query, it was entered into Google Academic. The review protocol and the implementation of the Inclusion and Exclusion criteria are presented in Fig. 1.

The resulting articles were checked, correcting references and duplicates. They were then fully read, extracting the information aimed at answering the research questions RQ1, RQ2 and RQ3, respectively.

## 4 Results and Discussion

After the systematic literature review, following the inclusion and exclusion criteria presented above, 16 articles were obtained. These articles are described and identified between P1 and P16 in Table 1.

### 4.1 What Are the Main Features of Wearable Devices in I4.0?

One of the most comprehensive definitions of a wearable is "wearable devices are smart electronic devices available in various forms that are used near or on the

**Table 1.** Papers analysed

| # | Title | Authors | Year |
|---|---|---|---|
| P1 [4] | A Novel MQTT Security Framework In Generic IoT Model | Chintan Patel, Nishant Doshi | 2020 |
| P2 [15] | Towards Assembly 4.0: Graduation Intelligent Manufacturing System for Fixed-position Assembly Islands | Daqiang Guo et al. | 2019 |
| P3 [16] | Promoting operator's wellbeing in Industry 5.0: detecting mental and physical fatigue | Valeria Villani et al. | 2022 |
| P4 [3] | Thermal discomfort in the workplace: measurement through the combined use of wearable sensors and machine learning algorithms | Silvia Angela Mansi et al. | 2022 |
| P5 [17] | Monitoring of shop-floor workers postural stability through the use of smart soles | Diogo Teixeira et al. | 2022 |
| P6 [18] | Acceptance and perception of wearable technologies: A survey on Brazilian and European companies | Gislene Cássia S. Schwambach et al. | 2022 |
| P7 [11] | Assessing mental workload with wearable devices - Reliability and applicability of heart rate and motion measurements | Sebastian Mach et al. | 2022 |
| P8 [19] | Employee acceptability of wearable mental workload monitoring: exploring effects of framing the goal and context in corporate communication | Bram B. Van Acker et al. | 2021 |
| P9 [5] | Smart operators: How Industry 4.0 is affecting the worker's performance in manufacturing contexts | Di Pasquale Valentina et al. | 2021 |
| P10 [20] | Smart Protective Protection Equipment for an accessible work environment and occupational hazard prevention | Márquez Sánchez et al. | 2020 |
| P11 [13] | A sensorized glove for industrial safety based on Near-Field Communication | Francesco di Rienzo et al. | 2020 |
| P12 [1] | State-of-the-Art of Industrial Wearables: A Systematic Review | Ekaterina Svertoka et al. | 2020 |
| P13 [6] | Industrial wearable system the human-centric empowering technology in Industry 4.0 | Xiang T. R. Kong et al. | 2019 |
| P14 [2] | An Industrial IoT Solution for Evaluating Workers' Performance Via Activity Recognition | Abdur Rahim Mohammad Forkan et al. | 2019 |
| P15 [21] | Line Balancing Assessment Enhanced by IoT and Simulation Tools | Marcello Fera et al. | 2019 |
| P16 [22] | Enabling visual community learning analytics with Internet of Things devices | István Koren, Ralf Klamma | 2018 |

human body to sense and analyze different types of data via applications either installed on the device or on external devices, such as smartphones connected to the cloud" [1]. The use of wearables in industry has been increasing due to their characteristics. They are easy to use, have low costs and several purposes, allowing the collection of information from workers and, consequently, contributing to the improvement of efficiency, well-being and reduction of risk of injuries. Another characteristic of these wearables is that they can collect a large amount of data continuously, and with these large amounts of data artificial intelligence algorithms can be used pre-emptively to, e.g., ascertain a person's physiological state, classify emotions or develop biometric recognition [3].

Wearables can be grouped into five classes according to their design in the industrial context: (i) product design ergonomics; (ii) data interaction in the device; (iii) operational stability; (iv) external software integration and (v) external hardware integration. These are further characterised by being devices that interact with humans and also allow data collection (smart glasses, smart watches, ring scanner, smart glove, among others). Another key concept of this type of wearables is the ability to provide the appropriate information to the worker in the form of image, voice or vibration therefore allowing the operator to acquire only the critical and necessary information in the simplest, quickest, and most secure way, to ensure that workers continue to perform their work [6].

The operator-machine-process dynamics is a central aspect of I4.0 [5,6,11, 18,21,22], which seeks to ensure flexibility, decentralisation, resource efficiency and digitalisation. Patel and Doshi [4] propose the use of the MQTT protocol, as it is a simple communication model that fits the characteristics of small computing capacity. On the other hand, it is also necessary to ensure communication between wearable devices and Enterprise Information Systems (EIS), where Enterprise Resource Planning (ERP), Material Requirements Planning (MRP) and Manufacturing Execution System (MES) are included. As worker and material mobility increases and processes become more complex, it will be beneficial for wearable technologies to cooperate with already existing emerging technologies, namely artificial intelligence, virtual and augmented reality, IoT and Cloud Computing [2,6]. IoT technologies enable continuous monitoring of equipment, remote control, and optimisation, while Cloud Computing allows for the storage of large volumes of data and enables the creation of advanced algorithms for future data analysis [2,13].

Wearable technologies have played a crucial role in the evolution of I4.0, shifting the focus of workers from manual labour to decision-making, while leaving physical work to machines. As a result, monitoring the cognitive and psychological well-being of work teams is becoming increasingly important, given the mentally demanding nature of their tasks [2,11]. To this end, the authors Bram B. Van Acker et al. [19] proposed the implementation of wearable mental workload (MWL), a concept that aims to assess and improve workers' physical and mental well-being. In this sense, factories will become more "intelligent", flexible and collaborative [21].

### 4.2 What Are the Main Benefits and Challenges of Wearable's Application in I4.0?

One of the main benefits of wearables implementation in the industry is that they substantially facilitate workers' tasks. However, the isolated implementation of wearables is insufficient in this field. The development of an operator-centred strategy that supports the interaction between the operator and the equipment and the technological aspect, designated as Industrial Wearable System (IWS) has been proposed [6]. Valeria Villani et al. [16] propose the designation of Industry 5.0 for this concept, which promotes the human-centricity of factories, with the operators' well-being the main factor. In fact, the cooperation between the collaborative robots developed in the context of I4.0 and the soft skills of the workers who operate them provides another level of flexibility, precision and profitability. By not needing to perform physically demanding tasks, they can focus in monitoring the production processes themselves and the maintenance of equipment operation [11,16]. In addition, another great advantage of wearables lies in the possibility of data connection between different devices, allowing the formation and monitoring of large networks in real time [2–4,11]. The authors in [5] make a distinction between traditional operators and what authors name 4.0 operators:

– Augmented and virtual operators: refers to operators using Augmented Reality or Virtual Reality. Augmented operators use this technology in their daily tasks, while virtual reality serves as an ideal simulation environment for training.

– Collaborative operator: collaborates with robots and/or machines to achieve higher productivity. Despite bringing several advantages, some concerns are raised by workers for fear of losing their jobs by being replaced by machines.

– Healthy and super-resistant operators: wearables are used to monitor operators' health in real time, adapting their decisions based on the information received from the devices. This data allows for greater worker safety, as well as preserving their health and being able to improve the ergonomics of the work environments.

In this context, the role of workers will include making decisions. Consequently, it is important to monitor and evaluate the worker mental workload to be as productive as possible. The authors in [11] refer that this monitoring may be possible through a wearable. However, despite all the advantages that these devices may bring to the evolution of the industrial sector, their use has also raised some concerns among the working class, particularly in terms of privacy and protection of the collected data, since they are under continuous monitoring, which in turn causes them a sense of discomfort and pressure [4,18,19,22]. In [18,19] this same concern is mentioned. The authors of both studies mention several theoretical methods to study the acceptance of wearables, namely the Unified Theory of Acceptance and Use of Technology 2 (UTAUT2), which is an adaptation of the original UTAUT, including new factors that affect technology acceptance, such as social influence and user experience. It was found in [19] that wearables are more easily accepted when proposed to improve health and the work environment itself. The opposite occurs when they are implemented with the purpose of assessing workers' efficiency and productivity. In addition to data security and privacy issues, the correct use of these devices must also be ensured [18].

### 4.3 What Applications of Wearables Currently Exist in the Industrial Sector?

In [15] a model to organise and optimise "assembly islands" is proposed, called Graduation Intelligent Manufacturing System (GiMS), which is inspired by graduation ceremonies and adapted to the context of I4.0 and Assembly 4.0, and consists of an intelligent manufacturing system aided by advanced technologies. Assembly 4.0 is characterised by the integration of new technologies into the assembly system. On the other hand, the analysis of the remaining papers [13,16,17,20,21] provided insights about three wearable applications in the industry.

**Health.** As previously mentioned, wearables have been widely applied to the sports and health areas, among many others. In the industrial context, the use of these devices is useful for monitoring workers' health, through the analysis of indicators such as body temperature and vital signs. In fact, in [16] the authors proposed the use of a chest band to monitor in real time the fatigue of workers. However, a major drawback pointed out by the authors is the impossibility to also assess mental fatigue using only this wearable and the data collected by it. Therefore, it will be necessary, in the future, to develop and implement other kind of wearables that allow collecting the necessary data for a more complete monitoring. Still on this health side, [17] work suggested the use of intelligent soles as a means of assessing workers' posture in an attempt to correct possible anomalies and ensure their well-being. On the other hand, the monitoring of vital signs through wearables, such as heart rate, seems to be still in a preliminary stage, however, the results obtained are promising and will allow, in the near future, to achieve improvements in this area [11].

**Security.** In addition to their usefulness in maintaining the well-being of workers, the implementation of wearables in industry has also helped companies overcome some safety-related challenges. According to the authors of [20], wearable technologies can be adopted in the development of smart Personal Protective Equipment's (PPEs) by implementing sensors that allow the equipment to respond in real time to the conditions presented to them. In their work, the authors propose the use of sensors present in a helmet and a belt that communicate with each other, in order to collect data that allow to predict, in future circumstances, situations of potential risk for the user. The authors of [13] have also worked in this field, developing a sensorised glove that can be exploited to monitor the worker's action and report dangerous situations in advance, in order to prevent accidents and injuries. This approach combines Near-Field Communication (NFC), IoT and Cloud Computing technologies to enable the collection and storage of a large volume of data, as well as algorithms capable of analysing it.

**Productivity.** Finally, in [21] it was concluded that workers' fatigue was not considered during the design and engineering phases of the workstation, which highlights the importance of the use of new technologies, including Industrial Internet of Things (IIoT) devices such as wearables, and thereby making changes in order to improve the productivity of production lines contributing to the emergence of smart factories.

## 5 Conclusions

Indeed, wearables have the potential to revolutionise I4.0, offering several advantages, namely increased production efficiency, improved worker safety and the ability to adapt to new technologies. However, their use in industrial environments is still infrequent, which highlights the need to make advances in this aspect. Among the difficulties for the adoption of wearables in industry are concerns with data security and privacy, the need for integration of other systems and equipment. However, with the progression of technology and the growing demand for more efficient and innovative solutions, it is possible that wearables will become more common in industrial environments. Hence, it is necessary to invest in the research and development of more robust devices suitable for use in industrial environments, as well as strategies to ensure data security and integration with other systems.

**Acknowledgments.** This work was supported by the R&D Project "Continental Factory of Future, (CONTINENTAL FoF) / POCI-01-0247-FEDER-047512", financed by the European Regional Development Fund(ERDF), through the Program "Programa Operacional Competitividade e Internacionalização (POCI) / PORTUGAL 2020", under the management of aicep Portugal Global Trade & Investment Agency.

## References

1. Svertoka, E., Rusu-Casandra, A., Marghescu, I.: State-of-the-art of industrial wearables: a systematic review. In: 2020 13th International Conference on Communications (COMM), pp. 411–415 (2020)
2. Forkan, A.R.M., Montori, F., Georgakopoulos, D., Jayaraman, P.P., Yavari, A., Morshed, A.: An industrial IoT solution for evaluating workers' performance via activity recognition. In: 2019 IEEE 39th International Conference on Distributed Computing Systems (ICDCS), pp. 1393–1403 (2019)
3. Mansi, S.A., Cosoli, G., Pisello, A.L., Pigliautile, I., Revel, G.M., Arnesano, M.: Thermal discomfort in the workplace: measurement through the combined use of wearable sensors and machine learning algorithms. In: 2022 IEEE International Workshop on Metrology for Industry 4.0 & IoT (MetroInd4.0 & IoT), pp. 54–59 (2022)
4. Patel, C., Doshi, N.: A novel MQTT security framework in generic IoT model. Procedia Comput. Sci. **171**, 1399–1408 (2020)
5. Salvatore, M., Stefano, R.: Smart operators: how Industry 4.0 is affecting the worker's performance in manufacturing contexts. Procedia Comput. Sci. **180**, 958–967 (2021)

6. Kong, X.T., Luo, H., Huang, G.Q., Yang, X.: Industrial wearable system: the human-centric empowering technology in Industry 4.0. J. Intell. Manuf. **30**(8), 2853–2869 (2019)
7. Lasi, H., Fettke, P., Kemper, H.G., Feld, T., Hoffmann, M.: Industry 4.0. Bus. Inf. Syst. Eng. **6**(4), 239–242 (2014)
8. Da Li, X., He, W., Li, S.: Internet of things in industries: a survey. IEEE Trans. Industr. Inf. **10**(4), 2233–2243 (2014)
9. Miorandi, D., Sicari, S., De Pellegrini, F., Chlamtac, I.: Internet of things: vision, applications and research challenges. Ad Hoc Netw. **10**(7), 1497–1516 (2012)
10. Ometov, A., et al.: A survey on wearable technology: history, state-of-the-art and current challenges. Comput. Netw. **193**, 108074 (2021)
11. Mach, S., Storozynski, P., Halama, J., Krems, J.F.: Assessing mental workload with wearable devices - reliability and applicability of heart rate and motion measurements. Appl. Ergon. **105**, 103855 (2022)
12. Shrouf, F., Ordieres, J., Miragliotta, G.: Smart factories in Industry 4.0: a review of the concept and of energy management approached in production based on the internet of things paradigm. In: 2014 IEEE International Conference on Industrial Engineering and Engineering Management, pp. 697–701 (2014)
13. di Rienzo, F., Virdis, A., Vallati, C., Carbonaro, N., Tognetti, A.: A sensorized glove for industrial safety based on near-field communication. In: 2020 IEEE International Conference on Smart Computing (SMARTCOMP), pp. 256–258 (2020)
14. Kitchenham, B.: Procedures for performing systematic reviews. Keele University, Keele, UK, 33(2004), 1–26 (2004)
15. Guo, D., et al.: Towards assembly 4.0: graduation intelligent manufacturing system for fixed-position assembly islands. IFAC-PapersOnLine **52**(13), 1513–1518 (2019)
16. Villani, V., Gabbi, M., Sabattini, L.: Promoting operator's wellbeing in Industry 5.0: detecting mental and physical fatigue. In: 2022 IEEE International Conference on Systems, Man, and Cybernetics (SMC), pp. 2030–2036 (2022)
17. Teixeira, D., Ferreira, J., Gonçalves, R.: Monitoring of shop-floor workers postural stability through the use of smart soles. IFAC-PapersOnLine **55**(10), 2234–2239 (2022)
18. Schwambach, G.C.S., López, Ó.H., Sott, M.K., Tedesco, L.P.C., Molz, R.F.: Acceptance and perception of wearable technologies: a survey on Brazilian and European companies. Technol. Soc. **68**, 101840 (2022)
19. Van Acker, B.B., Conradie, P.D., Vlerick, P., Saldien, J.: Employee acceptability of wearable mental workload monitoring: exploring effects of framing the goal and context in corporate communication. Cogn. Technol. Work **23**, 537–552 (2021)
20. Sánchez, M., Rodriguez, C., Manuel, J.: Smart protective protection equipment for an accessible work environment and occupational hazard prevention. In: 2020 10th International Conference on Cloud Computing, Data Science and Engineering (Confluence), pp. 581–585 (2020)
21. Fera, M., Greco, A., Caterino, M., Gerbino, S., Caputo, F.: Line balancing assessment enhanced by IoT and simulation tools. In: 2019 II Workshop on Metrology for Industry 4.0 and IoT (MetroInd4.0 & IoT), pp. 84–88 (2019)
22. Koren, I., Klamma, R.: Enabling visual community learning analytics with internet of things devices. Comput. Hum. Behav. **89**, 385–394 (2018)

# Towards Industry 5.0: A Capacitation Approach for Upskilling and Technology Transfer

Rui Pinto[1(✉)], João Pinheiro[1], Gil Gonçalves[1], and Anabela Ribeiro[2]

[1] SYSTEC, ARISE, Faculdade de Engenharia, Universidade do Porto, Rua Dr. Roberto Frias, 4200-465 Porto, Portugal
{rpinto,jmotp,gil}@fe.up.pt

[2] Continental Advanced Antenna, Sociedade Unipessoal, Lda., Rua Professor Anton Kathrein, Parque Industrial Constantim, 5000-082 Vila Real, Portugal
anabela.ribeiro@continental-corporation.com

**Abstract.** Considering the smart manufacturing context and the numerous opportunities to transform businesses, manufacturing companies are urging for options to improve current workforce skills by proper training in these new industrial technologies, enabling the desired digital transformation in shop-floor processes. On the other hand, together with the training is the need for technical consulting by leveraging disruptive scientific research results in real-world industrial applications. The Self-Made Program aims to deliver innovative training courses and tailored technical consulting to manufacturing companies. This will enable the exploitation of educational assets in topics such as Digitalization, Cyber-Physical Systems and the IEC 61499 standard. Finally, the program enables the experimentation and deployment of disruptive technologies in a Teaching and Learning Factory and in real-world environments to support the digital transformation of such companies.

**Keywords:** Industry 5.0 · Upskilling · Digital Twin · Self-Made · Digital Transformation

This work was supported by multiple funding sources including the: Project Continental FoF: Continental AA's Factory of the Future, with reference POCI-01-0247-FEDER-047512, co-funded by European Regional Development Fund (ERDF), through the Operational Programme for Competitiveness and Internationalization (COMPETE 2020), under the PORTUGAL 2020; and project "Self-Made: Self-management & device digitalization in manufacturing" that received funding from the European Institute of Innovation & Technology Manufacturing (EIT-M); and base funding (UIDB/00147/2020) and Programmatic funding (UIDP/00147/2020) of the SYSTEC - Center for Systems and Technologies and ARISE - Associate Laboratory for Advanced Production and Intelligent Systems (LA/P/0112/2020), both funded by national funds through the FCT/MCTES (PIDDAC).


R. Mehmood et al. (Eds.): DCAI 2023, LNNS 741, pp. 342–351, 2023.
https://doi.org/10.1007/978-3-031-38318-2_34

## 1 Introduction

Industry 5.0 (I5.0) is a relatively new concept that builds upon Industry 4.0 (I4.0). On the one hand, I4.0 is focused on the integration of digital technologies into industrial processes. I5.0, on the other hand, focuses on the integration of human intelligence and collaboration into the manufacturing process and seeks to find a balance between automation and human involvement [1].

I5.0 recognizes that certain tasks are better suited for automation, while others require human input and decision-making. By leveraging the strengths of both humans and machines, I5.0 aims to create more efficient and effective manufacturing processes that produce higher-quality products. I5.0 also emphasizes the importance of sustainability and social responsibility in manufacturing. This includes minimizing waste, reducing environmental impact, and creating products that are designed for recyclability and reuse.

In order to shift towards a more human-centred approach to manufacturing, where human skills and knowledge are combined with digital technologies to create innovative and sustainable solutions, three key pillars are often mentioned in discussions around I5.0 [12]:

- Human-centricity: I5.0 emphasizes the importance of human skills and knowledge in the manufacturing process, and seeks to find a balance between automation and human involvement. This involves designing manufacturing processes that can accommodate human creativity and decision-making, while still leveraging the efficiency and precision of machines.
- Sustainability: I5.0 places a strong emphasis on sustainability and social responsibility in manufacturing. This includes reducing waste, minimizing environmental impact, and creating products that are designed for recyclability and reuse.
- Resiliency: I5.0 emphasizes the importance of flexibility in the manufacturing process, allowing for quick adaptations to changing market demands and customer needs. This involves using technologies such as additive manufacturing and modular production to create more agile and responsive manufacturing processes.

I5.0 can be seen as an initiative for manufacturing companies to adopt I4.0, which tackles the necessity for improved collaboration between humans and technological systems. On the one hand, I5.0 emphasizes the importance of collaboration and communication between humans and machines, as well as between different human workers. This includes creating an environment where workers can share knowledge and ideas, and work together to solve problems. On the other hand, I5.0 recognizes the importance of digital technologies in manufacturing, and seeks to leverage these technologies to support the pillars mentioned before. This includes the use of technologies such as Cyber-Physical Systems (CPS), Internet of Things (IoT), Machine Learning (ML), and automation standards, such as the IEC 61499 [18], to create more efficient manufacturing processes.

Thus, there is a phenomenon where two industrial revolutions are currently ongoing. While I4.0 deals with digital transformation and the operationalization of new intelligent systems in manufacturing, I5.0 focuses on rethinking what engineering education should be, considering the skills required for future manufacturing professionals [11].

As a result, some key aspects of the human-centricity pillar of I5.0 include augmentation, education and training, workplace design and employee engagement. I5.0 recognizes the importance of educating and training workers to develop the skills and knowledge necessary to work alongside digital technologies. This includes training workers to operate and maintain machines, as well as to develop the skills necessary to work collaboratively with machines [5].

In this study we propose a methodology for blended education and training. This methodology is materialized in a training program offered to Small And Midsize Enterprises (SMEs), designated Self-Made Program, to assess their current digital maturity, and identify the required steps and knowledge needs to accelerate their digital transformation. This is done in a joint consulting and training journey with the individual SMEs, implemented through a series of co-design digital disruption workshops, where needs are identified and plans for the capacitation of the organization are defined. The Self-Made Program focuses, on the one hand, on pedagogical topics such as Digitalization, CPS design and creation using a Model-based Engineering (MBE) approach to foster SMEs' digital transformation. On the other hand, it focuses on an experimentation ecosystem, where Teaching and Learning Factories (TLF) infrastructures are used for hands-on and real-world experiences, to close the gap between the Universities/research centres and the manufacturing SMEs.

The paper is organized into three more sections. Section 2 provides the technological context and literature review on education and training programs for workforce upskilling. Section 3 describes in detail the training program offered, more specifically the training component and the TLF available for experimentation. Finally, Sect. 4 concludes the paper, stating final remarks about the study performed while further discussing the proposed approach.

## 2 Literature Review

Education and training in upskilling involve providing workers with the skills and knowledge necessary to work alongside digital technologies in the manufacturing process. This includes both technical training to operate and maintain machines, as well as soft skills training to develop the social and creative intelligence necessary. As I4.0 continues to transform the way we work, education and training usually target key competencies that will be important for workers in all industries, such as digital literacy, analytical thinking and problem-solving, adaptability and flexibility, collaboration and communication, continuous learning, emotional intelligence and entrepreneurial mindset. To develop the competencies needed for the future workforce in the context of I4.0, organizations can use a variety of training methodologies, such as:

- Online training: Online training can be a cost-effective way to deliver training on digital tools and technologies, such as data analytics, artificial intelligence,

and robotics. These training programs can be delivered through Learning Management Systems (LMS) or other online platforms.
- Blended learning: Blended learning combines online and in-person training, which can be particularly effective for developing interpersonal skills like collaboration and communication.
- Hands-on training: Hands-on training can be particularly effective for developing skills related to machinery and equipment. This type of training can be delivered through apprenticeships, on-the-job training, or simulation-based training programs.
- Experiential learning: Experiential learning involves learning through real-world experiences, such as internships, job shadowing, or project-based learning. This approach can be particularly effective for developing skills related to problem-solving, adaptability, and entrepreneurship.
- Microlearning: Microlearning involves delivering training in short, bite-sized modules that can be easily consumed on the go. This approach can be particularly effective for developing skills related to digital literacy, as it allows workers to learn new tools and technologies at their own pace.
- Gamification: Gamification involves using game-like elements, such as points, badges, and leaderboards, to motivate learners and reinforce learning. This approach can be particularly effective for developing skills related to continuous learning, as it encourages workers to engage in ongoing learning and development.

A combination of these training methodologies can be used to design training programs that are tailored to the specific needs of the workforce and that provide opportunities for hands-on practice and real-world application of skills. Blended learning can be particularly effective for developing interpersonal skills like collaboration, communication, and emotional intelligence, as well as for reinforcing learning related to digitalization and technologies, such as Digital Twin (DT), CPS and the IEC 61499 standard. These training programs can be offered by universities or technology-provider companies.

Siemens Digital Industries Academy [16] is a blended learning approach offered by Siemens to help learners develop competencies related to digitalization and I4.0. Festo Didactic [9] also offers a similar blended learning approach to develop competencies in topics such as (DT) and IEC 61499 standard. Their approach offers online courses, classroom training, e-learning modules, simulation-based training and hands-on labs to help learners develop skills.

The Institute for Manufacturing (IfM) [4] at the University of Cambridge and the Sustainable Manufacturing and Advanced Robotic Technologies at the University of Buffalo [3] offer similar blended learning approaches to help learners develop competencies related to digitalization, DT and MBE. Their approach includes online courses, in-person workshops, e-learning modules, virtual coaching and group projects to provide learners with a range of learning experiences.

The existing blended learning approaches consider only a training component, failing to address the need for technical consulting by leveraging disruptive scientific research results in real-world industrial applications, especially in the user's own industrial context. They also lack user support for technology experimentation in TLFs, allowing users to stay up-to-date with the latest technologies and techniques related to digitalization. This consulting component would support the facilitation and intermediation of an innovation ecosystem, where SMEs would benefit from technology prospecting and transference.

## 3 Self-Made Program

The Self-Made Program [15] is a European initiative that leverages a roadmap for SME digital transformation, including learning paths to support training and technology transfer. This program is framed in the Self-Made project, financed by EIT-M, which relies on a 6-partner consortium from 5 different hubs, Portugal, Finland, Czech Republic, Lithuania and France. The roadmap consists of 5 main phases.

First, the SME technological maturity is assessed in topics related to digitalization, which will lead to phase 2, the execution of online training actions. Online training relies on Skills.move [8] LMS offered by EIT-M to manage microlearning and learning path outline. Phase 3, focuses on TLF and onsite training by providing users with hands-on experience in the manufacturing industry. They can test technologies in a controlled environment, to be later deployed in their own manufacturing environments to refine and improve production processes.

After the training, phases 4 and 5 are related to technological consulting, technology transfer and adoption. This will have a very relevant contribution to the digital transformation of SMEs and will allow solution scale-up and new digital products. Phase 5 also includes an assessment of the training actions execution and the return of investment from the SME perspective.

### 3.1 Skills and Training

The pedagogical offer for the training component in the Self-Made program lays down four learning paths in two main pillars, as represented in Fig. 1. In the *Digitalization & Human-Centric Manufacturing* pilar, we offer the *Cyber-Physical Production Systems* and *Human-Machine Interaction* learning paths. On the other hand, in the *Data-driven Decision Making* pilar, we offer the *Equipment Life Cycle Assessment* and *Tools to Support Managerial Decision-Making* learning paths.

In the Portugal Hub, the focus is to deliver the *Cyber-Physical Production Systems* learning path that includes topics such as DT and the IEC 61499 standard for CPS design and creation using an MBE approach. This is a 45-hour training program, with 15 h of contact time, where users will address mainly the DINASORE framework [7,13] and Function Blocks (FB)-oriented programming.

The main purpose of the DINASORE framework is to act as a doorway to the incorporation of advanced analytics and data processing into a CPS, based on the IEC 61499 standard. DINASORE enables simple construction of FBs and Pipelines using the *Python* coding language. Recent trends reveal that *Python*

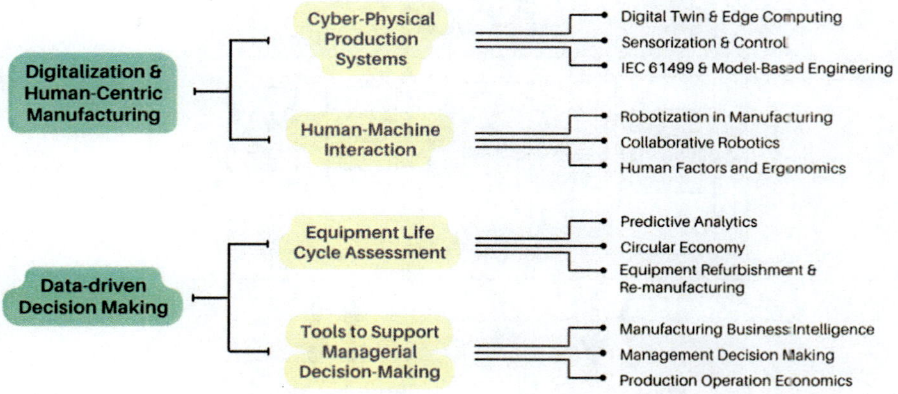


**Fig. 1.** Training component of the Self-Made Program.

is one of the most commonly used programming languages for constructing ML applications since it supports multiple frameworks for data science The aim behind adopting *Python* in DINASORE is to allow the most recent advancements in ML to be integrated at the Edge layer with current shop-floor equipment. Figure 2 depicts the general DINASORE architecture.

The general DINASORE architecture includes Edge devices that run DINASORE instances and the 4DIAC, which is used by a user to create FBs Pipelines and deploy the new configuration to all devices in the CPS. This means that DINASORE can be used to create digital models of a factory shop floor, enabling data collection, sensor integration, standardized connectivity and all the features that Python makes available, such as data science modules.

In this sense, the *Cyber-Physical Production Systems* learning path introduces users to the concepts and applications of digitalization in industry and DT technologies, illustrating how these concepts are used to increase the efficiency, adaptability and resilience of modern industrial production systems. The main goal is to develop knowledge in topics of DT and Edge Computing, sensorization, the IEC 61499 standard for automation control and MBE approaches for system design and implementation, using the DINASORE framework. As learning outcomes, at the end of the learning path, the users can:

1. Apply IEC 61499 for the design and development of CPS.
2. Create Function Blocks using the DINASORE framework.
3. Demonstrate how to develop and operate a Digital Twin in an Edge device.

### 3.2 Experimentation Ecosystem

The experimentation component consists of a European ecosystem of higher education institutions and research & technology organizations that concentrate and facilitate the dissemination of knowledge and technology transfer. This ecosystem intends to provide state-of-the-art TLFs infrastructures that enable the

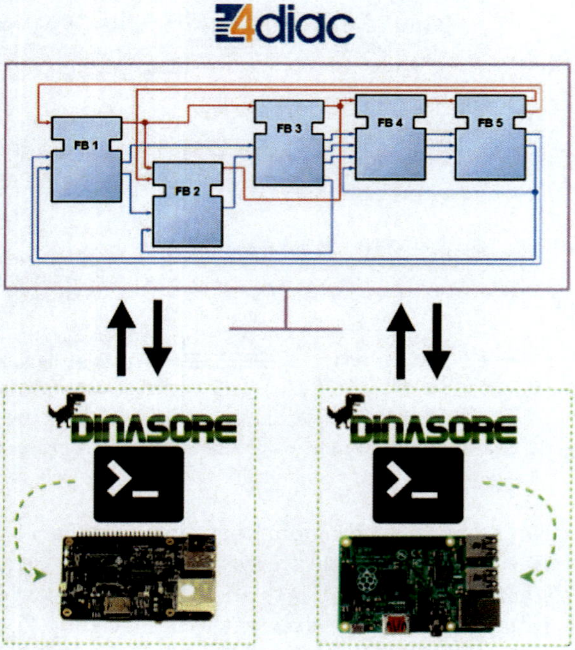


**Fig. 2.** General DINASORE architecture.

development and application of disruptive technology and new systems under the topics mentioned before, aiming for the digital transformation of SMEs.

In the Portugal Hub, there are two TLF, represented in Fig. 3, which target the experimentation with systems related to the design and creation of CPS, relying on the DINASORE framework and FBs-oriented programming for Pipeline deployment. The TLFs emulate sequential production lines where a user at a given workstation executes an operation. In these production lines, it is considered that, between workstations, parts can either move forward or be discarded. The production lines comprise several processing stations through which the parts must go through to get a finished product.

In the first TLF, represented in Fig. 3a, there are 3 workstations and users interact with the system through industrial button consoles. At each workstation, the user signals the beginning and end of the task execution and whether the task was successfully achieved or not. In this TLF, participants are asked to create a FB Pipeline for automatic data logging, equipment cycle time calculation, based on the button signals, and data representation in real-time in a dynamic Yamazumi chart [2,14].

In the second TLF, represented in Fig. 3b, there are 5 sequential workstations using the FESTO Modular Production System [10,17], emulating distribution, testing, processing, transporting and sorting tasks. We have also included a robotic arm in the transporting station, and a vision system to analyse prod-

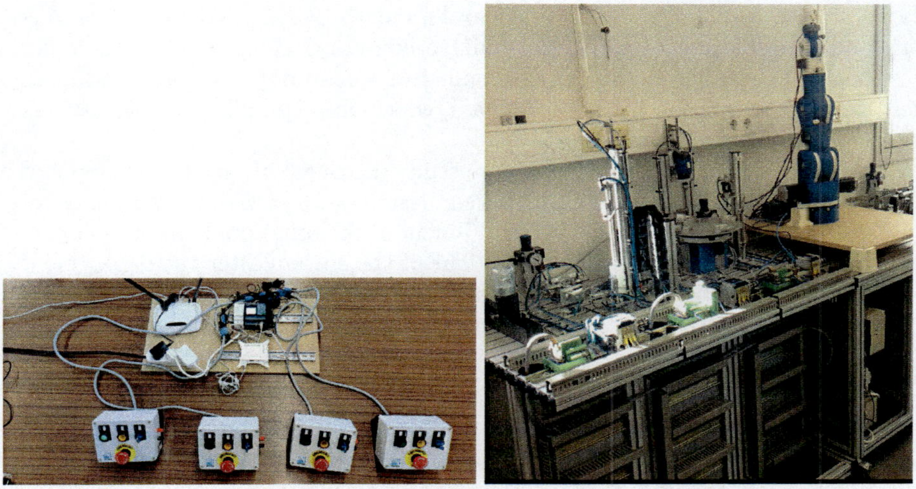

(a) Cycle time analysis and line optimization. (b) Festo MPS for system automation.

**Fig. 3.** TLF in Portugal Hub.

uct quality. The entire process is automatic, thus raw materials are fed into the first workstation to be processed, and a finished piece is expected at the end. Participants are asked to create a FB Pipeline that manages the M2M communication between workstations and integrates external sensors in the production line. Finally, data processing can be done in the Edge layer, using the sensor data for decision-making and actuator trigger.

An early validation of the Self-Made Program was achieved using the TLFs available in the Portugal Hub. Actually, the TLF represented in Fig. 3 takes inspiration from an actual shop floor case study. The tackled manufacturing company is Continental Advanced Antenna (CAA) [6], a car antenna manufacturer in Vila Real, Portugal. CAA production process includes Printed Circuit Boards (PCB) and electronic component assembly, which consists mainly of semi-automatic tasks, such as the insertion of cables, the screwing of different components, or the coupling of the module electronic connection to the plastic structure of the antenna.

At least two workshops were organized involving engineers project managers and IT technicians working in the field of industrial automation from CAA. The participants were asked to consume the available materials in Skills.move asynchronously, where they learn what the DINASORE framework is, how to install it, and how to develop FB Pipelines. Later, the participants were able to experiment with the technology in the TLFs facilities, by developing and deploying FB Pipelines.

Regarding assessment, a combination of methods was used to provide a well-rounded assessment of learners' understanding and competencies. There were online quizzes in Skills.move to assess learners' understanding of the key concepts related to IEC 61499, the design and development of CPS, DINASORE framework and the creation of FBs. The quizzes contained single and multi-choice questions. The grading method corresponds to the sum of grades achieved in all questions. There is a specific grade to pass, which corresponds to a 55% ratio of correct answers in the quiz.

Then, considering the cycle time case study, learners' ability to develop and deploy a FB Pipeline to manage the system described in each TLF was assessed. The assessment method considers the fulfilment of the functional requirements in each TLF, based on the empirical behaviour of the system after the deployment of a new FB Pipeline. This allowed participants to demonstrate their ability to analyze complex situations, relate different concepts and apply their knowledge to develop a CPS solution, considering the IEC 61499 standard and the DINASORE framework.

## 4 Conclusions

Within the I5.0 paradigm, where IoT and CPS technologies are perhaps the main innovations for digital business and the economic growth of the industrial fabric in recent years, several jobs will change and the overwhelming majority of today's employees will need to develop new skills. We propose a blended learning approach methodology to encourage the creation of unique pedagogical conditions, which integrate several scientific areas that are now highly desirable in the profiles of professionals in the industry.

Professionals working in the manufacturing industry who are interested in leveraging digitalization to improve production processes and optimize operations can benefit from the Self-Made Program. This program tackles both training and technical consulting, allowing participants to acquire knowledge, and apply this knowledge in real-world industrial contexts. Moreover, participants are supported later in the technology transfer and adoption. In the end, the trainee will be familiar with the new technologies that transform the future of intelligent production, as it involves areas of knowledge transversal to computer engineering, especially the digital transformation of modern industrial production systems using the DINASORE framework.

## References

1. Alexa, L., Pîslaru, M., Avasilcăi, S.: From Industry 4.0 to Industry 5.0–an overview of European Union enterprises, pp. 221–231. Springer, Singapore (2022). https://doi.org/10.1007/978-981-16-7365-8_8
2. Barbosa, V., Pinto, R., Pinheiro, J., Gonçalves, G., Ribeiro, A.: Towards a digital twin simulation for cycle times analysis in a cyber-physical production system. In: International Conference on Simulation and Modeling Methodologies, Technologies and Applications (SIMULTECH) (2023)
3. Buffalo, U.: Sustainable manufacturing and advanced robotic technologies (2023). https://www.buffalo.edu/smart.html. Accessed April 2023

4. Cambridge University: IFM engage (2023). https://engage.ifm.eng.cam.ac.uk/courses/. Accessed April 2023
5. Carayannis, E.G., Morawska-Jancelewicz, J.: The futures of Europe: Society 5.0 and Industry 5.0 as driving forces of future universities. J. Knowl. Econ. **13**(4), 3445–3471 (2022)
6. Continental: Advanced antenna solutions (2023). https://www.continental-automotive.com/Passenger-Cars/Vehicle-Networking/5G-Connectivity-Solutions/Advanced-Antenna-Solutions-(1). Accessed April 2023
7. DIGI2-Lab: Digi2-feup/dinasore. https://github.com/DIGI2-FEUP/dinasore
8. EIT: Skills.move (2023). https://www.skillsmove.eu/my/. Accessed April 2023
9. FESTO: Learning systems (2023). https://www.festo.com/us/en/c/technical-education/learning-systems/industrial-automation-and-industry-4-0-id_FDID_01_02/. Accessed April 2023
10. FESTO: Mps - mechatronics learning factories (2023). https://www.festo.com/us/en/e/technical-education/educational-concepts/highlights/learning-factories/mps-mechatronics-learning-factories-id_31963/. Accessed April 2023
11. Gürdür Broo, D., Kaynak, O., Sait, S.M.: Rethinking engineering education at the age of Industry 5.0. J. Ind. Inf. Integr. **25**, 100311 (2022). https://doi.org/10.1016/j.jii.2021.100311
12. Leng, J., et al.: Industry 5.0: prospect and retrospect. J. Manuf. Syst. **65**, 279–295 (2022). https://doi.org/10.1016/j.jmsy.2022.09.017
13. Pereira, E., Reis, J., Gonçalves, G.: Dinasore: a dynamic intelligent reconfiguration tool for cyber-physical production systems. In: Eclipse Conference on Security, Artificial Intelligence, and Modeling for the Next Generation Internet of Things (Eclipse SAM IoT), pp. 63–71 (2020). http://ceur-ws.org/Vol-2739/#paper_9
14. Pinheiro, J., Pinto, R., Gonçalves, G., Ribeiro, A.: Lean 4.0: a digital twin approach for automated cycle time collection and Yamazumi analysis. In: 3rd International Conference on Electrical, Computer, Communications and Mechatronics Engineering (ICECCME) (2023)
15. Self-Made: Self-made program (2023). https://selfmadeeit.com/self-made-program/. Accessed April 2023
16. Siemens: The future of learning starts now (2023). https://new.siemens.com/us/en/products/services/digital-enterprise-services/training-services/sitrain.html. Accessed April 2023
17. Thramboulidis, K.: Design alternatives in the IEC 61499 function block model. In: 2006 IEEE Conference on Emerging Technologies and Factory Automation, pp. 1309–1316 (2006). https://doi.org/10.1109/ETFA.2006.355411
18. Vyatkin, V.: The IEC 61499 standard and its semantics. IEEE Ind. Electron. Mag. **3**(4), 40–48 (2009). https://doi.org/10.1109/MIE.2009.934796

# Special Session on Artificial Intelligence for Enhanced Cyber Security (AI4CS)

# TestLab: An Intelligent Automated Software Testing Framework

Tiago Dias(✉), Arthur Batista, Eva Maia, and Isabel Praça

Research Group on Intelligent Engineering and Computing for Advanced Innovation and Development (GECAD), Porto School of Engineering (ISEP), 4200-072 Porto, Portugal
{tiada,1171006,egm,icp}@isep.ipp.pt

**Abstract.** The prevalence of software systems has become an integral part of modern-day living. Software usage has increased significantly, leading to its growth in both size and complexity. Consequently, software development is becoming a more time-consuming process. In an attempt to accelerate the development cycle, the testing phase is often neglected, leading to the deployment of flawed systems that can have significant implications on the users daily activities. This work presents TestLab, an intelligent automated software testing framework that attempts to gather a set of testing methods and automate them using Artificial Intelligence to allow continuous testing of software systems at multiple levels from different scopes, ranging from developers to end-users. The tool consists of three modules, each serving a distinct purpose. The first two modules aim to identify vulnerabilities from different perspectives, while the third module enhances traditional automated software testing by automatically generating test cases through source code analysis.

**Keywords:** Automated Software Testing · Artificial Intelligence · Testing Framework

## 1 Introduction

Software systems have become a crucial asset from a professional and personal standpoint. What was previously considered superfluous has now become much more ingrained to a point where it has become a dependency. Consequently, avid users have become compelled to trust these systems, which can be dangerous if they are faulty. The trustiness of software directly relates to its quality [22]. Higher software quality is typically associated with more reliable and secure systems, which in turn boosts users' confidence in interacting with them. The quality and effectiveness of software testing directly impacts the security of the software application. Incomplete or insufficient testing can leave vulnerabilities undetected, compromising the overall security of the system. Therefore, it is crucial to ensure the quality of software used in real-world tasks. One way to achieve this is through software testing, which plays a vital role in the software development cycle by enhancing the quality of the software.


R. Mehmood et al. (Eds.): DCAI 2023, LNNS 741, pp. 355–364, 2023.
https://doi.org/10.1007/978-3-031-38318-2_35

Software testing focuses on evaluating software to determine its correctness and ensure it meets the project requirements. The main objective of software testing is to identify defects within the software, that when correctly addressed may improve the software's quality, reliability, and longevity [22]. It consists of executing the Software Under Test (SUT) and monitoring it for errors, bugs, and other issues, and is usually performed by developers or testers. Despite the availability of automation and management tools, software testing is often skipped or rushed due to its time-consuming nature and the perceived increase in development costs. However, the long-term cost associated with the lack of testing and subsequent issues can also be significant.

According to recent reports, the total cost of poor software quality in the United States (US) was around 2 trillion dollars [11]. Additionally, the National Institute of Standards and Technology (NIST) found that an average bug detected in early development stages takes approximately five hours to be fixed, whereas, in a post-product release, it takes around 15.3 h. Therefore, in addition to improving the quality of the Software, testing decreases development time and costs. Regarding the use of automated testing tools, 44% of Information Technology companies automated 50% of testing in the year of 2020, with 24% seeing an increase in Return Of Investment [3]. Additionally, automated tools have proven to be helpful in uncovering vulnerabilties often missed by testers because of their randomness [22]. Despite the integration of automated software testing tools as part of the software testing process, during the first quarter of 2022, over 8000 vulnerabilities were discovered and documented [1].

The concerning number of software defects recently reported, underscore the need for improved software testing processes. As such, this work introduces the TestLab framework, an automated software testing tool designed to address multiple levels of software testing from different scopes ranging from developers to end-users. The framework aims to ensure the delivery of high-quality software systems throughout the software development cycle. By covering testing from different perspectives, TestLab enhances the efficiency and effectiveness of the software testing process, ultimately improving the overall quality of software systems.

This paper is organized as follows. Section 2 reviews the fundamentals of software testing and related work presented in recent literature. Section 3 describes the conceptualization of the proposed framework. Finally, Sect. 4 summarises and concludes the work, describing new research lines to be explored in the future.

## 2 State of the Art

Software testing is the verification and validation process of software, whose goal is to answer two respective questions: (i) is the right system being built?, and (ii) is the system being rightfully built? [21]. The validation question is answered by the product owner and end-users that validate the user requirements and the satisfaction of the developed solution, respectively [22]. The verification question

**Fig. 1.** Software Testing Process Model

can be answered by ensuring, to a certain extent, that the software was correctly built and tested.

In this sense, software testing plays a critical role in the software development process, as it usually appears as the phase that evaluates the correctness, completeness and accuracy of the developed work, revealing its defects. Nonetheless, compliance with best engineering practices is essential for ensuring the testability of software under development. Testability, which refers to the ease of testing a system, is a critical software quality attribute. Neglecting testability, along with insufficient testing, can lead to the failure of major projects. Therefore, prioritizing testability and conducting comprehensive testing are crucial for delivering high-quality software systems [23]. Errors and defects raised by software testing should not be disregarded nor considered an adversity towards the software being developed [23]. Software testing involves the creation of test cases, which are executed using test data to generate testing reports, as depicted in Fig. 1. However, this process can be arduous and time-intensive, particularly for large and intricate systems under test. Hence, automation is necessary to efficiently ensure software quality while minimizing the resources required [16].

Automated Software Testing (AST) can be applied in many different perspectives and at various levels, with the goal of testing in different ways multiple parts of the system. AST is often divided in three different types: (i) black-box, (ii) white-box, and (iii) grey-box. The former is used to evaluate the SUT from a end user's perspective by providing only information regarding the interaction with the system, without revealing its inner-workings [13,15]. White-box testing evaluates the SUT from the developers point of view by examining the internal structure and logic of a software program [12,15,18]. Grey-box testing combines the former types described, leveraging from their techniques to design more effective and efficient test cases [17]. Moreover, AST can be applied in multiple testing levels ranging from fine-grained testing, where the system is tested in units, to more general testing simulating the actions of a user. Testing a system in multiple levels is relevant, as the lack of it can jeopardise the quality of the software being developed [7]. The levels of testing are the following [23]:

1. **Unit Testing** - The system is broken down into units, and these are tested independently.
2. **Integration Testing** - The integration between units is tested to ensure it is defect-free.
3. **System Testing** - The system is tested in its entirety, focusing on multiple testing aspects inherent to publicly accessible systems.
4. **Acceptance Testing** - The last level of testing, which is performed by the end-users, reflects the acceptability of the system.

Testing a system in multiple levels ensures the correctness of the system at different granularity levels. Though, combined with multiple testing types, the produced test cases can simulate testing performed from different perspectives, which will ultimately provide different insights regarding the software's quality.

### 2.1 Related Work

The ever-growing interest, size and complexity of software systems have increased the testing effort. As a result, this process is many times skipped, in order to fasten the software development cycle. However, the release of faulty software can also produce unpleasant outcomes, creating an urge to invest in AST. Nowadays, most used and well-known AST tools rely on scripts to automate the execution of test scripts and the generation of reports to deliver faster and less prone to error testing. Additionally, they may even be integrated into Continuous Integration/Continuous Delivery pipelines to ensure the quality of the software under development throughout its development. Nonetheless, these tools still require human effort to write the test cases into the scripts and to function as oracle [21]. In this sense, the integration of Artificial Intelligence (AI) in the software testing process is promising, as it may be capable of fulfilling the human tasks that are found in the current state of AST.

Table 1 presents a comprehensive comparison of the most common AST tools based on the findings of multiple recent reviews [6,9,10,20,24], describing the license, testing levels considered and if they leverage AI.

Even though these tools are capable of automating software testing, they still require some degree of human expertise during the test case generation, as their use of AI is merely for assistance rather than automation. Moreover, most of them focus on high-level testing of the SUT, performing only at the system testing and acceptance testing levels.

Regarding the integration of AI in the AST process to automate the test case generation, multiple works have been developed and reviewed [4,5,25]. Vitorino et al. [25] present the challenges associated with the automatic test case generation faced in different testing type conditions. Their systematic review concludes that in a white-box setting most approaches resort to metaheuristics to constraint the input generation in order to achieve different test cases. Whilst in a black-box or grey-box setting, since the lack of data is the root problem,

**Table 1.** Common automated software testing tools.

| Tool and Frameworks | License | Testing Levels | AI-Aided |
|---|---|---|---|
| Selenium | Open Source | System Testing and Acceptance Testing | Yes |
| Unified Functional Testing | Licensed | System Testing | Yes |
| Test Complete | Licensed | System Testing and Acceptance Testing | Yes |
| Ranorex | Licensed | System Testing and Acceptance Testing | Yes |
| Watir | Open Source | System Testing | No |
| Load Runner | Licensed | System Testing | Yes |

most approaches rely on the information regarding the SUT provided by the testers. Vinicius et al. [5] reviews multiple works that achieve successful results using Machine Learning (ML) to further automate software testing, focusing on test case generation, refinement and evaluation, with some works mentioning the necessity for explainable algorithms in this context.

In conclusion, the findings show that even though there has been recent development towards software testing automation resorting to more intelligent methods, no prior work has addressed the development of a comprehensive framework comprising multiple testing methods that operate at various levels and in different settings (e.g. white-box, grey-box, and black-box). Such a framework would be capable of performing fully automated continuous testing of a SUT while integrating and analyzing the results from different testing methods.

## 3 Proposed Framework

TestLab is an automated and intelligent software testing framework that comprises multiple testing methods at different levels and various perspectives, such as white-box, black-box, and grey-box. This paper presents TestLab and its automated testing methods.

TestLab consists of three modules, as depicted in Fig. 2: FuzzTheREST, VulnRISKatcher, and CodeAssert. The first two aim to automate the process of analyzing the SUT for potential vulnerabilities in order to verify the software's security. FuzzTheREST is a fuzzer that focuses on testing Representational State Transfer (REST) Application Programming Interfaces (APIs) from a black-box point of view, where only data unrelated to the system's inner workings are utilized by the testing tool. VulnRISKatcher is a ML-based tool that has the capability to test source code developed in various programming languages in order to identify vulnerabilities. Lastly, CodeAssert is the part of the framework that attempts to automate the test scripts writing process by analyzing the source code in a white-box fashion. The subsequent subsections provide a more detailed description of each of these modules.

**Fig. 2.** TestLab architecture.

### 3.1 FuzzTheREST

As previously described in Sect. 2, software testing is composed of multiple methods capable of evaluating the quality of different software characteristics. Fuzzy testing is an automated software testing technique that consists in generating random malformed input and testing it against the SUT to find vulnerabilities and code defects [19]. This technique is usually applied in a black-box or grey-box setting leading to different input generation techniques. Typically, black-box fuzzers are quite naïve, as their testing input is randomly generated. This makes the vulnerability discovery an indefinitely time consuming task, rendering this kind of fuzzer unfeasible considering the size and complexity of modern day software systems [8]. On the other hand, grey-box fuzzers leverage existing information regarding the SUT's inner workings. Even though these fuzzers are usually capable of computing faulty input faster, they require more in-depth knowledge about the SUT.

FuzzTheREST is a fuzzer that performs intelligent fuzzy testing on REST APIs in a black-box fashion, with the goal of finding software defects. The fuzzer proposed relies on Reinforcement Learning (RL) to solve the search problem associated with discovering the right input values and combinations to uncover vulnerabilities in the software. The goal of the tool is to find vulnerabilities in web APIs in a platform-independent way, hence the need for the black-box approach. To tackle the naïveness associated with black-box fuzzers, the tool is equipped with a RL algorithm capable of driving the input that is generated relying on multiple mutation methods and using the API's feedback as guidance. Additionally, the tool can be fine-tuned to include grey-box information by integrating code execution analysis tools to obtain real-time execution metrics, such as coverage.

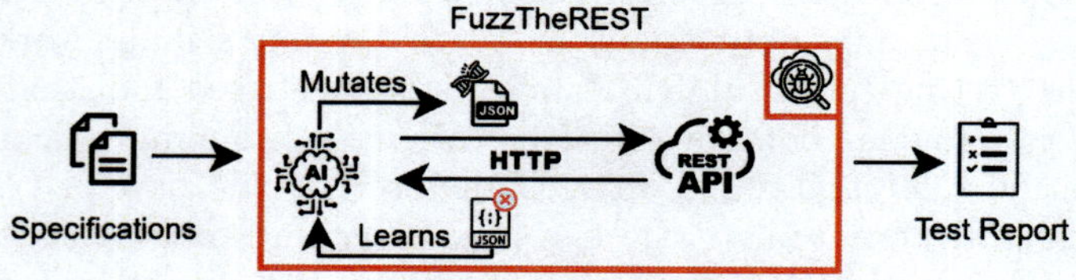


**Fig. 3.** FuzzTheREST pipeline.

Figure 3 describes the execution flow of the system. The tester provides all necessary information regarding the SUT, which includes the OpenAPI specification file, the scenarios file and the parameters of the RL algorithm. The OpenAPI specification file containing information to establish interaction with the target API, and the scenarios file describing user-defined combinations between functionalities to find vulnerabilities that may result of state changes. Afterwards, the tool starts fuzzying the API under test, iterating through each user-defined scenario. At the beginning of each scenario, a RL algorithm is instantiated with the chosen parameters and the initial population composed of the necessary input values is randomly generated. The fuzzer then starts sending the requests to the API to evaluate the generated input, using its feedback to mutate it accordingly. Throughout the fuzzy testing process, the algorithm records vulnerabilities, generated input values, and respective mutations to balance exploration and exploitation. This balance ensures that the search for faulty input values does not get stuck in a suboptimal solution (*local optimum*) allowing for a more efficient exploration of the solution space. After testing all scenarios, the fuzzer exports a test report file containing multiple metrics and the vulnerabilities found.

### 3.2 VulnRISKatcher

Every year, the number of vulnerabilities found contaminating software systems is rising, consequence of the ever-growing digital transformation. Their early detection and identification is crucial to ensure quality of the software and to significantly reduce the software development and maintenance costs.

Currently, code review tools mainly rely on static approaches and are typically introduced at the system test level. However, these tools are often less effective and may require a comprehensive understanding of the code context, resulting in a complex and time-consuming process [2]. As mentioned in Sect. 2, the potential of ML in software testing is significant [14], as it can overcome the limitations of static code review tools. By analyzing software metrics such as code complexity and code coverage, ML algorithms can learn patterns in the code that may suggest potential vulnerabilities, thereby improving the accuracy and efficiency of identifying them.

VulnRISKatcher is designed to accurately identify and classify vulnerabilities in source code and their associated risks. This tool uses ML techniques for code verification, which do not require the full context of the code, making it applicable at different levels of testing. VulnRISKatcher utilizes multiple ML techniques and leverages lexical code from diverse programming languages to train the models. This approach enables the solution to identify and classify various types of vulnerabilities accurately.

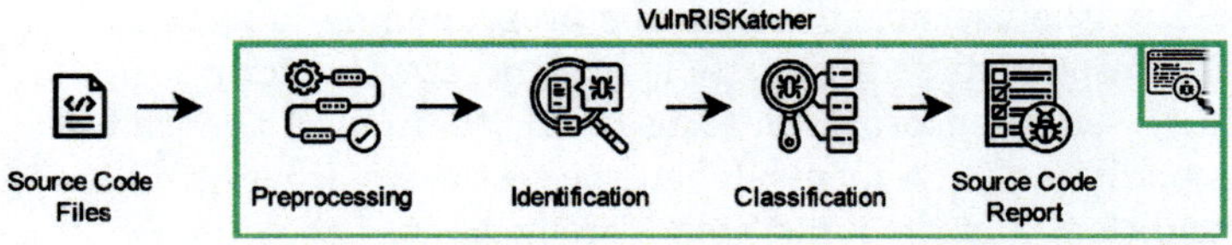


**Fig. 4.** VulnRISKatcher proposed pipeline.

Figure 4 illustrates pipeline of VulnRISKatcher. The user provides either all or part of the code, along with the programming language for accurate data processing. Following this, the tool will preprocess the input data by cleaning the code and converting it into discrete units. The preprocessing stage entails breaking down the code provided by the user into smaller units that can be analyzed. The preprocessed code is then analyzed to identify patterns and characteristics associated with vulnerabilities. Then, a classification of the potential vulnerabilities in the provided code is performed. Finally, a source-code report identifying and classifying the code vulnerabilities is presented to the user.

### 3.3 CodeAssert

The most commonly utilized AST tools attempt to automate the software testing process via scripting, as described in Sect. 2. These scripts define multiple tests which attempt to validate a software at multiple levels. However, they still require the manual effort of analyzing the code to create test cases, test data, and the expected results based on the tester's understanding of the SUT.

CodeAssert builds on top of the traditional AST process model by adopting the tools that already use scripting as their AST method. However, it further expands this model by automating the generation of the test scripts, which include all information that is traditionally provided by testers except for the expected result. Naturally, this is a white-box software testing tool that leverages Natural Language Processing (NLP) to analyze source code files and extract all possible test cases. It achieves 100% code coverage in an automated manner, ensuring the quality of the System Under Test (SUT) at both the unit and integration level.

**Fig. 5.** CodeAssert pipeline.

Figure 5 describes a simplified execution pipeline of CodeAssert. The user provides the testing tool access to the source code files and selects the type of testing (i.e. unit or integration) that should be conducted. Since it analyzes the source code files, the user must specify the programming language of the source code and the respective scripting tool, to generate the scripts in the correct format. CodeAssert then proceeds to use NLP to analyze the source code and identify all conditional statements. Then the test cases are written based on the identified conditions and the input values are generated accordingly. Afterwards, the generated testing scripts are reviewed by the tester, who proceeds to populate the expected value for each identified test case. Finally, the test cases are computed with the chosen scripting AST tool and a testing report is presented to the tester.

## 4 Conclusions

Considering the impact that software has in the people's daily lives, it is essential to ensure its quality in order to avoid disastrous events caused by faulty or vulnerable software. This work presents TestLab, an intelligent AST framework. Its purpose is to improve software testing in multiple testing levels from different testing perspectives relying on AI-driven methods to achieve automation. TestLab was designed to integrate with the software development cycle and enable continuous, comprehensive testing of the SUT, with the objective of ensuring high-quality output throughout the entire process.

Automated software testing has demonstrated its efficiency in saving time, resources, and costs by automating the most monotonous tasks, enhancing software testing, and elevating the overall quality of software. The authors strongly believe that the AI-driven proposed framework has the potential to further improve testing efficiency and effectiveness, reducing the risk of human error, and enabling a more comprehensive testing coverage.

The framework's future work includes the implementation and experimentation of all its modules to assess their performance and effectiveness, as well as to explore additional testing methods that can benefit from AI automation.

**Acknowledgements.** The present work was partially supported by the Norte Portugal Regional Operational Programme (NORTE 2020), under the PORTUGAL 2020 Partnership Agreement, through the European Regional Development Fund (ERDF), within project "Cybers SeC IP" (NORTE-01-0145-FEDER-000044). This work has also received funding from UIDB/00760/2020.

## References

1. 25+ cyber security vulnerability statistics and facts of 2022. https://www.comparitech.com/blog/information-security/cybersecurity-vulnerability-statistics/
2. Clang static analyzer. https://clang-analyzer.llvm.org/
3. Software testing statistics - truelist 2022. https://truelist.co/blog/software-testing-statistics/
4. Battina, D.S.: Artificial intelligence in software test automation: a systematic literature review. Int. J. Emerg. Technol. Innov. Res. **6**, 1329–1332 (2019)
5. Durelli, V.H.S., et al.: Machine learning applied to software testing: a systematic mapping study. IEEE Trans. Reliab. **68**(3), 1189–1212 (2019). https://doi.org/10.1109/TR.2019.2892517
6. Gamido, H.V., Gamido, M.V.: Comparative review of the features of automated software testing tools. Int. J. Electr. Comput. Eng. **9**(5), 4473 (2019). https://doi.org/10.11591/ijece.v9i5.pp4473-4478
7. Garousi, V., Zhi, J.: A survey of software testing practices in Canada. J. Syst. Softw. **86**(5), 1354–1376 (2013). https://doi.org/10.1016/j.jss.2012.12.051
8. Godefroid, P.: Random testing for security. In: ASE 2007. ACM, November 2007. https://doi.org/10.1145/1292414.1292416
9. Hanna, M., et al.: Automated software testing framework for web applications. Int. J. Appl. Eng. Res. **13**(11), 9758–9767 (2018)
10. Kakaraparthy, D.: Overview and analysis of automated testing tools: ranorex, test complete, selenium. Int. Res. J. Eng. Technol. **4**(10), 1575–1579 (2017)
11. Krasner, H.: The cost of poor software quality in the US: a 2020 report (2020). https://www.it-cisq.org/cisq-files/pdf/CPSQ-2020-report.pdf
12. Martin-Lopez, A., et al.: Black-box and white-box test case generation for restful APIs: enemies or allies? In: ISSRE 2021, pp. 231–241 (2021). https://doi.org/10.1109/ISSRE52982.2021.00034

13. Murnane, T., Reed, K.: On the effectiveness of mutation analysis as a black box testing technique. In: Proceedings 2001 Australian Software Engineering Conference, p. 0012. IEEE Computer Society, Los Alamitos, CA, USA, August 2001. https://doi.org/10.1109/ASWEC.2001.948492
14. Nazim, M.T.B., et al.: Systematic analysis of deep learning model for vulnerable code detection. In: COMPSAC 2022, pp. 1768–1773. Institute of Electrical and Electronics Engineers Inc. (2022). https://doi.org/10.1109/COMPSAC54236.2022.00281
15. Nidhra, S.: Black box and white box testing techniques - a literature review. Int. J. Embed. Syst. Appl. **2**(2), 29–50 (2012). https://doi.org/10.5121/ijesa.2012.2204
16. Rafi, D.M., et al.: Benefits and limitations of automated software testing: systematic literature review and practitioner survey. In: 2012 7th International Workshop on Automation of Software Test (AST), pp. 36–42 (2012). https://doi.org/10.1109/IWAST.2012.6228988
17. Rajamanickam, L., et al.: Software testing: the generation tools. Int. J. Adv. Trends Comput. Sci. Eng. **8**(2), 231–234 (2019). https://doi.org/10.30534/ijatcse/2019/20822019
18. Saglietti, F., et al.: White and grey-box verification and validation approaches for safety- and security-critical software systems. Inf. Secur. Tech. Rep. **13**(1), 10–16 (2008). https://doi.org/10.1016/j.istr.2008.03.002
19. Shen, Q., et al.: A systematic review of fuzzy testing for information systems and applications. In: CECIT 2021. IEEE, December 2021. https://doi.org/10.1109/cecit53797.2021.00035
20. Singh, I., Tarika, B.: Comparative analysis of open source automated software testing tools: selenium, sikuli and watir. Int. J. Inf. Comput. Technol. **4**(15), 1507–1518 (2014)
21. Sommerville, I.: Software Engineering, 9th edn. Pearson, Upper Saddle River, NJ (2010)
22. Torkar, R.: Towards automated software testing - techniques, classifications and frameworks, January 2006
23. Uddin, A., Anand, A.: Importance of software testing in the process of software development, pp. 141–145, January 2019
24. Umar, M.A., Zhanfang, C.: A study of automated software testing: automation tools and frameworks. IJCSE **6**, 217–225 (2019)
25. Vitorino, J., et al.: Constrained adversarial learning and its applicability to automated software testing: a systematic review (2023). https://doi.org/10.48550/ARXIV.2303.07546

# Data Privacy with Homomorphic Encryption in Neural Networks Training and Inference

Ivone Amorim[1], Eva Maia[2(✉)], Pedro Barbosa[2], and Isabel Praça[2]

[1] PORTIC - Porto Research, Technology & Innovation Center, Polytechnic of Porto (IPP), 4200-374 Porto, Portugal
ivone.amorim@portic.ipp.pt

[2] Research Group on Intelligent Engineering and Computing for Advanced Innovation and Development (GECAD), Porto School of Engineering, Polytechnic of Porto (ISEP-IPP), 4200-072 Porto, Portugal
{egm,pmbba,icp}@isep.ipp.pt

**Abstract.** The use of Neural Networks (NNs) for sensitive data processing is becoming increasingly popular, raising concerns about data privacy and security. Homomorphic Encryption (HE) has the potential to be used as a solution to preserve data privacy in NNs. This study provides a comprehensive analysis on the use of HE for NN training and classification, focusing on the techniques and strategies used to enhance data privacy and security. The current state-of-the-art in HE for NNs is analysed, and the challenges and limitations that need to be addressed to make it a reliable and efficient approach for privacy preservation are identified. Also, the different categories of HE schemes and their suitability for NNs are discussed, as well as the techniques used to optimize the accuracy and efficiency of encrypted models. The review reveals that HE has the potential to provide strong data privacy guarantees for NNs, but several challenges need to be addressed, such as limited support for advanced NN operations, scalability issues, and performance trade-offs.

**Keywords:** Homomorphic Encryption · Neural Networks · Data Privacy · Privacy Preserving · Machine Learning · Data Sharing · Cybersecurity

## 1 Introduction

The emergence of digital technology and the ever-increasing significance of data in the economy have made it crucial to protect personal and sensitive information. As the amount of data being collected and stored grows, the risk of

This work was partially supported by the Norte Portugal Regional Operational Programme (NORTE 2020), under the PORTUGAL 2020 Partnership Agreement, through the European Regional Development Fund (ERDF), within project "Cybers SeC IP" (NORTE-01-0145-FEDER-000044). This work has also received funding from the project UIDB/00760/2020.


R. Mehmood et al. (Eds.): DCAI 2023, LNNS 741, pp. 365–374, 2023.
https://doi.org/10.1007/978-3-031-38318-2_36

privacy breaches and data misuse also increases. Furthermore, the use of intelligent techniques, like Neural Networks (NNs), to extract insights from this data presents another potential vulnerability. Considering this, many privacy-preserving techniques have been proposed, including differential privacy, secure multi-party computation, and Homomorphic Encryption (HE). Among these, HE has attracted significant attention because it allows computations to be performed on encrypted data without requiring decryption, which can enhance data privacy while still allowing for data analysis. In recent years, HE has been increasingly applied to NNs. By training NNs on encrypted data using HE, sensitive information can be kept private, as the NN can operate on the encrypted data without needing to decrypt it. This can be particularly useful in scenarios where data privacy is a major concern, such as healthcare or finance. However, there are still challenges and limitations to the use of HE in NNs for data privacy, including computational overhead and potential security vulnerabilities.

Some scientific works have already proposed the use of HE to preserve data privacy in NNs. One of the most relevant works in this area was introduced by Dowlin et al. [3], who modified a trained NN to operate on encrypted data, creating what they called a CryptoNet. However, this work assumes that the model is trained based on plain data and then this model is adapted for classifying encrypted instances. Other works have emerged to address the limitations of this method and propose new ones [14], but few have explored the use of HE for NN training and classification. Furthermore, the current literature in this area is almost exclusively addressed by practitioners looking for suitable implementations. To the best of our knowledge, there are only two publications that provide an overview of the existing approaches that use HE to preserve data privacy in NNs: Poulido-Gaytan et al. [14] and Podschwadt et al. [13]. The former focuses mainly on works that use HE solely in the classification phase, which is the most prevalent approach in the literature. In fact, only one reference in their work addresses the problem of using HE in NN training and inference. On the other hand, Podschwadt's work is centered on deep learning architectures for privacy-preserving machine learning with Fully HE, without including other types of HE schemes like Partially or Somewhat HE. Therefore, in this work, we aim to provide a comprehensive analysis of the use of HE for NN training and classification. We will review and analyse the current state-of-the-art in this field and identify limitations of using homomorphically encrypted data in NNs. We will also analyse existing approaches aimed at addressing these limitations and discuss future research directions for this topic.

## 2 Background

Neural Networks (NNs) are weighted directed graphs that are inspired by the way the human brain works [9]. The nodes of these graphs are usually called neurons. These neurons are grouped in layers and are represented as being stacked. Each of the neurons computes a function over the values of the layer beneath it. This function is usually named Activation Function (AF) and is applied to the output of each neuron in a NN, to determine whether the neuron should be activated (produce an output) or not, based on the input it receives. Common activation functions include the sigmoid function, ReLU (Rectified Linear Unit), and tanh (hyperbolic tangent) function. The loss function is then used to measure NN performance, by calculating the difference between the predicted output and the actual output. The goal of training a NN is to minimize the loss function, typically using an optimization algorithm such as stochastic gradient descent. Common loss functions include mean squared error, cross-entropy, and binary cross-entropy. Deep Learning (DL) [1] is a subfield of NNs that utilizes multiple hidden layers to extract hierarchical features that can be used for tasks such as pattern recognition and classification. Two types of supervised learning approaches for DL are Deep Neural Networks (DNNs) and Convolutional Neural Networks (CNNs). DNNs consist of multiple layers of artificial neurons that adjust the strengths of their connections to learn complex patterns and relationships in data. CNNs use a combination of convolutional layers, pooling layers, and fully connected layers to extract features from images and classify them into different categories.

Homomorphic Encryption (HE) is a type of encryption scheme that allows computations to be performed on encrypted data without decrypting it first. HE can be divided into three categories: Partially Homomorphic Encryption (PHE), Somewhat Homomorphic Encryption (SWHE) and Fully Homomorphic Encryption (FHE). In PHE schemes, the homomorphic property is satisfied by only one operation, such as addition or multiplication, which can be performed an unlimited number of times. SWHE supports more than one operation but a limited number of times Finally, FHE schemes allow a set of operations to be performed an unlimited amount of times. The first FHE scheme was introduced by Gentry in 2009 [6]. Gentry also introduced the concept of bootstrapping in FHE, which reduces the noise generated during operations and ensures a correct decryption output. However, this bootstrapping process is computationally expensive.

There are several research branches on the topic of HE which are usually defined by the mathematical problems in which the security of the proposed models rely on. For instance, the FHE schemes BGV, BFV, and TFHE are based on the hardness of the Learning With Error over Rings (Ring-LWE) problem [10]. Both schemes BGV and BFV operate on integers, while TFHE operates in binary data. Additionally, the CKKS scheme is based on a variant of the Ring-LWE problem called the approximate number field sieve, and it is meant for floating point arithmetic [4]. Finally, the SWHE scheme YASHE is founded on the Short Integer Solution problem, and it also operates on integers [2].

Several popular libraries provide implementations of the HE schemes mentioned above. For instance, the SEAL library[1] provides implementations of the BGV, BFV, and CKKS schemes. Earlier versions of SEAL also included an implementation of the YASHE scheme. Another library, HElib[2], implements the

[1] https://github.com/microsoft/SEAL.
[2] https://github.com/homenc/HElib.

BGV and CKKS schemes. Finally, the TFHE library[3] offers an implementation of the TFHE scheme.

## 3 Methodology

The aim of this work is to present a comprehensive and well-structured overview of the current research literature concerning the use of HE in NN for preserving data privacy. For this we formulated the following research question: "How has HE been used to preserve data privacy in NNs training and inference?". To address this broader question, four narrower sub-questions were defined:

**(RQ1):** What types of NNs are most used in studies about the use of HE to preserve data privacy in NNs?
**(RQ2):** What are the commonly used HE schemes for preserving data privacy in NNs?
**(RQ3):** What are the major limitations identified in the literature regarding the use of HE to preserve data privacy in NNs?
**(RQ4):** What are the most common approaches used to address the limitations of HE in NNs, and what trade-offs need to be considered?

To achieve transparent, replicable and complete answers to these questions, the guidelines of PRISMA methodology [15] were followed. The search terms "Neural Network(s)", "Homomorphic" and "Privacy" were defined and used in several reputable bibliographic databases. Since HE is commonly used in conjunction with other privacy-preserving techniques, we also defined a set of exclusion terms that were omitted from the results: "Federated", "Mining", "Blockchain", "Mobile", "Multi-Party", "Edge", "Multiparty", "Distribut*", "Communication", where '*' denotes any sequence of characters.

The search was then executed in four different bibliographic databases: Web of Science, ACM, Scopus and IEEE Explore, and a total of 289 records were retrieved which resulted in 163 unique records. Following an initial screening of the titles and abstracts of the identified publications, a total of 126 papers were excluded. The full texts of the remaining 37 papers were then evaluated for eligibility, and only 6 were considered suitable for this study. This process is illustrated in Fig. 1.

## 4 Results and Discussion

The first known work that provides a solution for training NN models using HE was published in 2017 by Hesamifard et al. [8]. The authors implemented NNs over homomorphically encrypted data where the Sigmoid and ReLU activation functions are replaced by low-degree polynomial approximations to enable the use of NNs with HE. The authors also proposed that when the ciphertext noise reaches a given threshold, it should be sent back to the client so that a fresh

[3] https://www.tfhe.com/.

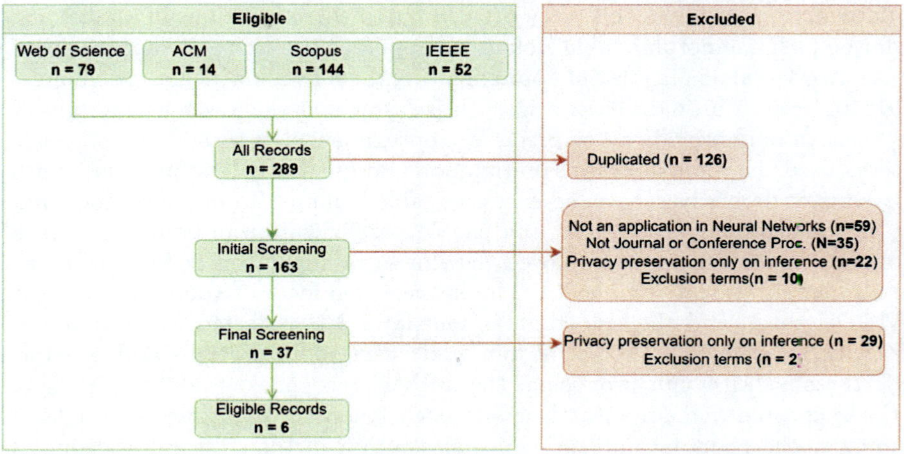


**Fig. 1.** Eligibility process

ciphertext can be returned, which they claim to be a more efficient alternative to bootstrapping. Experiments were conducted on three different datasets using five NNs configurations, and the reported accuracies are above 99%, outperforming two state-of-the-art approaches. One of those approaches also used HE, but only for classification, assuming that the model was trained in plaintext. The other approach used Secure Multi-party Computation (SMC) techniques to ensure privacy preservation in NN training and classification. The number of communications needed for training is also referred as an advantage over the state-of-the-art SMC approach, due to the significant reduction in the required number of interactions between the server and the client. The authors used the BGV scheme and its implementation in the HElib library.

In the following year, Ghimes et al. [7] implemented a NN where all data was encrypted, and all operations were performed over encrypted data using the YASHE scheme provided by the SEAL library. They utilized the simplest activation function possible, the identity function, to reduce the cost of computing over encrypted data. Their primary goal was to enable institutions to outsource NN predictions to third parties while ensuring data security and privacy. However, as recognized by the authors, their approach was not very promising, as it produced high error values and required long computation times. It is worth noting that the authors did not reference the previous work of Hesamifard et al. in their paper.

In 2019, Nandakumar et al. [11] published the first attempt to train a DNN on data encrypted using FHE, which can also be used for inference. Similar to Hesamifard et al.'s work, they used the HElib library to encrypt the data using the BGV scheme and to perform the computations on encrypted data. In this work, the authors identified two challenging steps in using HE, namely computing the activation function and its derivative, and computing the loss function and its derivative. To address this, they pre-computed these functions in a table and performed a homomorphic table lookup, an approach similar to the one proposed by Crawford et al. [5], instead of approximating them by a low-degree polynomial as Hesamifard et al. did. Their approach is faster and shallower when applicable, but can only provide a low-precision approximation of these functions. The authors used the sigmoid activation function and quadratic loss function, which have simpler derivatives, to avoid excessive table lookups. To minimize the number of bootstrapping operations and enable parallelization of computations, a data-packing strategy was employed, leading to a significant reduction of the overall computational complexity. The authors conducted experiments on the MNIST dataset, and the less complex four-layer NN reported an accuracy of 96% with an execution time of 40 min while using optimizations and multiple CPU threads. Although promising, the authors recognized that training data in the encrypted domain with their approach is four to five times slower than training in the plaintext domain, which shows that further research is required to improve efficiency.

After the work of Nandakumar et al., Lou et al. [10] published, in 2020, a faster alternative, that they called "Glyph", to train DNNs on encrypted data by combining TFHE and BGV schemes. Their approach employs a scheme switching technique that introduces only small computing overhead, but allows leveraging the use of TFHE for nonlinear activations, such as ReLU and softmax, and BGV for multiplicity-accumulations. This is possible because their scheme switching technique maps the plaintext spaces of BGV and TFHE to a common algebraic structure using natural algebraic homomorphisms, enabling, in this way, homomorphic switching between different plaintext spaces. The authors performed several experiments and have compared their approach with the previous one. To achieve similar accuracies, 97.8% for the approach of Nandakumar et al. and 98.6% for Glyph, they estimated a training time of 13.4 years for the first approach and 8 days for the second one. Overall, their experimental results show that Glyph obtains state-of-the-art accuracy, and reduces training latency by 69% to 99% over prior FHE-based privacy-preserving techniques on encrypted datasets.

In 2021, Onoufriou et al. [12] published a study which focus on the use of HE to preserve privacy in NN training and classification for milk yield forecasting in the agri-food sector. The authors employed a one-dimensional CNN (1D CNN) with encrypted data using the CKKS scheme. Since certain operations cannot be performed on fully homomorphically encrypted data, the authors implemented a set of measures to address this issue. First, they performed the forward pass in the encrypted domain by replacing the Sigmoid activation function by a polynomial approximation proposed by Chen et al. [3], in a similar approach to the one proposed by Hesamifard et al. This helped to avoid the necessity of computing divisions in the standard Sigmoid. Then, the backward pass and weight updating were done in plaintext to reduce computational cost. With this approach, the authors achieved an average accuracy of 87.6% and a Mean Absolute Percentage Error (MAPE) of 12.4% for milk yield prediction, demonstrating that FHE allows for absolutely private predictions. However, their approach has a flaw which is acknowledged by the authors, since it does not use encrypted data in all stages of training. This limitation requires further research to be addressed. It is worth mentioning that the authors used the MS-SEAL library for their experiments, and a dataset with 30 years of breeding, feeding, and milk yield data. The study showed that although FHE incurs a high spatial complexity cost, the run time remains within reasonable bounds.

In the same year, Yoo et al. [16] proposed a novel framework, named "t-BMPNet", for designing a foundational model of deep learning, namely a Multi-layer Perceptron Neural Network (MLPNN), over an FHE scheme. This framework allows for training in the encrypted domain without the need to replace activation functions with polynomial approximations, unlike the approach taken by Onoufriou et al. The authors achieved this by using the TFHE scheme, which enables building complex operations over encrypted data using primitive gates and fundamental bitwise homomorphic operations. They utilized the sigmoid function as the activation function and explained how to define the four main operations needed in the encrypted domain, namely: addition, two's complement, exponential function, and division. The authors conducted several experiments to validate their approach and confirmed that "t-BMPNet" achieved a highly accurate design of the nonlinear sigmoid function compared to other works that use polynomial approximations. However, the authors also recognized that their approach had low time performance and needs further research.

### 4.1 Discussion

After analysing the state-of-the-art studies on the use of HE to preserve data privacy in NN training and inference, in this section we will discuss the main findings, considering the research questions that were defined.

**RQ1 - What types of NNs are most used in studies about the use of HE to preserve data privacy in NNs?**

Based on the analysis of the reviewed literature, it is evident that DNNs and CNNs are the most commonly used types of NNs in studies that employ HE for preserving data privacy during training and inference. Out of the six works analysed, two of them utilized CNNs, while the other five employed DNNs. In fact, in the research conducted by Lou et al. [10], their approach was applied to both a DNN and a CNN.

**RQ2 - What are the commonly used HE schemes for preserving data privacy in NNs?**

From our analysis of the literature, it is clear that FHE is the preferred type of scheme. This is not surprising, given that FHE allows for multiple operations to be performed an unlimited number of times. BGV stands out as the most commonly used in this context, with three out of the six reviewed works using it [8,10,11]. It is worth noting that this scheme was prevalent in the first works published, probably because an implementation of it was available at the time in the HElib library. More recently, other FHE schemes have been explored. Two out of the six reviewed works employed the TFHE scheme [10,16], and the authors utilized the TFHE library for their experiments. On the other hand, one work uses the CKKS scheme [12]. Lou et al. [10] approach is also interesting since both BGV and TFHE are combined to leverage the best characteristics of each scheme. Specifically, they used TFHE for nonlinear activations and BGV for multiplicity-accumulations. Regarding SMHE, only a single work was found, namely one that uses the YASHE scheme [7].

**RQ3 - What are the major limitations identified in the literature regarding the use of HE to preserve data privacy in NNs?**

One of the key limitations is related with the fact that HE only supports basic arithmetic operations, such as addition and multiplication, as explained in Sect. 2. Thus, other commonly used operations in NNs, such as division, exponential, comparison, and maximum, cannot be directly supported by HE or they are computationally expensive. Consequently, some restrictions are introduced on the characteristics of the NNs that can be trained with homomorphically encrypted data. For instance, activation functions like the sigmoid, which involve division, cannot be directly used. Similarly, loss functions may also be affected by unsupported operations. Additionally, pooling techniques, such as max pooling and average pooling, which are commonly used in CNNs, can also be affected because they involve comparing and selecting the maximum or average value within a set of values, which is not possible in the majority of HE schemes.

Another significant limitation of homomorphically encrypted NNs is their computational overhead. The operations performed on encrypted data are computationally expensive, and their complexity increases with the size of the data and the depth of the NNs. Consequently, training and inference times can become significantly longer when compared with non-encrypted NN.

Finally, the use of HE with NNs can also introduce limitations on the way data is represented. HE schemes typically operate on encrypted data in a fixed-point representation, which is different from the floating-point representation used in traditional NNs. This difference can lead to numerical instability and can limit the range of values that can be represented, which may negatively affect the accuracy and performance of the model.

**RQ4 - What are the most common approaches used to address the limitations of HE in NN, and what trade-offs need to be considered?**

One of the most important limitations is related with the NN operations, as mentioned in the previous research question. To address this issue, most of the studied works replaced the activation and loss functions by a polynomial approximation [7,8,12]. However, this approach has shown to negatively affect the performance of the NNs, as low-degree approximations become progressively worse for input values farther from zero. To mitigate this problem, some authors have suggested normalizing input values, but this can be difficult to control, especially for NN with several layers. To avoid using polynomial approximations, other approaches have been proposed, such as pre-computing activation and loss functions and performing homomorphic table lookups [11] or implementing complex operations using fundamental bitwise operations [16]. However, each of these approaches also has trade-offs. For instance, while the table lookup approach is faster than using polynomial approximations, it can only provide low-precision approximations for the activation and loss functions. On the other hand, implementing complex operations using fundamental bitwise operations can lead to better approximations, but it can also result in low time performance. Finally, Lou et al. [10] proposed a combined approach using the schemes TFHE and BGV, that slightly increased the accuracy of this system.

Regarding the computation overhead limitation, one common strategy is to use a low degree polynomial approximation for the activation and loss functions, as explained before. This reduces the number of operations required, thereby reducing the computational overhead. Additionally, another way to reduce the computational overhead is to use NNs with fewer layers. In fact, most of the works reviewed have used DNN with just three layers [7,11,16]. However, some other studies have explored different NN configurations, such as varying the number of hidden layers from 1 to 5 [8], or utilizing a 1D CNN [12]. Since each homomorphic operation performed in encrypted data is slower than its counterpart, by reducing the layers, the number of homomorphic operations that need to be used can be reduced, which can lead to a significant reduction in the computational overhead. However, this reduction in the number of layers may also result in a reduction in the accuracy of the model. Therefore there is a clear trade-off between model accuracy and computational overhead when using HE in NN training and classification.

In relation to the limitation which may be introduced by HE regarding the way data is represented, most of the approaches use encoding strategies to make it possible to work with low-precision integers and fixed-point representation. By the other hand, CKKS, used in the work of Onoufriou et al. [12], uses a floating-point representation for plaintext and ciphertext data, requiring minimal adaptations. Even though these strategies are needed to use HE in NNs, they also come with a trade-off, such as potential accuracy loss with fixed-point representation and the need for additional computation with low-precision integer representation.

## 5 Conclusion

In this work, we have reviewed and analysed the state-of-the-art on the use of HE to preserve data privacy in NN training and inference. Our study showed that BGV scheme is the most commonly used in this context, which highlights the trend towards the use of FHE schemes. The Glyph and t-BMPNet approaches stand out for leveraging the characteristics of the TFHE scheme to reduce training times while maintaining high accuracy levels, without replacing nonlinear functions with polynomial approximations. Nevertheless, current approaches are still slow and may not provide high levels of accuracy. We believe that the selection of the most appropriate strategy depends on the specific application scenario, which may have varying requirements in terms of accuracy and computational complexity. Therefore, future research should focus on exploring the application of TFHE in more complex NNs and investigating other HE schemes. Additionally, improving computational performance through parallelization may be worth exploring to make HE more reliable and scalable for this type of architecture. In summary, the use of HE for NN training is a promising approach for preserving privacy, but optimizing its accuracy and efficiency requires further research and experimentation with novel techniques and strategies.

## References


1. Alom, M.Z., et al.: A state-of-the-art survey on deep learning theory and architectures. Electronics **8**(3) (2019)
2. Bos, J.W., Lauter, K., Loftus, J., Naehrig, M.: Improved security for a ring-based fully homomorphic encryption scheme. Cryptology ePrint Archive, Paper 2013/075 (2013)
3. Chen, H., et al.: Logistic regression over encrypted data from fully homomorphic encryption. Cryptology ePrint Archive, Paper 2018/462 (2018)
4. Cheon, J.H., Kim, A., Kim, M., Song, Y.: Homomorphic encryption for arithmetic of approximate numbers. In: Takagi, T., Peyrin, T. (eds.) ASIACRYPT 2017. LNCS, vol. 10624, pp. 409–437. Springer, Cham (2017). https://doi.org/10.1007/978-3-319-70694-8_15
5. Crawford, J.L.H., et al.: Doing real work with FHE: the case of logistic regression. In: WAHC Proceedings, pp. 1–12. WAHC '18, Association for Computing Machinery, New York, NY, USA (2018)
6. Gentry, C.: A Fully Homomorphic Encryption Scheme. PhD Thesis, Stanford University, Stanford, CA, USA (2009). iSBN: 9781109444506
7. Ghimes, A.M., Vladuta, V.A., Patriciu, V.V., Ioniţă, A.: Applying neural network approach to homomorphic encrypted data. In: ECAI 2018, pp. 1–6 (2018)
8. Hesamifard, E., Takabi, H., Ghasemi, M., Jones, C.: Privacy-preserving machine learning in cloud, pp. 39–43. Association for Computing Machinery, Inc. (2017)
9. Lippmann, R.P.: An introduction to computing with neural nets. SIGARCH Comput. Archit. News **16**(1), 7–25 (1988)
10. Lou, Q., Feng, B., Fox, G.C., Jiang, L.: Glyph: fast and accurately training deep neural networks on encrypted data. Neural Inf. Process. Syst. Found. (2020)
11. Nandakumar, K., et al.: Towards deep neural network training on encrypted data, vol. 2019-June, pp. 40–48. IEEE Computer Society (2019)
12. Onoufriou, G., Mayfield, P., Leontidis, G.: Fully homomorphically encrypted deep learning as a service. Mach. Learn. Knowl. Extr. **3**, 819–834 (2021)
13. Podschwadt, R., et al.: A survey of deep learning architectures for privacy-preserving machine learning with fully homomorphic encryption. IEEE Access **10**, 117477–117500 (2022)
14. Pulido-Gaytan, B., et al.: Privacy-preserving neural networks with homomorphic encryption: challenges and opportunities. Peer-to-Peer Netw. Appl. **14**, 1666–1691 (2021)
15. Shamseer, L., et al.: Preferred reporting items for systematic review and meta-analysis protocols (PRISMA-P) 2015: elaboration and explanation **349** (2015)
16. Yoo, J.S., Yoon, J.W.: T-BMPNet: trainable bitwise multilayer perceptron neural network over fully homomorphic encryption scheme **2021** (2021)

# A Privacy Risk Analysis of Identity Federation Topologies in Single Sign-On (SSO) Web Domain

Reem Al-Saidi(✉) and Ziad Kobti

University of Windsor, 401 Sunset Avenue, Windsor, ON N9B 3P4, Canada
{alsaidir,kobti}@uwindsor.ca

**Abstract.** Single Sign-On (SSO) allows users to authenticate once and gain access to different web services. Thus, users do not need to provide their credential information again. Three main entities are involved in the SSO: the user, Identity Provider (IdP), and Service Provider (SP) or Relying Party (RP). These entities are linked through different relationships under various IdP-SP association models, mainly decentralized, explicit, federated, and bridge. Utilizing SSO under these models or their overlapping in a specific context raises various privacy risks for all entities involved. This paper identifies a set of privacy requirements and outlines the privacy risks for the involved entities in simple SSO scenario and topologies. With regard to topology, it is vital to consider a combination of different association models with respect to privacy requirements. To address the common privacy risks with a focus on user identities, attributes, and changing behavior, the present study proposes a fuzzification system to infer the user's privacy risk during both the authentication and access processes under semi-honest and malicious models. Analyzing privacy risks is crucial in ensuring the web cloud environment's security as it helps reduce the likelihood of attacks and identify potential vulnerabilities.

**Keywords:** Single Sign On · Privacy risk · Privacy requirements · Federated identity management · Fuzzy inference system · Topologies

## 1 Introduction

In SSO [8], users access various web services located in different domains without needing to re-authenticate at each one individually. The main entities in web SSO are the IdP, SP/RP, and web user. The primary function of the IdP is to authenticate web users and confirm their identities to SP/RP. Trust [8] is the main relationship between IdP and SP, allowing SP/RP to delegate the authentication on the users' behalf through their IdP.

The present study focuses on privacy risk analysis and assessments for SSO users' during authentication and access processes against set of pre-defined privacy requirements. The access and authentication users' privacy risk includes


R. Mehmood et al. (Eds.): DCAI 2023, LNNS 741, pp. 375–385, 2023.
https://doi.org/10.1007/978-3-031-38318-2_37

threats from other entities, mainly IdP and SP, under malicious or semi-honest models, in which we rely on the probabilistic traditional risk assessment methods [3,6]. However, it becomes challenging to use the same techniques to assess the risk derived from user-changing behavior and rules during the authentication and access processes.

Fuzzy logic inference handles such dynamic changes in user behavior and uncertainty in the web domain for better risk analysis and assessment [4]. The fuzzy logic inference system [5] includes fuzzification, in which essential parameters for risk assessment are determined. The parameters should be dynamic and have uncertain values in their nature. The frequency and user rules are the dynamically changing parameters for the access privacy risk. The identity token sharing and user attribute disclosure are the primarily changing parameters for authentication privacy risk assessment. Then, a set of fuzzy rules is constructed based on a fuzzy set that defines the membership functions. Finally, the inference engine will make the corresponding decision based on the defined fuzzy rules known as defuzzification.

The main contributions of this paper are: First, identify the set of privacy requirements concerning user interaction and behavior in the web SSO domain. Second, construct different topologies demonstrating user-IdP-SP interactions under various association models. Third, outline the corresponding privacy risks for each entity across SSO simple scenario and topologies, considering the entities' interactions against privacy requirements. Finally, develop a fuzzy inference system to calculate the access privacy risks corresponding to frequency and roles attributes and the authentication privacy risk regarding attribute disclosure and token sharing attributes. The rest of this paper is organized as follows: In Sect. 2, we identify privacy requirements concerning web user data and behavior. In Sect. 3, we analyze the privacy risk for all the involved entities in a simple SSO scenario focusing on users' privacy risk from IdP and SP. In Sect. 4, we create different topologies utilizing distinct association models, present a scenario for each, and outline the related privacy risk for every entity involved. In the last Sect. 5, we develop a fuzzy inference system to calculate the access and authentication users' privacy risks.

## 2 Privacy Requirements in SSO

Identifying privacy requirements taking risk metrics into account, data sensitivity, the involved actors' roles, and various entities' interactions, is essential for better risk analysis and assessment in SSO [8]. However, most of the existing metrics for risk quantification [3] do not consider user behavior and access patterns, so our privacy requirements aim to address this gap. Other related work treats privacy requirements as instances of compliance, traceability, access control, verification, or usability [2]. Data privacy and user privacy are the two main privacy requirements in the SSO web application.

- **Data privacy:** The primary goals of data privacy are maintaining the confidentiality, integrity, and availability of the users' data and sensitive attributes

**Table 1.** Data and users privacy requirements in SSO.

| Data Privacy (DP) | User Privacy (UP) |
|---|---|
| **D.P.R1:** Confidentiality | **U.P.R1:** Authentication privacy |
| **D.P.R2:** Integrity | **U.P.R1.a** Identity privacy |
| **D.P.R3:** Availability | **U.P.R1.b** User profile privacy |
| **D.P.R4:** Tracking and transparency | **U.P.R1.c** Sensitive attribute privacy |
| **D.P.R5:** Un-linkability | **U.P.R1.d** Biometric privacy |
| **D.P.R6:** Data minimization | **U.P.R2:** Access privacy |
| | **U.P.R2.a** Access pattern privacy |
| | **U.P.R2.b** Access token privacy |
| | **U.P.R2.c** Behavioral and interaction privacy |
| | **U.P.R2.d** Roles and permission privacy |

[10]. The traceability and linkability of users' data and identities are other crucial data privacy requirements [3]. Tracking the interactions of various users allows IdP or SP to correlate/link their behavior to their identities. The primary objective is to guarantee a minimum data and attribute disclosure and only what is immediately relevant to fulfill a specified task [10]. For that purpose, the user should control the attributes revealed to SP to what is needed to accomplish a particular task. However, if more attributes need to be disclosed to SP, it should be upon user consent. Table 1 outlines the data privacy requirements in the SSO web domain.

- **User privacy:** Preserving web users' privacy is the main privacy requirement in the SSO domain, including their identities, sensitive attributes, profiles, and biometrics, while they are authenticating through IdP or accessing various SPs. The likelihood of users accessing different web services to be monitored by malicious and colluded entities is still high in a distributed web domain. Revealing users' access patterns, behavior, and interaction poses various risks which could threaten users' lives. Table 1 shows user privacy requirements, including authentication and access privacy.

## 3 Risk Analysis in Simple SSO Scenario

This section illustrates the corresponding user privacy risk from IdP and SP in a simple SSO scenario. The risk is calculated under semi-honest and malicious models using traditional risk techniques [3,6] against a set of privacy requirements. Moreover, the risk for IdP and SP is introduced, which is common among simple SSO scenarios and the upcoming topologies.

The simple SSO scenario involves three main entities: the IdP, SP, and user. The users first register their credentials to a specific IdP. The SP that users want to access and benefit from its services has an explicit trust relationship with the

IdP with which the user registered and subscribed. When a user access a service in SP; an IdP can authenticate the user and send an authentication token for SP, known as authentication delegation through the user IdP [8]. IdPs and SP can behave under **semi-honest** or **malicious** models [9].

We assumed that the users would not notice such IdP or SP illegal behavior in semi-honest as they do in the malicious model. In the malicious model, users know about the IdP breaches through news and alerts [7].

However, under both models, IdP and SP poses various threats to users, putting them at risk during the authentication and access processes in a specific context. The risk assessment is based on the traditional risk approach [3,6] as shown in Eqs. (1) and (2).

$$R = P \times I \tag{1}$$

The Eq. (1) shows that $R$ is proportional to both the probability of an undesirable event occurring ($P$) and the impact of this event ($I$) [3,6].

Usually the risk consist of a series of risks, as shown in Eq. (2):

$$Risk_{total} \sum_{i=1}^{n} R_i \tag{2}$$

The total risk $Risk_{total}$ can be calculated as the sum of individual risks ($R_i$) where $i$ ranges from 1 to $n$.

The following subsections, 3.1, 3.2, and 3.3, outline the potential privacy risk for the involved entities in a simple SSO scenario focusing on analyzing users' risks based on the threats from IdP and SP. The risks are assessed from a general perspective without relying on any specific SSO protocol using traditional risk techniques [3,6]. We assign probabilities for each model based on the user notice: 0.6 for the semi-honest and 0.4 for the malicious model. Malicious behavior is less likely to occur as most IdPs care about their reputation and building trust towards their users [7].

The impact is assessed against a set of privacy requirements where each privacy requirement has a weight of 1.

### 3.1 Users Risk from IdP($URisk_{IdP}$)

- **Semi-honest IdP:**
  1. IdP can monitor users' behaviors and interactions during their access to extract a specific access pattern and the corresponding privileges and roles for the user. This will affect the following privacy requirements $\{U.P.R1.a, U.P.R1.c, U.P.R2.a, U.P.R2.c, U.P.R2.d\}$.
  2. IdP tracks users' behavior and data access by linking their identities and registered profiles to their access revealing users sensitive information, which will affect $\{D.P.R4,\ D.P.R5, U.P.R1(a)\}$ privacy requirements.
  Based on Eq. (1), the risk value is 3, 1.8 on each, with a total of 4.8 of users risk under semi-honest IdP, based on Eq. (2).

- **Malicious IdP:**
  1. IdP discloses users registered information and their corresponding attributes, users' profile, to unauthorized entities or provide more information than required during the user interaction life-cycle. This will affect the $\{D.P.R6, U.P.R1(b), U.P.R1(c)\}$ privacy requirements.
  2. IdP manipulates users' data and their access while issuing the corresponding authentication or access token, which will impact $\{D.P.R2, U.P.R1, U.P.R2(b)\}$ privacy requirements.
  3. IdP colludes with SP during user access or authentication process to reveal more users' attributes to SP without user consent or to extract user access pattern and their access intention which will impact $\{D.P.R6, D.P.R6, U.P.R1(a), U.P.R1(c), U.P.R2(a), U.P.R2(c), U.P.R2(d)\}$ privacy requirements.

  Based on Eq. (1), the risk value is 1.2, 1.2 and 2.8 on each, with a total of 5.2 of users risk under malicious IdP, based on Eq. (2).

### 3.2 Users Risk from SP($URisk_{SP}$)

- **Semi-honest SP:**
  1. SP accesses more data and attributes than required without user consent and with the collude of IdP, which will affect $\{D.P.R6, U.P.R1(c)\}$ privacy requirements.
  2. SP links user attribute during a sequence of user access to their identities, which will impact the unlinkability $\{D.P.R5\}$ privacy requirement

  Based on Eq. (1), the risk value is 1.2, 0.6 on each, with a total of 1.8 of users risk under semi-honest SP, based on Eq. (2).
- **Malicious SP:**
  1. SP legitimate users access token and prevent them from access other web services. SP knows other user preferences by monitoring their behavior and access. This will affect the $\{D.P.R4, U.P.R2(a), U.P.R2(c)\}$ privacy requirements.

  Based on Eq. (1), the risk value is 1.2, with a total of 1.2 of users risk under malicious SP, based on Eq. (2).

### 3.3 IdP and SP Risk

IdP and SP will be at risk due to other entities' semi-honest or malicious behaviors.

- **IdP risk:** IdP will be at risk when a malicious user colludes with SP to corrupt or get illegal access to IdP or when SP asks IdP to access extra user attributes without their consent
- **SP risk:** SP will be at risk if a malicious user performs a specific attack or behaves illegally toward SP. As well, IdP can give access to an unauthorized user, which seems to be a legal entity to SP.

Note that the corresponding risks for IdP and SP will be the same through the upcoming topology discussions.

## 4 Topologies: User-IdP-SP Relationships Within Different Association Models

This section explores different architectural design scenarios for SSO with varying user-IdP-SP relationships, showcasing their interactions across four association models [1] and their overlapping: decentralized, explicit, federated, and bridge. We refer to the relationships between entities in a specific design with a particular association model as a **"Topology"**. Moreover, the term "Topology" used for trust requirements identification in SSO [8]. The purpose of introducing topologies is to demonstrate the complexity of the underlying SSO infrastructure in the web domain and to give a future road-map for assessing risk on various topologies under particular SSO protocols; the protocols differ in their use of protection techniques and the corresponding trust establishment [8]. For that, it is become difficult to assess the user privacy risk by the traditional risk techniques on these topologies.

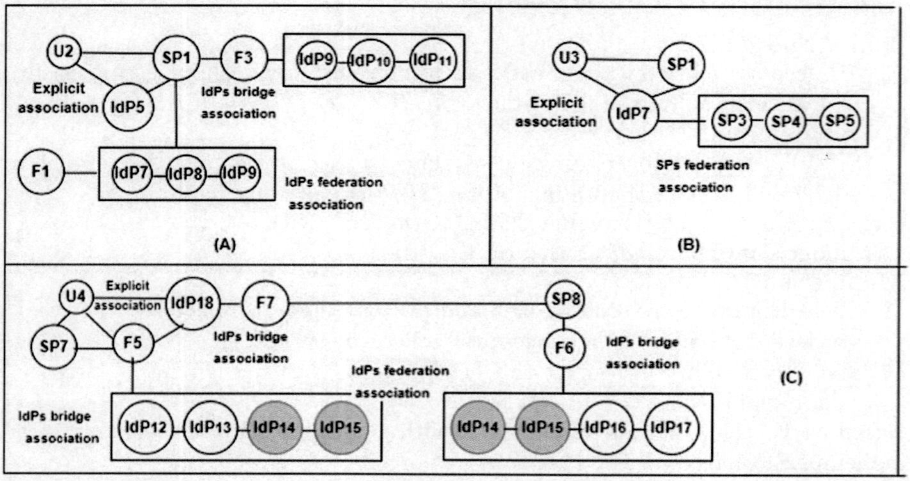


**Fig. 1.** Topology models (A) Explicit, bridge, and IdPs federation association models. (B) Explicit and SPs federation association models. (C) Explicit, bridge, and IdPs federation association models

### 4.1 Topology 1: Explicit, IdPs Bridge and Federation Association Models

The U2 has an explicit association with IdP5 and wants to access SP1. However, SP1 doesn't have a static trust relationship with IdP5. Rather, it belongs to F1. It is evident that IdP5 isn't part of F1. Thus, SP1 can't rely on any of the IdPs in the federation. If the user registers with IdP7, which is part of the F1 members, then SP1 can rely on IdP7 to delegate user authentication to it. The corresponding actions on that case are: **(1)** SP1 can explicitly ask U2 for registration to sign and access the requested service, (No SSO). **(2)** SP1 denies user access. **(3)** SP1 can start a dynamic discovery to build on-fly trust with IdP5, which users register on ending with an implicit association between SP1 and IdP5. Figure 1(A) shows an explicit, federated and bridge association models.

- **Users risk from IdP**($URisk_{IdP}$)
  1. **Semi-honest IdP:**
     - IdP5 knows user access upon SP1's success in the discovery of IdP5. Not only the access pattern but also more information could be revealed if IdP5 colludes with SP1, which will affect $\{U.P.R1.a, U.P.R1.c, U.P.R2.a, U.P.R2.c, U.P.R2.d\}$ privacy requirements.
     - IdP5 colludes with SP1 during U2 access to extract the access pattern and its intention, which will affect $\{D.P.R6, U.P.R1.a, U.P.R1.c, U.P.R2.a, U.P.R2.c, U.P.R2.d\}$ privacy requirements.
  2. **Malicious IdP:**
     - IdP5 authenticates U2 but issue a token for U1, which will impact $\{D.P.R2, U.P.R1\}$ privacy requirements.
     - IdP5 discloses more information than required to SP1 without user consent, which will affect $\{D.P.R6, U.P.R1.b, U.P.R1.c\}$ privacy requirements.
- **Users risk from SP**($URisk_{SP}$)
  1. **Semi-honest SP:**
     - SP1 asks to access more information about U2. However, if U2 approve such an access, there is no risk. Otherwise, it will be against $\{D.P.R6, U.P.R1(a)\}$ privacy requirements.
  2. **Malicious SP:**
     - SP1 act on legitimate users' behalf to prevent them from accessing other web services.

### 4.2 Topology 2: Explicit and SPs Federation Association Models

U3 explicitly registers to IdP7, which has a federation association with different service providers $\{SP3, SP4, SP5\}$. U3 wants to access the two related services offered by SP3 and SP5.

U3 can authenticate through IdP7. Noted that IdP7 relies on all federated members. However, user access, in that case, will add an extra complexity due to the service orchestration in the federation. The users risks from IdP and SP ($URisk_{IdP}$) ($URisk_{SP}$) would be the same as in topology 1. Figure 1(B) demonstrate the topology 2 and corresponding entities relationships.

### 4.3 Topology 3: Explicit, SP Bridge and IdPs Federation Association Models

U4 explicitly registers to IdP18 and wants to access SP8. As SP8 and IdP18 are in the same federation. Then, U4 can authenticate through IdP18 and access the SP8. If U4 wants to access SP7 and SP8, which don't belong to the same federation but have common IdPs.

Neither SP7 nor SP8 knows about the common IdPs. On that case, U4 can access both SP7 and SP8 and authenticate either through IdP14 or IdP15. Neither SP7 nor SP8 will know about the common IdPs. This topology model carries the same potential risks as those identified in the topology 1 and 2, except for the fact that SPs will not cooperate as they are unaware of the shared IdPs. Figure 1(C) demonstrates topology 3.

## 5 Authentication and Access Privacy Risks Analysis

In this section, we assess access and authentication privacy risks using the fuzzy inference system to calculate the likelihood of each risk. The risk calculated derived from user behavior which is difficult to assess by traditional risk techniques due to the dynamic nature of user behavior and rules. Fuzzy logic inference handles such dynamic changes in user behavior and uncertainty in the web domain.

### 5.1 Users Access and Authentication Privacy Risks

The access and authentication privacy risk defines under semi-honest and malicious models. The risk during user access and authentication is derived from IdP and SP interacting entities (see Sect.3) and users' behavior. Assessing the risk from user behavior and other entities helps better protect users and their data.

The following formula shows the access privacy risk under semi-honest or malicious models

$$User\ Access\ Privacy\ Risk_{[semihonest,malicious]} \\ = User\ behaviour_{[Roles,frequency]} + URisk_{IdP} + URisk_{SP} \tag{3}$$

We will calculate the user access privacy risk based on their changing behavior taking the following as primary parameters. **(1) Roles:** users' roles and interests change over time, which makes it more challenging for semi-honest and malicious entities to derive a specific user role and intention.

**(2) Frequency:** how often a user interacts and accesses different web services. The more frequent access the more semi-honest and malicious are able to derive more information about user access.

For the authentication privacy risk, we will rely on the two dynamic parameters: **(1) Identity token sharing:** the amount of time users rely on SSO. Regardless of the type of protocol that may be deployed, there will be a token that allows users to access SPs without the need to re-authenticate or enter their credentials again **(2) User attribute disclosure:** the user's awareness about the consent to grant SP access to a particular attribute.

The following formula calculated the authentication privacy risk under semi-honest or malicious models

$$User\ Authentication\ Privacy\ Risk_{[semihonest,malicious]} \\ = User\ behaviour_{[tokensharing,attributesdisclosure]} + \boldsymbol{URisk_{IdP}} + \boldsymbol{URisk_{SP}} \quad (4)$$

$\boldsymbol{URisk_{IdP}}$ and $\boldsymbol{URisk_{SP}}$ are calculated in Sect. 3 under semi-honest and malicious models.

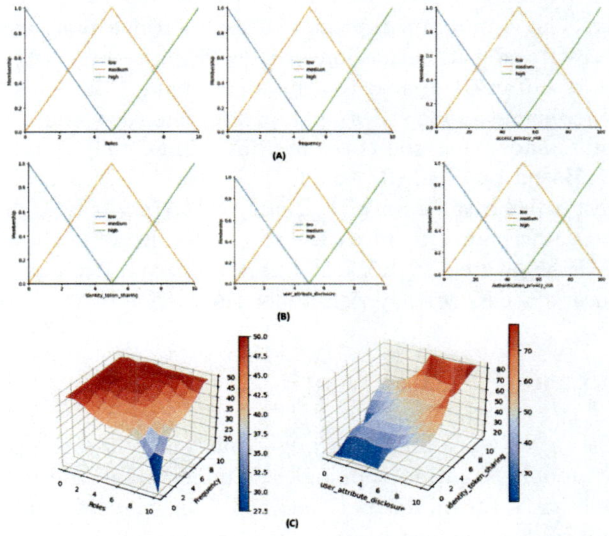


**Fig. 2.** Implementation: (A) frequency, roles and access privacy risk triangular membership functions (B) Identity token sharing, user attribute disclosure and authentication privacy risk triangular membership functions (C) Simulation of access and authentication privacy risks

## 5.2 Users Access and Authentication Privacy Implementation

When a user frequently accesses data, it is more likely that a semi-honest or malicious IdP and SP entities will monitor and expose the user's access pattern and reveal their access purpose. However, as users' roles changes, it becomes difficult to determine the exact users roles and intention. Based on changing of these parameters, we developed a fuzzy inference system to assess the corresponding access privacy risk derived from user behavior including the following steps: **(1)** We define the frequency and roles as input variable and access privacy risk as an output variable. **(2)** We generate triangular membership functions for input and output variables. Figure 2(A) shows the triangular membership functions for the roles, frequency, and access privacy risk. **(3)** We create nine main fuzzy rules based on the input variables to get the access privacy risk as an output. **(4)** We design a simulation to calculate the corresponding privacy access risk based on roles and frequency. The implementation is based on Python with the scikit-fuzzy library.

Same previous steps are followed for the authentication privacy risk with two different changing parameters: identity token sharing and user attribute disclosure. Figure 2(B) shows the triangular membership functions for identity token sharing, user attribute disclosure, and authentication privacy risk. The simulations show all the possible combinations of roles and frequency for access privacy risk. Identity token sharing and user attribute disclosure for authentication privacy risk. The simulation represented as a surface plot, computing the corresponding access and authentication privacy risk, see Fig. 2(C). The results indicate less risk is obtained due to the dynamic changes in users' roles under semi-honest and malicious models. For example, for the roles and frequency with 8 and 1, the results show that the corresponding users' risk derived from their behavior is 0.38. Based on Eqs. (3) and (4), the total user privacy access risk under semi-honest will be the sum of $URisk_{IdP}$, $URisk_{SP}$, and user behavior risk from the fuzzy inference system with a total 6.98. In a malicious model, the $URisk_{IdP}$ and $URisk_{SP}$ are 5.2, 1.5, and the risk derived from user behavior is 0.38. The total user's access privacy risk under the malicious model will be 6.78.

## 6 Conclusion and Future Work

The main entities in SSO are linked through different relationships under various association models, creating topologies. Under these different topologies, SSO raises privacy risks for all entities involved. The present study considered user privacy concerns from a general point of view without considering any specific protocol to behave under these topologies. Based on this, a set of privacy requirements was identified, and the privacy risks for each entity were outlined in simple SSO scenario and on topologies with respect to the privacy requirements, focusing on all involved entities. Under the assumption that IdP or SP will behave under semi-honest or malicious modes which cause threats to the users, the study proposes a fuzzification system to address user authentication and access privacy risks based on changing parameters that is difficult to assess by traditional risk techniques. While this study has provided valuable insights into the authentication and access privacy risks among different topologies, there is still much to be explored in this area. For example, future research should explore SSO protocols within these topologies and comprehensively analyze the associated privacy risks for all entities involved.

**Acknowledgments.** We acknowledge the support of the Natural Sciences and Engineering Research Council of Canada (NSERC [funding reference number 03181]).

## References


1. Alaca, F., Oorschot, P.C.V.: Comparative analysis and framework evaluating web single sign-on systems. ACM Comput. Surv. (CSUR) **53**(5), 1–34 (2020)
2. Anthonysamy, P., Rashid, A., Chitchyan, R.: Privacy requirements: present & future. In: 2017 IEEE/ACM 39th International Conference on Software Engineering: Software Engineering in Society Track (ICSE-SEIS), pp. 13–22. IEEE (2017)
3. Arias-Cabarcos, P., Almenárez-Mendoza, F., Marín-López, A., Díaz-Sánchez, D., Sánchez-Guerrero, R.: A metric-based approach to assess risk for "on cloud" federated identity management. J. Netw. Syst. Manag. **20**, 513–533 (2012)
4. Atlam, H.F., Azad, M.A., Alassafi, M.O., Alshdadi, A.A., Alenezi, A.: Risk-based access control model: a systematic literature review. Future Internet **12**(6), 103 (2020)
5. Azimi, S.R., Nikraz, H., Yazdani-Chamzini, A.: Landslide risk assessment by using a new combination model based on a fuzzy inference system method. KSCE J. Civ. Eng. **22**, 4263–4271 (2018)
6. Boehm, B.W.: Software risk management: principles and practices. IEEE Softw. **8**(1), 32–41 (1991)
7. Brown, A.J.: "Should I stay or should I leave?": exploring (dis) continued Facebook use after the Cambridge analytica scandal. Soc. Media+ Soc. **6**(1), 2056305120913884 (2020)
8. Grüner, A., Mühle, A., Gayvoronskaya, T., Meinel, C.: A comparative analysis of trust requirements in decentralized identity management. In: Barolli, L., Takizawa, M., Xhafa, F., Enokido, T. (eds.) AINA 2019. AISC, vol. 926, pp. 200–213. Springer, Cham (2020). https://doi.org/10.1007/978-3-030-15032-7_18
9. Hazay, C., Lindell, Y.: A note on the relation between the definitions of security for semi-honest and malicious adversaries. Cryptology ePrint Archive (2010)
10. Kumar, R., Bhatia, M.: A systematic review of the security in cloud computing: data integrity, confidentiality and availability. In: 2020 IEEE International Conference on Computing, Power and Communication Technologies (GUCON), pp. 334–337. IEEE (2020)

# Special Session on the Next Generation of IoT Infrastructures, Cybersecurity, and Applications (IoTalentum)

# ML for Attack and Defense of PUFs: Current Status and Future Directions

Mieszko Ferens(✉), Edlira Dushku, and Sokol Kosta

Aalborg University, Copenhagen, Denmark
{mjfm,edu,sok}@es.aau.dk

**Abstract.** The integration of IoT devices is becoming increasingly inevitable in the development of next-generation systems and applications. Due to such a wide adoption, IoT devices handle large quantities of private and sensitive data, and operate safety-critical systems. As such, failure to comply with security requirements would prove to be catastrophic. However, the resource-constrained nature of IoT devices is a fundamental limitation in designing their security features. To tackle the problem of implementing lightweight security functionalities that enable trusted communications, Physical Unclonable Functions (PUFs) have been proposed. Exploiting the manufacturing variations of Integrated Circuits (ICs), these primitives aim to give devices a unique identifier that no attacker can violate or clone. That said, in the past decade many studies have shown the great threat that Machine Learning (ML) poses to the security of Physical Unclonable Functions. In this paper, we provide an up-to-date situation of this field of research, as well as our current work and future directions.

**Keywords:** Physical Unclonable Function · Machine Learning · IoT · hardware security

## 1 Introduction

Be it industrial, automotive, energy, or any other applications, with each passing year the deployment of IoT devices is becoming more pervasive and ubiquitous [3]. The reason for such rapid development is simple, IoT devices can serve as a large and highly granular base for data collection, monitoring, and actuation. While this proliferation brings enormous opportunities for new and improved services, it comes at the expense of increased susceptibility to attacks.

A system composed of many small and potentially more vulnerable devices is more prone to exploitable vulnerabilities [3]. For example, in an industrial setting, a compromised sensor could lead to not detecting an overheated component. In such a scenario, a single small device that has been successfully attacked

This work is supported by the IoTalentum project, funded by the European Union Horizon 2020 research and innovation program within the framework of Marie Skłodowska-Curie Actions ITN-ETN with grant number 953442.


R. Mehmood et al. (Eds.): DCAI 2023, LNNS 741, pp. 389–398, 2023.
https://doi.org/10.1007/978-3-031-38318-2_38

can lead to a catastrophic event with broader implications beyond the individual device. This is one reason that makes security in IoT a very challenging topic, since each small device has a large responsibility in their system.

Because IoT devices are sought to be very cost-effective, designers must make them very limited in resources. However, traditional security features do not scale with this cost. If designers are to use such features, they are faced with a hard choice. Either the cost of the device must go up significantly, or security features have to be removed. Unfortunately, recent history has shown that many developers choose the former to be competitive in the market [3]. Seemingly, there is no good solution to the problem, but one may be found if the security of IoT devices is reevaluated from the ground up.

To fix security vulnerabilities in IoT, the main idea is to replace traditional security primitives with more lightweight ones [7,9]. This avoids the omission of security features while keeping costs down. To this end, a key proposal is the Physical Unclonable Function (PUF). Originally envisioned back in 2002 [7,16], its core idea is the possibility to exploit random physical variations of materials and manufacturing processes. This means that given a device with a specific PUF design implemented, multiple devices that are manufactured will posses essentially unique PUFs. These can then be used as a source of identity on the device. Once such a feature is present, other fundamental security functionalities like authentication, intellectual property protection, or even secure storage become possible.

That said, since the original PUF proposal, many research works have shown an everlasting vulnerability of PUFs. This vulnerability is the fact that PUFs have known mathematical models, where the parameters of the model are randomly generated for each instance [7]. Given such models, Machine Learning (ML) algorithms can be applied to attempt to learn the unknown random parameters [17,18]. This is possible due to open communications between PUF-enabled devices, which lack cryptography (due to resource constrains), leaking data that can be used to train ML algorithms.

In this paper, we study the evolution of the field of PUFs, highlighting the most significant attack and defense mechanisms proposed in the literature. We first show how PUF proposals have evolved due to new and improved attack methods which show the lack of security of previous PUFs. Then, we present recent advances showing the currently believed secure PUFs and the direction in which countermeasures to attacks have evolved. Moreover, considering the trend of attacks through the years, we consider the potential vulnerabilities that might soon be exposed. As such, the contributions of this paper are as follows:

- We provide a clear historical evolution of both PUF defense mechanisms and ML attack techniques. This allows us to see the direction in which the field has been taken, identifying unexplored ideas as well as their probable future evolution.
- We explain our current efforts towards more secure PUF designs.
- We provide future research directions with potential for developing more secure and applicable PUFs.

The rest of this paper is organized as follows. Section 2 gives an overview of how PUFs work and how they can be classified. In Sect. 3, the ideas behind ML modeling attacks on PUFs are presented. Placing both attack and defense research in parallel, Sect. 4 provides an explanatory historical evolution of the field. Following, Sect. 5 showcases how we are dealing with PUF vulnerabilities. Section 6 gives ideas on how to best proceed to improve PUFs. Finally, Sect. 7 concludes the paper.

## 2 Physical Unclonable Function

### 2.1 Key Concepts

The basic idea of a Physical Unclonable Function (PUF) is the use of random variations in materials and manufacturing processes. Consider a generic PUF that takes some input and provides an output. The key factor is that each unique input provides a specific output, and this output is not necessarily shared between different instances of the same PUF.

Depending on the PUF design, the input and output can be very different. For example, one could create an Optical PUF [16] which is stimulated with a laser at a specific intensity and angle of incision. The output then is the light scatter pattern, randomized by the micro-structure of the material composing the PUF. It is easy to see that, since this micro-structure cannot be replicated by any current manufacturing process with enough precision, no two Optical PUFs will give the exact same output to the same input.

In the previous example, it is very likely that the output needs to be processed into a binary sequence (e.g., a cryptographic key). Since PUFs are designed to be lightweight, it is more convenient to use designs intrinsic to Integrated Circuits (ICs), which can generate output bits directly. This is the case for the Arbiter PUF [7], the first of many PUF designs directly compatible with standard IC manufacturing technologies (such as CMOS) that require no specialized hardware.

As illustrated in Fig. 1, The Arbiter PUF relies on the fact that two wires of the same length on the same circuit are not guaranteed to have the same propagation delay. By creating two unique paths in a circuit and measuring their delays, the comparison yields a binary output (based on which path had lower delay). This can be done by creating a chain of switch elements (e.g., multiplexers) each controlled by an input bit. The input bits form an input vector called the challenge, and each unique challenge will generate two symmetrical and unique paths of wires. Finally, An Arbiter (e.g., a latch) at the end of the chain compares the delays of both paths, generating a single bit output.

As will be shown in Sect. 4, PUF research has mostly evolved to develop designs based on the original Arbiter PUF. For this reason we focus on this group of PUFs, as they all work similarly. However, there are many more PUF designs which we do not cover in this paper. Interested readers are referred to [11].

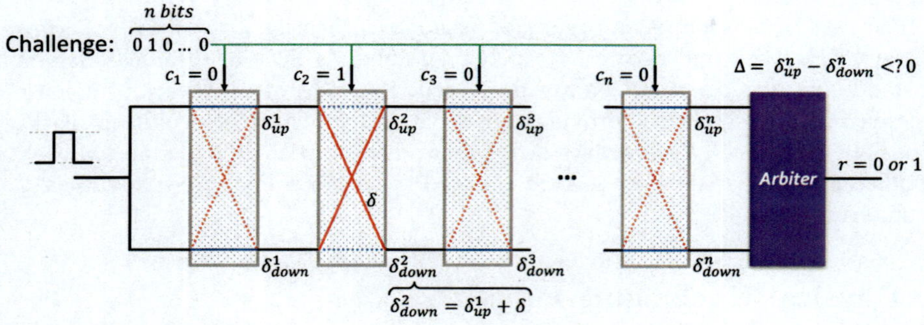


**Fig. 1.** Example of Arbiter PUF in action.

### 2.2 Use Case

In a typical PUF use case scenario, two phases can be distinguished: (i) enrollment phase and (ii) authentication phase. Firstly, in the enrollment phase (performed once after manufacturing), a large number of input-output pairs are collected from the device. These are commonly referred to as Challenge-Response Pairs (CRPs) and the interface to obtain them should be permanently disabled after this phase. The collected CRPs can be used in the repetitive authentication phase by a server provided with them. This is done by having the server "challenge" the device with inputs to which it already knows the output. Since only the authentic device can recreate the adequate responses to the challenges, an authentication is possible. The server simply checks if the given responses match the expected ones with some tolerance.

The amount of CRPs that are used per authentication query varies depending on the application scenario. It is obvious that for a single-bit output more that one must be collected to securely authenticate a device. The number of CRPs that are required will depend on parameters such as the number of devices in the system or the tolerance, and has already been estimated for Arbiter PUFs [10].

## 3 Machine Learning Modeling Attacks

To understand how the previously described PUF scheme can be attacked we must first understand its vulnerabilities. Consider the use case described in Sect. 2.2. In this scenario, two fundamental security requirements exist.

Firstly, any abnormal behavior from the device must be detected. This can come as a result of natural breakdown or random failure, but also due to malicious tampering. As long as one can detect that the device is no longer operating as expected, certain measures can be taken to solve the issue (e.g., replacing the device). In this regard, PUFs have been shown to be tamper-evident [7], since their input-output mapping is heavily tied to the device's physical condition. Taking the Arbiter PUF as an example, the wire delays are highly susceptible to changes in supply voltage, temperature, resistance, capacitive coupling, etc. This means that physical tampering or degradation of the device is almost guaranteed to change the PUF's outputs and be detected by the server.

Secondly, only legitimate devices should be authenticated by the server. This means that a malicious party neither should be able to deploy their own device in the system, nor impersonate an existing device. If attackers are unable to predict the responses to the challenges that the server gives, they cannot perform the previous. Here, the unclonability property of PUFs comes into play.

On paper, PUF schemes provide security guarantees regarding their unclonability, however in reality they can fall victim to various attacks. In particular, due to the lack of secure communications in resource-constrained devices, the information between the devices and the server is usually transmitted in the clear. This means that an attacker can passively observe these exchanges and collect the CRPs that were used. To prevent this from being an issue the server is programmed to never reuse the same CRP in an authentication, and discards them. However, this is not enough to ensure security, as a software model of the PUF might be able to be built based on the exchanged CRPs.

By defining a mathematical model of a PUF, where some randomized parameters define each physical instance, an attacker can attempt to approximate the correct model for any given instance. This is done with well-known Machine Learning (ML) algorithms with the collected CRPs as training data. If enough training data is provided, it is provable that the model can be replicated and the original device impersonated [1,2,17,18].

## 4 Historical Evolution

In this section, we give an overview of the evolution of the landscape surrounding PUF research, as depicted in Fig 2.

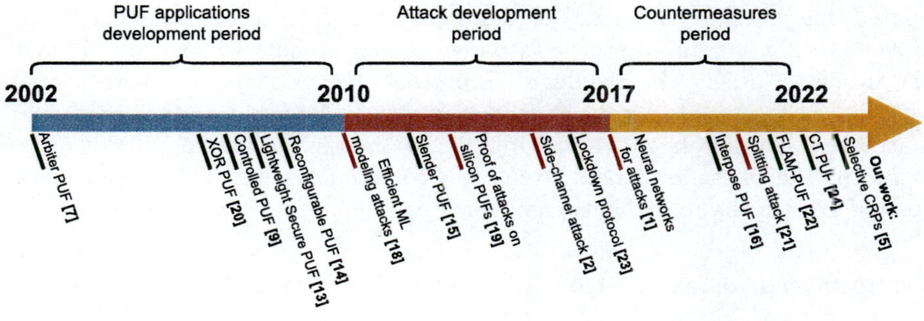


**Fig. 2.** Evolution of PUF research.

### 4.1 PUF Application Development Period

The concept of PUF appeared in 2002 with the proposal of the Optical PUF [16] and the Arbiter PUF [7][1]. It quickly became apparent how an intrinsic PUF design like the Arbiter PUF was more appealing due to its extremely low-cost implementation. Stemming from this, the research works that followed focused on developing the use of Arbiter PUFs in device identification applications [9,10,19]. However, even back then, researchers were already considering the possibility of software models being built based on the exchanged CRPs between server and device [7,9]. That said, it was yet to be shown that such a model could be built. Still, certain proposals for improving the security of PUFs were made. Most notably, the concept of controlled PUF (originally designed to ensure reliability [6]) was established for security purposes [8], and more expensive designs based on the Arbiter PUF started appearing (e.g., XOR PUF [19], Lightweight Secure PUF [12], or Reconfigurable PUF [13]). This marks a key point in the history of PUF research, as the original lightweight Arbiter PUF proposal was being turned into a more heavyweight design for security concerns.

### 4.2 Attack Development Period

By deviating from the lightweight implementation by adding additional control logic (controlled PUFs) or making heavier designs, the problem of resource-constrainedness reappeared. To make matters worse, in 2010 an extensive paper succeeded in cloning various PUFs in realistic conditions [17], and was extended to real PUFs implemented in FPGAs in 2013 [18]. Moreover, 2015 saw the exposure of a serious side-channel vulnerability of Arbiter PUFs regarding their reliability, when in [2] a method for learning PUF model parameters based on their reliability was shown to scale linearly with the complexity of the PUF (instead of exponentially), outperforming previous attacks.

During this advent of successful attacks, some preventive measures where proposed, emphasizing the need for more costly designs to ensure security. In this context, some notable protocols based on controlled PUF designs showed great promise. Namely, the Slender PUF protocol [14] and the Lockdown protocol [22] both attempted to prevent modeling attacks by avoiding the core issue: the attackers had access to training data. This marked a new idea for defense mechanisms which saw designers attempting to hide the true input-output mappings of PUFs in the most lightweight form possible.

### 4.3 Countermeasures Period

Unfortunately, using controlled PUFs can be costly and bring other potential vulnerabilities due to the additional modules [4]. For that reason, a similar methodology was adopted in [15], which proposed the Interpose PUF. This design was

[1] Technically, the concept of PUF can be traced back even further [11]. However, it was not properly defined yet and its use for resource-constrained devices was not envisioned.

not a controlled PUF, but still managed to poison the training data available to attackers. However, [20] showed that a specialized attack could still break it. This attack was based on the use of neural networks, which since 2017 [1] showed their potential in this field.

Ultimately, a pure (uncontrolled) and secure PUF has not been proven to exist as of today. Fortunately, some designs have gotten close by learning some key lessons from past PUFs and attacks. These are mainly: (i) avoid complex control logic that requires synchronization between server and device, or vulnerable modules that could be exploited, and (ii) prevent the attacker from obtaining quality training data (CRPs) instead of attempting to create a heavier and more complex PUF model. Currently, the FLAM-PUF [21] and the CT PUF [23] are considered to be secure. Both implement some simple additional control logic over an Arbiter PUF that does not significantly change the basic use case for PUFs, but prevent attackers from obtaining good training data for their models.

## 5 Our Work

The original design principle of PUFs was an extremely lightweight security primitive, however due to successful attacks the direction of new designs and defense mechanisms has deviated from this. To this end, our research focuses on designing protocols with no hardware requirements that can still ensure security against ML modeling attacks.

For this purpose, in our latest work [5], we studied a form of adversarial attack on the ML algorithms used to model XOR PUFs (derived from Arbiter PUFs). We note that up until now, previous works have assumed that the collection of CRPs from the device during the enrollment phase is performed at random. We question this assumption and attempt to use selective CRP subsets that leak less information about the underlying PUF model. By having the server use CRPs that are similar to one another in the short-term, but change more in the long-term, we avoid giving attackers information about the PUF model's parameters. This limits any adversary in how fast they can obtain enough data to fully model the PUF. Our results show that our best mechanism, called Binary Shifted Pattern (BSP) CRPs, can reduce the modeling accuracy of attacks by 10 to 15%, which is enough to prevent most attacks from succeding in impersonating a device.

An extension of this research is being worked on, where we are developing a mathematical model for estimating the modeling accuracy of attacks based on the training CRPs available to the attacker. We plan to elaborate on the quantitative aspects of how secure PUFs are when our proposed mechanism is used, in regards to the time it would take an attacker to break it. Additionally, we see our future efforts leading to a PUF proposal with certain security guarantees, while not using any additional control logic. Finally, we will consider the possibility of an attacker directly querying the PUF for responses to specific challenges, in the event they know the selective CRP subset.

## 6 Future Directions

Based on the historical review of PUF literature we can identify two main directions for future research to improve PUFs. Firstly, it is clear that the current preferred method of defense against ML modeling attacks is the use of controlled PUFs [21,23]. Pursuing this direction further, one can attempt to reduce the hardware overhead of the additional control logic. To do this, a new controlled PUF design which avoids heavy modules like hash functions or random number generators is required. We think that such a research direction is very likely to be explored in the future, especially due to the possibility of new attacks succeeding in breaking novel PUFs.

Until now, most proposals for new PUF designs test them against known attacks, and the FLAM-PUF [21] and the CT PUF [23] are no exception. The issue with such a methodology is that all modern controlled PUFs require tailor-made attacks to have any significant chance at creating a clone of the PUF [20]. By testing ML algorithms in a naive manner (i.e., the algorithm is given CRPs of the whole PUF that is treated as a blackbox), the security of new PUFs is not truly guaranteed. Authors tend to give some proof of the security of their PUFs concerning what the attacker can realistically do [23], but this is still not a certainty. This makes it so that after the publication of a new PUF, an attack that proves its vulnerability is usually discovered within a year or two [20]. That said, looking from another perspective, researching viable ML attacks against the FLAM-PUF and the CT PUF is an interesting future direction.

Finally, a not-so-popular idea for securing PUFs is the use of pure (uncontrolled) PUFs with some defense mechanism. This is the case of our work, and has the advantage of being more lightweight. The reason why this is not usually explored is that such ideas have failed in the past [1,2,17,18,20]. Mainly, evolving the Arbiter PUF to the XOR PUF, the Lightweight Secure PUF, or even the Interpose PUF, did not have the desired result. More efficient attacks were discovered that could still break these constructions despite their increased hardware overhead. However, similar to what our work tries to accomplish, we think that it is worth researching supporting mechanisms to already existing PUF designs that require no additional hardware on the devices.

## 7 Conclusion

This paper analyzes the state-of-the-art and evolution of research on Physical Unclonable Functions (PUFs). These security primitives exploit random material and manufacturing process variations to implement lightweight identification capabilities, which are especially useful in resource-constrained devices like IoT devices. However, even when the concept of PUF was first proposed, researchers were already aware of the possibility of creating software clones. If such a thing could be done, then the unclonability of PUFs would be undermined and their security broken.

By reviewing the literature, we can see that this is exactly what happened. The past decade has been a constant arms race where new and improved PUFs were proposed, just to be proven vulnerable to novel ML attacks. Currently, the preferred method for securing PUFs is the use of additional control logic, making a controlled PUF. This has led to the FLAM-PUF and the CT PUF, which are considered secure by today's standard. However, based on previous trends, we expect new attacks to appear within two years proving the opposite.

Creating a truly secure PUF that does not deviate from the original Arbiter PUF lightweight design principle seems unrealistic. However, current designs (FLAM-PUF and CT PUF) are fairly lightweight for controlled PUFs, and further improvement is still possible. Additionally, defense mechanisms that can be placed on top of pure (uncontrolled) PUFs, like the Arbiter PUF, that do not require additional hardware are promising. Specifically, our focus is on ML adversarial attacks that can prevent attackers from training accurate models of these PUFs.

## References


1. Alkatheiri, M.S., Zhuang, Y.: Towards fast and accurate machine learning attacks of feed-forward arbiter PUFs. In: 2017 IEEE Conference on Dependable and Secure Computing, pp. 181–187 (2017). https://doi.org/10.1109/DESEC.2017.8073845
2. Becker, G.T.: The Gap between promise and reality: on the insecurity of XOR arbiter PUFs. In: Güneysu, T., Handschuh, H. (eds.) CHES 2015. LNCS, vol. 9293, pp. 535–555. Springer, Heidelberg (2015). https://doi.org/10.1007/978-3-662-48324-4_27
3. Cheruvu, S., Kumar, A., Smith, N., Wheeler, D.M.: Demystifying Internet of Things Security. Apress, Berkeley, CA (2020). https://doi.org/10.1007/978-1-4842-2896-8
4. Delvaux, J., Gu, D., Schellekens, D., Verbauwhede, I.: Secure lightweight entity authentication with strong PUFs: mission impossible? In: Batina, L., Robshaw, M. (eds.) CHES 2014. LNCS, vol. 8731, pp. 451–475. Springer, Heidelberg (2014). https://doi.org/10.1007/978-3-662-44709-3_25
5. Ferens, M., Dushku, E., Kosta, S.: Securing PUFs against ml modeling attacks via an efficient challenge-response approach. In: IEEE International Workshop on the Security, Privacy, and Digital Forensics of Mobile Systems and Networks (Mobisec) (2023, in press)
6. Gassend, B., Clarke, D., van Dijk, M., Devadas, S.: Controlled physical random functions. In: 18th Annual Computer Security Applications Conference, 2002. Proceedings, pp. 149–160 (2002). https://doi.org/10.1109/CSAC.2002.1176287
7. Gassend, B., Clarke, D., van Dijk, M., Devadas, S.: Silicon physical random functions. In: Proceedings of the 9th ACM Conference on Computer and Communications Security, pp. 148–160. CCS '02, Association for Computing Machinery, New York, NY, USA (2002). https://doi.org/10.1145/586110.586132
8. Gassend, B., Dijk, M.V., Clarke, D., Torlak, E., Devadas, S., Tuyls, P.: Controlled physical random functions and applications **10**(4) (2008). https://doi.org/10.1145/1284680.1284683
9. Gassend, B., Lim, D., Clarke, D., van Dijk, M., Devadas, S.: Identification and authentication of integrated circuits. Concurr. Comput. Pract. Exp. **16**(11), 1077–1098 (2004). https://doi.org/10.1002/cpe.805

10. Lim, D., Lee, J., Gassend, B., Suh, G., van Dijk, M., Devadas, S.: Extracting secret keys from integrated circuits. IEEE Trans. Very Large Scale Integr. (VLSI) Syst. **13**(10), 1200–1205 (2005). https://doi.org/10.1109/TVLSI.2005.859470
11. Maes, R., Verbauwhede, I.: Physically Unclonable Functions: A Study on the State of the Art and Future Research Directions, pp. 3–37. Springer, Berlin, Heidelberg (2010). https://doi.org/10.1007/978-3-642-14452-3_1
12. Majzoobi, M., Koushanfar, F., Potkonjak, M.: Lightweight secure PUFs. In: 2008 IEEE/ACM International Conference on Computer-Aided Design, pp. 670–673 (2008). https://doi.org/10.1109/ICCAD.2008.4681648
13. Majzoobi, M., Koushanfar, F., Potkonjak, M.: Techniques for design and implementation of secure reconfigurable PUFs. ACM Trans. Reconfigurable Technol. Syst. **2**(1) (2009). https://doi.org/10.1145/1502781.1502786
14. Majzoobi, M., Rostami, M., Koushanfar, F., Wallach, D.S., Devadas, S.: Slender PUF protocol: a lightweight, robust, and secure authentication by substring matching. In: 2012 IEEE Symposium on Security and Privacy Workshops, pp. 33–44 (2012). https://doi.org/10.1109/SPW.2012.30
15. Nguyen, P.H., Sahoo, D.P., Jin, C., Mahmood, K., Rührmair, U., van Dijk, M.: The interpose PUF: secure PUF design against state-of-the-art machine learning attacks. IACR Trans. Cryptogr. Hardw. Embed. Syst. **2019**(4), 243–290 (2019). https://doi.org/10.13154/tches.v2019.i4.243-290
16. Pappu, R., Recht, B., Taylor, J., Gershenfeld, N.: Physical one-way functions. Science **297**(5589), 2026–2030 (2002). https://doi.org/10.1126/science.1074376
17. Rührmair, U., Sehnke, F., Sölter, J., Dror, G., Devadas, S., Schmidhuber, J.: Modeling attacks on physical unclonable functions. In: Proceedings of the 17th ACM Conference on Computer and Communications Security, pp. 237–249. CCS '10, Association for Computing Machinery, New York, NY, USA (2010). https://doi.org/10.1145/1866307.1866335
18. Rührmair, U., et al.: PUF modeling attacks on simulated and silicon data. IEEE Trans. Inf. Forensics Secur. **8**(11), 1876–1891 (2013). https://doi.org/10.1109/TIFS.2013.2279798
19. Suh, G.E., Devadas, S.: Physical unclonable functions for device authentication and secret key generation. In: 2007 44th ACM/IEEE Design Automation Conference, pp. 9–14 (2007)
20. Wisiol, N., et al.: Splitting the interpose PUF: a novel modeling attack strategy. IACR Trans. Cryptogr. Hardw. Embed. Syst. **2020**(3), 97–120 (2020). https://doi.org/10.13154/tches.v2020.i3.97-120
21. Wu, L., Hu, Y., Zhang, K., Li, W., Xu, X., Chang, W.: FLAM-PUF: a response-feedback-based lightweight anti-machine-learning-attack PUF. IEEE Trans. Comput. Aided Des. Integr. Circuits Syst. **41**(11), 4433–4444 (2022). https://doi.org/10.1109/TCAD.2022.3197696
22. Yu, M.D., Hiller, M., Delvaux, J., Sowell, R., Devadas, S., Verbauwhede, I.: A lockdown technique to prevent machine learning on PUFs for lightweight authentication. IEEE Trans. Multi-Scale Comput. Syst. **2**(3), 146–159 (2016). https://doi.org/10.1109/TMSCS.2016.2553027
23. Zhang, J., Shen, C., Guo, Z., Wu, Q., Chang, W.: CT PUF: configurable tristate PUF against machine learning attacks for IoT security. IEEE Internet Things J. **9**(16), 14452–14462 (2022). https://doi.org/10.1109/JIOT.2021.3090475

# A Framework for Next Generation Cloud-Native SDN Cognitive Resource Orchestrator for IoTs (NG2CRO)

Hafiza Kanwal Janjua, Ignacio de Miguel(✉), Ramón J. Durán Barroso, Maryam Masoumi, Soheil Hosseini, Juan Carlos Aguado, Noemí Merayo, and Evaristo J. Abril

Universidad de Valladolid, Valladolid, Spain
{hafizakanwal.janjua,ignacio.demiguel,ramon.duran,maryam.masoumi, soheil.hosseini,juancarlos.aguado.manzano,noemer,ejabril}@uva.es

**Abstract.** SDN (Software Define Networking) and NFV (Network Function Virtualization) are the key enablers for 5G systems and also open many doors in the cloud-native application. Besides, it invites new challenges to the efficiency and scalability of resource management. This work aims to provide a cognitive framework for 5G resource and service orchestration in a cloud-native SDN environment. The proposed NG2CRO framework resource orchestrator is designed to adapt the network's self-learning capabilities and dynamicity while taken on to account the network's Markovian properties and diverse service requirements. We consider incorporating AI (Artificial Intelligence) techniques specifically RL (Reinforcement Learning) methodologies because literature has shown that these techniques can efficiently address and comply with the current dynamic behaviors and heterogeneity of 5G services and applications. In conclusion, both benefits and liabilities are discussed of incorporating AI specifically RL into resource orchestration practices that provide us with future research challenges.

**Keywords:** Next Generation 5G Network · Artificial Intelligence · Network Automation · Software-Define Networking · Cloud-Native SDN · Resource Orchestration

## 1 Introduction

SDN orchestration facilitates energy and resource optimization for the Internet of Things (IoT) industry. SDN allows efficient, cost-effective, and dynamic scalability of resources in cloud and edge computing services [1,12]. The evolution

This work is part of the IoTalentum project, which has received funding from the EU H2020 research and innovation programme under the MSCA grant agreement No 953442. It is also supported by Consejería de Educación de la Junta de Castilla y León and the European Regional Development Fund (Grant VA231P20), and the Spanish Ministry of Science of Innovation and the State Research Agency (Grant PID2020-112675RB-C42 funded by MCIN/AEI/10.13039/501100011033).


R. Mehmood et al. (Eds.): DCAI 2023, LNNS 741, pp. 399–407, 2023.
https://doi.org/10.1007/978-3-031-38318-2_39

of the IoT industry has raised the demand for cloud data centers where tens of thousands of compute nodes are connected through thousands of networks, creating avenues for manageability and performance issues. Cloud data centers need dynamic provision computing and network resources to adapt to the fluctuating demands of customers [17].

With advents in software-defined clouds (SDC) that integrate SDN and cloud resource management where the programmable controller provides dynamic and autonomic configuration and management of the underlying resources; cognition of such scenario can predict the requirement and increase the efficiency in such scenarios [2,10]. Despite the increasing popularity of studies for joint resource optimization in the cloud environment with SDN technology, the realization is still limited for developing integrated management platforms providing simultaneous controllability and energy optimization of computing and networking infrastructures. The testing cost and energy efficiency in real-time are required for migration to new technologies [4,6].

Moreover, 5G networks are complex, large, and diverse with heterogeneous services which require high flexibility and efficiency to meet the SLA and QoS agreed upon by the services provider. However, due to the lack of flexibility in traditional 4G networks, when comes to the dynamics of 5G systems due to the heterogeneous services. Therefore, the current resource optimization techniques are not suitable for the current 5G infrastructure, which, is faced with challenges of SLA and QoS violation due to the incapability to handle the system dynamicity.

Hence, there is a need to have advanced cognitive control methods that can efficiently maneuver and orchestrate the dynamicity of the network demands. Consequently, studies extrapolated that RL-based cognitive approaches are beneficial to learn and exploring the dynamics of the network autonomously which is capable to support service flexibility and can provide an enhanced user experience in terms of SLA and QoS fulfillment [8].

### 1.1 Aims and Objectives

The aim of this project is an association with the IoTalentum project [7] specifically to the topic ESR7: "Cognitive orchestrator of MEC and network resources". Hence, the vision is to design a Cloud-Native SDN joint resource orchestration system for the network and service resources. Also, it should optimize the energy and cost profile via pivoting the ability of self-learning and self-management to guarantee performance in the 5G systems.

Moreover, the integration of Cloud and SDN platforms should leverage the framework functioning at the controller via intelligent network management, optimal resource utilization, load balancing, and service orchestration to provide IaaS (Infrastructure-as-a-Service) and SaaS (Software-as-a-Service) to users that will operate the MEC and the underlying IoT network of IoTalentum architecture.

We aim to leveraged the objectives of ESR7 by integrating the efficient, optimized, and proactive performance-aware Deep Reinforcement Learning techniques to develop the advanced decision-making pattern for energy monitoring, protection, and control mechanisms to deliver optimal SLA and QoS in IaaS and SaaS environments.

We designed some research questions with the consideration of 5G network resources orchestration using artificial intelligence specifically reinforcement learning and Machine learning along with the major problems in current network resource management. Table 1 provides an overview of the research questions in relation to the problems.

**Table 1.** Problem and Research Questions

| Problem (P) | Research Questions (RQ) |
|---|---|
| **P1:** Dynamicity in traffic demands and users' mobility | **RQ1:** How AI techniques has performed in literature orchestration in 5G Network? |
| **P2:** Dynamic systems render markovian properties due to heterogeneous network devices | **RQ2:** How RL-based methods can be useful for resource orchestration to to address the dynamicity? |
| **P3:** Need for safe orchestration due to constraints of the physical system even in virtualized SDN NFV infrastructure | **RQ3:** What will be use of constraint-aware RL methods having an impact on SLA and QoS of the 5G slicing network scenario? **Safe Orchestration is Required!!!** which is termed as making the resource orchestration policies which should consider the physical system constraint such as Bandwidth, CPU, RAM and Memory |

## 2 Background Studies

In the research area of 5G resource and service orchestration, we will specifically discuss the current state-of-the-art projects related to Cloud and SDN management and orchestration likewise IoTalentum.

### 2.1 Projects Related to 5G Resource Orchestration

Virtual Network Functions as a Service over virtualized infrastructures (T-NOVA) focuses on network functions as a service (NFaaS) [9]. This project provides a solution for deploying and managing NFaaS, which allows operators to create and deliver new services efficiently on virtualized infrastructure. 5GEx focuses on creating an ecosystem that allows the exchange of resources and services to and from different 5G experimental infrastructures [3]. The aim of 5GEx is to facilitate cross-domain experimentation and validate 5G technologies. Sonata [13] focuses on developing a service-oriented 5G architecture that supports the dynamic creation and management of network services, including smart cities, Industry 4.0, and virtual/augmented reality. Vital [16] is a 5G-related project that focuses on the development of an integrated framework for the design, deployment, and management of 5G networks and services. It addresses the challenges associated with the deployment of 5G networks, including network slicing, orchestration, and automation. 5G Transformer [11] aims to develop a flexible and scalable 5G architecture that can support the dynamic creation and management of network slices. The project's focus is on developing an SDN-based architecture that allows the efficient and flexible orchestration of 5G network resources. 5G Growth presents an ecosystem for the co-creation of 5G services and applications [14]. It provides a platform to allow collaboration between different stakeholders to develop and deploy new 5G services and applications. Inspire 5G Plus [15] project aims to develop an integrated platform to create, deploy, and manage 5G services and applications to enable the seamless integration of several 5G components and services, such as network slicing, edge computing, and AI. 5G Zorro aims to develop a security framework for 5G networks [5]. The project aims to address the security challenges associated with the deployment of 5G networks by proposing a security architecture that can ensure the confidentiality, integrity, and availability of 5G network resources.

**Table 2.** Projects related to 5G resource Orchestration

| Project | Domain | Resource Orchestration Goals |
|---|---|---|
| T-NOVA [9] | Cloud, NFV, SDN | Enable end-to-end service provisioning via NFVaaS, VNFaaS, NFVIaaS |
| 5GEx [3] | Cloud, NFV, SDN | Enable cross-domain orchestration of 5G services |
| Sonata [13] | Cloud, NFV SDN | Develop an integrated service platform |
| Vital [16] | Cloud, NFV, SDN, IoT | Develop a framework for IoT and smart city |
| 5G Transformer [11] | Cloud, NFV, SDN | Develop SDN service-oriented 5G network architecture |
| 5G Growth [14] | NFV | Develop 5G infrastructure and services for rural areas |
| Inspire 5G Plus [15] | NFV, SDN, DLTs, IoT | Develop 5G infrastructure and services for IIoT |
| 5G Zorro [5] | DLTs | Develop an integrated platform for 5G services |

## 3 Proposed NG2CRO Framework

We designed a cloud-native SDN-based framework addressing cognitive capabilities using RL methods. Also, the proposed architecture is designed to achieve an intelligent, autonomous, and self-managed network and resource orchestration system. The design of the proposed system is separated into four logical layers that are; the user plane (UP), the data plane (DP), and the control plane

(CP) (further CP is subdivided into the following: management, orchestration, and provisioning), management plane (MP) and application plane (AP). On the top of Fig. 1 we described the use cases of NG2CRO. Also, we aimed to solve the problems related to network orchestration addressed in Sect. 1 by separating each part of the network (NVF-I, VNF-O, and slice orchestration & management) by separating each module in master-slave architecture. Here in master-slave architecture, each part will do its resource management on its own but get the updates from the main controller, which can lower the delays and latency when operations are directly performed by the main controller.

The main aim of the four logical layers is to adapt the convergence of DP into CP and MP that leads towards the AP. The data plane performs data transfer to upper abstract layers in terms of infrastructure and configuration information. The control plane performs the following tasks that are, monitoring of network function virtualization management (NFVI-M), Virtual network function orchestration (VNF-O), VNF life cycle management, and Slice management. Resource provisioning and orchestration are the capabilities of CP, but in contrast, the convergence of CP operations must be adaptable to DP as the MEC environment is dynamic and heterogeneous. Moreover, to maintain service continuity the CP convergence to DP must be addressed in terms of infrastructure adaptability, service flexibility, dynamic updates, and resource provisioning.

Following, the process flow of NG2CRO is depicted in Fig. 1 where the flow is labeled with numbers. At 1. The compatible application scenarios are discussed and also the application services related to the scenarios are given. 2. The enabling fronthaul and backhaul technologies are given for 5G applicability. 3. Next the information flows goes toward and controller in the control plane. Here the information flow will be in back fourth manner between the controller and applications. 4. Here we have the cloud-native SDN controller to support the cloud-native 5G applications.

Further on, the resource orchestration operations are carried out by the controller, here in step number 10 the policy optimization, and model aggregation of AI methods is performed. The data fed into the main brain is acquired from the modules which are VIM (5) and NFVO (7). At level 10 NG2CRO will make use of RL techniques to build the resource orchestration policies and send these policies to NFV and VNf-O orchestrator components. Next, it incorporates the TL (transfer learning) techniques also to build future predictions, but here we can face the problems with data privacy issues which is an area to address.

At level (9) bandwidth management is carried out among 5G slices, we have carried out the experimentation of this module here [8]. Our future work includes further experimentation of modules 7 and 6. Lastly, the main brain is a main controller that manages all of the operations based on the given states as traffic information and sends back the required action based on the current state and optimal policies.

The Proposed NG2CRO is the subpart of the IoTalentum [7] project and it aimed to solve the issue in current state-of-the-art resource orchestration projects by adhering to the standards used by existing projects. Furthermore, we plan to perform the experimentation of each module of NG2CRO with RL techniques and integrate all of them with the main controller with the aim of optimizing the RL methods compliant with network management.

### 3.1 Discussion

The platforms given in the Table 2 have distinct goals and sets of capabilities, but they all seek to make it possible to manage and orchestrate SDN resources effectively in a scalable and flexible way. However, certain platforms could have restrictions or gaps in their SDN-specific functionality, or they might not support setups with many vendors. For instance, certain systems could lack sophisticated SDN-specific capabilities like effective flow management, network slicing, or network function chaining.

A platform's capacity to serve complex, multi-vendor setups may also be constrained, which can make integration and administration more difficult. It's crucial to assess the specific requirements of the network and consider the advantages and disadvantages of each platform before selecting the best SDN management and orchestration platform. We aimed to design a platform that tried to suit the needs of current and help to efficiently manage and orchestrate the SDN resources by carefully considering these criteria.

The integration of RL into SDN and NFV resource orchestration has the potential to have beneficial as well as detrimental impacts on the interoperability, flexibility, and scalability of the infrastructure. Also, we strive to integrate RL algorithms into our work. The initial part to integrate RL into this work is been published in [8].

On the plus side, adding RL to SDN and NFV resource orchestration can increase the infrastructure's automation and intelligence, which can further boost interoperability and adaptability. For instance, RL algorithms may be applied to forecast traffic patterns, optimize resource allocation, and detect abnormalities in real time, allowing for more effective use of network resources.

Additionally, RL can provide a dynamic response to shifting network settings, enhancing the infrastructure's adaptability. RL algorithms may make SDN and NFV resource orchestration more sensitive to changes in traffic patterns, network topology, and user behavior by continually learning and adjusting to new scenarios.

The integration of RL into SDN and NFV resource orchestration might hinder scalability and interoperability, which is a serious drawback. For instance, for RL algorithms to be effective, huge volumes of data must be collected and processed instantly. This can make scaling difficult, especially in large-scale networks where there is a lot of data to manage.

The lack of a generally adopted standard for RL integration into SDN and NFV resource orchestration might also cause compatibility and interoperability problems. This can make it challenging for network operators to combine various systems and parts, which can affect the infrastructure's overall interoperability and adaptability.

Overall, integrating RL into SDN and NFV resource orchestration has the potential to have beneficial as well as adverse impacts on scalability, flexibil-

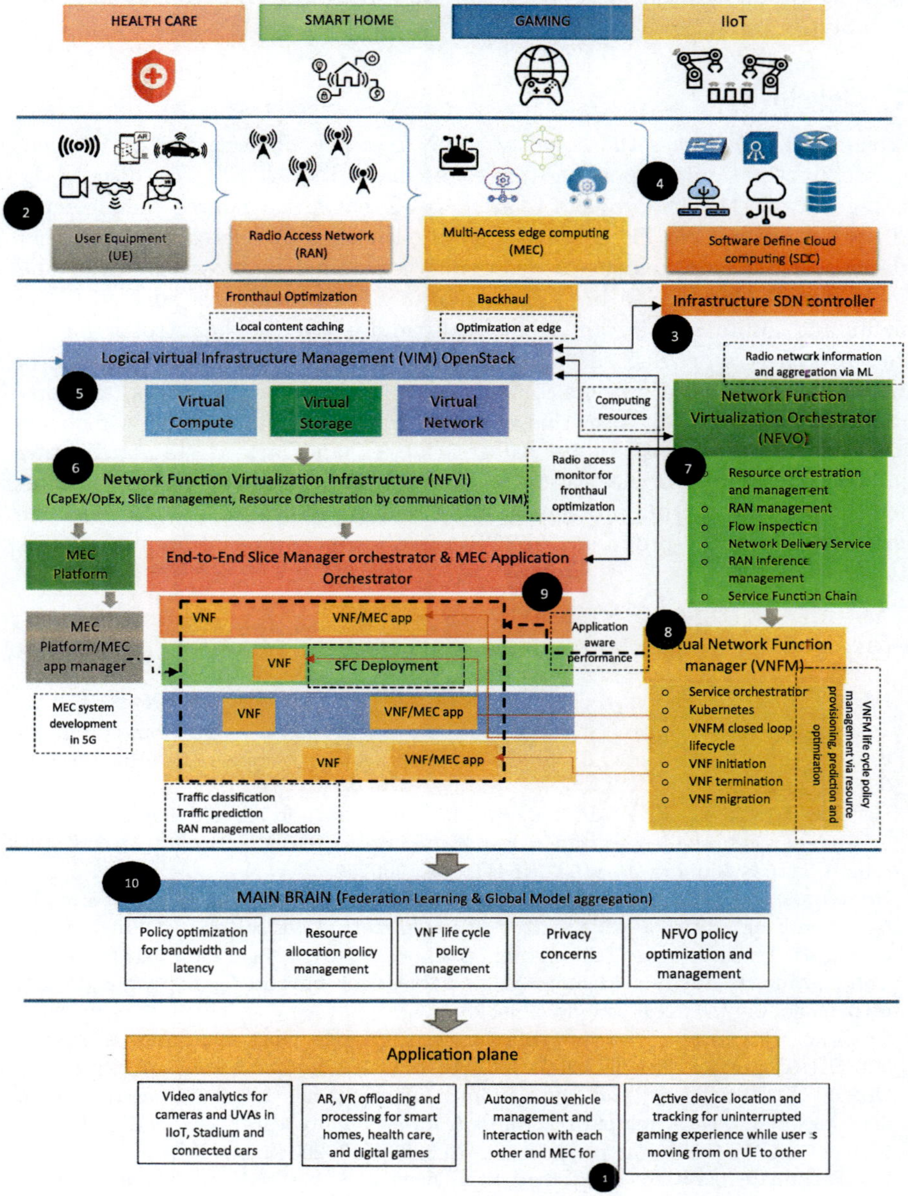


**Fig. 1.** Proposed NG2CRO Framework

ity, and interoperability. It is crucial to carefully weigh the possible advantages and challenges of integrating RL into SDN and NFV resource orchestration as well as to build standards and best practices to guarantee interoperability and scalability.

### 3.2 Conclusions

In this work, we adress the cognitive SDN resource orchestrator of IoTalentum architecture and proposed a framework named NG2CRO. We discuss the existing state-of-the-art projects related to 5G network resource management and orchestration. We discuss and analyze the existing listed projects that address network management using cognitive techniques across the cloud and SDN domains. Moreover, a brief discussion is given providing a curated list of benefits and limitations for an AI-enabled cloud-native SDN controller for the resource orchestrator. Such as RL techniques will be very useful in terms of improved efficiency and automation, real-time analytics and insights, and predictive maintenance. But also, this invites crucial points to consider which are privacy issues, AI explainability, Adversarial AI, and lack of standardization regarding RL resource orchestration frameworks. We aimed to address the complexity and cost-incurring problem of implementing RL techniques for resource orchestration and the proposed NG2CRO framework is the initial step to this path.

## References


1. Bega, D., Gramaglia, M., Perez, R., Fiore, M., Banchs, A., Costa-Pérez, X.: Ai-based autonomous control, management, and orchestration in 5g: from standards to algorithms. IEEE Netw. **34**(6), 14–20 (2020). https://doi.org/10.1109/MNET.001.2000047
2. Benzaid, C., Taleb, T.: Ai-driven zero touch network and service management in 5g and beyond: challenges and research directions. IEEE Netw. **34**(2), 186–194 (2020). https://doi.org/10.1109/MNET.001.1900252
3. Bernardos, C.J., Gerö, B.P., Di Girolamo, M., Kern, A., Martini, B., Vaishnavi, I.: 5gex: realising a Europe-wide multi-domain framework for software-defined infrastructures. Trans. Emerg. Telecommun. Technol. **27**(9), 1271–1280 (2016)
4. Camps-Mur, D., et al.: AI and ML - enablers for beyond 5g networks (2021). https://doi.org/10.5281/zenodo.4299895
5. Chainho, P., et al.: Overview of the security and trust mechanisms in the 5GZORRO project. EUCNC 2021 (2021)
6. Duan, Q.: Intelligent and autonomous management in cloud-native future networks-a survey on related standards from an architectural perspective. Future Internet **13**, 42 (2021). https://doi.org/10.3390/fi13020042
7. IoTalentum: Cognitive orchestrator of mec and network resources. http://iotalentum.eu/esrs. Accessed 8 May 2023
8. Janjua, H.K., et al.: Efficient optimization of actor-critic learning for constrained resource orchestration in ran with network slicing. In: 2023 26th Conference on Innovation in Clouds, Internet and Networks and Workshops (ICIN), pp. 100–104 (2023). https://doi.org/10.1109/ICIN56760.2023.10073489

9. Kourtis, M.A., et al.: T-nova: an open-source MANO stack for NFV infrastructures. IEEE Transa. Netw. Serv. Manage. **14**(3), 586–602 (2017)
10. Latah, M., Toker, L.: Artificial intelligence enabled software defined networking: a comprehensive overview. IET Netw. **8**, 79–99 (2018)
11. Mangues-Bafalluy, J., et al.: 5G-TRANSFORMER service orchestrator: design, implementation, and evaluation. In: 2019 European Conference on Networks and Communications (EuCNC), pp. 31–36. IEEE (2019)
12. Manias, D.M., Shami, A.: The need for advanced intelligence in NFV management and orchestration. IEEE Netw. **35**(1), 365–371 (2021). https://doi.org/10.1109/MNET.011.2000373
13. Martrat, J.: SONATA approach towards DevOps in 5G networks. In: SDN World Congress (2017)
14. Nguyen, T.H., Yoo, M.: A VNF descriptor generator for tacker-based NFV management and orchestration. In: 2018 International Conference on Information and Communication Technology Convergence (ICTC), pp. 260–262. IEEE (2018)
15. Ortiz, J., et al.: INSPIRE-5Gplus: intelligent security and pervasive trust for 5G and beyond networks. In: Proceedings of the 15th International Conference on Availability, Reliability and Security, pp. 1–10 (2020)
16. Trichias, K., et al.: Vital-5G: innovative network applications (NetApps) support over 5G connectivity for the transport & logistics vertical. In: 2021 Joint European Conference on Networks and Communications & 6G Summit (EuCNC/6G Summit), pp. 437–442. IEEE (2021)
17. Zhang, S., Zhu, D.: Towards artificial intelligence enabled 6G: state of the art, challenges, and opportunities. Comput. Netw. **183**, 107556 (2020). https://doi.org/10.1016/j.comnet.2020.107556

# Decentralized Identifiers for IoT Systems: Current Status and Future Vision

Muhammad Waleed(✉), Knud Erik Skouby, and Sokol Kosta

Aalborg University, Copenhagen, Denmark
{muhammadw,skouby,sok}@es.aau.dk

**Abstract.** Traditional blockchain technology poses challenges to implementing practical and sustainable decentralized identity solutions, such as scalability issues, high transaction fees, and energy consumption. However, Decentralized Identifiers (DIDs) with IOTA's Tangle technology offer a promising solution to these challenges. This paper explores IOTA's recent version Chrysalis, which provides a more energy-efficient and scalable infrastructure for implementing DIDs, particularly for heterogeneous IoT devices. The system enables these devices to create tamper-proof digital identities that are easily verified and used in various IoT applications. Additionally, the system is highly secure and cost-effective, with low transaction fees and minimal resource requirements. We discuss the current status of the work and provide a future vision for the IOTA-based DID system. This technology offers a more secure, efficient, and sustainable approach to decentralized identity management than traditional systems. The potential of IOTA's DID system for IoT devices is significant, and we anticipate that it will lead to innovative use cases and solutions in the future.

**Keywords:** IoT · Distributed Ledger Technology · IOTA Chrysalis · Decentralized Identifiers · Scalability and Data Security

## 1 Introduction

Decentralized Identifiers (DIDs) have emerged as a promising solution for secure and privacy-preserving identity management in the Internet of Things (IoT) ecosystem [7]. DIDs enable the creation of unique and immutable identities that can be owned and controlled by individuals or organizations without the need for intermediaries such as governments or corporations. However, using DIDs also poses several challenges due to the distributed ledger technologies (DLTs) as a foundation layer, including the need for scalable and efficient identity resolution, secure and private identity verification, and interoperability between different identity systems [11]. To address these challenges, DLTs have been proposed

This work was supported by the IoTalentum research program funded by the European Union Horizon 2020 research and innovation program within the framework of Marie Skłodowska-Curie Actions (MSCA) ITN-ETN with grant number 953442.


R. Mehmood et al. (Eds.): DCAI 2023, LNNS 741, pp. 408–417, 2023.
https://doi.org/10.1007/978-3-031-38318-2_40

as a potential solution for implementing decentralized identity systems. DLTs, such as Blockchain, provide a tamper-evident and decentralized infrastructure for identity management, enabling secure and transparent identity transactions without relying on central authorities. However, traditional blockchain-based identity systems also suffer from several limitations, such as high transaction fees, slow processing speeds, limited scalability and required high computation power to solve the complex Proof of Work (PoW) that make it infeasible for low-Power IoT devices [10,16].

These problems with the existing identity management system create a need to develop a lightweight system for the authentication and identification of IoT devices. An efficient and lightweight DLT is required to achieve the decentralization goal in DIDs systems to enable secure communication of IoT devices. The new version of the IOTA[1] (i.e., IOTA Chrysalis) protocol was introduced in 2021 as a novel DLT designed and developed, especially for IoT applications. Unlike traditional blockchain systems, IOTA is based on a Directed Acyclic Graph (DAG) data structure known as the Tangle, allowing faster and more scalable transactions without requiring miners or transaction fees. The transaction handling is exciting as it allows several transactions simultaneously to add to the Tangle. Using IOTA's leverage as a decentralized technology, we will use digital identities as a standard to develop our system for uniquely identifying IoT devices. Integrating IOTA Chrysalis with DIDs can provide a more efficient and cost-effective solution for identity management in the IoT ecosystem.

In this paper, we introduce a new idea based on integrating DIDs with the IOTA Chrysalis for heterogeneous devices and explain how it can help to overcome limitations faced by traditional DLTs such as Blockchain by providing a feeless and scalable infrastructure for identity management. We perform initial experiments with different amounts of data on two devices (a workstation PC and raspberry pi) to analyze the performance and feasibility of DIDs creation and publishing to the Tangle. Furthermore, we provide the current status of the work and explore the direction towards future goals. For future goals, our system will use the IOTA Tangle to create a decentralized and scalable DID registry that can store and manage DIDs for IoT devices. The idea for proposed system provides a robust and efficient solution for managing decentralized identities in IoT environments. The system at the current stage ensures secure and efficient interactions between IoT devices.

### 1.1 Contributions

The contributions of the paper are as follows:

- We design a system based on IOTA Chrysalis, which accommodates and enables the exchange of secure information between resource-constrained IoT devices.

[1] https://chrysalis.iota.org/.

- We introduce the idea of IOTA Chrysalis for creating a scalable and efficient DID system for heterogeneous IoT devices.
- We provide a comprehensive overview of the work's current status and outline the future vision for an IOTA-based DID system. This will help the development of future research and efforts in this area.

The rest of the paper is organized as follows: Sect. 2 shows the related work, focusing on traditional approaches and approaches based on DLTs. Section 3 demonstrates the DIDs with IOTA and details of its components, followed by Sect. 4 that presents the system model in detail. Section 5 illustrates current work by providing system architecture details and implementation procedures. Section 6 concludes the paper with a discussion and future direction.

## 2 Related Work

Securing data is always a challenge when it comes to low-power IoT devices. Below discussion is based on traditional approaches with no use of DLTs and approaches based on DLTs for the authentication of IoT devices.

### 2.1 Traditional Approaches

For identity management systems, several traditional models were developed for the authentication of IoT devices. A public key infrastructure (PKI) is one of the early approaches providing various security services such as authentication, data integrity and confidentiality, data fairness, and timestamp [15]. The approach addresses most of the security requirements of IoT systems. However, there are several critical issues with the traditional PKI, such as a single point of failure as a central authority controls PKI, providing an easy way for attacks leading to substantial security risks. Digital identity models were applied to current IoT systems. However, these centralized models lead to the same scalability and single-point failure issues [2,9].

### 2.2 Approaches Based on Distributed Ledger Technologies

Researchers are nowadays more interested in decentralized ledger technologies. DLTs are taking place in recent IoT environments to enable the secure exchange of information. One motivation behind bringing attention to DLTs is the problem of a single point of failure with the centralized authority; however, these models also pose various issues, such as scalability and performance are the critical ones [10,16]. A decentralized PKI was proposed based on blockchain to manage identities and their public keys to address the registration of users' identities if they have already registered [4]. A scheme based on the blockchain Ethereum was proposed, termed a bubble of trust [5]. The scheme makes a secure bubble before establishing communication in IoT applications. However, the proposed scheme was highly demanding in terms of resources and vulnerable to attacks like DDoS and man in the middle.

Following that, [6] used the Ethereum smart contract to manage the RSA public key and the private key stored by a separate device, which can effectively control the identity of the IoT device. However, the schemes still have limitations for IoT applications, such as throughput and scalability. In a nutshell, all the mentioned approaches use blockchain to achieve decentralization without considering the massive energy consumption and scalability issues that lead to limitations in the IoT environment. Further, the transaction fee is also involved due to miners utilizing their resources.

The IOTA Chrysalis is the latest version of the IOTA DLT that has undergone a significant overhaul in terms of performance, reliability, and user experience. The Chrysalis network boasts a range of new features and improvements, making it a more streamlined and efficient platform for IoT systems. One of the most notable improvements in IOTA Chrysalis is adopting a new UTXO (Unspent Transaction Output) model, which replaces the previous account-based model used in earlier IOTA versions [3]. This new model provides a more efficient and scalable approach to transaction processing, enabling faster and more secure transactions on the network. Another significant improvement is the new client libraries (e.g., iota.rs), which enable to build more robust and secure applications on the network [1]. Furthermore, the Chrysalis network features a new and improved consensus mechanism called Fast Probabilistic Consensus (FPC), which enables faster and more secure transaction confirmation [3]. The FPC consensus mechanism provides a more efficient and lightweight approach to transaction validation, reducing the network's energy consumption and improving its overall scalability.

The purpose of this work is to provide an idea to develop a system based on DIDs with IOTA for IoT systems and to analyze the performance of IOTA Chrysalis as a foundation layer for decentralization. We also analyze the feasibility of IOTA on low-power IoT devices with different amounts of data to see the performance overhead.

## 3 Decentralized Identifiers with IOTA

As the critical level of trust and privacy is authentication and identification, new decentralized digital identity standards were proposed by W3C [13], where both parties (i.e., individuals and devices) take part in communication, trust each other, and provide proof of their identities. Various identity management models have been proposed to address the issues of unique identification. Self-Sovereign Identity (SSI) assembled a trusted environment while giving more authority to the holders that can decide what type of data and with whom the data should be shared [12]. These identity management models have mainly focused on the subject that is usually human; however, the proliferation of IoT systems in our daily lives has created a need for a trustworthy environment where device identity is a crucial aspect to be considered [17]. In IoT systems, DIDs can fulfil the need for verifiable information managed and provided by IoT devices. With DIDs in IoT, the devices can decide what information to share and where

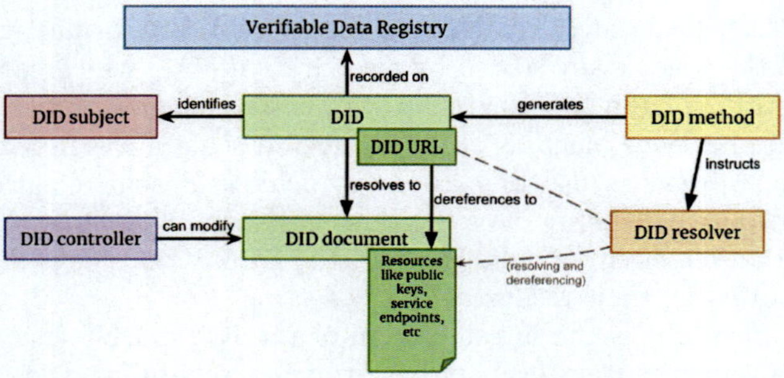


**Fig. 1.** Different basic components of DID (Source: [14]).

to maintain and improve the IoT environment's security and privacy [8].Different DID components are shown in Fig. 1 and further discussed in detail below.

A DID consists of three main components: the method, the identifier, and the DID document.

- Method: The method component of a DID specifies the method or scheme used to create and manage the DID. This could be a DLTs-based method. The method defines the rules and procedures for generating and managing DIDs within a specific context.
- Identifier: The identifier component of a DID is a unique string of characters that identifies a specific entity or object. The identifier is generated using the rules defined by the method component. For example, a DID generated using an IOTA-based method might include a public key or a public key hash.
- DID Document: The DID document is a JSON document that contains information about the entity or object identified by the DID. The DID document typically includes public keys, service endpoints, and other metadata that enable verification of the entity's identity and enable interactions with the entity or object and is stored on a DLT.

In addition, several other components are essential for managing DIDs in IoT applications. These include:

- Resolver: A DID resolver is a software component that enables the resolution of a DID to its associated DID document. The resolver receives a DID as input and returns the corresponding DID document.
- Verifiable Credentials: Verifiable credentials are a type of digital credential that enable the issuance, verification, and revocation of identity claims in a decentralized and interoperable manner. Verifiable credentials are typically stored in a DID document and can be used to provide proof of identity, authority, or ownership in IoT applications.
- Trust Framework: A trust framework is a set of rules and procedures that establish and maintain trust between entities or objects in a decentralized and interoperable manner. Trust frameworks define identity verification, access control, and data-sharing requirements in IoT applications.

# 4 System Model

We will discuss the in-depth detail of integrating DIDs with IOTA and provide the structure of our system in the following discussion step by step.

## 4.1 Define DID Standard for IOTA Chrysalis

The first step for developing the system based on DIDs integrated with IOTA for IoT systems is to define the DID standard for IOTA Chrysalis. This standard would specify the structure of a DID on the IOTA network and the methods for generating and verifying DIDs. We can leverage existing DID standards, such as W3C DID, to create a customized version of DID for IOTA Chrysalis.

## 4.2 Create Layer for Managing DIDs

The next step is to create a layer for managing DIDs on IOTA Chrysalis. This layer would include a set of APIs for generating, updating, and verifying DIDs. It would also include a mechanism for revoking DIDs in case of security breaches or device loss. To implement this layer, we can also use IOTA smart contract technology such as IOTA Smart Contract (ISC) or IOTA Contracts to create and manage smart contracts for DIDs. These smart contracts would be deployed on the IOTA network and would interact with the IoT devices through APIs.

## 4.3 Generate DIDs for IoT Devices

The third step is to generate DIDs for IoT devices. Each IoT device would generate its own unique DID using the standard defined in Step 1. The device would store the DID on its own secure storage and use it to authenticate and authorize its actions on the IOTA network.

## 4.4 Authenticate IoT Devices Using DIDs

The fourth step is to authenticate and authorize IoT devices using DIDs. The IoT devices would use their unique DIDs to authenticate and authorize their actions on the IOTA network. For example, an IoT device could generate a DID and use it to send data to another device on the network. The receiving device could then verify the sender's identity using the DID and process the data accordingly.

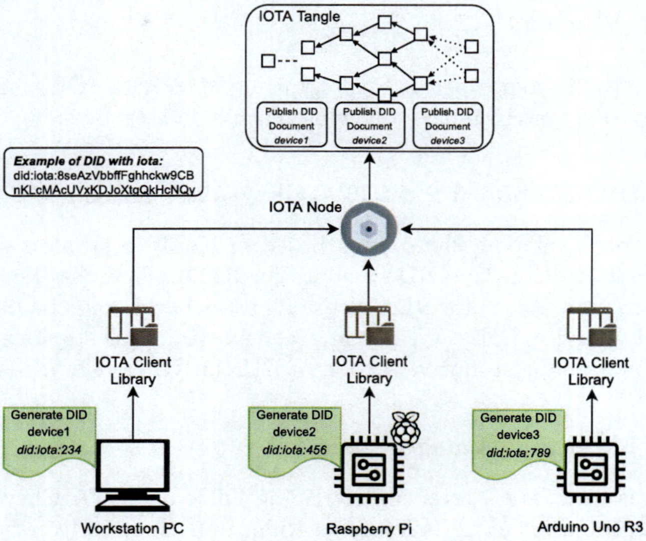


**Fig. 2.** System architecture for decentralized identifiers with IOTA.

### 4.5 Device Discovery and Management Using DIDs

The fifth step is to use DIDs for device discovery and management on the IOTA network. For example, an IoT device could broadcast its DID to other devices on the network to establish connections and form peer-to-peer networks. DIDs could also be used to manage access control and permissions for IoT devices on the network.

Overall, the system for integrating DIDs with IOTA Chrysalis for IoT devices would provide a secure and decentralized identity management solution for the IoT ecosystem.

## 5 Current Work Status

### 5.1 System Architecture

Initially, we propose a design on lightweight DLT IOTA Chrysalis to enable a secure exchange of information between different devices. In the first stage, the developed system aims to evaluate the performance of the IOTA on different devices by creating and broadcasting different amount of information. In the next stage, the devices will create their own DIDs, resolve these DIDs to their respective DID documents containing the keys, and publish them to the Tangle. The DIDs stage will be initiated as part of future work. The system architecture based on different devices is shown in Fig. 2.

### 5.2 Implementation Procedures

The system incorporates necessary libraries such as IOTA client library `iota.rs`[2] for IoT devices to facilitate communication with the IOTA network. The library is used by binding it with node.js. To achieve the full node capabilities, we use the Chrysalis public Hornet node[3] on devices that provide access to the IOTA Chrysalis devnet. For testing purposes, we generate different amounts of data of different lengths ranging from 10 characters to 30,000 characters. Next, we broadcast the generated data from the workstation PC and raspberry pi to IOTA Chrysalis Network using the IOTA client libraries. These tests are imperative to perform an in-depth analysis of the performance of both devices for sending data to the IOTA Chrysalis network. This will explore the computation overhead from different devices for publishing DID documents associated with them to the IOTA Tangle.

To implement DIDs on IOTA for IoT devices, we describe the process in steps in detail below:

1. *Step 1:* The first step is to define the DID method and standard for IoT devices. This involves selecting an appropriate method that suits the requirements of IoT applications and defining the standard for generating and managing DIDs. The standard defines a DID's structure and specifies the methods for generating, storing, and verifying DIDs.
2. *Step 2:* The next step is implementing the DID method and standard on IoT devices. This involves integrating the DID generation and management functionality into the IoT devices, enabling them to generate their own unique DIDs.
3. *Step 3:* The third step is to implement the DID resolver. The resolver would enable the resolution of a DID to its associated DID document, which contains the public keys and other metadata required for identity verification and interaction with the IoT device. The resolver is implemented as a cloud-based service or a component of the IoT device.
4. *Step 4:* The fourth step is implementing verifiable credentials for IoT devices. This would involve defining the requirements for issuing, verifying, and revoking verifiable credentials and implementing the necessary functionality on the IoT devices. Verifiable credentials provide proof of identity, authority, or ownership of a particular IoT device.

Integrating DIDs with IOTA Chrysalis provides a promising secure, decentralized identity management solution in the IoT ecosystem. By leveraging the unique features of the IOTA protocol, DIDs can be used to create a scalable, low-cost, and fast solution for managing identities on the network. This integration has the potential to address the security and privacy challenges of the IoT ecosystem, where billions of devices generate massive amounts of data that require secure and decentralized identity management. Further, DIDs enable a

[2] https://github.com/iotaledger/iota.rs.
[3] https://wiki.iota.org/hornet/welcome.

decentralized and self-sovereign identity model that gives devices control over their identity and data. By implementing it on IoT devices and establishing trust frameworks, we can create a trusted and secure environment for IoT applications that enables data sharing and collaboration between different entities. It can also help reduce the risk of data breaches and unauthorized access to IoT devices by providing a secure and decentralized identity management solution.

## 6 Conclusion and Future Work

DIDs have emerged as a promising solution for secure and privacy-preserving identity management in the IoT ecosystem. In this paper, we have provided an idea for integrating DIDs with IOTA Chrysalis for IoT devices that offers a unique opportunity for secure and decentralized identity management in the IoT ecosystem. The proposed system model provides a step-by-step approach for building such a solution, taking into account the unique features of both technologies. This helps address the security and privacy challenges of the IoT ecosystem, where billions of devices generate massive amounts of data that require secure and decentralized identity management. We have provided details of current work where we performed experiments on the initially developed system by creating and broadcasting different amounts of data to the Chrysais Tangle to see its performance on different devices and feasibility for future work related to DIDs.

Future directions for this integration include pursuing the work towards implementing DIDs with IOTA for IoT devices. We will also research how to optimize the performance and scalability of the system model and its comparative analysis with other decentralized identity solutions such as Sovrin or uPort. We are also interested in performing simulation-based work for which we will explore the VANETS and NS3 to extract the use of vehicles as devices and see how to implement DIDs with IOTA in a simulated environment.

## References

1. Bernstein, D.J., Duif, N., Lange, T., Schwabe, P., Yang, B.Y.: High-speed high-security signatures. J. Cryptogr. Eng. **2**(2), 77–89 (2012)
2. Camp, J.: Digital identity. IEEE Technol. Soc. Mag. **23**(3), 34–41 (2004). https://doi.org/10.1109/MTAS.2004.1337889
3. Conti, M., Kumar, G., Nerurkar, P., Saha, R., Vigneri, L.: A survey on security challenges and solutions in the IOTA. J. Netw. Comput. Appl. **203**, 103383 (2022). https://doi.org/10.1016/j.jnca.2022.103383, https://www.sciencedirect.com/science/article/pii/S1084804522000467
4. Fromknecht, C., Velicanu, D., Yakoubov, S.: A decentralized public key infrastructure with identity retention. Cryptology ePrint Archive (2014)
5. Hammi, M.T., Hammi, B., Bellot, P., Serhrouchni, A.: Bubbles of trust: a decentralized blockchain-based authentication system for IoT. Comput. Secur. **78**, 126–142 (2018). https://doi.org/10.1016/j.cose.2018.06.004, https://www.sciencedirect.com/science/article/pii/S0167404818300890

6. Huh, S., Cho, S., Kim, S.: Managing IoT devices using blockchain platform. In: 2017 19th International Conference on Advanced Communication Technology (ICACT), PyeongChang, Korea (South), pp. 464–467 (2017). https://doi.org/10.23919/ICACT.2017.7890132
7. Huo, R., et al.: A blockchain-enabled trusted identifier co-governance architecture for the industrial internet of things. IEEE Commun. Mag. **60**(6), 66–72 (2022). https://doi.org/10.1109/MCOM.001.2100448
8. IOTA Foundation: The Case for a Unified Identity (2021). https://files.iota.org/comms/IOTA_The_Case_for_a_Unified_Identity.pdf. Accessed 12 Apr 2023
9. Masiero, S., Bailur, S.: Digital identity for development: the quest for justice and a research agenda. Inf. Technol. Dev. **27**(1), 1–12 (2021)
10. Okegbile, S.D., Cai, J., Alfa, A.S.: Performance analysis of blockchain-enabled data-sharing scheme in cloud-edge computing-based IoT networks. IEEE Internet Things J. **9**(21), 21520–21536 (2022). https://doi.org/10.1109/JIOT.2022.3181556
11. Saidi, H., Labraoui, N., Ari, A.A.A., Maglaras, L.A., Emati, J.H.M. DSMAC: privacy-aware decentralized self-management of data access control based on blockchain for health data. IEEE Access **10**, 101011–101028 (2022). https://doi.org/10.1109/ACCESS.2022.3207803
12. Soltani, R., Nguyen, U.T., An, A.: A survey of self-sovereign identity ecosystem. Secur. Commun. Netw. **2021**, 1–26 (2021)
13. W3C: Decentralized Identifiers (DIDs) v1.0 (2021). https://www.w3.org/TR/did-core/. Accessed 10 Apr 2023
14. W3C: Decentralized Identifiers (DIDs) v1.0 (2021). https://www.w3.org/TR/2021/WD-did-core-20210131/. Accessed 02 May 2023
15. Wang, S., Li, H., Chen, J., Wang, J., Deng, Y.: Dag blockchain-based lightweight authentication and authorization scheme for IoT devices. J. Inf. Secur. Appl. **66**, 103134 (2022). https://doi.org/10.1016/j.jisa.2022.103134, https://www.sciencedirect.com/science/article/pii/S2214212622000242
16. Xie, J., et al.: A survey of blockchain technology applied to smart cities: research issues and challenges. IEEE Commun. Surv. Tutor. **21**(3), 2794–2830 (2019). https://doi.org/10.1109/COMST.2019.2899617
17. Zwitter, A.J., Gstrein, O.J., Yap, E.: Digital identity and the blockchain: universal identity management and the concept of the "self-sovereign" individual. Front. Blockchain **3**, 26 (2020)

# Network Automation: From Intent-Based Networking to Cloud-Native Networking

Mahmoud Abbasi[1(✉)], Javier Prieto[1], and Juan Manuel Corchado[1,2]

[1] BISITE Research Group, University of Salamanca, Salamanca, Spain
{mahmoudabbasi,javierp,corchado}@usal.es

[2] AIR Institute, IoT Digital Innovation Hub, Salamanca, Spain
https://bisite.usal.es, https://www.innovationhub.es

**Abstract.** This paper presents a review of recent advancements in network automation, such as Intent-based Networking (IBN), network orchestration, automation tools, network analytics, Software-defined Networking (SDN), and cloud-native networking, and examines their potential impact on various applications. It highlights the benefits, challenges, and potential applications of these developments and notes that the adoption of network automation has grown significantly in recent years and is projected to continue to do so. Nevertheless, to optimize outcomes, organizations must cautiously evaluate their network automation requirements and select appropriate tools and techniques that align with their business goals and objectives. The paper concludes by considering future possibilities for network automation and research directions that can help to enhance this field.

**Keywords:** Network automation · NTMA · Intent-based Networking (IBN) · Software-defined Networking (SDN)

## 1 Introduction

The rapid growth of network infrastructure and the increasing complexity of network operations have highlighted the need for automation in network management [1]. Network automation refers to the use of software tools and processes to automate the configuration, management, and monitoring of network devices and services. It has become increasingly important as networks have grown more complex and dynamic, with the rise of cloud computing, the Internet of Things (IoT), and other trends. There are several benefits to network automation, including increased efficiency, greater reliability, and improved security. By automating routine tasks such as device configuration and software updates, network engineers can save time and reduce the risk of errors. Automation also allows for faster deployment of new services and better management of network resources. In recent years, significant developments in network automation have emerged, such as Intent-based Networking (IBN), network orchestration, and automation tools [2]. These developments have enabled organizations to automate network


R. Mehmood et al. (Eds.): DCAI 2023, LNNS 741, pp. 418–427, 2023.
https://doi.org/10.1007/978-3-031-38318-2_41

tasks and workflows, streamline network operations, and improve business outcomes. This paper reviews the recent developments in network automation and their potential applications in various contexts. We begin with a literature review that summarizes the current state of network automation and identifies any gaps or opportunities for further research. Next, we discuss the recent developments in network automation, such as IBN, network orchestration, automation tools, network analytics, Software-defined Networking (SDN) [3], and cloud-native networking. We analyze the benefits and challenges of these developments and discuss their potential applications. The contribution of this paper is to provide a overview of the recent developments in network automation and their potential impact on network management and operations. We conclude with a discussion of the future of network automation and the research directions that can further advance this field.

## 2 Background and Motivations

Networks are the backbone of modern society, enabling communication, commerce, and innovation across the globe [4]. However, the increasing complexity and scale of network infrastructure and operations have made it challenging for organizations to manage and maintain their networks effectively. Manual network management processes are time-consuming, error-prone, and often unable to keep up with the pace of technological change [5]. Therefore, organizations have been exploring automation as a means to overcome these challenges and improve network performance, reliability, and security. Network automation involves using software and tools to automate network management tasks, such as configuration, deployment, monitoring, and troubleshooting. The key aspects of network automation are including: 1) Programmable infrastructure, 2) Orchestration tools, 3) Configuration management, 4) Network monitoring, and 5) Compliance management (see Fig. 1).

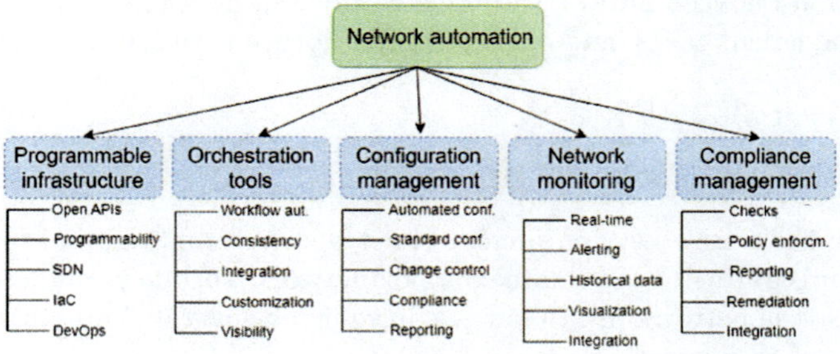


**Fig. 1.** Fundamental components of network automation.

### 2.1 Programmable Infrastructure

A programmable infrastructure is a network infrastructure that can be controlled and managed through Application Programming Interfaces (APIs) [6]. This means that network administrators can automate network management tasks by writing scripts or programs that interact with the network infrastructure through APIs. This approach allows network administrators to automate routine tasks, reduce errors, and improve network reliability. Below are some of the essential attributes of a programmable infrastructure.

- ***Open APIs:*** A programmable infrastructure has open APIs that allow network administrators to interact with network devices and services. These APIs provide a standardized way to communicate with network devices and enable automation tools to interact with the network infrastructure.
- ***Network programmability:*** Network programmability refers to the ability to use programming languages to configure and manage network devices. With network programmability, network administrators can automate complex network workflows and create custom applications that interact with the network infrastructure.
- ***SDN:*** SDN is a key component of a programmable infrastructure. SDN separates the network's control plane from the data plane, allowing network administrators to manage network devices centrally from a single controller. SDN also enables network administrators to create custom network policies and automate network workflows.
- ***Infrastructure as Code (IaC):*** IaC is an approach to managing network infrastructure through code [7]. IaC allows network administrators to define network configurations using code, and then deploy these configurations to network devices automatically. This approach can reduce errors and improve network reliability.
- ***DevOps:*** DevOps is a cultural approach to software development and IT operations that emphasizes collaboration, automation, and continuous improvement. DevOps can be applied to network automation to improve collaboration between network administrators and developers, automate network management tasks, and continuously improve network performance and reliability.

### 2.2 Orchestration Tools

Orchestration tools are a key component of network automation, and they enable network administrators to automate complex network workflows and ensure consistency across the network [8]. Below, we have listed several characteristics that are indicative of orchestration tools.

- ***Workflow automation:*** Orchestration tools can automate complex network workflows by defining a sequence of tasks to be executed in a specific order. These workflows can include tasks such as configuring network devices, deploying network services, and monitoring network performance.

- ***Consistency:*** Orchestration tools are able to ensure consistency across the network by enforcing standard configurations and policies. By defining a set of standard configurations and policies, network administrators can ensure that network devices and services are configured in a consistent and standardized manner.
- ***Integration:*** Orchestration tools can integrate with other systems, such as cloud platforms and IT service management tools. This allows network administrators to automate workflows that span multiple systems and provides a more seamless user experience.
- ***Customization:*** An organization can adapt orchestration tools to fulfill their unique requirements. Customization may involve creating custom workflows, defining new tasks, or integrating with other systems.
- ***Visibility:*** By using orchestration tools, network administrators can gain insight into network workflows and performance. This visibility can help administrators identify bottlenecks, troubleshoot issues, and optimize network performance.

### 2.3 Configuration Management

Configuration management is an essential component of network automation [9]. It involves managing the configuration of network devices and services to ensure that they are configured correctly and consistently across the network. Configuration management tools can automate configuration tasks, enforce standard configurations, and ensure that changes are made in a controlled and consistent manner. Configuration management tools offer several important features that make them valuable assets for managing network resources. These tools provide a centralized platform for configuring and maintaining devices in a network, enabling administrators to make changes quickly and efficiently. We have identified several key characteristics of configuration management tools, including:

- ***Automated configuration:*** Configuration management tools have the ability to automate configuration tasks, such as deploying new network devices, configuring network interfaces, and updating network configurations. By automating these tasks, configuration management tools can reduce errors and ensure consistency across the network.
- ***Standard configuration:*** Configuration management tools can enforce standard configurations across the network. By defining standard configurations and policies, network administrators can ensure that network devices and services are configured in a consistent and standardized manner.
- ***Change control:*** To ensure changes are implemented in a controlled and consistent manner, configuration management tools can offer change control mechanisms. Change control mechanisms can include approval workflows, change tracking, and rollback capabilities.
- ***Compliance:*** Configuration management tools can ensure compliance with regulatory requirements and industry standards. By defining compliance policies and automating compliance checks, configuration management tools can

ensure that the network is compliant with regulatory requirements and industry standards.
- ***Reporting:*** Reporting capabilities of configuration management tools can track configuration changes, enforce compliance, and detect configuration issues. Reporting capabilities can provide network administrators with the visibility they need to manage network configurations effectively.

### 2.4 Network Monitoring

Network monitoring is a crucial aspect of network management [10]. It involves the continuous monitoring and analysis of network traffic, performance, and security to identify issues, troubleshoot problems, and optimize network performance. Network monitoring tools can provide real-time visibility into network performance, detect and alert administrators about issues, and provide historical data for analysis and optimization. In the following, we outline several critical characteristics of network monitoring:

- ***Real-time monitoring:*** Network monitoring tools enable real-time monitoring of network traffic, performance, and security. Real-time monitoring allows administrators to detect and respond to issues as they occur, reducing the risk of downtime and improving network performance.
- ***Alerting:*** Network monitoring tools can alert administrators about issues or events that require attention. Alerts can be sent via email, SMS, or other means, and can be customized to suit the needs of the organization.
- ***Historical data:*** Network monitoring tools can offer historical data on network performance, allowing administrators to analyze trends and identify patterns that may impact network performance. Historical data can be used for capacity planning, performance optimization, and troubleshooting.
- ***Visualization:*** Network monitoring tools are capable of presenting visual representations of network performance, such as graphs, charts, and heat maps. Visualization allows administrators to quickly identify issues and understand network performance at a glance.
- ***Integration:*** Integration with other systems such as configuration management tools, orchestration tools, and IT service management tools is possible through network monitoring tools. The integration enables administrators to automate workflows and improve the efficiency of network management.

### 2.5 Compliance Management

Compliance management is the process of ensuring that an organization complies with applicable laws, regulations, and industry standards [11]. In the context of network management, compliance management involves ensuring that the network infrastructure and services are configured, operated, and maintained in accordance with applicable regulations and standards. Compliance management tools can automate compliance checks, track compliance status, and provide reporting capabilities. We have discovered several crucial attributes of compliance management tools during our research, which are summarized below:

- ***Compliance checks:*** Compliance checks can be automated using compliance management tools, such as security scans, vulnerability assessments, and configuration audits. Automated compliance checks can reduce the risk of non-compliance and improve the efficiency of compliance management.
- ***Policy enforcement:*** By using compliance management tools, organizations can implement and enforce policies and standards like security policies, access control policies, and configuration standards. Policy enforcement can ensure that the network infrastructure and services are configured and operated in accordance with applicable regulations and standards.
- ***Reporting:*** Compliance management tools can offer capabilities for generating reports to track compliance status, identify compliance issues, and generate compliance reports. Reporting capabilities can provide network administrators with the visibility they need to manage compliance effectively.
- ***Remediation:*** Compliance management tools can effectively address compliance issues by providing remediation capabilities. These remediation capabilities typically include automated workflows, change management processes, and integration with other tools.
- ***Integration:*** Compliance management tools can be integrated with various other systems, including configuration management tools, network monitoring tools, and IT service management tools.

## 3 Literature Review

In recent years, there has been an increase in the adoption of network automation, with numerous organizations implementing automation tools and techniques to enhance their network management. The adoption rate of network automation is expected to continue to grow, with the global network automation market expected to reach $29.3 billion by 2025. The adoption of network automation is not limited to large enterprises or service providers; small and medium-sized businesses can also benefit from network automation by reducing the time and resources required to manage their networks. This section reviews the existing literature on network automation and identifies the key trends and challenges in this field.

Several studies have highlighted the benefits of network automation in improving network performance, reducing operational costs, and enhancing security. For example, a study by International Data Corporation (IDC) found that organizations that implemented network automation saw a 43% reduction in network downtime and a 45% reduction in network operation costs [13]. Similarly, a survey conducted by Network Computing found that 91% of respondents cited improved network performance as a benefit of network automation, and 66% cited reduced network downtime [14].

Recent developments in network automation include IBN, network orchestration, automation tools, network analytics, SDN, and cloud-native networking. These developments have enabled organizations to automate more complex network tasks and workflows, providing more efficient and effective network management. IBN, for example, is a paradigm shift in network management that enables network administrators to specify the desired network behavior in natural language rather than configuring the network devices manually [15]. The work in [16] discusses the potential application of IBN in 6G networks. The authors present the key features of IBN, including its ability to provide automatic network configuration, policy enforcement, and continuous optimization. They then discuss the unique requirements of 6G networks, such as ultra-high reliability and low latency, and how IBN can help address these challenges. The paper highlights the need for a standardized IBN architecture and the development of machine learning algorithms to support IBN in 6G networks. Overall, the paper provides valuable insights into the potential of IBN for 6G networks and the challenges that need to be addressed to realize its full potential.

Network orchestration [17], on the other hand, is the process of automating the coordination and management of multiple network devices and services. A survey by Nathan et al. [18] provides a comprehensive review of the existing literature on Network Service Orchestration (NSO) and its key concepts. The authors define NSO as the process of automating the deployment and management of network services and resources in a coordinated and dynamic manner. The paper presents an overview of the different NSO architectures and technologies, including network function virtualization (NFV), SDN, and cloud computing. The authors also discuss the various NSO frameworks and tools that have been developed to facilitate NSO. Finally, the paper identifies some of the current challenges and future directions in NSO, such as service composition, network security, and network slicing. Automation tools have also become more sophisticated and powerful in recent years. These tools can automate various network tasks, such as configuration management, network monitoring, and security management. Network analytics, which is the process of using data analysis tools to gain insights into network performance and behavior, is another significant development in network automation [19]. The paper by Eva Papadogiannaki and Sotiris Ioannidis [20] provides a comprehensive review of the state-of-the-art techniques, applications, and countermeasures for encrypted network traffic analysis. The authors discuss the main challenges and opportunities of encrypted traffic analysis and the tools and techniques used to analyze it, including statistical analysis, machine learning, and deep learning. The paper also presents different applications of encrypted traffic analysis, such as intrusion detection, malware detection, and network performance monitoring. Finally, the authors discuss countermeasures to protect against encrypted traffic analysis, including encryption and anonymity techniques. This paper provides valuable insights into the current trends and challenges in encrypted traffic analysis and can serve as a useful reference for researchers and practitioners in the field.

SDN and cloud-native networking are two other recent developments in network automation [21]. SDN separates the network control plane from the data plane, enabling centralized network management and programmability. One of the first study that investigate SDN for network management conducted by Hyojoon Kim and Nick Feamster in [22]. The paper provides an overview of SDN and its potential to improve network management, including automation, agility, and scalability. The authors discuss the architecture, protocols, and applications of SDN and highlight the benefits and challenges of adopting SDN in real-world network environments. They also present a case study of an SDN implementation in a large-scale data center network to demonstrate the benefits of SDN in improving network management. Cloud-native networking, on the other hand, is the use of cloud-native technologies, such as containers and microservices, to deploy and manage network services [23]. These developments have led to more agile and flexible network management and have enabled organizations to scale their networks more effectively.

## 4 Trends and Challenges

As computer networks continue to evolve and become more complex, network automation is encountering new trends and challenges. Below are some potential future advancements and obstacles in network automation that we have identified through our investigation:

- ***Artificial intelligence and machine learning:*** Artificial intelligence and machine learning are expected to play a major role in network automation in the future. These technologies can be used to automate network management tasks, such as fault detection, capacity planning, and performance optimization.
- ***Intent-based networking:*** IBN is an emerging approach to network automation that focuses on the intent of network operations rather than the underlying configuration. IBN aims to simplify network management by abstracting network operations to higher-level policies and intentions, making it easier for administrators to manage the network.
- ***Multi-cloud environments:*** As more organizations adopt multi-cloud strategies, network automation tools will need to support the management of networks across multiple clouds. This includes the ability to automate network provisioning, security, and performance across different cloud environments.
- ***Security automation:*** With the increasing threat of cyber attacks, network automation tools will need to include security automation capabilities. This includes the ability to detect and respond to security threats in real-time, automate security policy enforcement, and integrate with security analytics tools.
- ***Skill gaps:*** The adoption of network automation requires a skilled workforce with expertise in network automation technologies, programming, and software development. The shortage of skilled network automation professionals is expected to be a major challenge in the future.

In summary, the significance of network automation in network management is growing. The future of network automation will witness the integration of emerging technologies such as artificial intelligence and machine learning, the implementation of intent-based networking, and the management of networks across multiple cloud platforms. Nevertheless, the scarcity of skilled personnel and the necessity for security automation will continue to be significant obstacles.

## 5 Conclusion

In conclusion, this study has highlighted the significant potential of network automation to improve network performance, reduce operational costs, and enhance security. The recent developments in network automation, such as intent-based networking, network orchestration, automation tools, network analytics, software-defined networking, and cloud-native networking, have enabled organizations to automate more complex network tasks and workflows, providing more efficient and effective network management. However, despite the benefits, organizations face challenges in implementing automation, such as the need for skilled personnel and the complexity of integrating multiple automation tools. It is crucial for organizations to carefully evaluate their network automation needs and choose appropriate tools and techniques that align with their business goals and objectives. All in all, network automation is a rapidly evolving field that holds great promise for improving the efficiency, reliability, and security of network operations. By staying informed about the latest developments and best practices in network automation, network engineers can help their organizations stay ahead of the curve and achieve their business goals.

**Acknowledgement.** This work has been performed under the IoTalentum project within the framework of Marie Skłodowska-Curie Actions ITN-ETN (grant number 953442) which is funded by the European Union Horizon 2020 research and innovation program.

## References

1. Ayoubi, S., et al.: Machine learning for cognitive network management. IEEE Commun. Mag. **56**(1), 158–165 (2018)
2. Aboubakar, M., Kellil, M., Roux, P.: A review of IoT network management: current status and perspectives. J. King Saud Univ.-Comput. Inf. Sci. **34**(7), 4163–4176 (2022)
3. Mendoza-Cardenas, F., Aparcana-Tasayco, A.J., Diaz-Ataucuri, D.: Network monitoring system for software-defined networking: experimental evaluation in a real SDN environment. IEIE Trans. Smart Process. Comput. **11**(5), 361–367 (2022)
4. Shahraki, A., et al.: A comprehensive survey on 6G networks: applications, core services, enabling technologies, and future challenges. arXiv preprint arXiv:2101.12475 (2021)
5. Clemm, A., Zhani, M.F., Boutaba, R.: Network management 2030: operations and control of network 2030 services. J. Netw. Syst. Manage. **28**(4), 721–750 (2020)
6. Campbell, A.T., et al.: A survey of programmable networks. ACM SIGCOMM Comput. Commun. Rev. **29**(2), 7–23 (1999)
7. Morris, K.: Infrastructure as Code. O'Reilly Media (2020)
8. Chiang, Y., et al.: Management and orchestration of edge computing for IoT: a comprehensive survey. IEEE Internet Things J. (2023)
9. Ismail, H., Hamza, H.S., Mohamed, S.M.: Semantic enhancement for network configuration management. In: 2018 IEEE Global Conference on Internet of Things (GCIoT). IEEE (2018)

10. D'Alconzo, A., et al.: A survey on big data for network traffic monitoring and analysis. IEEE Trans. Netw. Serv. Manage. **16**(3), 800–813 (2019)
11. Talesh, S.A.: Data breach, privacy, and cyber insurance: how insurance companies act as "compliance managers" for businesses. Law Soc. Inq. **43**(2), 417–440 (2018)
12. LNCS Homepage. http://www.springer.com/lncs. Accessed 4 Oct 2017
13. Cisco, U.: Cisco annual internet report (2018-2023) white paper. Cisco: San Jose, CA, USA, vol. 10, no. 1, pp. 1–35 (2020)
14. Arzo, S.T., et al.: A theoretical discussion and survey of network automation for IoT: challenges and opportunity. IEEE Internet Things J. **8**(15), 12021–12045 (2021)
15. Leivadeas, A., Falkner, M.: A survey on intent based networking. IEEE Commun. Surv. Tutor. (2022)
16. Wei, Y., Peng, M., Liu, Y.: Intent-based networks for 6G: insights and challenges. Digit. Commun. Netw. **6**(3), 270–280 (2020)
17. Tam, P., et al.: Graph neural networks for intelligent modelling in network management and orchestration: a survey on communications. Electronics **11**(20), 3371 (2022)
18. Sousa, D., Saraiva, N.F., et al.: Network service orchestration: a survey. Comput. Commun. **142**, 69–94 (2019)
19. Abbasi, M., Shahraki, A., Taherkordi, A.: Deep learning for network traffic monitoring and analysis (NTMA): a survey. Comput. Commun. **170**, 19–41 (2021)
20. Papadogiannaki, E., Ioannidis, S.: A survey on encrypted network traffic analysis applications, techniques, and countermeasures. ACM Comput. Surv. (CSUR) **54**(6), 1–35 (2021)
21. Shah, S.D.A., Gregory, M.A., Li, S.: Cloud-native network slicing using software defined networking based multi-access edge computing: A survey. IEEE Access **9**, 10903–10924 (2021)
22. Kim, H., Feamster, N.: Improving network management with software defined networking. IEEE Commun. Mag. **51**(2), 114–119 (2013)
23. Duan, Q.: Intelligent and autonomous management in cloud-native future networks-a survey on related standards from an architectural perspective. Future Internet **13**(2), 42 (2021)

# Enhancing the Security of Software Defined Networks via Quantum Key Distribution and Post-Quantum Cryptography

Carlos Rubio García[1(✉)], Simon Rommel[1], Juan Jose Vegas Olmos[2], and Idelfonso Tafur Monroy[1]

[1] Department of Electrical Engineering, Eindhoven University of Technology, Eindhoven, The Netherlands
{c.rubio.garcia,s.rommel,i.tafur.monroy}@tue.nl
[2] Software Architecture, NVIDIA Corporation, Yokneam, Israel
juanj@nvidia.com

**Abstract.** This paper investigates the integration of quantum key distribution (QKD) and post-quantum cryptography (PQC) into software defined networks (SDNs) to maintain their security even against quantum computers which threaten practically all current cryptography. We discuss the relevance of securing SDNs, the threats and limitations of classical cryptography against upcoming quantum computing, and how QKD and PQC can be used for mitigating potential attacks featuring quantum computers on both SDN control and data planes. Finally, we propose an SDN architecture that allows for end-to-end quantum-safe communications.

**Keywords:** SDN · QKD · PQC · Security · Control plane · Data plane

## 1 Introduction

As we approach the era of 6G networks, the security of critical information in current networks relies on a mix of asymmetric public key cryptography (PKC) and symmetric encryption. However, the emergence of sufficiently large quantum computers poses a significant threat to these security schemes. This is particularly concerning for software defined networks (SDNs), where the control of the network is centralized in a single entity, commonly known as the network controller [23]. Although the introduction of a network controller signifies increased flexibility, scalability and programmability, it also introduces a potential single point of failure and a high-value target for attack [1]. In this scenario, it becomes critical that the network controller remains secure against any (quantum-enabled) attack [23].

This work was supported by the EC H2020 MSCA ITN-ETN IoTalentum (GA 953442) project and the Dutch Ministry of Economic Affairs and Climate Policy (EZK), as part of the Quantum Delta NL programme.


R. Mehmood et al. (Eds.): DCAI 2023, LNNS 741, pp. 428–437, 2023.
https://doi.org/10.1007/978-3-031-38318-2_42

Quantum key distribution (QKD) and post-quantum cryptography (PQC) are the first cyber-security technologies that allow the network controller to remain secure in any scenario, even with the advent of quantum computing. While the aforementioned technologies use different cryptographic assumptions for securing communications against quantum computers, integrating both of them into SDN can ensure the reliability of control plane communications at any time. If the control plane remains quantum-safe, the same strategy can be followed towards ensuring the safety of data plane communications. We propose a quantum-enabled SDN architecture in which QKD and PQC could be combined to ensure end-to-end quantum-safe communications, both in control and data planes, thus fulfilling the safety requirements of next-generation SDN networks.

The remaining of the paper is structured as follows: Sect. 2 explains the concept of upcoming SDN networks, why is relevant to secure control plane communications, how TLS is currently used for that purpose and why quantum-safe technologies could be integrated into TLS as a way to secure SDN networks. Section 3 explains the two technologies that enable quantum-safe communications: QKD and PQC. Section 4 explains a quantum-enabled SDN architecture that can make use of both QKD and PQC for providing end-to-end quantum-safe communications, both in control and data planes. Finally, Sect. 5 is devoted to conclusions.

## 2 Software Defined Networks

Software defined networking (SDN) has emerged in response to the limitations of conventional networks [18]. It relies on the idea of decoupling the hardware from the software by transferring network intelligence to a logically centralized software platform called the network controller [23]. This controller has complete visibility and authority over all network activity. Meanwhile, the data plane is dedicated solely to traffic forwarding, which is performed by executing instructions delivered from the network controller. The separation of control and data plane gives network operators efficient use of network resources and enough flexibility to provide new services so that operators can quickly and easily manage, configure and optimize the resources given by a network through dynamic, automated open-source network controllers and interfaces, while the role of data plane devices can be simplified to forwarding packets, thus reducing their complexity and the cost of the network.

### 2.1 Relevance of Securing SDN Control Plane

Despite the various benefits of SDN architecture, the centralization of the controller poses a potential failure point that could be exploited by an external attacker [10]. A vulnerable interface towards the network controller could potentially expose the full control of the network to a third party, causing serious consequences on the behavior of the networks or may even allow an attacker to damage the network and equipment attached to it, potentially causing harm beyond the network itself. That is the reason why SDN control plane security has emerged as a hot topic with the advent of 6G and beyond networks. Some of the issues that need to be addressed in regard to control plane security are [6]:

- Critical information traveling back and forth from the network switches to the controller.
- Introducing "network applications" that interact with the controller and can modify network behavior based on application demands.
- Introducing a centralized controller architecture is a potential single point of failure and attack.

As SDN networks become more geographically distributed, control plane packets that traverse insecure communication channels from the network controller to remote network elements are susceptible to interception and eavesdropping. Thus, the following interfaces within an SDN architecture have to be secured:

- The northbound interface (NB): from the controller(s) to the orchestrator or application layer.
- East-westbound interface (EWB): between controllers.
- The southbound interface (SB): From the data plane devices to the controller(s).

If the aforementioned communication interfaces are properly secured, then the controller can have a guarantee that the incoming communications are coming from trustworthy network elements such as switches, routers or programmable encryptors, and these devices can be equally sure to be correctly receiving unaltered commands from the controller. This means that the processes of both authentication and encryption happening between controller and network elements must guarantee: 1) The identity of the server side of the communications must be verifiable. 2) The plaintext data sent must only be visible to the rightful endpoints of the communications. 3) The encrypted data sent among parties cannot be modified without both endpoints noticing [15].

In the case of a fully secure and centralized SDN architecture, then the SDN network controller can use its flexibility, complete overview of the network and programmability to enforce different security measures across the data plane, enabling customized security solutions such as wireless network anomaly detection or dynamically enforcing security policies across network nodes, thus augmenting the overall security of the network [6]. With the emergence of quantum computing, it is crucial that the aforementioned security remains under control. This can be achieved by securing the different SDN communication interfaces by means of quantum-safe cryptography, namely QKD and PQC.

### 2.2 Transport Layer Security for Securing SDNs Control Plane Communications

With respect to how SDN can be secured, there are several encryption solutions such as IPsec [9], secure shell (SSH) [24] and transport layer security (TLS) [15] that could be employed for securing the communications on the control plane. While there is not a clear standard that states which network security protocol should be used for SDN control plane communications, TLS is one of such protocols that is recommended for securing SDN control plane communications [10,13,19,23].

TLS is a cryptographic protocol designed to provide secure communications over a computer network, primarily between a client and a server. It is widely used over the internet, as it allows the data transmitted between two parties to remain confidential, authentic and resistant to tampering. TLS is built at the transport layer of the TCP/IP protocol suite, and its main advantage is that it can be placed on top of any system allowing end-to-end communications, even in an unsafe environment. As TLS is proposed to secure the control and management communication for SDN in 5G and 6G networks [10,13,19,23], this paper studies how TLS can be enhanced so that it can comply with quantum-safe requirements of SDN networks.

### 2.3 Transport Layer Security for Securing SDNs - Related Work

Although TLS is one of the most recommended solutions for enforcing control plane security within SDN networks, some of the most prominent communication standards in SDN (e.g., OpenFlow, NETCONF [7], RESTCONF [4], etc.) do not enforce TLS implementations but rather define it as optional [12]. Nonetheless, extensive work has been carried out in the literature with a focus on improving the security of TLS for SDN networks. [14] states that the security of the certificate authority (CA) is a problem, which might have some consequences on the authentication phase of a TLS handshake. [23] investigates why TLS is not implemented in current SDN networks and reaches two conclusions: first, are performance concerns that exist when implementing any security mechanisms on any communication channel (authentication and encryption communication cost, complexity of configuration and maintenance, etc.). Second, data centers, which are the most prominent application for SDN, usually live in relatively controlled and restricted environments, where the attacks on the communication channels are not considered a problematic issue.

Additionally, and inconveniently for implementing TLS in SDN, in most the controller software implementations TLS cannot be activated or deactivated dynamically and must be activated before the controller is started and cannot be modified/deactivated while the controller is working [23]. This situation becomes impractical if a controller needs to work with different network devices that might support different versions of the protocol, be placed in secure or insecure locations (which generates different security demands) or even when working with devices that might not perform well with encryption. [16] further notes that the implementation of TLS adds complexity to the preparation and maintenance of SDN networks, in particular, regarding the preparation of certificates, signing of certificates, and the installation of these certificates (and private keys) in each device. When assuming a distributed topology with many locations, the installation and management of private keys and certificates generates an added complexity that is avoided in scenarios where private networks are distributed in different physical locations without access to the internet.

Focusing on the southbound communications, [11] explains that controllers that do not implement TLS are prone to MITM attacks, in which the malicious opponent can penetrate the communication channel while remaining undetected. The sum of all of the factors mentioned above have led industry to overlook the security of the control plane communication within SDN networks, which is low on the list of security priorities to be addressed [16,23]. However, as SDN becomes deployed in reality and is considered a enabler technology of 6G critical applications in which quantum-safe communications is a must, the combination of QKD and PQC cryptography, as well as classical cryptography will be the key for enabling next-generation, quantum-safe SDN networks.

## 3 Enabling Technologies for Quantum-Safe SDN Control Plane Communications

### 3.1 Quantum Key Distribution

Quantum key distribution (QKD) is a cryptographic technique that leverages the principles of quantum mechanics to securely exchange encryption keys between two parties, typically referred to as Alice and Bob. The primary goal of QKD is to allow Alice and Bob to exchange a series of quantum signals that can be used by each peer to distill the same secure, shared secret key. The security of QKD is given by some fundamental principles of quantum physics. First, any attempt by an attacker (eavesdropper) to obtain information about the state of the quantum signals exchanged between Alice and Bob is a generalized form of measurement. A well known fact within quantum physics is that the mere act of measuring a quantum object inevitably changes its state [17]. This guarantees that an attacker cannot eavesdrop on the quantum channel without being detected by Alice and Bob. The second quantum physics law that guarantees the security of QKD is the no-cloning theorem [22], which states that a quantum state cannot be cloned while keeping the initial state intact. This guarantees that an eavesdropper could not obtain a copy of the quantum signals exchanged between Alice and Bob without being detected.

After the transmission of the quantum signals, Alice and Bob can execute a series of key sifting and post-processing operations and communicate the results to each other through the classical service channel. This allows both transmitter and receiver to determine how much information about the exchanged quantum signals has been exposed to an eavesdropper. If the quantum signals have been overexposed to an external party, Alice and Bob must assume the presence of an eavesdropper and abort the protocol. Otherwise, after a defined key distillation process, the key is considered to be secure, and can be provided to requesting secure application entities (SAE) [8] for the implementation of a network security protocol of choice [20] as shown in Fig. 1.

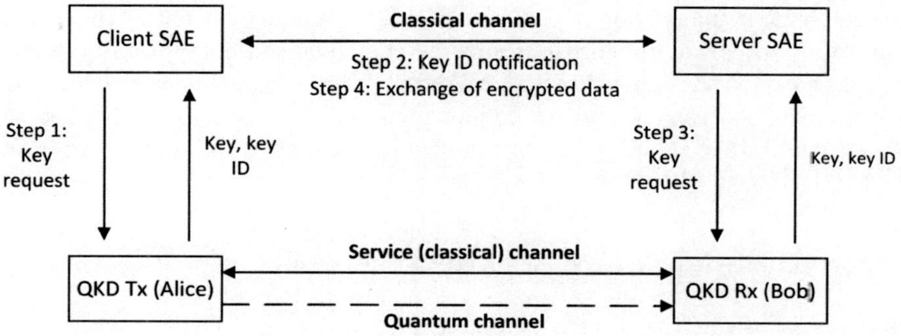


**Fig. 1.** Example QKD key retrieval process for securing communications among SAEs [8].

### 3.2 Post-Quantum Cryptography

Post-quantum cryptography (PQC) is a branch of cryptography that focuses on developing cryptographic algorithms that can resist the attacks of a quantum computer. As seen in classical cryptography, PQC provides a kind of public key cryptography (PKC) family of algorithms that allow for operations such as private-public key pair generation, digital signatures, digital signature verification, key exchange, and key encapsulation/decapsulation. These operations allow network security protocols to implement some key processes for encryption such as authentication and authenticated key exchange. The national institute of science and technology (NIST) is currently conducting a standardization to solicit, evaluate, and standardize one or more quantum-safe PKC algorithms [2]. The process is currently on the fourth round of submissions, and there are several algorithms that have already been selected for standardization: CRYSTALS-KYBER as a mechanism for public-key encapsulation (KEM). CRYSTALS-Dilithium, FALCON, and SPHINCS+ will be standardized for digital signatures, although CRYSTALS-Dilithium is recommended by NIST as the primary algorithm to be implemented [2]. Although PQC cryptography is designed to directly substitute classical cryptography algorithms, due to its recent nature, it is recommended to be used in a hybrid mode, as there have been examples during the NIST standardization project of several PQC-based algorithms that have been broken, such as the Rainbow algorithm [3,5].

### 3.3 Combining PQC and QKD

By enabling both QKD and PQC technologies into SDN networks, augmented quantum-resistance could be achieved. One side, PQC can be used for authentication, by digitally signing messages (e.g., certificates) and being able to verify the signature at the other endpoint. Regarding the key distribution, although both QKD and PQC could be used by themselves, the proposed system could enable a high-security protocol in which both technologies are combined for providing shared, secret keys of the highest security that can later on be integrated into a symmetric encryption algorithm of choice. It must be noted that some symmetric cryptography algorithms such as the advanced encryption standard (AES) do not face the same threat to quantum computing as asymmetric PKC, and key expansion can be used to countermeasure the advantage conferred by quantum computing. Thus, a symmetric encryption algorithm such as AES-256 can be considered to be quantum-safe [21].

## 4 Quantum-Safe SDN Architecture

To remain secure in the advent of quantum computing, SDN networks need to integrate a quantum-safe technology of choice as the base for encrypting control plane communications. In this regard, QKD and PQC appear as the most prominent choices. While QKD offers information-theoretically secure keys, PQC is more flexible, and can adapt to almost any possible network topology as well as offering a more direct drop-in replacement for current cryptographic solutions.

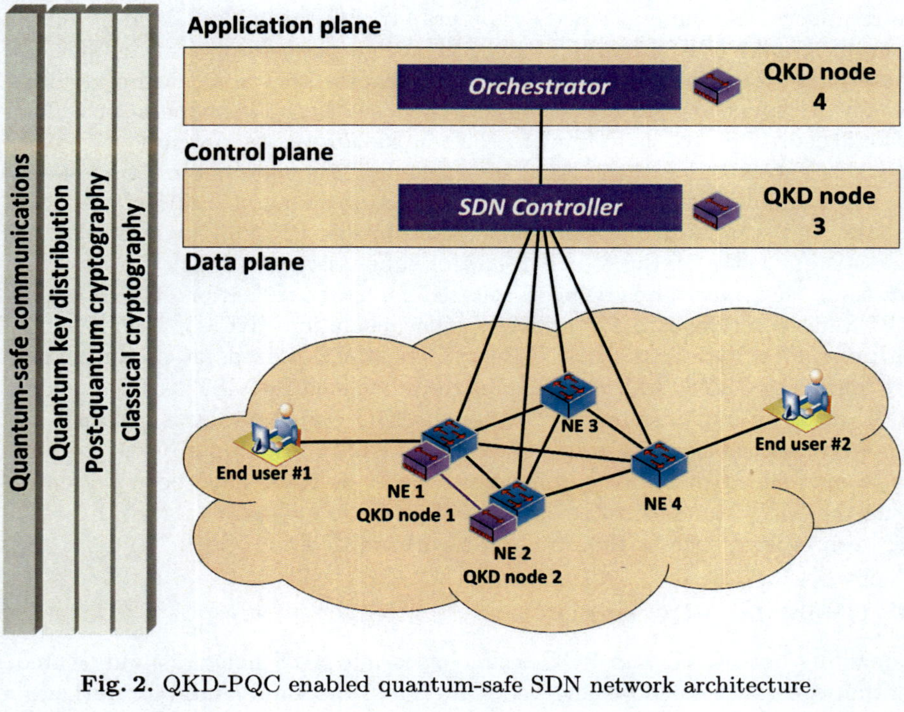


**Fig. 2.** QKD-PQC enabled quantum-safe SDN network architecture.

As can be seen in Fig. 2 SDN enabled networks can facilitate the integration of QKD and PQC for providing end-to-end quantum-safe communications in the control plane. For the parts of the network in which optical fiber infrastructure is already in place, SDN can be use to coordinate the QKD resources and allocate QKD-based keys for the requesting applications such as TLS-based communications among a network controller and a network element (e.g., programmable encryptor). In case that maximum security is required by the controller in a very specific scenario (e.g., renewal of the private key of a certain user), QKD-based keys could even be combined with PQC-based and classical-based keys, guaranteeing that three different cryptographic assumptions must be broken before the communications can be compromised.

A similar quantum-safe solution for encrypting communications could be as well applied to the SDN-controlled data plane. We could imagine the implementation of multiple-cryptography-based (QKD, PQC, classical cryptography) secure links among the most demanding communication channels of the networks, which are assumed to be implemented over optical fiber. On the parts of the networks where wireless links are implemented, PQC can be used in combination with classical cryptography for assuring that TLS communications remain quantum-safe up until the end-user. To properly guarantee that communications remain quantum-safe even in the case of disruption in the network, SDN must be dynamically aware of the status of the QKD network, having constant communication with the QKD nodes through the control plane to determine which links have higher or lower key consumption needs.

This becomes a different use case for SDN and QKD, in which the network controller can allocate and program the resources to ensure an optimum functioning of the QKD links, thus demonstrating a mutually beneficial relationship among SDN and QKD. In the case that some of the QKD links are overwhelmed by key consumption requests coming from applications, or in the face of possible attacks such as denial of service (DoS) or man-in-the-middle (MITM), the SDN controller must be able to dynamically reallocate the routes of the QKD network, avoiding those nodes that might be compromised, thus providing a minimum guarantee that the QKD communication remains operative. In a catastrophic scenario, in which the quantum channel among two QKD nodes becomes completely disrupted, the SDN controller must be able to anticipate and dynamically address the issue and halt communication – in case of extreme security requirements not met by PQC – or allocate PQC-based keys for the disrupted links, ensuring that communications remain safe until the quantum channel becomes operative again and QKD-based keys can be used again.

The use of quantum-safe cryptography for securing communications potentially allows the development of more distributed network architectures, in which applications such as edge/far edge computing, artificial Intelligence as a service (AIaaS) or confidential computing can be made available for end-users without the risk of exposing confidential information to potential (quantum-enabled) attackers.

## 5 Conclusion

This paper presents a novel architecture for enabling quantum-safe SDN communications. Within our architecture, we propose the integration of QKD and PQC into TLS to guarantee the safety of control plane communications. Moreover, a similar strategy can be followed towards securing data plane communications, enabling end-to-end quantum-safe communications, in which both QKD and PQC can inter operate to provide quantum-safe end-to-end communications for both the control and data planes.

## References

1. Aguado, A., et al.: Hybrid conventional and quantum security for software defined and virtualized networks. J. Opt. Commun. Netw. **9**(10), 819–825 (2017). https://doi.org/10.1364/JOCN.9.000819
2. Alagic, G., et al.: Status report on the third round of the nist post-quantum cryptography standardization process. Technical report NIST IR 8413, National Institute of Standards and Technology (2022). https://doi.org/10.6028/NIST.IR.8413
3. Beullens, W.: Breaking rainbow takes a weekend on a laptop. In: Dodis, Y., Shrimpton, T. (eds.) CRYPTO 2022. LNCS, vol. 13508, pp. 464–479. Springer, Cham (2022). https://doi.org/10.1007/978-3-031-15979-4_16
4. Bierman, A., Bjorklund, M., Watsen, K.: Restconf protocol. The Internet Engineering Task Force (IETF) (2017)
5. Shing Chen, M., et al.: Pqcrainbow (2023). https://www.pqcrainbow.org/
6. Dacier, M.C., König, H., Cwalinski, R., Kargl, F., Dietrich, S.: Security challenges and opportunities of software-defined networking. IEEE Secur. Priv. **15**(2), 96–100 (2017)
7. Enns, R., Bjorklund, M., Schoenwaelder, J., Bierman, A.: RFC 6241: network configuration protocol (netconf) (2011)
8. ETSI: ETSI GS QKD 014 v1.1.1 (2019). https://www.etsi.org/deliver/etsi_gs/QKD/001_099/014/01.01.01_60/gs_qkd014v010101p.pdf
9. Frankel, S., Krishnan, S.: IP security (IPsec) and internet key exchange (IKE) document roadmap. Technical report RFC 6071, Internet Engineering Task Force (2011). https://www.rfc-editor.org/rfc/rfc6071
10. Karakus, M., Durresi, A.: A survey: control plane scalability issues and approaches in software-defined networking (SDN). Comput. Netw. **112**, 279–293 (2017)
11. Lam, J.H., Lee, S.G., Lee, H.J., Oktian, Y.E.: Securing distributed SDN with IBC. In: 2015 Seventh International Conference on Ubiquitous and Future Networks, pp. 921–925. IEEE (2015)
12. Lopez, V., Pastor, A., Lopez, D., Aguado, A., Martin, V.: Applying QKD to improve next-generation network infrastructures. In: 2019 European Conference on Networks and Communications (EuCNC), pp. 283–288. IEEE (2019)
13. Nguyen, V.L., Lin, P.C., Cheng, B.C., Hwang, R.H., Lin, Y.D.: Security and privacy for 6G: a survey on prospective technologies and challenges. IEEE Commun. Surv. Tutor. **23**(4), 2384–2428 (2021). https://doi.org/10.1109/COMST.2021.3108618
14. Peng, Y., Wu, C., Zhao, B., Yu, W., Liu, B., Qiao, S.: QKDFlow: QKD based secure communication towards the OpenFlow interface in SDN. In: Yuan, H., Geng, J., Bian, F. (eds.) GRMSE 2016. CCIS, vol. 699, pp. 410–415. Springer, Singapore (2017). https://doi.org/10.1007/978-981-10-3969-0_45
15. Rescorla, E.: The transport layer security (TLS) protocol version 1.3. Request for Comments RFC 8446, Internet Engineering Task Force, p. 160 (2018). https://doi.org/10.17487/RFC8446, https://www.rfc-editor.org/rfc/rfc8446

16. Samociuk, D.: Secure communication between OpenFlow switches and controllers. In: AFIN 2015, vol. 39 (2015)
17. Scarani, V., Bechmann-Pasquinucci, H., Cerf, N.J., Dušek, M., Lütkenhaus, N., Peev, M.: The security of practical quantum key distribution. Rev. Mod. Phys. **81**, 1301–1350 (2009). https://doi.org/10.1103/RevModPhys.81.1301
18. Sezer, S., et al.: Are we ready for SDN? Implementation challenges for software-defined networks. IEEE Commun. Mag. **51**(7), 36–43 (2013). https://doi.org/10.1109/MCOM.2013.6553676
19. Sjoholmsierchio, M., Hale, B., Lukaszewski, D., Xie, G.: Strengthening SDN security: protocol dialecting and downgrade attacks. In: 2021 IEEE 7th International Conference on Network Softwarization (NetSoft), pp. 321–329 (2021). https://doi.org/10.1109/NetSoft51509.2021.9492614
20. Stan, C., Garcia, C.R., Cimoli, B., Olmos, J.J.V., Monroy, I.T., Rommel, S.: Securing communication with quantum key distribution: implications and impact on network performance. In: Optica Advanced Photonics Congress 2022, p. SpW2J.2. Optica Publishing Group (2022). https://doi.org/10.1364/SPPCOM.2022.SpW2J.2
21. Strand, M.: A status update on quantum safe cryptography. In: 2021 International Conference on Military Communication and Information Systems (ICMCIS), pp. 1–7 (2021). https://doi.org/10.1109/ICMCIS52405.2021.9486413
22. Wootters, W.K., Zurek, W.H.: A single quantum cannot be cloned. Nature **299**(5886), 802–803 (1982)
23. Yigit, B., Gur, G., Tellenbach, B., Alagoz, F.: Secured communication channels in software-defined networks. IEEE Commun. Mag. **57**(10), 63–69 (2019)
24. Ylonen, T., Lonvick, C.: The secure shell (SSH) protocol architecture. Technical report RFC 4251, Internet Engineering Task Force (2006). https://www.rfc-editor.org/rfc/rfc4251

# AI-Optimized Energy Management for More Efficient and Sustainable Microgrids

Sebastián López Flórez[1,2,3](✉), Guillermo Herniández[1,2,3,4], Alfonso Gonziález-Briones[1,2,3], and Fernando de la Prieta[1,3]

[1] BISITE Digital Innovation Hub, University of Salamanca, Edificio Multiusos I+D+i, Calle Espejo 2, 37007 Salamanca, Spain
{sebastianlopezflorez,guillehg,alfonsogb,fer}@usal.es
[2] Air Institute, IoT Digital Innovation Hub, 37188 Salamanca, Spain
[3] Universidad Tecnológica de Pereira, Cra. 27 N 10-02, Pereira, Risaralda, Colombia
[4] Institute for Artificial Intelligence and Big Data, Universiti Malaysia Kelantan, City Campus, Pengkalan Chepa, 16100 Kota Bharu, Kelantan, Malaysia

**Abstract.** The implementation of renewable energy sources reduces dependence on fossil fuels and greenhouse gas emissions, but it comes at a significant cost to the grid due to the need for coupling multiple energy sources. This necessitates the optimization of energy production using artificial intelligence techniques that can reflect the characteristics of distributed energy management from the storage capacity of the microgrid while considering the consumer load. As such, this study develops an energy management algorithm based on the Deep-Q-Network (DQN) with prioritized experience replay (PER) utility function to model the operator's risk-aversion or risk-seeking behavior, generating optimal exchange strategies in a dynamic environment. The reinforcement learning-based agent provides an understanding of energy storage capacity constraints in aggregate load/discharge energy decision making in the microgrid, using a discrete action space that depends on a reward related to the value of the optimal online objective function of the microgrid. By analyzing microgrid management under different system state definitions, the results show that substantially improved performance can be achieved compared to a traditional model that assumes deterministic information. The findings provide solid experimental validation of the potential of deep reward-based learning techniques to enable more efficient, flexible, and cost-effective energy management, particularly with the integration of more intermittent renewable energy sources.

**Keywords:** Smart grid technologies · Energy efficiency optimization · contingency reserve · Reinforcement learning

The IoTalentum Project within the Framework of Marie Skłodowska-Curie Actions Innovative Training Networks (ITN)-European Training Networks (ETN), which is funded by the European Union Horizon 2020 Research and Innovation.


R. Mehmood et al. (Eds.): DCAI 2023, LNNS 741, pp. 438–447, 2023.
https://doi.org/10.1007/978-3-031-38318-2_43

## 1 Introduction

Due to the exponential rise in energy demand, efforts are being made to increase the production of clean, sustainable energy while improving energy efficiency to lower consumption through the use of renewable energy sources including solar, eolian, and geothermal energy. This is due to the intermittent nature of the energy along with the stochasticity of the loads, the cost of network energy, and the availability of the network, which makes the generation of energy from renewable sources and the demand for energy a challenge as modern electrical systems become more complex [5].

The incorporation of distributed energy into the electrical system gives rise to the concept of a microgrid, an energy supply system that may function independently or in conjunction with an energy distribution network. When connected to the energy distribution network, the Microgrid can operate either in an isolated mode to provide local services with reliability guarantees or in a connected mode to fully utilize the distributed energy resources, subject to the characteristics of the main network and providing supplementary services like redundant energy sales to the network or purchasing additional energy from the network when needed. It is anticipated that a combination of several cooperating Microgrid autonomous systems will become the dominating configuration in the next-generation intelligent network [16].

The objective of the Microgrid energy management strategy is to use distributed energy efficiently and, at the same time, reduce long-term operating costs by buying and selling electricity on the distribution grid. In addition, due to the large number of sources distributed by Microgrid, it is necessary to perform statistical analysis, load consumption forecasting and data mining in order to establish benefits of optimal generation dispatch, energy savings, increased reliability and reduced costs due to losses. A number of studies have addressed this problem focusing on energy management considering various approaches to achieve the optimal operation of microgrids based on the expected load and renewable energy resources. Clarke et al. [2] studied a standalone desalination renewable energy system with the intention of sizing and power management. In the end, it was discovered that MOPSO's optimization produced greater outcomes in lowering NPC and CO2 emissions. TACCARI et al. [15] propose comparing two optimization approaches to deal with short-term operational planning of the energy systems, such as electric power plants, calderas, heat storage facilities, and cogeneration units. In considering the option optimized operation of multiple microgrids, reference [10] uses the random optimization method of generation and scene reduction to cope with renewable energy uncertainty. In ILR Gomes et al. [6] focuses on a management system to support and plan the operation of a microgrid by the new entrant in the electricity market, the microgrid aggregator This system is based on mixed-linear entera estocatic programming. Zhang et al. [17] took a step forward, which both includes electric turbines and photovoltaic panels. They developed an energy management model based on uncertain factors represented by typical scenarios to coordinate the operation of the electrical and thermal systems. However, these methods are not guaranteed to fall into the local optimum easily, In addition, the lack of an ideal target for the correction makes it difficult to apply the methods now available to make practical control decisions and in some cases generates a significant computational burden.

The challenges surrounding energy consumption scheduling, energy trading, and storage management in microgrids entail intricate interdependencies among various components and nonlinear dynamics under constant changes in system parameters, such as real-time pricing, contingency reserve requirements, and operator risk preferences. Traditional optimization and rule-based techniques rely on strong assumptions of precise predictions, static models, and linear relationships that are invalid for such complex cyber-physical energy systems [9,12].

In contrast, reinforcement learning acquires the optimal policy through continuous trial-and-error interactions with the dynamic system. The learned policy adapts to changes in conditions without requiring retraining, providing a flexible and optimal approach [4]. Deep reinforcement learning further enhances its capabilities by allowing the handling of high-dimensional state and action spaces with complex nonlinear interactions. Empirical results also suggest that deep Q-learning(DQN) techniques can outperform rule-based and optimization-based baselines for complex decision-making problems under uncertainty.

Specifically, we propose a deep neural network (DNN) model with prioritized experience replay (PER) that prioritizes experiences with large errors, which has shown to substantially improve the accuracy and speed up the training of deep neural networks in high-dimensional spaces by updating the weights of relevant experiences in the buffer memory as shown in the Fig. 1. This technique enables the algorithm to focus on the most informative data points, rapidly improving the optimal control policy compared to a Q-Network approach, also validated in this research for this use case.

This document is structured as follows: Sect. 2 details the microgrid energy management problem considering flexible option sources and describes the deep RL framework. Section 3 presents comparative results of the methods used in the use case. Section 4 provides conclusions

## 2 Deep RL for Energy Microgrids Management

Microgrids are interconnected local distribution power systems that can operate in isolation or connected to the main distribution grid. These smart grids integrate distributed renewable generation, energy storage systems and loads, and allow optimizing the use of resources at the local level, as well as improving reliability and security of supply. They facilitate the integration of renewable energies by providing operational flexibility and active management of supply and demand. They can include multiple generation technologies such as solar photovoltaic, wind, biomass or mini-hydro. Storage systems, such as batteries or flywheels, provide backup capacity, smooth the variability of renewables and enable new grid management services. Finally, this paper assumes that the microgrid is not connected to the grid, where the objective function is the energy

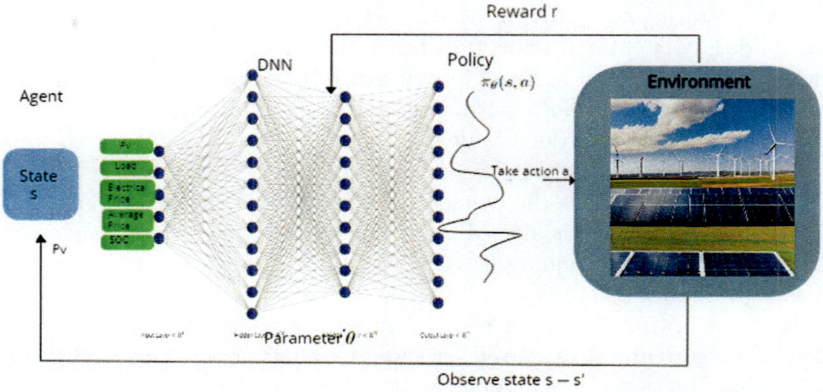


**Fig. 1.** Simple schematic of algorithm

management system that determines how the microgrid behaves in response to different scenarios. In these papers, the factors considered in the objective functions consist of an aggregate consumption load, a PV farm and an Energy Storage System, which is connected only to the main grid through a power distribution line. The microgrid is managed by a microgrid operator who has full control of all microgrid operations, including the power transaction operation with the main grid [1].

The following assumptions are made about microgrid systems each time interval is assumed to last 1 h. Therefore, a system divided into N time slots of 1 h duration each is considered. Each time slot, denoted as 4t, has a duration of 1 h, indicating that there are 24 power scheduling time slots per day. Each of these 24 time slots represents a 1-h period, so the optimal energy scheduling must be determined hour by hour over a full day. Second, a simplified (quasi-static) model of the dynamics of the microgrid energy system is assumed to facilitate its study and optimization, third, it is assumed that the generation and consumption of energy in the microgrid remains constant within each time interval (slot) considered, but changes from one slot to another to capture its variability, third, it is assumed that the generation and consumption of energy in the microgrid remains in equilibrium, with no power deficit or surplus [3].

## 2.1 Reinforcement Learning

Reinforcement learning (RL) is a field of artificial intelligence focused on the sequential interaction of decisions between an agent and an environment. The agent pursues an optimal policy of actions that maximizes a reward received from the environment over time.

Formally, we define a state space $S$, an action space A and a family of reward functions $r(s, a, s') : S \times A \times S \longrightarrow R$, where R are the reals. A sequence of states, actions and rewards forms a trajectory (also called episode, history or roll-out) $\tau = (s_0, a_0, r_0, s_1, a_1, r_1, ...)$ defined by the dynamics of the Markov decision model (MDP).Each transition has a likelihood of happening $p(s'|s,a)$ and offers a specific amount of benefit determined by $r(s,a,s')$. In episodic tasks, the horizon $\tau$ is finite, while in continuing tasks $\tau$ is infinite.

The agent's policy of actions is defined as $\pi : S \longrightarrow A$. Given a trajectory and a policy, the cumulative reward is computed as $G(\tau, \pi) = \sum_{k=0}^{\tau} \gamma r_t$. The optimal policy $\pi_*$ maximizes the expected reward $G\pi$. If the task is episodic ($\tau$ is finite, the trajectories ends after a finite number of transitions), $\gamma$ can be set to 1, but if the task is continuing ($\tau = \infty$, trajectories have no end), $\gamma$ must be chosen smaller than 1.

The iterative optimization process of the policy $\pi_*$ is known as reinforcement learning (RL) algorithm. It can be either value-based or policy-based. The former update a value function $V\pi(s)$, while the latter directly update the policy $\pi(s)$. In Policy-Based methods, we learn a policy function directly.

Learning takes place by exploring the space of states and actions to find the optimal paths, and exploiting the knowledge gained to obtain the highest possible reward. After many iterations, the policy converges to an optimal solution $\pi_*$.

## 2.2 Deep RL Solutions for Sequential Decision Making

This work uses a Deep Learning version of Q-Learning [13] called Deep Q-Network (DQN) and Prioritized Experience Replay (PER) as configured and conditioned in [5,14]. Another feature of some of these algorithms is the ability to work with continuous and discrete action spaces, but since DQN only works with discrete action spaces, these algorithms require discretization of continuous action space environments.

**Deep Q-Network (DQN).** To formally define the DQN algorithm, let $Q(s,a;\theta)$ be the $Q$ function that we estimate with a deep neural network with parameters $\theta$. The goal of the algorithm is to find the parameters $\theta*$ that minimize the difference between the estimated $Q$ function and the true $Q$ function. The algorithm updates the $Q$ function as follows.

$$q_\star(s,a) = E\Big[R_{t+1} + \gamma\ max_{a'} q_\star(s',a')\Big] \tag{1}$$

After calculating the loss, the weights are updated in the network using SGD and backpropagation, like any other normal network. This process is repeated for each state of the medium until the losses are sufficiently reduced and an approximately optimal Q function is obtained.

The Q-function is updated as follows:

$$q_\star(s,a) = q_\star(s,a) + \alpha[r + \gamma\ max_{a'} q_\star(s',a') - q_\star(s,a)] \tag{2}$$

**Prioritized Experience Replay (PER).** The Prioritized Experience Replay (PER) algorithm is an extension of the memory replay method used in the DQN algorithm. In PER, experiences stored in the replay memory are prioritized based on their importance for improving learning. Each experience is associated with a weight indicating its importance, and a non-uniform selection criterion is used to sample experiences from the replay memory, ensuring that the probability of being sampled is monotonic in the priority of a transition, while ensuring a non-zero probability even for the lowest priority transition. The most important experiences (i.e., those with higher weight) are sampled more frequently, helping the algorithm to focus on the most relevant experiences and improve learning more efficiently. The stochastic sampling method interpolates between purely greedy prioritization and uniform random sampling. PER prioritizes them according to their expected importance for future learning. Transitions with higher temporal difference (TD) errors, which indicate greater learning potential, are replayed more frequently. This allows learning to quickly focus on regions of the state-action space where the approximate value function differs most from the optimal.

The sampling probability of a transition is defined as follows:

$$P(i) = \frac{p_i^\alpha}{\sum_k p_k^\alpha} \tag{3}$$

Here, $p_i > 0$ represents the priority of the transition, and the exponent determines the extent of prioritization, where $\alpha = 0$ corresponds to the uniform case.

Importance sampling weights rectified the uncontrolled shift in sampling distribution under prioritized experience replay. By reweighting priorities, IS stabilized the solution that value estimates would converge to rather than diverging due to bias from prioritization.

$$w_i = \left(\frac{1}{N} \cdot \frac{1}{P(i)}\right)^\beta \tag{4}$$

which fully explains the non-uniform probabilities $P(i)$ if $\beta = 1$.

We use the absolute value of the magnitude of our TD error:

$$p_i = |\delta_i| + \epsilon \tag{5}$$

## 2.3 Application of RL in the Microgrid

**State Definition.** In the case of microrgrids, the state of the system can be described using a tuple of five components that captures the relevant details of the system's state for each time step. This state representation is then used to construct a state-based window, represented as *st*, which captures the system's current and past states for a given time interval.

$$s_t \in S = \{P_t^{WG}, P_t^{Load}, Pr_t, Pr_t^{avg}, SOC_t\}$$

where electrical load $P_t^{Load}$, wind power $P_t^{PV}$ and electrical price $Pr_t$ are obtained from the time series dataset, while SOC and average electrical price $Pr_t^{avg}$ for the past 24 h need to be calculated at each time step [11].

An excerpt from a dataset encompassing load consumption data and dynamic prices, sourced from [8] and employed for model evaluation, exhibits the trends of photovoltaic solar energy generation, electricity consumption, and prices over a 24-h period.

**Action Definition.** In the context of managing lithium-ion battery energy, it is crucial to determine the amount of energy to be charged or discharged using a reinforcement learning algorithm. This algorithm enables the agent's action at time t to be defined, incorporating fundamental aspects into the utility function [3].

- Maintaining a minimum level of contingency reserves or target state of charge (SOC) in the ESS to ensure critical operations. The utility function will be piecewise, depending on if the SOC is below, at or above this target level.
- Below the target SOC, the agent must charge the ESS as soon as possible regardless of price. And the price must be high enough for discharging at this level.
- Above the target SOC, the criteriachanges. At higher SOC levels, the agent should charge at lower prices and discharge at higher prices to better utilize the remaining storage capacity. Closer to the target SOC, the agent should do the opposite.
- A negative exponential multiplier (M) is used to incorporate these criteria into the utility function. Parameters can be tuned to control the agent's behavior below and above the target SOC.
- Two penalties are also introduced to further control the agent's actions:
  - PntESSt: Penalizes actions that violate ESS operating limits. This helps preserve the ESS.
  - PntPVt: Penalizes not storing surplus solar power when there is spare capacity. This aims to maximize use of free energy within limits.

The overall utility function (u) incorporates all these elements to guide the agent's trading strategies regarding the ESS in a way that maintains necessary reserves, utilizes storage effectively and prolongs the ESS lifespan.

## 3 Results

An energy storage system utilizing lithium-ion batteries (ESS) is employed, with a nominal power of 1000 kW and a nominal capacity of 5000 kWh. The ESS boasts a cycle life of 4996, a $\delta$ value of 1, and a respective $\eta_c$ and $\eta_d$ of 90%. It is established that $C_i$ is 171/kWh. The critical microgrid operations require a target SOC of 0.5, equivalent to 2500 kWh [7]. Other hyperparameters include (Table 1).

**Table 1.** Hyperarameters Reinforcement Learning Model

| $\rho$ | $d$ | err | $\alpha$ | $\beta$ |
|---|---|---|---|---|
| 0.6 | 0.0001 | 0.01 | 0.001 | 0.4 |

Simulations were conducted to assess the impact of risky and prudent behaviors of the MGO on the ESS operation decisions and performance. Furthermore, the performance of the proposed DQN-PER and DQN algorithm was compared with other algorithms. The numerical results of the simulations were compiled and tabulated in Table 2. The ESS operations and discharge patterns during a single day using different algorithms are depicted in Fig. 2a and 2b., respectively, under restricted conditions of surplus solar energy and ESS capacity reserve.

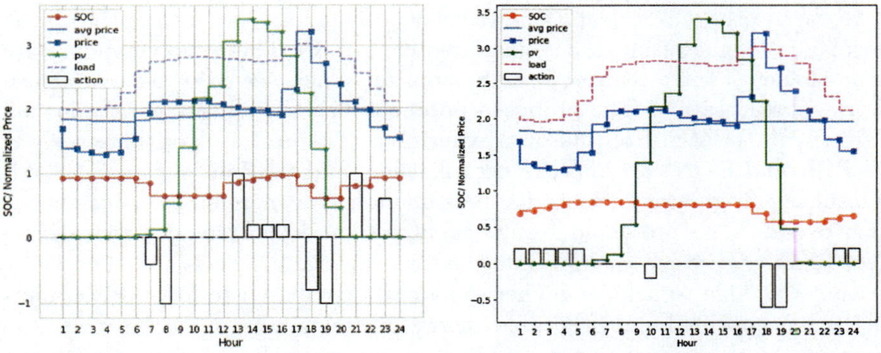


(a) PER ESS actions .

(b) DQN ESS actions

**Fig. 2.** Optimal trading strategy for a 24 h period

The average daily costs for DQN amount to $62.62. By taking perfect knowledge into account, DQN is capable of making more informed decisions when compared to relying solely on forecast knowledge. Conversely, the average daily costs for PER are $22.75, respectively. With a risk-averse perspective, the Microgrid tends to be less aggressive in energy trading compared to a risk-seeking perspective. This implies that, when considering potential risks, Microgrid takes a more cautious approach in its energy marketing activities.

**Table 2.** Hyperarameters Reinforcement Learning Model

| | DQN-PER | DQN | DQN [7] |
|---|---|---|---|
| Benefit($) | 22.75 | 62.62 | 53.45 |

There are two main observations from the results:

– The probability of the state of charge (SOC) falling below the target SOC is lowest using the DQN algorithm, and then the DQN-PER algorithm
– The daily monetary benefit obtained is highest using the DQN surpassing that of the state of the art

The ESS actions on how DQN-PER algorithm acted as an energy optimizer that made intelligent data-driven decisions. It charged the battery to maintain a high state of charge when the load was low and discharged the stored capacity during periods of high demand, thus fulfilling the reward function designed. Thus, ESS acts more meticulously in the DQN algorithm compared to the DQN-PER algorithm.

## 4 Conclusions

This research proposes a deep reinforcement learning approach to optimize the performance of microgrids under simulated conditions. The results demonstrate that an efficient control strategy can be applied to new scenarios using a neural network representation. The proposed approach has the advantage of allowing the algorithm to learn from its own experience. The use of algorithms such as DQN-PER and DQN can improve the efficiency and flexibility of energy trading, resulting in greater benefits for prosumers and a more sustainable energy system overall. The presented results highlight the importance of considering perfect knowledge when making informed energy trading decisions. The DQN algorithm was able to achieve higher daily costs compared to DQN-PER, indicating a more optimal and profitable energy trading strategy. Furthermore, the study revealed the impact of risk aversion on energy trading decisions, with the Microgrid taking a more cautious approach to energy marketing activities. These findings demonstrate the importance of considering risk in energy trading and the potential benefits of implementing adaptive algorithms that can respond to changing market conditions. Overall, these results provide valuable insights into the effective management of energy trading activities and the potential for AI-based approaches to enhance energy system sustainability and profitability.

As a future work, implementing a more descriptive model where the output is characterized by a distribution is a promising direction. This approach can provide a more accurate representation of the uncertainty involved in energy trading and enable more informed decision-making. Furthermore, incorporating probabilistic models can facilitate the development of risk-averse strategies that take into account the potential outcomes of different decisions. The use of distribution-based models can also enable the modeling of complex systems with multiple sources of uncertainty and provide a more comprehensive understanding of the underlying dynamics. Overall, the implementation of distribution-based models represents a promising avenue for enhancing the accuracy and robustness of energy trading algorithms.

## References


1. Adhikari, R.S., Paudyal, P.: Handbook on Microgrids for Power Quality and Connectivity, 1st edn. Elsevier, Amsterdam (2019)
2. Affi, S., Cherif, H., Belhadj, J.: Smart system management and techno-environmental optimal sizing of a desalination plant powered by renewables with energy storage. Int. J. Energy Res. **45**(5), 7501–7520 (2021)
3. Aih, H.C., Kumar, P.P.S., Srinivasan, P.D.: Energy management economic evaluation of grid-connected microgrid operation. Department of Electrical Computer Engineering, National University of Singapore (2020)
4. Du, Y., Li, F.: Intelligent multi-microgrid energy management based on deep neural network and model-free reinforcement learning. IEEE Trans. Smart Grid **11**(2), 1066–1076 (2019)
5. Gassi, K.B., Baysal, M.: Improving real-time energy decision-making model with an actor-critic agent in modern microgrids with energy storage devices. Energy **263**, 126105 (2023)
6. Gomes, I., Melicio, R., Mendes, V.: A novel microgrid support management system based on stochastic mixed-integer linear programming. Energy **223**, 120030 (2021)
7. Hau, C., Radhakrishnan, K.K., Siu, J., Panda, S.K.: Reinforcement learning based energy management algorithm for energy trading and contingency reserve application in a microgrid. In: 2020 IEEE PES Innovative Smart Grid Technologies Europe (ISGT-Europe), pp. 1005–1009. IEEE (2020)
8. Hong, T., Pinson, P., Fan, S., Zareipour, H., Troccoli, A., Hyndman, R.: Probabilistic energy forecasting: global energy forecasting competition 2014 and beyond. Int. J. Forecast. **32**(3), 896–913 (2016). https://EconPapers.repec.org/RePEc:eee:intfor:v:32:y:2016:i:3:p:896-913
9. Ji, Y., Wang, J., Xu, J., Fang, X., Zhang, H.: Real-time energy management of a microgrid using deep reinforcement learning. Energies **12**(12), 2291 (2019)
10. Kargarian, A., Rahmani, M.: Multi-microgrid energy systems operation incorporating distribution-interline power flow controller. Electr. Power Syst. Res. **129**, 208–216 (2015)
11. Liu, F., Liu, Q., Tao, Q., Huang, Y., Li, D., Sidorov, D.: Deep reinforcement learning based energy storage management strategy considering prediction intervals of wind power. Int. J. Electr. Power Energy Syst. **145**, 108608 (2023)
12. Nakabi, T.A., Toivanen, P.: Deep reinforcement learning for energy management in a microgrid with flexible demand. Sustain. Energy Grids Netw. **25**, 100413 (2021)
13. Ong, H.Y., Chavez, K., Hong, A.: Distributed deep Q-learning, arXiv preprint arXiv:1508.04186 (2015)
14. Saglam, B., Mutlu, F.B., Cicek, D.C., Kozat, S.S.: Actor prioritized experience replay, arXiv:2209.00532 (2022)
15. Taccari, L., Amaldi, E., Martelli, E., Bischi, A.: Short-term planning of cogeneration power plants: a comparison between MINLP and piecewise-linear MILP formulations. In: Computer Aided Chemical Engineering, vol. 37, pp. 2429–2434. Elsevier (2015)
16. Wu, N., Wang, H.: Deep learning adaptive dynamic programming for real time energy management and control strategy of micro-grid. J. Clean. Prod. **204**, 1169–1177 (2018)
17. Zhang, Y., Meng, F., Wang, R., Kazemtabrizi, B., Shi, J.: Uncertainty-resistant stochastic MPC approach for optimal operation of CHP microgrid. Energy **179**, 1265–1278 (2019)

# Efficient Protected VNF Placement and MEC Location Selection for Dynamic Service Provisioning in 5G Networks

Maryam Masoumi(✉), Ignacio de Miguel, Ramón J. Durán Barroso, Soheil Hosseini, Hafiza Kanwal Janjua, Noemí Merayo, Juan Carlos Aguado, and Rubén M. Lorenzo

Universidad de Valladolid, Valladolid, Spain
{maryam.masoumi,ignacio.demiguel,ramon.duran,soheil.hosseini, hafizakanwal.janjua,noemer,juancarlos.aguado.manzano, ruben.lorenzo}@uva.es

**Abstract.** Multi-Access Edge Computing (MEC) and Network Function Virtualization (NFV) have emerged as promising technologies for providing low-latency and high-bandwidth services to mobile users through Service Function Chains (SFCs) consisting of Virtual Network Functions (VNFs). However, the efficient utilization of network resources and optimal placement of MECs to host VNFs remain challenging issues in NFV-based dynamic networks. Since the proximity of MEC nodes to end-users and the availability of resources at the location impact the blocking ratio in VNF placement, in this paper we analyze the impact of the location of the set of MEC resources in the network. Thus, starting from the work previously done related to the VNF placement scheme with different protection methods to accommodate services in MEC nodes, we conduct a simulation to evaluate the impact of different MEC locations on the blocking ratio. Our results demonstrate that strategic placement of MEC locations can significantly reduce the blocking ratio.

**Keywords:** VNF Placement · MEC Location · Blocking Ratio

## 1 Introduction

Network Function (NF) refers to a component or application that performs a specific networking task within a larger network infrastructure. Despite the fact that these components are created to enhance network security and efficiency,

This work is part of IoTalentum project that has received funding from the EU H2020 research and innovation programme under the MSCA grant agreement No 953442. It is also supported by Consejería de Educación de la Junta de Castilla y León and the European Regional Development Fund (Grant VA231P20), and the Spanish Ministry of Science of Innovation and the State Research Agency (Grant PID2020-112675RB-C42 funded by MCIN/AEI/10.13039/501100011033).


R. Mehmood et al. (Eds.): DCAI 2023, LNNS 741, pp. 448–456, 2023.
https://doi.org/10.1007/978-3-031-38318-2_44

they come with a high price tag, limited adaptability, and pose operational challenges [9]. Network Function Virtualization (NFV) is introduced to address these challenges by virtualizing and consolidating network functions onto commodity servers, storage, and switches. This approach replaces traditional dedicated network hardware with software-based virtualized network functions (VNFs) running on standard IT infrastructure [10] and [14]. Multi-Access Edge Computing (MEC) is another technology that is predicted to have a significant influence on 5G networks by fulfilling the demand for ultra-low latency in specific applications and services and minimizing transport network congestion [11]. In order to have more services, Service Function Chaining (SFC) can be employed. This involves processing a sequence of VNFs in a specific order to provide a particular service. As a result, a crucial concern is determining the appropriate placement of the VNFs in the predefined sequence to fulfill the service requirements [9].

VNF placement refers to the process of selecting the optimal locations to deploy VNFs within a network infrastructure. To enhance customers' Quality of Experience (QoE), VNFs need to be dynamically relocated between network nodes. One of the challenges concerns the placement of VNFs in edge nodes consisting of selecting the most suitable edge (MEC) nodes to deploy VNFs- SFCs to meet service requests. Reliability is also an important criterion for selecting a service provider, which prompts them to seek NFV deployment algorithms that maintain reliability above specific standards. Sometimes, networks are faced with increased computing loads, hardware failures, or malicious attacks. To ensure a desired level of reliability, protection schemes are commonly employed, with dedicated or shared backup being the two types of protection methods [3]. This is necessary to ensure the reliability of each request. Therefore, in all the protection strategies, a primary SFC is set up to provide the related service in normal conditions and is protected by a backup SFC that has its VNFs located in distinct physical positions [1].

Network operators strive to boost the service acceptance ratio by accommodating as many SFC requests as possible and minimizing the blocking probability. Previous research has investigated different techniques for protecting SFCs for static traffic. However, our research focuses on the challenge of ensuring reliable SFC mapping in dynamic settings. In our previous study [7], we conducted an investigation into the VNF Placement problem, specifically focusing on VNF Placement at the Edge and dynamic control and resource allocation for CPU and RAM. Our approach involved identifying MEC locations that would minimize communication delays between nodes and MEC sites. We then implemented and compared different scenarios with varying numbers of MEC sites. For this aim, we compared 3-MEC, 5-MEC and 7-MEC scenarios.

The results indicated that concentrating computing resources in a smaller number of MEC sites rather than dispersing them across multiple locations results in decreased probabilities of blocking. To enhance network resiliency against a single failure, we proposed two protection methods: dedicated and shared protection. A dedicated approach can be effective in providing high levels of protection for VNFs, but can also be costly due to the need for additional resources while the shared approach can be more cost-effective than dedicated protection, but may not provide the same level of protection as dedicated resources. Since the location of MECs is of great importance in service blocking, in this current paper we aim at considering the impact of the locations of the MEC on the blocking ratio. Moreover, we focus on the significance of MEC node placement in relation to service blocking. The primary objective of this paper is to analyze and understand how the locations of MECs influence the blocking ratio. In our previous study, we made certain assumptions about the placement of 5 MEC nodes.

In this current research, we are taking a further step by exploring the implications of keeping the same number of MEC nodes but locating them in different positions. This is a contribution compared to our previous work. Additionally, we are also considering a more realistic assumption regarding service chains. Essentially, this paper has a two-fold objective. Firstly, it aligns with the IoTalentum project [5], specifically addressing one of the research topics focused on integrating NFV in a distributed computing resources in MEC by means of SDN-based control mechanism and determining the optimal locations for MEC nodes. Secondly, it aims to provide updates and enhancements to our previous research, which are being presented in this paper.

The main contributions of this paper can be summarized as follows: Several possibilities for the 5-MEC locations are taken into account and then to assess and compare the blocking rate of different locations for 5MEC configuration, a real-world network topology is employed.

The remainder of this paper is as follows: In Sect. 2, we provide a background on VNF placement and MEC location problems. Section 3 discusses the problem addressed in our previous paper. In Sect. 4 an overview of our previous work is given followed by an explanation of VNF placement algorithm in Sect. 4.1. We present the main contribution of this work by providing the different assumptions for MEC locations in Sect. 4.2. Section 5 is our numerical results obtained by simulation. Finally, Sect. 6 concludes the paper and explains the potential direction of future research.

## 2 Background Works

MEC utilizes the characteristics of NFV, SDN, and NS (Network Slicing) to provide delay-sensitive services dynamically as requested. As a result, MEC has become a critical component for future 5G networks and beyond [4]. Due to its importance as a 5G technology enabler alongside NFV and SDN, there has been interest from both industry and academia in addressing issues related to MEC technology. Dash et al. in [2] propose an efficient clustering-based approach to locate MEC servers and associate them with RANs. The approach uses a greedy-based algorithm called MEC Placement and Association (MEC PA) to minimize the number of required MEC servers while satisfying the propagation delay constraint of all RANs. Under this approach, each MEC server acts as a cluster head, and each RAN is treated as a member of the cluster that meets the propagation delay constraint. This placement strategy not only meets the propagation delay requirement but also balances the workload.

In [12], an algorithm is proposed for selecting a few MEC locations from a set of base stations in order to establish MEC infrastructure and address the user requirements of delay-sensitives services. In this paper, a set of BS nodes are selected as potential MEC locations based on high Closeness Centrality (CC) values. So, the high-CC value BSs are selected as potential MEC locations. However, the aspects of resiliency and service continuity have not been adequately explored in resource-limited MEC-enabled 5G networks to ensure that latency requirements and service continuity are maintained even in the event of failures. This section will examine previous studies that have focused on the deployment of VNFs in MEC-enabled networks with the aim of enhancing system performance and network efficiency. [6] addresses the problem of VNF placement and resource allocation in MEC networks enabled by NFV/SDN, with the objective of minimizing both the overall placement and resource cost. However, it does not take into account the latency requirements of users, failures of network components, or resiliency aspects.

The issue of VNF placement in a MEC-NFV environment is addressed in [13], with a focus on minimizing costs by taking into account both latency and resource availability. However, the study does not address backup methods or resiliency aspects, which are necessary to ensure service continuity in the case of failures. Furthermore, the study only considers a single VNF placement for slice creation. [8] is the most closely related prior studies to previous and current work. The issue of resilient VNF placement in 5G networks is addressed, with the objective of minimizing the impact caused by network component failures. This paper aims to explore how the blocking ratio is impacted by varying MEC node locations in a network that will have five MEC nodes.

## 3 Problem Description

In this paper, an analysis of a 5G network is presented, which employs a distributed MEC infrastructure with virtualization at each site to handle service requests. However, this network faces limitations in computational and storage capacities as well as bandwidth. VNF failures can lead to disruptions in VNF-dependent services, and multiple VNF operations are vulnerable to interruptions, causing various services to fail. Therefore, the necessity arises to identify a solution for mapping dynamic service requests onto MECs while considering protection schemes. To reduce communication delays between nodes and their nearest MEC site, strategically chosen MEC sites are utilized. The dynamic resource allocation problem is formulated in this paper without protection methods, and then dedicated VNF backup and shared VNF backup protection methods are applied to ensure the resilience, security, and availability of established services. The objective of this paper is to minimize the blocking ratio of a given service request while focusing on protection against single failures.

## 4 A Summary of Our Proposed VNF Placement Method

The network is represented as an undirected graph with base stations and physical links. MEC locations are set up by selecting a small subset of base stations and assigning a finite number of servers at each MEC location to accommodate services for user requests. Each server is equipped with computing capacity with a certain number of CPU cores and RAM. Physical links have a specific bandwidth capacity and introduce latency consisting of propagation and processing delay. Three scenarios are considered: unprotected operation, dedicated SFC protection, and shared VNF protection.

### 4.1 VNF Placement Algorithm

In this part, we provide a brief explanation of the method for VNF placement that we previously introduced in our work [7]. We assume dynamically arriving of SFC requests. In order to set up a service function chain, we must allocate all of its VNFs on MEC nodes and direct the traffic between them. Each SFC needs a certain amount of bandwidth and has a maximum latency limit. Additionally, the VNFs associated with each SFC require a particular amount of computational resources, which is measured in terms of CPU core usage and RAM. Each SFC is modeled based on some specifications including the source base station from which the user requests service, the number of VNF, the required bandwidth, the maximum allowed end-to-end latency, the time of arrival, and the lifetime of the SFC request.

Our algorithm includes two phases:

1. Phase I: The placement of the whole SFC in a single MEC site is considered for service provisioning. In the first scenario (unprotected operation), a primary SFC is established for each request (with no backup SFC), and the MEC node closest to the user requesting the service is selected to host the SFC. The service is provided by selecting the first server within the MEC site that has sufficient available CPU and RAM capacity, and the required CPU and RAM are reserved during the service duration. If no server is found with available resources in this MEC, the other nearest MECs are considered so as to find a server with enough capacity. The request will be blocked if no server can be found with enough resources (or if bandwidth or delay constraints are not met).
   In the second and third scenarios (when dedicated and shared protection are applied), the primary MEC is selected as the first nearest MEC to the user, and the backup MEC is considered to be the second closest one. If there are insufficient resources in the preferred MEC site, the algorithm considers the other nearest MEC. In dedicated protection, one backup SFC protects one primary SFC in a single MEC site using the first-fit policy. In contrast, shared protection allows one backup VNF to protect multiple primary VNFs located in different MEC sites. If a suitable backup VNF cannot be found for shared protection, a new instance of the VNF is created. When the lifetime

of a request expires, all resources are released except for shared protection, where backup VNF resources are released only when the last SFC using them is released.

2. Phase II: During this phase, the total time it takes for both the primary and backup routes to transmit data is computed, considering both the propagation and processing times. The end-to-end delay for the primary route and the backup route is calculated based on the distance between the source base station and the primary MEC and backup MEC respectively. Moreover, each VNF introduces an additional delay which should be taken into account in each SFC.
   After meeting the latency criteria, the SFC is provisioned and the computational resources are then allocated. The purpose of this mechanism is to ensure that the backup VNF is available for use in case of a failure of any primary VNF in any of the SFC instances that use it.

### 4.2 Different Assumptions for MEC Locations

In our previous study, we proposed concentrating computing resources in a smaller number of MEC sites, rather than spreading them across a larger number of locations, as it demonstrated better performance. Building upon that research, this paper investigates the influence of different MEC locations on the blocking ratio while utilizing a shared protection mechanism. We present a scenario with 5 MECs and introduce three distinct sets of MEC node locations to evaluate their impact on service availability.

## 5 Simulation Results

To obtain numerical results we develop a discrete event-driven simulator in Python. We consider a reference metro regional network of Italy similar with 51 nodes, out of which 5 nodes are MEC, each one has 315 servers which are equipped with 512 CPU cores and RAM. In this network, there are 61 bidirectional links enabling communication in both directions. The longest among these links measures 101.46 km in length. The SFCs considered in this paper, their VNFs as well as their latency and bandwidth requirements are depicted in Table 1 while computational requirements of VNFs are shown in Table 2. We consider that incoming SFC requests are selected with the possibility of 25% among those. SFC requests are generated at MEC nodes dynamically.

The SFC requests are generated with an inter-arrival rate $\lambda$ that follows a Poisson distribution and a lifetime according to negative-exponential distribution, with mean lifetime $\mu = 60$ s. All the results are obtained with a confidence level of 95% with at most 5% confidence interval on blocking probability. The load is determined based on the average lifetime of each request, the average inter-arrival time, and the number of nodes in the topology as follows:

$$load = \frac{\lambda * \mu}{N(N-1)} \tag{1}$$

**Table 1.** Services Function Chain Requirements [15].

| Services | Service-Chain-VNFs | Banddwidth (Mbps) | Delay (ms) |
|---|---|---|---|
| Cloud Gaming | 4 | NAT, FW, VOC, WO, IDPS | 80 |
| Augmented Reality | 200 | NAT, FW, TM, VOC, IDPS | 2 |
| MIoT | 100 | NAT, FW, IDPS | 5 |
| Video Streaming | 4 | NAT, FW, TM, VOC, IDPS | 100 |

**Table 2.** CPU Core and RAM Usage For Various VNFs [15].

| VNF | CPU | RAM |
|---|---|---|
| NAT | 2 | 4 |
| FW | 1 | 4 |
| IDPS | 1 | 2 |
| TM | 1 | 2 |
| VOC | 2 | 4 |
| WO | 1 | 2 |

The unit of load is typically expressed as Erlang. In order to measure the latency, the nodes are linked together using optical fibers, which leads to a propagational delay. Moreover, each VNF add a value of 0.050 ms to the total delay.

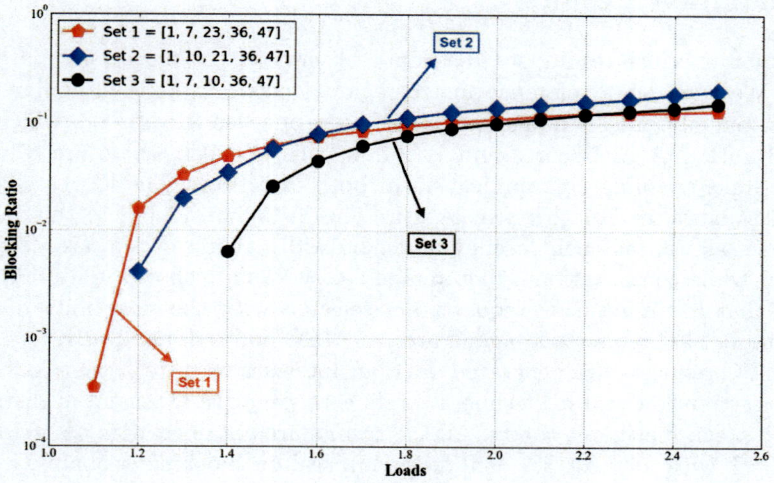


**Fig. 1.** Comparison in Blocking Ratio between Different Location Sets with Shared Protection

For the purpose of this study, we make the simplifying assumption that the constraints are only due to computing resources and delay, and we do not consider any bandwidth requirements. We consider a scenario with 5 MECs and introduce three sets of MEC node locations, denoted as location set 1, location set 2, and location set 3. Each set consists of 5 MEC locations. In location set 1, the MEC nodes are situated at nodes 1, 7, 23, 36, and 47. Location set 2 places the MECs at nodes 1, 10, 21, 36, and 47, while location set 3 consists of nodes 1, 7, 10, 36, and 47. The performance, specifically the blocking probability, varies depending on the MEC location.

The graph in Fig. 1 illustrates the blocking ratio for these proposed node sets. Location set 3, with MEC nodes placed at nodes 1, 7, 10, 36, and 47, exhibits superior performance in terms of blocking probability. On the other hand, configuring the network according to the location set 1 results in a higher blocking ratio, particularly for lighter loads. This discrepancy can be attributed to the distribution of the loads handled by each server, as MEC nodes in certain locations may serve a larger number of nodes. Location set 3 appears to have a more homogeneous distribution of workloads. As we transition from set 1 to set 2 and finally to set 3, the blocking rate decreases, indicating a more balanced workload in set 3, where MECs are positioned at nodes 1, 7, 10, 36, and 47.

## 6 Conclusion

To conclude, this paper provides an overview of the approach proposed in our previous work for dynamically placing VNFs in a MEC-enabled environment. The approach aims to minimize communication delays and introduces two protection methods to enhance network resiliency against single failures. We concluded that locating computing resources in fewer MEC sites rather than distributing them in a higher number of MEC locations shows better performance in terms of blocking ratio. In our previous work, we focused on the blocking rate with 5 MEC nodes. However, in this present paper, we investigate the placement of 5 MECs using realistic VNF specifications. We recognize the significance of the MEC location and its relationship with load distribution. Our next step involves analyzing the underlying factors contributing to different blocking rates, leveraging the static context, and ultimately identifying optimal methods for locating MECs. Furthermore, we are going to analyze the workload and traffic patterns of the network to identify the optimal MEC locations for specific use cases and traffic profiles.

## References


1. Askari, L., Tamizi, M., Ayoub, O., Tornatore, M.: Protection strategies for dynamic VNF placement and service chaining. In: 2021 International Conference on Computer Communications and Networks (ICCCN), pp. 1–9. IEEE (2021)
2. Dash, S., Khan, A.U., Swain, S.K., Kar, B.: Clustering based efficient MEC server placement and association in 5g networks. In: 2021 19th OITS International Conference on Information Technology (OCIT), pp. 167–172. IEEE (2021)

3. Ghazizadeh, A., Akbari, B., Tajiki, M.M.: Joint reliability-aware and cost efficient path AllocationFig and VNF placement using sharing scheme. J. Netw. Syst. Manag. **30**, 1–28 (2022)
4. Giust, F., et al.: MEC deployments in 4G and evolution towards 5G. ETSI White paper **24**(2018), 1–24 (2018)
5. IoTalentum: SDN for network virtualization in iotalentum infrastructure. iotalentum.eu/esrs. Accessed 8 May 2023
6. Kiran, N., Liu, X., Wang, S., Yin, C.: VNF placement and resource allocation in SDN/NFV-enabled MEC networks. In: 2020 IEEE Wireless Communications and Networking Conference Workshops (WCNCW), pp. 1–6. IEEE (2020)
7. Masoumi, M., et al.: Dynamic online VNF placement with different protection schemes in a MEC environment. In: 2022 32nd International Telecommunication Networks and Applications Conference (ITNAC), pp. 1–6. IEEE (2022)
8. Mohan, P.M., Gurusamy, M.: Resilient VNF placement for service chain embedding in diversified 5G network slices. In: 2019 IEEE Global Communications Conference (GLOBECOM), pp. 1–6. IEEE (2019)
9. Ren, H., et al.: Efficient algorithms for delay-aware NFV-enabled multicasting in mobile edge clouds with resource sharing. IEEE Trans. Parallel Distrib. Syst. **31**(9), 2050–2066 (2020)
10. Scazzariello, M., Ariemma, L., Di Battista, G., Patrignani, M.: Megalos: A scalable architecture for the virtualization of network scenarios. In: NOMS 2020-2020 IEEE/IFIP Network Operations and Management Symposium, pp. 1–7. IEEE (2020)
11. Tabatabaee Malazi, H., et al.: Dynamic service placement in multi-access edge computing: a systematic literature review. IEEE Access **10**, 32639–32688 (2022). https://doi.org/10.1109/ACCESS.2022.3160738
12. Thiruvasagam, P.K., Chakraborty, A., Murthy, C.S.R.: Latency-aware and survivable mapping of VNFs in 5G network edge cloud. In: 2021 17th International Conference on the Design of Reliable Communication Networks (DRCN), pp. 1–8. IEEE (2021)
13. Yala, L., Frangoudis, P.A., Ksentini, A.: Latency and availability driven VNF placement in a MEC-NFV environment. In: 2018 IEEE Global Communications Conference (GLOBECOM), pp. 1–7. IEEE (2018)
14. Yang, S., Li, F., Trajanovski, S., Yahyapour, R., Fu, X.: Recent advances of resource allocation in network function virtualization. IEEE Trans. Parallel Distrib. Syst. **32**(2), 295–314 (2020)
15. Zhang, Q., Liu, F., Zeng, C.: Online adaptive interference-aware VNF deployment and migration for 5G network slice. IEEE/ACM Trans. Netw. **29**(5), 2115–2128 (2021)

# Lossless Preprocessing of Floating Point Data to Enhance Compression

Francesco Taurone(✉), Daniel E. Lucani, Marcell Fehér, and Qi Zhang

DIGIT, Department of Electrical and Computer Engineering, Aarhus University, Aarhus, Denmark
{francesco.taurone,daniel.lucani,sw0rdf1shqz,qz}@ece.au.dk

**Abstract.** Data compression algorithms typically rely on identifying repeated sequences of symbols from the original data to provide a compact representation of the same information, while maintaining the ability to recover the original data from the compressed sequence. Using data transformations prior to the compression process has the potential to enhance the compression capabilities, being lossless as long as the transformation is invertible. Floating point data presents unique challenges to generate invertible transformations with high compression potential. This paper identifies key conditions for basic operations of floating point data that guarantee lossless transformations. Then, we show four methods that make use of these observations to deliver lossless compression of real datasets, where we improve compression rates up to 40%.

**Keywords:** Compression · Lossless · Floating point

## 1 Introduction

Data management is a key aspect of most IoT related applications, since data generated by sensors usually has to be transmitted, stored and analyzed. A common practice to store time-series datasets is to compress them before archival, so to limit the needed database space. Before compression, data is usually preprocessed, undergoing a transformation process where the content of the dataset is adjusted for compression, which can potentially enhance the compressibility of the dataset. There are various preprocessing approaches that the user could adopt, ranging from rounding the decimals, to changing domain of the signal with the discrete wavelet transform [2], or dropping bits that carry minimal information [8]. One key distinction among preprocessing transforms is whether they are lossy or lossless. Given a transformation function $f$ and a dataset DS, we say that $f$ is *lossless* when $f^{-1}(f(\mathrm{DS})) = \mathrm{DS}$, *lossy* otherwise. The goal of this paper is to characterize lossless operation on floating point data and to propose 4 lossless preprocessing techniques enhancing the compressibility for

This work was supported by the IoTalentum Project within the Framework of Marie Skłodowska-Curie Actions Innovative Training Networks (ITN)-European Training Networks (ETN), which is funded by the European Union Horizon 2020 Research and Innovation Program under Grant 953442.


R. Mehmood et al. (Eds.): DCAI 2023, LNNS 741, pp. 457–466, 2023.
https://doi.org/10.1007/978-3-031-38318-2_45

mono-dimensional floating point datasets, meaning that the preprocessed dataset $f(\text{DS})$, once compressed, is smaller in size than the compressed vanilla DS. We quantify compression in terms of compression ratio (CR), defined as

$$\text{CR} = \frac{\text{Compressed dataset size in bits} + \text{Compression metadata}}{\text{Uncompressed dataset size in bits}} \tag{1}$$

where the metadata might be needed to be able to undo the compression. Since these transformations operate on each individual floating point number, they are naturally compatible with compression methods with fine-grained random access capabilities, such as Generalized Deduplication [12], which are useful for computing analytics directly on the compressed data [6].

### 1.1 Compressing Floating Point Numbers

The most common standard for floating point numbers is the IEEE-754 [1], where a number $x$ is represented by three components: the sign (S), the exponent (E) and the mantissa (M). In this paper, we assume the use of double precision, namely 64 bits per number. When interpreted as unsigned integers, $x$ is

$$x = (-1)^S \cdot 2^{E-B} \cdot (1 + M \cdot 2^{-l}), \tag{2}$$

where $B = 1023$ is the bias of the exponent, and $l = 52$ is the length of the mantissa in bits. Out of the 64 bits, 1 is the sign bit, 11 are the exponent and the remaining 52 bits are the mantissa. Each mantissa bit is $m_i$, with $i \in \{1, 2, \ldots, 52\}$ and $m_1$ being the most significant one. From Eq. (2), we see that all $x$ such that $|x| \in [2^{E^*}, 2^{E^*+1})$ have the same exponent $E^* + B$.

As studied in [11], a possible idea to improve compressibility is to maximize the number of bits shared by all numbers in the collection we want to compress. We say that the i-th bit is *shared* when all numbers int eh dataset have the same value for the i-th bit, meaning that if all bits are shared, the dataset is composed of a series of equal numbers. The work in [11] empirically shows that having more shared bits generally results in better compression, with both standard compressors, like zlib [4], and ones based on deduplication, used in Sect. 4. However, the preprocessing technique $\bar{f}$ showed in [11] is lossy: in the following, we elaborate on it to propose lossless preprocessing techniques.

## 2 Key Observations

### 2.1 Lossless Operations on Floating Point Numbers

Numbers represented with a finite amount of bits will necessarily have finite precision. IEEE-754 can represent a subset of all numbers on the real line, while values that are not in this subset get approximated to one that is. In this context, we refer to the precision of a number $x = (-1)^{S_x} \cdot 2^{E_x-1023} \cdot (1 + M_x \cdot 2^{-52})$ as the distance between $x$ and the next representable floating point number. We call this quantity *unit in the last place*, or $\mathrm{ULP}(x)$, calculated as

$$\mathrm{ULP}(x) = 2^{E_x - 1023 - 52} \tag{3}$$

The reason why manipulating floating point numbers can result in losses is that $\mathrm{ULP}(x)$ is not a constant in floating point, but rather a function of the number's unbiased exponent. Transformations whose output results in $\mathrm{ULP}(x)$ different from the original could potentially lead to information losses. In order to explain this phenomenon, we introduce the operations $\oplus$, $\otimes$, $\ominus$, $\oslash$, which represent addition, multiplication, subtraction and division according to IEEE-754. For our purposes, we can think of them as being the theoretical operations followed by rounding to the nearest representable number.

An example of potential losses due to floating point precision can be shown by considering $x = 3.5$, the transform $f(x) = x \oplus 10^{16}$ with its inverse $g(x) = x \ominus 10^{16}$ and their theoretical counterpart $f'(x) = x + 10^{16}$ with its inverse $g'(x) = x - 10^{16}$. While $g'(f'(x)) = x, \forall x \in \mathbb{R}$, the same is not true in floating point, since $g(f(3.5)) = 4.0 \neq 3.5$.

While there are multiple strategies to limit the effects of this approximation error, as described in Section 5.3 of [10], this paper aims at devising strategies to avoid it completely. In Sect. 2.1, we discuss particular scenarios for which summation does not produce error, and in Sect. 2.1 we analyze multiplication.

**Lossless Addition.** In the scenario where $x \in [2^{E^*}, 2^{E^*+1})$ and $f(x) = A \oplus x$ is such that $f(x) \in [2^{E^*+1}, 2^{E^*+2})$ and $A \in [2^{E^*}, 2^{E^*+1})$, the operation is lossless as long as both $A$ and $x$ have the same least significant mantissa bit $m_{52}$. The reason is that during an addition under these conditions, a bit called *guard bit* has to be rounded away, since $\mathrm{ULP}(\cdot)$ goes from $\mathrm{ULP}(x) = 2^{E^*-52}$ to $\mathrm{ULP}(f(x)) = 2^{E^*-51}$. However, if both addends have the same $m_{52}$, the guard bit is guaranteed to be equal to zero, requiring no rounding. This result is summarized in Table 1 and more details on how addition works in floating point in Section 7.3 of [10].

**Table 1.** Mantissa least significant bits for lossless addition $y = x \oplus A$. The blue results indicate a lossless transformation, namely $y \ominus A = x$.

| $m^y_{52}$ | | $m^x_{51}m^x_{52}$ | | | |
|---|---|---|---|---|---|
| | | 00 | 01 | 10 | 11 |
| $m^A_{51}m^A_{52}$ | 00 | **0** | 0 | **1** | 0 |
| | 01 | 0 | **1** | 0 | **0** |
| | 10 | **1** | 0 | **0** | 0 |
| | 11 | 0 | **0** | 0 | **1** |

Another particular useful scenario is the addition $y = x \oplus A$ where $x, y \in [2^{E^*}, 2^{E^*+1})$ and $A \in [2^{\tilde{E}}, 2^{\tilde{E}+1}) < 2^{E^*}$. It can be shown that given the floating point addition process detailed in [9], this operation is lossless when

$$m_i = 0 \quad \text{for } i \in \left\{52 - \left(E^* - \tilde{E} + 1\right), \ldots, 52\right\}. \tag{4}$$

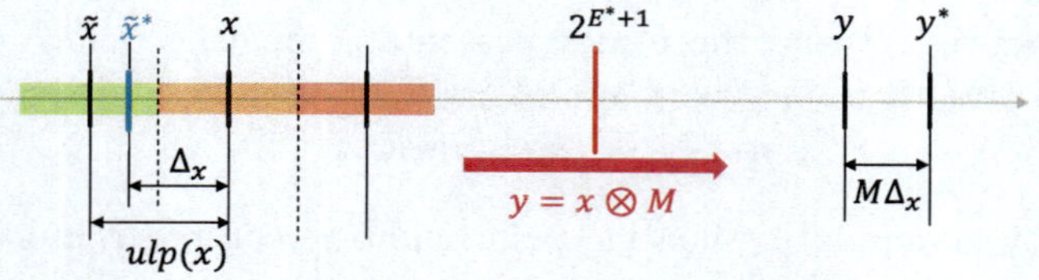


**Fig. 1.** Representation of the loss with $f(x) = y = x \odot M$ and $f^{-1}(x) = \tilde{x} = y \oslash M$. Assuming the use of the *round to nearest* approach, all real numbers in the same color region get rounded to the same floating point number.

**Lossless Multiplication.** In the scenario where $x \in [2^{E^*}, 2^{E^*+1})$ and $f(x) = x \otimes M$ is such that $f(x) \in [2^{E^*+1}, 2^{E^*+2})$, the operation is guaranteed to be lossless as long as $M \geq 2$. To prove it, we consider Fig. 1, where $y^* = x \cdot M$, $y = x \otimes M$, $\tilde{x}^* = y/M$ and $\tilde{x} = y \oslash M$. In order for this transformation to be lossless, we need

$$|x - \tilde{x}^*| = \Delta_x < \text{ULP}(x)/2 \quad (5)$$

since this would result in $\tilde{x} = x$ under the *round to nearest* rounding scheme. Since $y = \tilde{x}^* \cdot M = (x + \Delta_x) \cdot M$, we have$y - y^* = M \cdot \Delta_x$, which combined with Eq. (5), results in the condition for lossless multiplication being

$$y - y^* < M \cdot \frac{\text{ULP}(x)}{2}. \quad (6)$$

Since $\text{ULP}(y) = 2 \cdot \text{ULP}(x)$ and $y - y^* < \frac{\text{ULP}(y)}{2}$ because of the rounding method, we have $y - y^* < \text{ULP}(x) < M \cdot \frac{\text{ULP}(x)}{2}$, fulfilling Eq. (6) when $M \geq 2$.

## 3 Lossless Dataset Manipulations for Better Compression

In order to increase the number of shared mantissa bits in the dataset, our goal is to ensure that all numbers after preprocessing lie in the portion of the real line where the mantissa bits are guaranteed to be as we desire. Given a dataset DS, in order to guarantee that the $D$ most significant mantissa bits are shared, the datapoints $x$ would all have to be such that

$$|x| \in \left[2^E, 2^E + 2^{E-D}\right] \quad \text{with } E \in [-1022, 1023] \quad \forall x \in \text{DS}. \quad (7)$$

Next, we present four techniques to losslessly place the numbers in the dataset in these preferred regions, supposing without loss of generality and for clarity sake to have a dataset where all numbers have the same exponent. This can be easily generalized by storing as metadata the information on the original exponent of each sample. We named these techniques *compact bins, multiply and shift, shift and separate even from odd, shift and save evenness.*

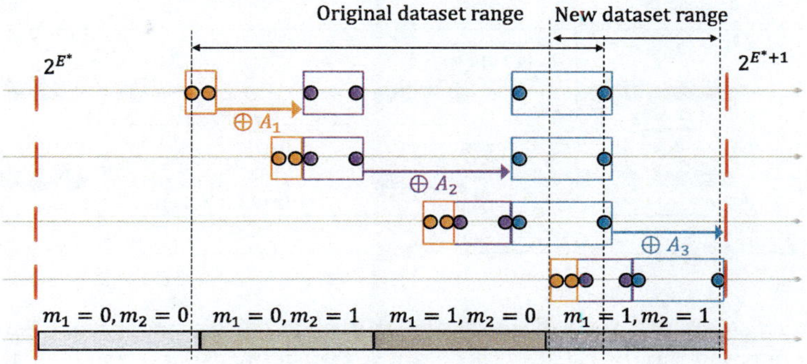


**Fig. 2. Illustration of method *compact bins*.** In this example, by using 3 bins all numbers in the new dataset are guaranteed to have $m_1 = 1$ and $m_2 = 1$, while there was no such guarantee in the original dataset.

### 3.1 Compact Bins

In the scenario where $x \in [2^{E^*}, 2^{E^*+1})$, if we cluster this set of values into bins and shift the bins so they are closer to each other, we are effectively reducing the range of values of the transformed dataset, possibly making it fit to a region where some mantissa bits are guaranteed to be shared. An example of this binning process is in Fig. 2, where we use 3 bins to ensure $m_1 = 1$ and $m_2 = 1$ $\forall x \in \text{DS}$. The optimal amount of bins depends on the distribution of values in the real axis. Two conditions are needed in order for this process to be lossless. First, supposing to use $k$ bins on a dataset with $\ell$ unique values, we have to store $(k-1) \cdot \lceil \log_2(\ell) \rceil$ bits as metadata representing the boundaries of the bins, as well as the $k$ values $\{A_1, \ldots, A_k\}$ used to shift the bins. The size of metadata in bytes is $\text{Z} = (k \cdot 64 + (k-1) \cdot b)/8$. Secondly, the $A_i$ have to be chosen so that the operation $y = x \oplus A_i$ is lossless. We can use Eq. (6) to select the correct $A_i$ values.

### 3.2 Multiply and Shift

As per Eq. (7), there is one portion of the real axis per exponent region guaranteeing that the $D$ most significant bits are equal to 1. Using the lossless operations in Subsect. 2.1, we can manipulate the dataset so that the modified DS has all values lying in those regions. The *multiply and shift* transform we propose, represented in Fig. 3, involves one multiplication and one addition to move to the next exponent region, applied interatively until all modified numbers lie on the portion of the real axis where the most significant $D$ mantissa bits are guaranteed to be shared. Assuming that $x \in [2^{E^*}, 2^{E^*+1})$ $\forall x \in \text{DS}$, the operation is

$$f(x) = (2.0 \otimes x) \oplus A, \quad \text{with } A = 2^{E^*-D+1} - 2 \cdot \text{ULP}\left(2^{E^*+1}\right), \tag{8}$$

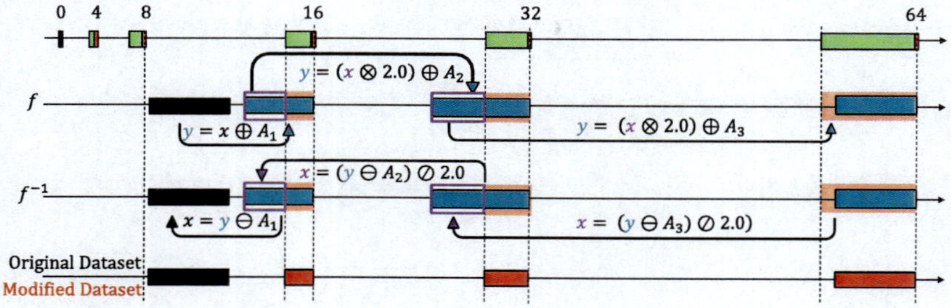


**Fig. 3. Illustration of method *multiply and shift*.** From the original dataset in black, we iteratively apply Eq. (8) to obtain a transformed dataset in red lying on regions in green, where all numbers are guaranteed to share at least the first $D$ mantissa bits. We can recover the original dataset with the inverse transform since all operations are lossless.

where $A$ has to be rounded down if necessary to the first value fulfilling Eq. (4). As we can see in Fig. 3, after an iteration, only the values out of the region with desired common bits have to go through another one, progressively shaving the dataset under analysis. A limitation of this algorithm is that the multiplication by a factor of 2.0 effectively enlarges the window of values of the transformed portion of the dataset, increasing the number of needed iterations. However, $M = 2$ is the smallest lossless factor, as per Eq. (6). In terms of metadata, we need to store $D$ and $A_1$, since all $A_i$ with $i \neq 1$ can be computed by knowing the exponent of the original dataset and $D$. In order to invert the transformation, given the transformed dataset, we can iteratively apply the inverse transformation $f^{-1}(y) = (y \ominus A) \oslash 2.0$ on each element belonging to the rightmost exponent region, concluding with the inverted shift stored as metadata.

### 3.3 Shift and Separate Even from Odd

In order to alleviate the drawbacks of *multiply and shift* in Subsect. 3.2, we can substitute the multiplication with an addition, while carefully selecting the addendum so that it fulfills the conditions in Sect. 2.1. Specifically, given the operation $y = x \oplus A$ with $x, A \in [2^{E^*}, 2^{E^*+1})$, their mantissas $M_x$ and $M_A$ have to have the same evenness, namely $m^x_{52} = m^A_{52}$. We do that by ensuring that all $y$ resulting from $x$ with even mantissas lie in a portion of the real axis that does not overlap with the one for the odds. In this way, we can distinguish the two during the inverse transformation. As depicted in Fig. 4, the transformation differs depending on the evenness of the mantissas as input, resulting in

$$f(x) = \begin{cases} x \oplus A_i^{\text{even}} & \text{if} \quad \text{mod}\ (M_x, 2) = 0 \\ x \oplus A_i^{\text{odd}} & \text{if} \quad \text{mod}\ (M_x, 2) = 1. \end{cases} \tag{9}$$

$A_i^{\text{even}}$ and $A_i^{\text{odd}}$ at the i-th iteration are computed so to fulfill Eq. 4. In particular, at th i-th iteration, with $x, A_i^{\text{even}}, A_i^{\text{odd}} \in [2^{E^*}, 2^{E^*+1})$ and $D$ desired shared most

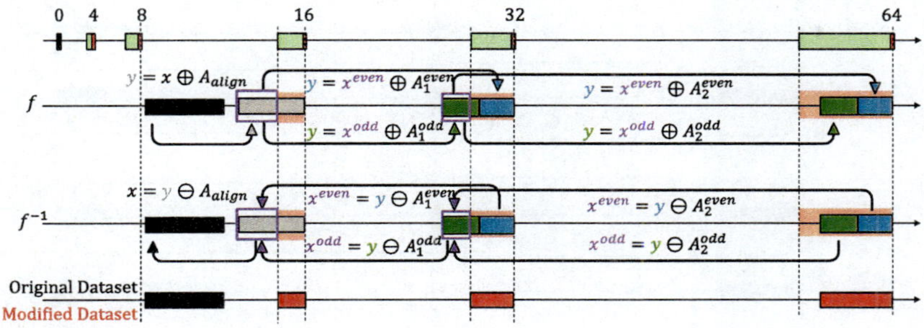


**Fig. 4. Illustration of method *shift and separate even from odd*.** We iteratively apply Eq. (9), ensuring that the output of $x$ with even and odd mantissas lie in different portions of the real axis ath every iteration.

significant mantissa bits, they are

$$A_i^{\text{even}} = 2^{E^*+1} + 2^{E^*-D}, \quad A_i^{\text{odd}} = A_i^{\text{even}} - W_i, \tag{10}$$

where $W_i$ represents the length of the portion of the dataset to be processed at the i-th iteration. It is calculated starting from the original dataset DS with $W_{i+1} = 2 \cdot W_i - 2^{E^*-D}$, where $W_0 = \max(\text{DS}) - \min(\text{DS})$. In terms of metadata, we need to store the initial shift $A_{align}$, $D$ and $W_0$, since all $A_i$ can be reconstructed from those. The i-th step of the inverse transformation of $y \in \left[2^{E^*}, 2^{E^*+1}\right]$ is

$$f^{-1}(y) = \begin{cases} y \ominus A_i^{\text{even}} & \text{if } y > 2^{E^*+1} - W_i \\ x \ominus A_i^{\text{odd}} & \text{if } y \leq 2^{E^*+1} - W_i. \end{cases} \tag{11}$$

With this method we strategically choose $A_i^{\text{even}}$ and $A_i^{\text{odd}}$ so that we can guess the evenness of the original $x$ by checking the position of $y$ on the real axis. The drawback is that we need to enlarge the range of values of the transformed dataset at every iteration, similarly to what happened with the multiplication by 2.0 in Subsect. 3.2.

### 3.4 Shift and Save Evenness

In order to limit the scaling up of the transformed dataset at every iteration in *shift and separate even from odd*, we could store the evenness information as a single metadata bit per sample, meaning $n$ bits per iteration for a dataset of $n$ elements. The increased size of metadata is compensated by a drastic reduction in the number of iterations needed to end the procedure. The direct transformation formulas is the same as per Eq. (9) and it is depicted in Fig. 5 where the only difference is that the calculation of $A_i^{\text{odd}}$ becomes $A_i^{\text{odd}} = A_i^{\text{even}}$.

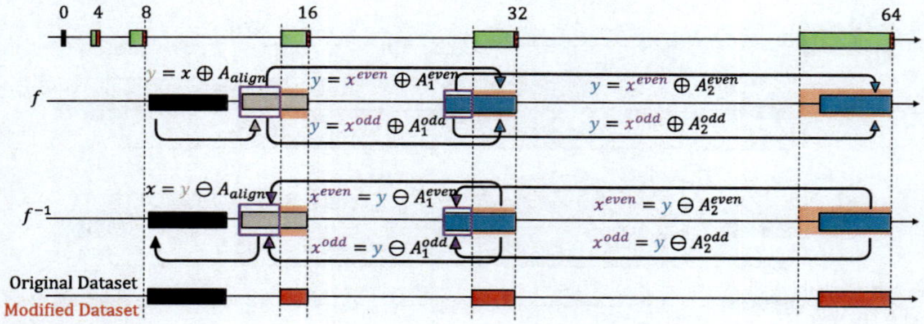


**Fig. 5. Illustration of method *shift and save evenness*.** While similar in concept to Fig. 4, we use metadata to distinguish whether to apply the inverse transformation with $A_i^{\text{odd}}$, namely having odd mantissas, or with $A_i^{\text{even}}$, which has it even.

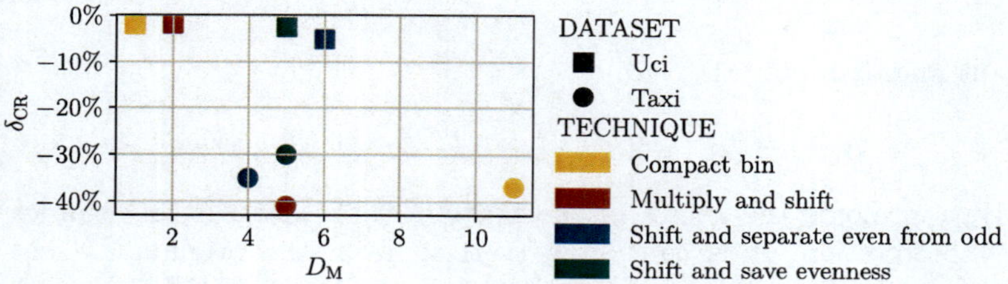


**Fig. 6.** Comparison of the best results from Fig. 7a and Fig. 7b, where lower $\delta_{\text{CR}}$ are better.

## 4 Results

In this section we analyze the benefits of the four proposed techniques for lossless preprocessing of floating point datasets, namely *compact bins, multiply and shift, shift and separate even from odd, shift and save evenness.* In order to quantify their effectiveness, we compare the compression ratio achieved with preprocessing ($\text{CR}_{\text{PREP}}$) and without preprocessing ($\text{CR}_{\text{NO-PREP}}$), using the metric

$$\delta_{\text{CR}} = \frac{\text{CR}_{\text{PREP}} - \text{CR}_{\text{NO-PREP}}}{\text{CR}_{\text{NO-PREP}}} \tag{12}$$

where negative values mean that our techniques resulted in better compressed datasets. To measure the impact of metadata on the final size, we define Z as

$$Z = \frac{\text{Metadata size in bytes}}{\text{Compressed dataset size in bytes}}. \tag{13}$$

As compressor, we use Greedy-GD [7], since the number of common bits is particularly beneficial for its effectiveness in terms of CR. Here, $D_M$ is the number of desired most significant guaranteed shared mantissa bits, whereas the total number of shared mantissa bits is $S_M$, $S_E$ is for shared exponent bits and $S_{\text{TOT}}$ for the total number shared bits. As input, we use the first 1000 elements of the

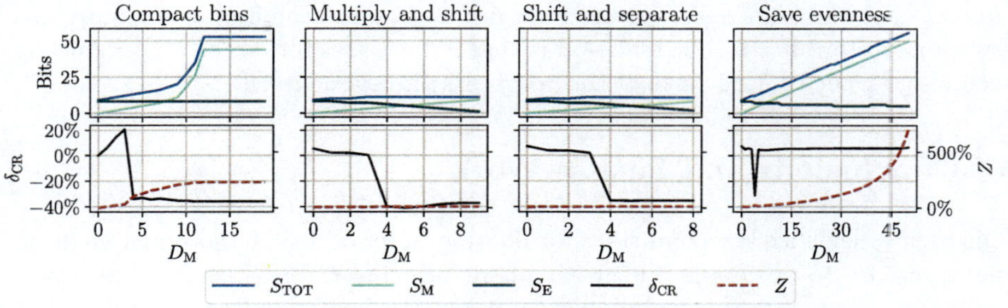


(a) Chicago-taxi-trips-fares [3]

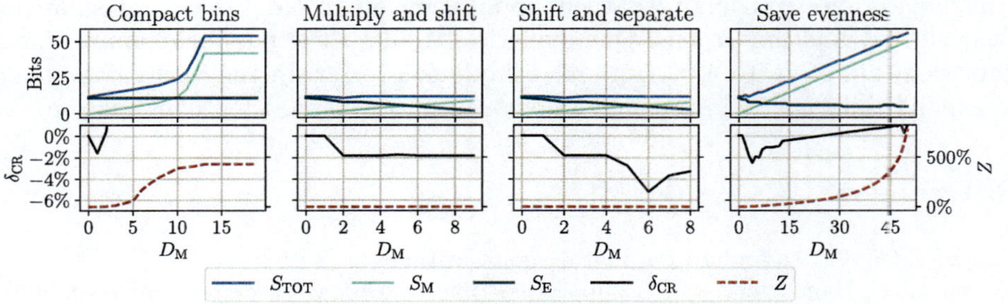


(b) UCI-gas-turbine-emissions [5].

**Fig. 7.** Performances of the 4 techniques for lossless preprocessing in terms of CR on the datasets Chicago-taxi-trips-fares and UCI-gas-turbine-emissions.

datasets Chicago-taxi-trips-fares [3] and UCI-gas-turbine-emissions [5], having a single dimension of non-negative floating point values: these technique could be extended to mixed signed datasets. In Fig. 6, we summarize the best results of all 4 techniques with both datasets, while in Fig. 7 we report more detailed performances.

Looking at Fig. 6, we see that in all scenarios we were able to find a transformation that improved compression, since all optimal results are below zero. We also notice that the best transformation was able to achieve an improvement of 40% in CR, meaning that the final size of the compressed dataset was 40% smaller in size than the compressed vanilla version.

Regarding the *compact bins* technique, we notice that $S_{\text{TOT}}$ is always larger than $D_M$. It plateaus when all 52 mantissa bits are shared, causing CR to plateaus as well. We also see that $S_E$ is constant, since this technique preserves the original exponents. We see a plateau in $S_{\text{TOT}}$ also with *multiply and shift* and *shift and separate even from odd.* Here, after a certain $D_M$ value, for every additional mantissa bit we want to be shared, we lose a shared exponent bit, making $S_{\text{TOT}}$ constant: as a consequence, CR plateaus as well.

We also notice that with *Compact bins* and *shift and save evenness* we can impose a much higher number of common mantissa bits compared to both *multi-*

*ply and shift* and *shift and separate even from odd*, since the first two require less iterations to terminate. The trade-off is that $Z$ increases for larger $D$, potentially becoming larger that the size of the compressed dataset itself.

## 5 Conclusions and Future Work

This paper identifies key conditions for floating point addition and multiplication operations to be invertible, i.e., without introducing errors. Using these observations, we proposed four techniques to preprocess arrays of floating point data in a lossless fashion to enhance their compression potential. We discussed their implementations and analyzed their application on real datasets and showed their efficacy and limitations. Our numerical results show possible compression improvements of up to 40%, without introducing losses. In the future, we plan to expand their use to mixed sign datasets and investigate their combination.

## References


1. IEEE 754-2019 standard for floating-point arithmetic (2019)
2. Batal, I., Hauskrecht, M.: A supervised time series feature extraction technique using DCT and DWT. In: 2009 International Conference on Machine Learning and Applications (2009)
3. City of Chicago: Taxi trips dataset (2022). https://tinyurl.com/4rypurjp
4. Jean-loup Gailly, M.A.: Zlib compressor. https://www.zlib.net/
5. Kaya, H.: Gas turbine CO and NOx emission dataset (2019). https://tinyurl.com/2ubk63ra
6. Hurst, A., Lucani, D.E., Assent, I., Zhang, Q.: Glean: generalized-deduplication-enabled approximate edge analytics. IEEE Internet of Things J. **10**, 4006–4020 (2023)
7. Hurst, A., Lucani, D.E., Zhang, Q.: GreedyGD: enhanced generalized deduplication for direct analytics in IoT (2023). https://arxiv.org/abs/2304.07240
8. Klower, M., Razinger, M., Dominguez, J.J., Duben, P.D., Palmer, T.N.: Compressing atmospheric data into its real information content. Nat. Comput. Sci. **1**, 713–724 (2021)
9. Hill, M.D.: Notes on floating point arithmetic, part of course CS/ECE 354 at university of Wisconsin. https://tinyurl.com/mvzcyc3x
10. Muller, J.M., et al.: Handbook of Floating-Point Arithmetic (2010)
11. Taurone, F., Lucani, D.E., Fehér, M., Zhang, Q.: Change a bit to save bytes: compression for floating point time-series data. In: IEEE ICC (2023). https://arxiv.org/abs/2303.04478
12. Vestergaard, R., Lucani, D.E., Zhang, Q.: A randomly accessible lossless compression scheme for time-series data. In: IEEE Conference on Computer Communications, IEEE INFOCOM 2020 (2020)

# Multi-band Elastic Optical Networks: Overview and Recent Contributions from the IoTalentum Project

Soheil Hosseini[1](✉), Ignacio de Miguel[1], Ramón J. Durán Barroso[1], Noemí Merayo[1], Óscar González de Dios[2], Maryam Masoumi[1], Hafiza Kanwal Janjua[1], Juan Carlos Aguado Manzano[1], and Patricia Fernández[1]

[1] Universidad de Valladolid, Valladolid, Spain
soheil.hosseini@uva.es
[2] Telefónica I+D, Madrid, Spain

**Abstract.** Space division multiplexing (SDM) technology and band division multiplexing (BDM) are potential candidates for the bandwidth extension of the existing optical transport networks. Contrary to the SDM technology, whose activation requires a huge CAPEX cost, BDM is viewed as a practical solution in the short or middle term. However, the complete upgrade of a network to a multi-band elastic optical network (MB-EON) would incur significant expenses due to the costly components like erbium doped fiber amplifiers (EDFAs). Therefore, an efficient solution may come from performing a partial upgrade of the EON, lighting up the L-band of only some of the fibers, as an intermediate or a set of intermediate steps towards the final full upgrade. On the other hand, capacity extension by using other spectral bands in BDM also requires survivability provisioning. As providing connections with protection leads to additional spectrum consumption, by benefitting from service level agreement (SLA) differentiated protection, resources for connection protection can be more efficiently utilized. In this paper, we provide an overview of multi-band elastic optical networks and a summary of our previous works in the framework of the EU MSCA IoTalentum project, with the main focus on the partial migration of a C-band only network to C + L band system, and on SLA-differentiated survivability in MB-EONs.

**Keywords:** Network Planning · Partial Network Upgrade · Elastic Optical Network · Space Division Multiplexing · Multi-Band

## 1 Introduction

The continuous rapid growth of IP traffic causes the exhaustion of the available capacity of optical transport networks. Therefore, attempts are made to increase the capacity of optical fibers for supporting the accelerated network traffic demands. To this end, and to postpone the relative cost of space division multiplexing (SDM) technology realization, network operators put the spotlight on multi-band elastic optical network (MB-EON).


R. Mehmood et al. (Eds.): DCAI 2023, LNNS 741, pp. 467–472, 2023.
https://doi.org/10.1007/978-3-031-38318-2_46

[1–3]. In MB-EON, getting around the cost of capacity enhancement is achieved by taking the advantage of different spectral bands of already installed single mode fibers (SMFs) [4], which maximizes the value of return on investment for these fibers [5].

As it is mentioned, thanks to the usage of already installed optical fibers, MB-EON would be a more practical solution for increasing the network capacity in medium-term period compared to the SDM technology. However, the deployment of costly network components like doped fiber amplifiers and reconfigurable optical add-drop multiplexers (ROADMs) are necessary for the implementation of MB-EON. As ROADMs and erbium doped fiber amplifiers (EDFAs) are already commercialized for lighting up the L-band, in our previous research, the focus was set on C + L bands.

Several architectures have been proposed for the implementation of C + L bands [6]. In the most common architecture, which is depicted in Fig. 1(a), separate EDFAs are used to perform optical amplification over different spectral bands. In this model, a bandwidth of around 400 GHz is considered as the guardband between the C-band and the L-band. Figure 1(b) shows an alternative model for the implementation of MB-EON in such a way that bandwidth waste due to the guardband is removed. In this model, a semiconductor optical amplifier (SOA) is used to achieve a wideband amplifier architecture [6].

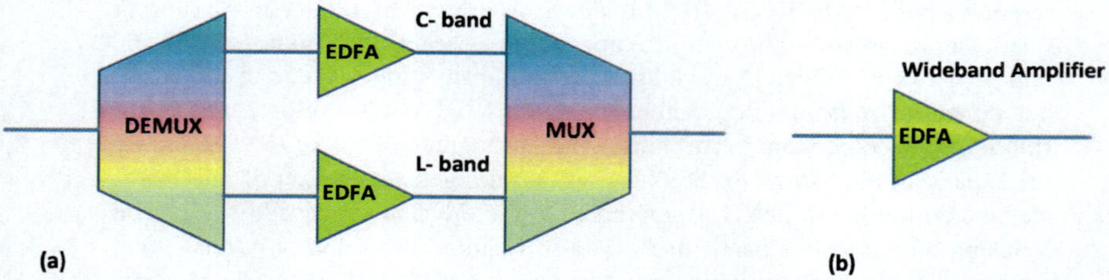


**Fig. 1.** Architecture of amplifiers for the implementation of C + L line system, a) separate amplifiers architecture, b) wideband amplifier architecture.

Even in the domain of C + L bands, a fully upgraded network requires the deployment of EDFAs and ROADMs for all of the links within the network. In other words, fully upgrading a network would require a significant investment. Hence, researchers have focused on partially upgrading a network to take the advantage of additional resources that the L-band provides while keeping the total cost of the upgrade as minimum as possible. In this way, several research works have focused either on how to partially upgrade a network in which only some of the links work over C + L bands or on control algorithms for a hybrid C/C + L network. For instance, Ahmed *et al.* [7] have introduced several techniques to select potential links for the migration, and they have also investigated the proper timing for performing the migration to achieve a partially upgraded network. Another example is the work by Yao *et al.* [8], who have presented a routing, band, modulation level, and spectrum assignment (RBMLSA) algorithm as well as different spectrum assignment policies for a partially upgraded network.

On the other hand, as moving towards the L-band provides the network operators with more resources, network failures in C + L band systems, for instance due to fiber cuts or EDFA failures [9] would result in enormous data loss. Therefore, the concept of survivability should get more emphasis for launching a MB-EON.

In this paper we provide a summary of our recent contributions in the domain of MB-EON in the framework of the EU MSCA IoTalentum project. As the partial migration of an EON reduces the cost of the implementation of C + L band systems, Sect. 2 includes a brief explanation of our proposed network planning strategies to perform partial migration over a C-band only network. Section 3 presents a review on our previous works about enabling the C + L band line system with different levels of protection for different class of services in terms of priority.

## 2 Strategies for Partial Migration Towards MB-EON

Our first work on strategies for the partial migration towards MB-EON can be found in [10]. There we proposed a heuristic approach to perform an incremental migration towards a C + L MB-EON determining which fibers to upgrade, as well as an RBMLSA algorithm to serve incoming connections dynamically. In the phase of network planning, optical fibers are sorted from "most" to "least" used in terms of their appearances in the primary path between each source and destination pair. In other words, the links with more probability of resource exhaustion owing to the congestion are prioritized to be migrated to C + L bands. Once the network has been partially upgraded based on that policy, the RBMLSA algorithm is executed to serve connection requests. When a connection is entered into the network, it would be routed using the $K$-shortest paths algorithm. We proposed two strategies, called length-based and hop-based, to route a certain connection request. One of the main constraints of spectrum allocation in EONs is called as spectrum continuity, which guarantees that the same range of spectrum slots should be assigned to the connection along the whole path. Therefore, taking this into account, in the phase of spectral band selection, the L-band is selected for the spectrum allocation if, and only if, all of the links within the selected path have been upgraded to C + L bands. Therefore, in the length-based approach, if at least one of the links of the selected shortest path has not been migrated to the L-band, the path with the equal length, which can work over C + L bands would be prioritized. As finding different paths with the same distance is infrequent in occurrence, the hop-based approach is proposed. In the hop-based approach, different paths with the same number of hops are analyzed to route the connection through the path in which the L-band is lit up for all the links.

The assessment of network performance was done using the NSFNet topology, which has 14 nodes and 21 bidirectional links. In that paper [10], we demonstrated that it is not necessary to equip all fibers of the network with multiband devices to have significant performance improvements. Instead, a trade-off between the number of upgraded bidirectional links and an acceptable level of bandwidth blocking ratio (BBR) should be found. Simulation results showed that by upgrading roughly half of the links (12 links), the bandwidth blocking ratio is reduced in almost two orders of magnitude or, equivalently, by upgrading 12 links the maximum load supported by the network with BBR $< 10^{-3}$ doubles compared with the use of only the C-band. Furthermore, another conclusion of that work is related to the performance of proposed routing strategies. As the hop-based approach always outperforms the length-based approach, it is concluded that for C-band only network or fully upgrade one, hop-based approach should be used in the RBMLSA algorithm.

Then, in [11], we introduced a novel Integer Linear Programming (ILP) formulation to determine the set of fibers to equip with multiband devices for a partial upgrade of the network. In that paper, the objective of the proposed ILP formulation is to maximize the number of precomputed shortest paths (which will be used to route the connections) that benefit from the upgrade. An interesting point of the presented method is that, despite being an ILP formulation, which are generally slow, the network operator can be informed about the potential fibers for the migration very fast, in less than 1 min in the considered topologies. To make a comprehensive assessment of the performance of a partially upgraded network using the ILP formulation, three different network topologies have been considered, namely NSFNet, JPN12, and Deutsche Telekom. Simulation results of the paper demonstrated that in 16 of the 17 upgrade scenarios analyzed in the paper, the fibers for the migration suggested by the ILP formulation led the network to perform better (or at least similar) than the method that we proposed in [12] in terms of BBR.

## 3 Strategies for Differentiated Protection in MB-EON

We have also performed some studies on survivability, assuming that a network has been fully migrated to a C + L MB-EON. Thus, in [12], we have analyzed different strategies for providing protection against, at least, EDFA failures considering the use of the separate amplifiers architecture and the wideband amplifier architecture previously mentioned. For the separate amplifiers architecture, we have proposed a protection scheme for the employed optical amplifiers over each spectral band. In this scenario (called basic approach) the primary path is assigned to the C-band, and the L-band is considered for protection path provisioning, or vice versa. In order to have survivability in both fiber and optical amplifier over the wide-band amplifier architecture, we have introduced another approach (called protected wideband amplifier architecture approach), in which for accommodating a connection request, two link-disjoint shortest paths are selected to work as the primary path and backup path. In this scenario, the transmission media for both of the selected paths can be C + L bands. We have extended the basic protection scenario to have the survivability against amplifier failures for all of the incoming connections and also survivability against link failures for some of them, which is called as the hybrid method. In the hybrid approach, survivability of the deployed EDFA over a single spectral band is always achieved because accessible spectrum along the same lightpath for the primary and backup paths belongs to different spectral bands. Hence, for this scenario, serving the connections is always independent from the EDFA's failures. When a connection is going to be blocked due to the unavailability of resources in the C-band, the algorithm of the hybrid approach will go to the second phase. In this case, one working path, which is link disjoint from the backup path would be chosen. The working path uses the spectrum resources of the L-band, while the backup path is able to use the more resources provided by C + L bands. Therefore, protection against the fiber failure and also EDFA failure can be obtained for those connections that are served in the second phase of the hybrid approach.

Network performance evaluation for the proposed survivability methods show that the application of the hybrid approach has the best performance compared to the other proposed methods. On the other hand, the classical method of protection in the wideband amplifier architecture not only is the best method in terms of providing resiliency against both fiber and EDFA failures, but also leads to an acceptable level of blocking probability.

In survivable MB-EONs, classical path protection techniques can be applied by reserving additional resources to establish a backup lightpath when solving the routing, band, modulation level, and spectrum assignment (RBMLSA) problem for each incoming request. The application of these techniques leads to a higher use of resources and, therefore, to higher blocking ratios. However, not all users/services require the same level of protection against failures. Therefore, in [13], we split the connections in "gold" and "silver" categories, the former having more stringent requirements on survivability. It was assumed that gold connections are prioritized over silver connections. This priority is related to the level of the protection that must be provided for the connections. Under the separated amplifiers architecture, silver connections would be able to tolerate single EDFA failure (but not fiber failures). This is because working paths of the silver connections are established over the C-band, and at the same time the L-band will be selected for backup path assignment. On the other hand, gold connections have protection against EDFA and fiber failures. To this end, the edge disjoint paths are considered for working path and protection path.

Simulation results showed that the application of SLA differentiated protection in the separate amplifiers architecture results in blocking ratio reduction for the considered SLAs compared to the usage of dedicated protection for all of the incoming connections regardless of their required type of service. Consideration of different levels of protection for different class of service would enable the network operators to manage efficiently the available bandwidth of optical fibers.

## 4 Conclusions and Future Directions

In this paper, we have briefly reviewed our contributions in multi-band elastic optical networks (MB-EONs). As fully upgrading a network is a considerably costly process, we have focused on partial upgrade of EONs. In this way, a trade-off between the number of upgraded fibers and the level of capacity upgrade can be achieved. We invite the interested reader to consult [10] and [11] for details. The issue of survivability in MB-EONs is discussed in [12] in which different techniques have been introduced to provide a survivable C + L band system against at least erbium doped fiber amplifiers (EDFAs). Differentiating between different incoming requests by providing them with different levels of protection is the main goal of [13].

Regarding the partial upgrade, a direction for the future work includes considering that the cost of upgrading different optical fibers of the network is different. For instance, the number of amplifiers in a link depends on its length, and thus links of different lengths translate in different upgrade costs. Therefore, rather than partially upgrading a network subject to the maximum number of fibers that should be equipped with multiband devices, as done in our previous work and in other related works in the literature, a more pragmatic approach should focus on determining how to upgrade the network ensuring that the affordable cost by the network operator is not exceeded. Additionally, in SLA differentiated protection, research directions include the analysis of more classes of service as well as the use of multipath routing or split spectrum techniques, where by splitting a whole traffic demand to be served using different spectral bands might lead to new opportunities to increase the efficiency in differentiated protection.

**Acknowledgements.** This work is part of the IoTalentum project, which has received funding from the EU H2020 research and innovation programme under the MSCA grant agreement No 953442. It is also supported by Ministerio de Ciencia e Innovación / Agencia Estatal de Investigación (Grant PID2020-112675RB-C42 funded by MCIN/AEI/https://doi.org/10.13039/501100011033), and Consejería de Educación de la Junta de Castilla y León and the European Regional Development Fund (Grant VA231P20).

## References

1. Napoli, A., et al.: Towards multiband optical systems. In: Photonic Networks and Devices, p. NeTu3E. 1: Optica Publishing Group (2018)
2. Napoli, A., et al.: Perspectives of multi-band optical communication systems. In: 2018 23rd Opto-Electronics and Communications Conference (OECC), pp. 1–2. IEEE (2018)
3. Fischer, J.K., et al.: Maximizing the capacity of installed optical fiber infrastructure via wide-band transmission. In: 2018 20th International Conference on Transparent Optical Networks (ICTON), pp. 1–4. IEEE (2018)
4. Sambo, N., et al.: Provisioning in multi-band optical networks. J. Lightwave Technol. **38**(9), 2598–2605 (2020)
5. Correia, A., et al.: Power control strategies and network performance assessment for C+ L+ S multiband optical transport. J. Opt. Commu. Netw. **13**(7), 147–157 (2021)
6. Cantono, M., et al.: Opportunities and challenges of C+ L transmission systems. J. Lightwave Technol. **38**(5), 1050–1060 (2020)
7. Ahmed, T., et al.: C+ L-band upgrade strategies to sustain traffic growth in optical backbone networks. J. Opt. Commun. Netw. **13**(7), 193–203 (2021)
8. Yao, Q., et al.: SNR re-verification-based routing, band, modulation, and spectrum assignment in hybrid C-C+L optical networks. J. Lightwave Technol. **40**(11), 3456–3469 (2022)
9. Wiatr, P., et al.: Device reliability performance awareness: Impact of RWA on EDFA failure reparation cost in optical networks. In: 2017 9th International Workshop on Resilient Networks Design and Modeling (RNDM), pp. 1–5. IEEE (2017)
10. Hosseini, S., et al: Migration Strategy from C-Band Elastic Optical Network to C+ L Multiband Optical Network. In: 2022 32nd International Telecommunication Networks and Applications Conference (ITNAC), pp. 204–206. IEEE (2022)
11. Hosseini, S., et al.: An ILP Formulation for Partially Upgrading Elastic Optical Networks to Multi-Bands. In: 7th International Conference on Optical Network Design and Modelling (ONDM 2023), Coimbra, Portugal (8–11 May, 2023)
12. Hosseini, S., et al.: Survivability against amplifier failures in multi-band elastic optical networks. In: 2022 Asia Communications and Photonics Conference (ACP), pp. 1299–1302. IEEE (2022)
13. Hosseini, S., et al.: SLA-Differentiated Protection in Multi-Band Elastic Optical Networks. In: 2022 IEEE Photonics Conference (IPC), pp. 1–2. IEEE (2022)

# Computation Offloading in Beyond 5G/6G Networks with Edge Computing: Implications and Challenges

Catalina Stan[1](✉), Simon Rommel[1], Juan José Vegas Olmos[2], and Idelfonso Tafur Monroy[1]

[1] Department of Electrical Engineering, Eindhoven University of Technology, Eindhoven, The Netherlands
{c.i.stan,s.rommel,i.tafur.monroy}@tue.nl
[2] NVIDIA Corporation, Ofer Industrial Park, Yokneam, Israel
juanj@nvidia.com

**Abstract.** The emerging beyond 5G/6G networks come with novel, latency-sensitive and computation-intensive applications that require enhanced network performance and infrastructure to meet the expected quality of experience for end users. To cope with this challenge, computation offloading leverages the benefits of multi-access edge computing to migrate the application tasks requiring additional computing resources for reduced execution delay. Although the benefits of introducing offloading mechanisms into the network might be straightforward, the implementation is not trivial due to various communication and computation trade-offs that must be made to obtain optimal offloading decisions. In this paper, we provide an overview of computation offloading with highlight on the networking perspective by looking at different offloading decisions, current research efforts, as well as the challenges that may be encountered while building an efficient and robust offloading mechanism. In addition, we provide our view on the evolution of computation offloading in 6G networks to support novel applications through enriched infrastructure and powerful artificial intelligence techniques.

**Keywords:** Beyond 5G · 6G · edge computing · computation offloading

## 1 Introduction

The integration of edge computing into the existing networks aims to bring additional computing, storage and networking resources to support the emerging 5G/6G applications. In contrast to the traditional network architectures that

This work was supported by the EC H2020 MSCA ITN-ETN IoTalentum (grant no. 953442), ECSEL JU BRAINE (grant no. 876967) and EC KDT JU CLEVER (grant no. 101097560) projects.


R. Mehmood et al. (Eds.): DCAI 2023, LNNS 741, pp. 473–479, 2023.
https://doi.org/10.1007/978-3-031-38318-2_47

mainly rely on the centralized mobile cloud computing (MCC) to handle complex computational tasks, edge computing introduces an intermediate layer, i.e., the edge layer, into the current infrastructure where the computing capabilities are distributed due to the dynamic deployment of the edge nodes [13]. In the mobile networks context, the edge computing paradigm is often referred to as multi-access edge computing (MEC) [3], as introduced by the European Telecommunications Standards Institute (ETSI), and it will be used throughout the rest of this paper. With this promising solution, the computing capabilities are brought closer to the end users, therefore reducing the distance between the source and destination of the data. Given this context, offloading intensive processing to edge nodes through computation offloading has been employed increasingly at a time when the latency requirements coming from applications such as augmented reality/virtual reality (AR/VR), autonomous driving or interactive gaming are more demanding with each generation of mobile networks. Although computation offloading is not a new technique, with its origins dating back to the early 2000s [7], it has been studied during the MCC environment deployment and more recently in the MEC environment in order to alleviate network congestion and improve quality of experience (QoE) for end users.

This paper gives an overview of the advantages of computation offloading by comparing MCC with MEC, describes different offloading decisions and their trade-offs, discusses the main challenges that arise during an offloading procedure, as well as presenting our view on how offloading will fit into the future 6G networks. The remainder of this paper is organized as follows: Sect. 2 presents an overview of computation offloading, Sect. 3 provides a closer look at the different challenges faced in offloading, Sect. 4 presents our view on offloading in 6G networks, and Sect. 5 summarizes and concludes the paper.

## 2 Overview of Computation Offloading

In the computation offloading process, tasks coming from computation-intensive applications are offloaded to remote servers such as MEC or MCC servers for faster computing. To meet the expected QoE, which often translates to low latency in terms of network response [2], offloading processing to MEC rather than MCC has become a preferred solution due to shorter distance between the end users and edge nodes. In addition, this decentralized data processing method reduces the number of data flows transmitted in the network backhaul, therefore saving network bandwidth [7]. In the case of latency-tolerant applications, offloading computing to MCC is more appropriate, increasing the availability of resources at MEC for the latency-sensitive applications since edge resources are limited compared to cloud servers in terms of storage and computing capabilities. The aforementioned MEC and MCC aspects are summarized in Table 1 and are the basis of understanding the benefits of computation offloading to different elements of the network.

**Table 1.** Comparison between cloud and edge computing [9,10].

| Criteria | Cloud computing | Edge computing |
|---|---|---|
| Deployment | Centralized | Distributed |
| Configuration and planning | Sophisticated | Lightweight |
| Distance to end user | Large | Small |
| Latency | High | Low |
| Backhaul usage | Frequent | Infrequent |
| Computational capabilities | High | Limited |
| Storage capacity | High | Limited |

### 2.1 Computation Offloading Mechanisms

When building a computation offloading algorithm, one key aspect is to analyze the type of applications involved in the offloading process since it provides information about whether an application task can be partitioned or not and, as a consequence, if it can be included in the transmission queue for edge processing. With this in mind, the output of an offloading algorithm may result in one of the following decisions [9]:

- Local execution where application tasks are executed on the device.
- Full offloading to MEC where the entire computation is migrated at the edge.
- Partial offloading where the application is partitioned in multiple tasks that can be executed part on the device and part on the MEC host.

In addition to the above offloading decisions, tasks can be sent to the cloud for processing, but such solution may come at the expense of lower QoE. In Fig. 1, the three offloading decisions are depicted in the context of 5G networks where the wireless and optical network (front-, mid- and backhaul) are the main components of the communication path when transmitting tasks to MEC/MCC.

Making optimal offloading decisions in time-varying 5G networks is not trivial, therefore various trade-offs must be made in the process. Some of these trade-offs may include the following network parameters: wireless and optical communication time, execution and queuing time at MEC/MCC, energy consumption for communication and computation, MEC load or MEC placement. Numerous efforts have been dedicated to addressing the optimal offloading problem in edge computing taking into account the implications of making these trade-offs between two or more network parameters. In the rest of this subsection, we will be looking at some of them.

In [8], computation offloading is placed in a vehicle edge computing network where tasks can be computed locally, on a fixed edge server or a vehicular edge server. The offloading decision is made using deep reinforcement learning (DRL), also including allocation for the computing and communication resources in the action space, while considering the delay of the computation task. In [16], a

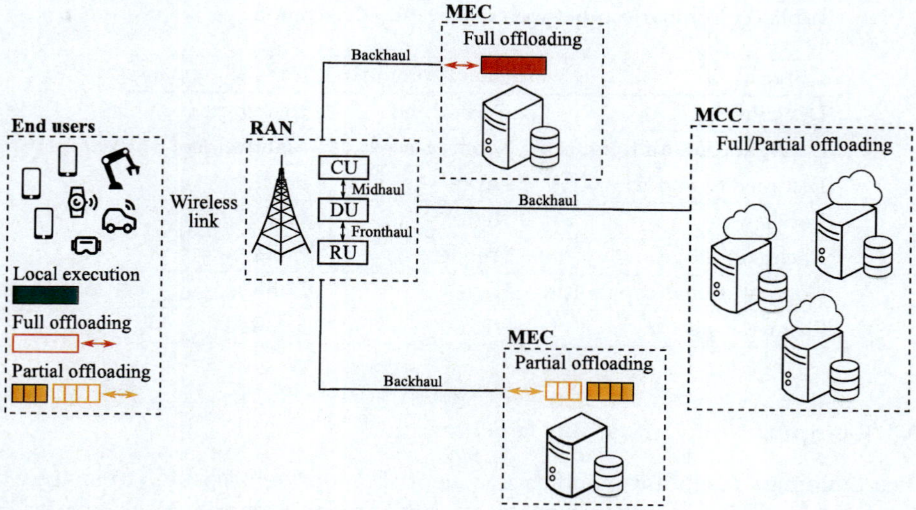


**Fig. 1.** Computation offloading decisions (local execution, full offloading and partial offloading) in 5G network with MEC and MCC.

computation offloading scheme for dependent Internet of Things (IoT) applications is proposed, where the target is to obtain the offloading decision (local or edge). In the scheme, the decision is made with DRL by investigating the dependency between tasks, as well as the overhead, i.e., latency and energy consumption. In [17], an intelligent ultradense edge computing framework is proposed to solve the offloading decision problem together with resource allocation and service caching placement. DRL was employed to tackle the combined optimization problems, while federated learning was introduced for end user privacy preservation in the model training phase. A DRL-based online offloading method is proposed in [5], where the problem is divided into two sub-problems: offloading decision solved with DRL and resource allocation in terms of time required for energy harvesting and offloading, solved with one-dimensional bi-section search. Model-free DRL is used in [14] to solve the offloading challenge by considering non-divisible and delay-sensitive tasks as well as the load level dynamics of the edge nodes. The decision to offload a task and the selection of which edge node to offload it to was indicated in the DRL action space. A multilevel vehicular edge-cloud computing network is considered in [6] where both computation offloading and resource allocation are included in the optimization problem. Reinforcement learning was employed to solve the problem by minimizing the vehicle's consumption in terms of communication and computation time, as well as energy consumption. Similar network parameters were used in [4], where the aim was to maximize the long-term system utility in an edge computing assisted power IoT scenario.

## 3 Challenges in Computation Offloading

Offloading comes with a wide set of challenges that may be considered while building a computation offloading framework. Examples of such challenges are described as follows:

- The type of offloading decision, presented in Sect. 2, is the center of the algorithm and it is usually the first problem to be addressed.
- In partial offloading, the dependability between application tasks is difficult to approach and implement due to the execution order (one task may require the output of another task) and software/hardware restrictions (some tasks may require local execution, such as the camera input) [10].
- Definition of the network model and input parameters – it refers to the choices regarding the offloading environment (e.g., single or multi-user, static or mobile users, single or multi-MEC, MEC placement) and its parameters: communication (e.g., transmit power, available bandwidth) and computation capabilities (e.g., number of CPUs).
- Resource allocation for both communication (e.g., radio resources) and computation (e.g., number of cores per task) to achieve the desired offloading trade-off in terms of execution delay, transmission delay, energy consumption, etc.
- Management of user mobility as well as edge mobility (e.g., when deploying vehicle edge computing, unmanned aerial vehicles (UAVs)) since network parameters are time-varying and service continuity may be difficult to guarantee [9].

The examples above are proof that making optimal offloading decisions, especially in dynamic and dense 5G networks, is challenging, therefore employing advanced artificial intelligence (AI) techniques is a natural step to solve this problem, as seen lately in the literature [12].

## 4 6G Vision for Computation Offloading

The range of applications is expanding with each generation of mobile networks, each application requiring its specific network infrastructure model, techniques and minimum performance indicators in order to provide the expected QoE. Offloading to edge has been one of the techniques used for improving the performance of latency-sensitive 5G applications such as AR/VR, face recognition, gaming, video analytics that require enhanced computing capabilities. In future 6G networks more applications are expected to emerge, requiring AI on both application and networking sides of the deployment. Therefore, AI-enabled computation offloading is expected to become one of the key drivers in providing real-time user experience for future applications including telepresence (mixed reality co-design, merged reality game/work), AI-assisted vehicle-to-everything (V2X), to name but a few [15]. With the increasing number of end devices, far edge computing can become more relevant in future 6G networks since the resources are placed even closer to the end users [1], creating a computing continuum spanning across the network. As a consequence, the latency can be reduced even more and the interaction and data sharing with external parties is avoided for privacy reasons, but coming at the expense of complex offloading mechanisms due to the increased number of computing nodes and offloading possibilities. In addition, 6G architectures are set to employ distributed AI mechanisms such as federated learning to preserve user privacy and reduce security risks by decoupling the model training from the need to access the raw training data generated and stored on end user devices [11].

## 5 Conclusion

This paper presented an overview of computation offloading starting with a comparison between MEC and MCC and highlighting the benefits of these computing resources. Then, different offloading decisions were introduced, followed by a short description of current computation offloading works and the challenges faced when building an offloading framework. Finally, we described our vision on the evolution of computation offloading in future 6G networks by looking at novel applications, enhanced computing resources through far edge, and federated learning as an AI training model solution for user data protection.

## References

1. 5GPPP Architecture Working Group: The 6G Architecture Landscape (2023). https://doi.org/10.5281/zenodo.7313232
2. Bhattacharya, A., De, P.: A survey of adaptation techniques in computation offloading. J. Netw. Comput. Appl. **78**, 97–115 (2017). https://doi.org/10.1016/j.jnca.2016.10.023
3. ETSI GS MEC 003: Multi-access edge computing (MEC); Framework and Reference Architecture (2022). v3.1.1
4. Hu, J., Li, Y., Zhao, G., Xu, B., Ni, Y., Zhao, H.: Deep reinforcement learning for task offloading in edge computing assisted power IoT. IEEE Access **9**, 93892–93901 (2021). https://doi.org/10.1109/ACCESS.2021.3092381
5. Huang, L., Bi, S., Zhang, Y.J.A.: Deep reinforcement learning for online computation offloading in wireless powered mobile-edge computing networks. IEEE Trans. Mob. Comput. **19**(11), 2581–2593 (2020). https://doi.org/10.1109/TMC.2019.2928811
6. Khayyat, M., Elgendy, I.A., Muthanna, A., Alshahrani, A.S., Alharbi, S., Koucheryavy, A.: Advanced deep learning-based computational offloading for multilevel vehicular edge-cloud computing networks. IEEE Access **8**, 137052–137062 (2020). https://doi.org/10.1109/ACCESS.2020.3011705
7. Lin, L., Liao, X., Jin, H., Li, P.: Computation offloading toward edge computing. Proc. IEEE **107**(8), 1584–1607 (2019). https://doi.org/10.1109/JPROC.2019.2922285
8. Liu, Y., Yu, H., Xie, S., Zhang, Y.: Deep reinforcement learning for offloading and resource allocation in vehicle edge computing and networks. IEEE Trans. Veh. Technol. **68**(11), 11158–11168 (2019). https://doi.org/10.1109/TVT.2019.2935450

9. Mach, P., Becvar, Z.: Mobile edge computing: a survey on architecture and computation offloading. IEEE Commun. Surv. Tutor. **19**(3), 1628–1656 (2017). https://doi.org/10.1109/COMST.2017.2682318
10. Mao, Y., You, C., Zhang, J., Huang, K., Letaief, K.B.: A survey on mobile edge computing: the communication perspective. IEEE Commun. Surv. Tutor. **19**(4), 2322–2358 (2017). https://doi.org/10.1109/COMST.2017.2745201
11. McMahan, H.B., Moore, E., Ramage, D., Hampson, S., y Arcas, B.A.: Communication-efficient learning of deep networks from decentralized data (2023)
12. Shakarami, A., Ghobaei-Arani, M., Shahidinejad, A.: A survey on the computation offloading approaches in mobile edge computing: a machine learning-based perspective. Comput. Netw. **182**, 107496 (2020). https://doi.org/10.1016/j.comnet.2020.107496
13. Shi, W., Cao, J., Zhang, Q., Li, Y., Xu, L.: Edge Computing: Vision and Challenges. IEEE Internet Things J. **3**(5), 637–646 (2016). https://doi.org/10.1109/JIOT.2016.2579198
14. Tang, M., Wong, V.W.: Deep reinforcement learning for task offloading in mobile edge computing systems. IEEE Trans. Mob. Comput. **21**(6), 1985–1997 (2022). https://doi.org/10.1109/TMC.2020.3036871
15. The 5G Infrastructure Association: European Vision for the 6G Network Ecosystem (2021). https://doi.org/10.5281/zenodo.5007671
16. Xiao, H., Xu, C., Ma, Y., Yang, S., Zhong, L., Muntean, G.M.: Edge intelligence: a computational task offloading scheme for dependent IoT application. IEEE Trans. Wirel. Commun. **21**(9), 7222–7237 (2022). https://doi.org/10.1109/TWC.2022.3156905
17. Yu, S., Chen, X., Zhou, Z., Gong, X., Wu, D.: When deep reinforcement learning meets federated learning: intelligent multitimescale resource management for multiaccess edge computing in 5G ultra-dense network. IEEE Internet Things J. **8**(4), 2238–2251 (2021). https://doi.org/10.1109/JIOT.2020.3026589

# Quantum-Inspired Network Optimization in 6G: Opportunities, Challenges and Open Research Directions


Oumayma Bouchmal[1(✉)], Bruno Cimoli[1], Ripalta Stabile[1], Juan Jose Vegas Olmos[2], and Idelfonso Tafur Monroy[1]

[1] Department of Electrical Engineering, Eindhoven University of Technology, Eindhoven, The Netherlands
{o.bouchmal,b.cimoli,R.Stabile,i.tafur.monroy}@tue.nl

[2] Software Architecture, NVIDIA Corporation, Yokneam, Israel
juanj@nvidia.com



**Abstract.** The sixth-generation (6G) wireless networks are envisioned to deliver a paradigm shift in performance, capacity, and complexity compared to the previous generations. The increased complexity, ultra-low latency, and higher data rate requirements of 6G networks require advanced computational capabilities for network optimization, including resource allocation, dynamic slicing, and traffic routing. In this paper, we discuss the need of integrating quantum computing and quantum machine learning for network optimization in 6G.



**Keywords:** 6G · QC · QML · optimization · routing · resource allocation · dynamic slicing


## 1 Introduction

The emergence of 6G networks is expected to bring to the table unprecedented performance goals, diverse service requirements, and advanced technologies and applications such as machine-to-machine (M2M) communication and extended reality (XR) [30]. However, the increased complexity of 6G networks and the strict quality of service (QoS) requirements of 6G applications bring significant challenges in resource allocation, dynamic network slicing, and traffic routing, which are crucial aspects to maintaining optimal network performance and user experience [3]. Classical machine learning (ML) based on conventional computation computers have been implemented in the previous generation, however, their effectiveness is becoming limited due to the exponential growth in the number of devices, connections, and network nodes expected in 6G networks.


This work was supported by the EC H2020 MSCA ITN-ETN IoTalentum (GA 953442) project, the ECSEL-JU project BRAINE with grant number 876967 and the Dutch Ministry of Economic Affairs and Climate Policy (EZK), as part of the Quantum Delta NL programme.





R. Mehmood et al. (Eds.): DCAI 2023, LNNS 741, pp. 480–488, 2023.
https://doi.org/10.1007/978-3-031-38318-2_48

Moreover, network optimization in 6G will involve solving complex combinatorial problems, such as routing optimization and resource allocation, which are typically NP-hard. Additionally, ultra-low latency requirements of 6G networks make classical optimization techniques insufficient for real-time optimization of resources and traffic routing in 6G networks [11,16].

Quantum Machine Learning (QML) leverages the principles of quantum computing (QC) to perform ML tasks more efficiently than classical methods. Since its appearance, QML has shown promise in solving complex optimization problems such as NP-Hard problems, providing the potential to overcome the challenges faced in resource allocation, dynamic slicing, and traffic routing in 6G networks. Accordingly, this paper explores and discusses the use of QML to address these challenges. The rest of the paper is structured as follows: fundamentals concepts of QC and QML are presented in Sect. 2. Section 3 is devoted to the challenges of resource allocation, dynamic slicing and traffic routing in 6G networks and how QML algorithms can be used in order to solve these challenges. Section 4 presents the challenges of applying QML and some open research directions. Finally, Sect. 5 presents conclusions.

## 2 Fundamentals of Quantum Computing and Quantum Machine Learning

Quantum computing is a field that focuses on developing and utilizing quantum computers, which are based on the principles of quantum physics (superposition and entanglement). The field includes various aspects, such as designing and building quantum hardware, creating quantum algorithms, understanding the fundamental limits of quantum information processing, and exploring applications of quantum computing across various domains, including cryptography, optimization, and machine learning. When compared to classical computers, a fault-tolerant quantum computer has the potential to achieve a speed-up of 100 million times for certain types of problems [7].

Quantum machine learning (QML) is a subfield of quantum computing that applies quantum algorithms to machine learning programs. By utilizing the inherent properties of quantum computing, such as superposition, entanglement, and quantum parallelism, QML can potentially improve the efficiency and capabilities of machine learning algorithms for certain types of problems [14].

Before delving into the details of how QML can be employed for network optimization in 6G, it is essential to first review some fundamental concepts of quantum computation.

### 2.1 Qubits

A quantum bit (qubit) is the fundamental unit of quantum information. It is the quantum analog of a classical bit, which can have a value of either 0 or 1. Qubits, however, exhibit unique quantum properties that enable them to exist in a superposition of both 0 and 1 states simultaneously. The state of a qubit can be represented as a linear combination of its basis states, $|0\rangle$ and $|1\rangle$:

$$|\psi\rangle = \alpha|0\rangle + \beta|1\rangle \tag{1}$$

Here, $\alpha$ and $\beta$ are complex numbers, and the sum of their squared magnitudes is equal to 1.

The ability of qubits to exist in a superposition of states is one of the key features that enable quantum computers to perform certain calculations significantly faster than classical computers. Moreover, when multiple qubits are entangled, which is another important quantum property, they can be used to represent an exponentially large number of classical states simultaneously, allowing quantum computers to perform parallel computations (known as quantum parallelism) and solve certain problems more efficiently and faster than conventional computers [15].

### 2.2 Quantum Data

Quantum data refers to information encoded and stored using quantum systems, such as qubits in a quantum computer. Unlike classical data, which consists of binary bits (0 or 1), quantum data takes advantage of the unique properties of quantum mechanics (e.g., superposition and entanglement) to represent and process information more efficiently for certain types of problems. There are several techniques that can be used to encode/translate classical data into quantum data. Some popular encoding methods include computational basis encoding, angle encoding, amplitude encoding, Hamiltonian encoding, and QSample encoding [20,21,27]. The choice of encoding method depends on the specific problem, the quantum algorithm being employed, and the requirements for error resilience.

### 2.3 Quantum Algorithms

Quantum algorithms are computational procedures designed to run on quantum computers or simulators in order to solve problems more efficiently and faster than classical algorithms in certain cases. Quantum algorithms are usually expressed using quantum circuits, which consist of a sequence of quantum gates that operate on qubits. There are two types of quantum algorithms:

- Full quantum algorithms: these algorithms are designed to run entirely on quantum devices without the integration of any classical hardware. While some of these algorithms can achieve exponential speed-ups and may demonstrate "quantum advantage" when executed on fault-tolerant quantum computers, not all fully quantum algorithms guarantee exponential improvements. Some notable examples include Shor's algorithm, Grover's algorithm, and the quantum phase estimation algorithm. A comprehensive list of these algorithms can be found in [22].

- Variational quantum algorithms (VQAs): are a specific subclass of quantum algorithms that utilize a hybrid quantum-classical approach to find approximate solutions to problems. In these algorithms, a parameterized quantum circuit (ansatz) is optimized using a classical optimizer to minimize or maximize a given cost function. These types of algorithms are well suited for Noisy Intermediate scale quantum computers (NISQ) [4]. An example of these algorithms is the quantum approximate optimization algorithm (QAOA) [8] and variational quantum eigensolver (VQE) [25].

## 2.4 Types of Quantum Computation

The range of problems that can be solved using QML algorithms depends on the type of quantum computer and the specific architecture involved in the process. Currently, there are two primary categories of quantum computing models: universal and specialized quantum computers.

- Universal quantum computers: these are general-purpose quantum computers designed to perform a wide range of computations by implementing all the types of quantum algorithms (mentioned in Sect. 2.3 ). They contain a large number of qubits and a universal set of quantum gates. In theory, a universal quantum computer is capable of solving any problem that can be solved by a classical computer and can provide significant speed-ups for certain types of problems. Both academia and industry are investing in the development of this type of quantum hardware. However, the currently available quantum devices are NISQ computers, which have limited numbers of qubits and lack error correction (e.g., IBM's Osprey with 433 qubits) [23].
- Specialized quantum computers: also known as non-universal quantum computers. These are designed to solve specific types of problems. Examples of specialized quantum computers include quantum annealers (e.g., D-Wave systems with 5640 qubits and Fujitsu with 8192 qubits) and analog quantum simulators (e.g., NIST trapped ion systems). These devices can be more easily built compared to universal quantum computers and can provide computational advantages for certain problems, but their applicability is more limited [6].

## 2.5 Processing Classical Data on a Quantum Computer

The procedure of processing classical data on a quantum computer can be simplified in six steps as depicted in Fig. 1 [12,14]:

- Data preprocessing: classical data is often preprocessed to reduce noise, standardize features, and transform it into a suitable format for encoding into quantum states.
- Data encoding: the preprocessed classical data is translated into quantum states, or qubits, by employing a suitable encoding technique (e.g., Hamiltonian encoding).

- Quantum algorithm: a quantum algorithm, designed for the specific problem, is applied to the encoded quantum states to process the information.
- Measurement: the resulting quantum state is measured, resulting in a classical bit string or probability distribution as an output.
- Post-processing: to extract relevant information, classical post-processing algorithms are applied to the measurement outcomes.
- Interpretation: the final processed data is interpreted to provide insights or make decisions based on the problem requirements.

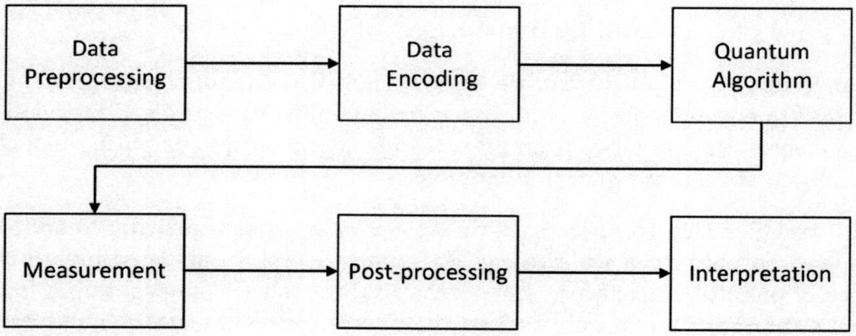


**Fig. 1.** Simplified procedure of processing classical data on a quantum computer

# 3 Quantum Machine Learning for Network Optimization

## 3.1 Resource Allocation

Resource allocation refers to the process of efficiently distributing and managing network resources, such as radio spectrum, computing power, and energy. An effective resource allocation is essential to minimize interference, prevent congestion, and maintain the desired level of service quality across the network [28]. However, due to the high degree of uncertainty and dynamicity in network conditions, resource allocation is considered a very challenging optimization problem. Moreover, as networks become more complex and heterogeneous, the number of variables that need to be optimized increases, leading to a larger search space and potentially making the problem more difficult to solve [29]. In this regard, the power provided by QC techniques and the speed-up offered by quantum algorithms can efficiently improve resource allocation in these types of networks [7]. One of the algorithms that can be employed is the Grover algorithm, which is a quantum search algorithm that belongs to the fully quantum category mentioned in Sect. 2.3. It is considered one of the most powerful computational procedures that can reduce the time complexity from $O(N)$ to $O(\sqrt{N})$ where $N$ is the number of items in the database being searched [9]. Another approach that can be used is quantum annealing (QA). It is a quantum computing technique used to solve NP-Hard optimization problems by finding the lowest energy state that presents the solution of a given Hamiltonian cost function. QA can be implemented using various types of quantum devices including D-Wave quantum annealers.

### 3.2 Dynamic Network Slicing

Dynamic network slicing (dynamic-NS) refers to the process of creating and managing multiple virtual and independent networks (slices) on a single physical infrastructure, with the ability to adapt and reconfigure these slices in real-time based on changing demands and network conditions. In the context of 6G, dynamic-NS faces several challenges. Developing efficient algorithms for real-time adaptation and reconfiguration becomes increasingly difficult due to the stringent latency requirements and complex network conditions. Moreover, accommodating the heterogeneity and ensuring scalability in 6G networks adds to the complexity of implementing dynamic-NS [5]. In this respect, QML can potentially address these challenges. For example, one can use quantum algorithms such as quantum neural networks (QNN) and quantum support vector machine (QSVM) to efficiently process and optimize large-scale problems, enabling better scalability. Furthermore, by exploiting the inherent parallelism and entanglement properties of QC, QML can process large amounts of data and learn patterns for efficient dynamic-NS. Not to mention that QML can potentially enable faster optimization algorithms that can keep up with the stringent latency requirements and facilitate real-time adaptation of network slices. In this context, Quantum Deep Reinforcement Learning (QDRL) can also be used to learn and optimize decision-making processes related to network configurations and resource allocations in dynamic-NS [18].

### 3.3 Traffic Routing

Similar to resource allocation and dynamic-NS, traffic routing will also face several challenges in 6G networks, starting from supporting the scalability, dynamicity and heterogeneity of the complex networks to ensuring strict QoS requirements for 6G applications and services [1,24]. Furthermore, routing optimization with multiple constraints can be a difficult problem to solve. In particular, when multiple constraints are involved, such as optimizing for both network latency and energy efficiency, the problem becomes NP-hard, meaning that finding the optimal solution may be computationally infeasible [17]. VQAs can be considered a possible way to overcome these challenges. The Quantum Approximate Optimization Algorithm (QAOA) is a type of VQA specifically designed to solve combinatorial optimization problems [8]. As a VQA, it is particularly suited to run on current NISQ devices. By utilizing a hybrid classical-quantum approach, QAOA provides a scalable and practical method to optimize problems, even when limited by the current noisy and error-prone nature of quantum hardware. Due to its compatibility with NISQ devices, QAOA has attracted considerable attention from the quantum computing community and has demonstrated promising results for solving various NP-Hard optimization problems such as MaxCut [26], vehicle-routing problem (VRP) [2], and the traveling salesman problem (TSP) [19]. Another approach that can be also employed for traffic routing is QA [13].

## 4 Challenges and Open Research Directions

While QML has the potential to optimize network performance in 6G, there are several challenges and limitations associated with its employment [7,16]:

- Quantum hardware limitations: Current quantum computers have a limited number of qubits and are prone to errors due to noise and decoherence. This limits the size and complexity of problems that can be solved using QML algorithms. Research in both academia and industry is carried out to build fault-tolerance quantum computers by 2030 [10].
- Integration with existing systems: Integrating QML algorithms into 6G network optimization frameworks could be complex, as it may necessitate reworking existing infrastructure and adapting to new paradigms of computation.
- Training data and complexity: QML algorithms may require large amounts of training data and computational resources to achieve satisfactory performance. Collecting, preprocessing, and storing the necessary data for 6G network optimization tasks can be challenging. To tackle this problem, High-performance computing (HPC)-enabled quantum computers can be considered a promising approach for the training phase.

Researchers and industry experts are actively working to address these challenges, and it is expected that the applicability and effectiveness of QML in 6G network optimization will continue to improve in the coming years.

## 5 Conclusions

This paper has provided a comprehensive investigation of the potential of QML in addressing the challenges associated with resource allocation, dynamic network slicing, and traffic routing in the emerging 6G networks. We have presented an overview of the fundamentals of QC and QML and discussed their applicability to complex optimization problems encountered in 6G network optimization. Furthermore, we have mentioned various QML algorithms that can be applied for network optimization and identified the challenges associated with integrating QML techniques within 6G networks as well as some research directions.

## References

1. Amin, R., Rojas, E., Aqdus, A., Ramzan, S., Casillas-Perez, D., Arco, J.M.: A survey on machine learning techniques for routing optimization in SDN. IEEE Access **9**, 104582–104611 (2021). https://doi.org/10.1109/ACCESS.2021.3099092
2. Azad, U., Behera, B.K., Ahmed, E.A., Panigrahi, P.K., Farouk, A.: Solving vehicle routing problem using quantum approximate optimization algorithm. IEEE Trans. Intell. Transp. Syst., 1–10 (2022). https://doi.org/10.1109/TITS.2022.3172241

3. Banafaa, M., et al.: 6g mobile communication technology: requirements, targets, applications, challenges, advantages, and opportunities. Alex. Eng. J. **64**, 245–274 (2023). https://doi.org/10.1016/j.aej.2022.08.017
4. Cerezo, M., et al.: Variational quantum algorithms. Nat. Rev. Phys. **3**(9), 625–644 (2021). https://doi.org/10.1038/s42254-021-00348-9
5. Chochliouros, I.P., Spiliopoulou, A.S., Lazaridis, P., Dardamanis, A., Zaharis, Z., Kostopoulos, A.: Dynamic network slicing: challenges and opportunities. In: Maglogiannis, I., Iliadis, L., Pimenidis, E. (eds.) AIAI 2020. IAICT, vol. 585, pp. 47–60. Springer, Cham (2020). https://doi.org/10.1007/978-3-030-49190-1_5
6. Das, A., Chakrabarti, B.K.: Colloquium: quantum annealing and analog quantum computation. Rev. Mod. Phys. **80**, 1061–1081 (2008). https://doi.org/10.1103/RevModPhys.80.1061
7. Duong, T.Q., Nguyen, L.D., Narottama, B., Ansere, J.A., Huynh, D.V., Shin, H.: Quantum-inspired real-time optimization for 6G networks: opportunities, challenges, and the road ahead. IEEE Open J. Commun. Soc. **3**, 1347–1359 (2022). https://doi.org/10.1109/OJCOMS.2022.3195219
8. Farhi, E., Goldstone, J., Gutmann, S.: A quantum approximate optimization algorithm. arXiv preprint arXiv:1411.4028 (2014). https://doi.org/10.48550/arXiv.1411.4028
9. Grover, L.K.: A fast quantum mechanical algorithm for database search. In: Proceedings of the Twenty-Eighth Annual ACM Symposium on Theory of Computing. Association for Computing Machinery, New York, NY, USA (1996). https://doi.org/10.1145/237814.237866
10. IBM: The IBM Quantum Development Roadmap (2022). https://www.ibm.com/quantum/roadmap
11. Kato, N., Mao, B., Tang, F., Kawamoto, Y., Liu, J.: Ten challenges in advancing machine learning technologies toward 6G. IEEE Wirel. Commun. **27**(3), 96–103 (2020). https://doi.org/10.1109/MWC.001.1900476
12. Khan, T.M., Robles-Kelly, A.: Machine learning: quantum vs classical. IEEE Access **8**, 219275–219294 (2020). https://doi.org/10.1109/ACCESS.2020.3041719
13. Krauss, T., McCollum, J.: Solving the network shortest path problem on a quantum annealer. IEEE Trans. Quantum Eng. **1**, 1–12 (2020). https://doi.org/10.1109/TQE.2020.3021921
14. Martín-Guerrero, J.D., Lamata, L.: Quantum machine learning: a tutorial. Neurocomputing **470**, 457–461 (2022). https://doi.org/10.1016/j.neucom.2021.02.102
15. McMahon, D.: Quantum Computing Explained. Wiley, Hoboken (2007)
16. Nawaz, S.J., Sharma, S.K., Wyne, S., Patwary, M.N., Asaduzzaman, M.: Quantum machine learning for 6G communication networks: state-of-the-art and vision for the future. IEEE Access **7**, 46317–46350 (2019). https://doi.org/10.1109/ACCESS.2019.2909490
17. Puri, A., Tripakis, S.: Algorithms for routing with multiple constraints. In: AIPS-02, p. 7 (2002)
18. Rezazadeh, F., Kahvazadeh, S., Mosahebfard, M.: Towards quantum-enabled 6G slicing. arXiv preprint arXiv:2212.11755 (2022)
19. Ruan, Y., Marsh, S., Xue, X., Liu, Z., Wang, J., et al.: The quantum approximate algorithm for solving traveling salesman problem. Comput. Mater. Continua **63**(3), 1237–1247 (2020)
20. Sergioli, G.: Quantum and quantum-like machine learning: a note on differences and similarities. Soft. Comput. **24**(14), 10247–10255 (2019). https://doi.org/10.1007/s00500-019-04429-x

21. Sierra-Sosa, D., Telahun, M., Elmaghraby, A.: TensorFlow quantum: impacts of quantum state preparation on quantum machine learning performance. IEEE Access **8**, 215246–215255 (2020). https://doi.org/10.1109/ACCESS.2020.3040798
22. Jordan, S.: Quantum Algorithm Zoo (2011). https://quantumalgorithmzoo.org/
23. Tacchino, F., Chiesa, A., Carretta, S., Gerace, D.: Quantum computers as universal quantum simulators: state-of-the-art and perspectives. Adv. Quant. Technol. **3**(3), 1900052 (2020). https://doi.org/10.1002/qute.201900052
24. Tang, F., Mao, B., Kawamoto, Y., Kato, N.: Survey on machine learning for intelligent end-to-end communication toward 6G: from network access, routing to traffic control and streaming adaption. IEEE Commun. Surv. Tutorials **23**(3), 1578–1598 (2021). https://doi.org/10.1109/COMST.2021.3073009
25. Tilly, J., et al.: The variational quantum eigensolver: a review of methods and best practices. Phys. Rep. **986**, 1–128 (2022). https://doi.org/10.1016/j.physrep.2022.08.003
26. Wang, Z., Hadfield, S., Jiang, Z., Rieffel, E.G.: Quantum approximate optimization algorithm for MaxCut: a fermionic view. Phys. Rev. A **97**, 022304 (2018). https://doi.org/10.1103/PhysRevA.97.022304
27. Weigold, M., Barzen, J., Leymann, F., Salm, M.: Data encoding patterns for quantum computing. In: Proceedings of the 27th Conference on Pattern Languages of Programs. The Hillside Group (2022). https://doi.org/10.5555/3511065.3511068
28. Xu, Y., Gui, G., Gacanin, H., Adachi, F.: A survey on resource allocation for 5G heterogeneous networks: current research, future trends, and challenges. IEEE Commun. Surv. Tutorials **23**(2), 668–695 (2021). https://doi.org/10.1109/COMST.2021.3059896
29. Zhang, S., Zhu, D.: Towards artificial intelligence enabled 6G: state of the art, challenges, and opportunities. Comput. Netw. **183**, 107556 (2020). https://doi.org/10.1016/j.comnet.2020.107556
30. Zhang, Z., et al.: 6G wireless networks: vision, requirements, architecture, and key technologies. IEEE Veh. Technol. Mag. **14**(3), 28–41 (2019). https://doi.org/10.1109/MVT.2019.2921208

# Telefónica I+D's Roadmap for Integrated SDN Orchestration of Multi-domain Multi-layer Transport Networks

Liesbeth Roelens(✉), Alejandro Muñiz Da Costa, and Óscar González de Dios

Telefónica I+D, Madrid, Spain
{liesbeth.roelens,alejandro.muniz,oscar.gonzalezdedios}@telefonica.com

**Abstract.** With increasing demand for faster, more reliable connectivity, next-generation transport networks must evolve toward autonomous and customer-driven network management. Software-Defined Networking (SDN) is widely used to dynamically provide high-availability, low-latency paths. At the same time, however, transport networks are heterogeneous, consisting of multiple layers (L3-L0) and operator domains, and must be able to interoperate smoothly. To meet these requirements in multi-vendor, multi-domain end-to-end network slicing, it is important to hide customer-specific implementation details from different operators. Similarly, "traffic intent modelling" is proposed as a means to abstract the required services from the customers. In this paper, we explore a standardised intent-driven multi-layer network management architecture that has been proposed by Telefónica I+D.

**Keywords:** Software-Defined Networking (SDN) · Multi-Layer Network Planning · Traffic Intent Modelling

## 1 Introduction

In 5G and beyond (B5G) transport networks, the demand for a diverse range of applications has significantly increased. Not only the progressively stricter service level agreements of applications such as high-definition video streaming, but also the exponential growth of Internet of Things (IoT) devices, exemplify this trend. Consequently, the conventional one-size-fits-all approach for providing network connectivity for these distinct requirements has become inadequate [1]. As a response, network slicing emerged, which involves the partitioning of a physical network infrastructure into multiple virtual networks, each tailored to an individual service. However, telecommunications networks are composed of dedicated hardware. The design, integration and operation of individual network elements require specialised skills and expertise. Such a rigid infrastructure poses limitations on the adaptability and agility needed to accommodate the dynamic requirements of network slicing. In this context, the role of B5G transport networks goes beyond providing connectivity and they must guarantee


R. Mehmood et al. (Eds.): DCAI 2023, LNNS 741, pp. 489–496, 2023.
https://doi.org/10.1007/978-3-031-38318-2_49

a coordinated allocation of networking resources. To do so, the management and orchestration plane serves as a mediator between the service and infrastructure plane.

This paper presents the functional architecture of the management and orchestration plane in a multi-domain multi-layer network. This work is considered as an ongoing effort towards its implementation within the IoTalentum[1] and Int5Gent[2] projects. The primary aim of this paper is to address key challenges in the field of network planning and optimisation. First, we propose the use of common data models and standard, open APIs for extracting layer or domain-specific information. This approach will facilitate interoperability and information exchange between different layers of the network stack. Second, we emphasise the need for better alignment of the network layers, particularly between the IP and optical layer. Subsequently, we advocate for a multi-layer path computation approach, wherein common knowledge of both the IP and optical layers adds more flexibility and leads to smarter and more efficient planning decisions.

This paper is structured as follows. Section 2 presents an overview of the functionalities of the proposed network slice orchestrator. We then delve into the detailed discussion of the various modules comprising this orchestrator in Sect. 3. Section 4 concludes our discussion and provides future research directions.

## 2 Network Orchestrator Architecture

Figure 1 presents the functional architecture of the multi-layer network slice orchestrator, illustrating the components and their interactions. The architecture focuses on network planning and visualisation and consists of several modules, each playing a role in achieving efficient network management and resource utilisation. The presented architecture serves two primary objectives: streamlining operations across the IP and optical layer and standardising interfaces and interactions between the modules.

The core of the architecture is the *multi-layer network planning* module. This module encompasses strategic decision-making, resource allocation, and optimisation techniques across multiple network layers. By considering the requirements and constraints of different layers and domains, the network planner aims to maximise the overall performance of the network. Working in tandem with the network planner is the *transport slice intent* module. This module captures the high-level network slice requests. By translating these intents into practical instructions for the planner, the module facilitates efficient provisioning of network resources.

To obtain network topology information, the architecture relies on a *multi-layer topology database*. This comprehensive database stores detailed data about the physical and logical connectivity of network elements across different layers. To collect this information, the *IP* and *optical connectors* play crucial roles.

[1] https://iotalentum.eu/.
[2] https://5g-ppp.eu/int5gent/.

These modules act as interfaces for retrieving information from the IP and optical network layers, respectively. They make use of standardised interfaces, ensuring vendor-independent operation. Finally, the *slice visualisation* module permits the network operator to observe and analyse established slices in the network.

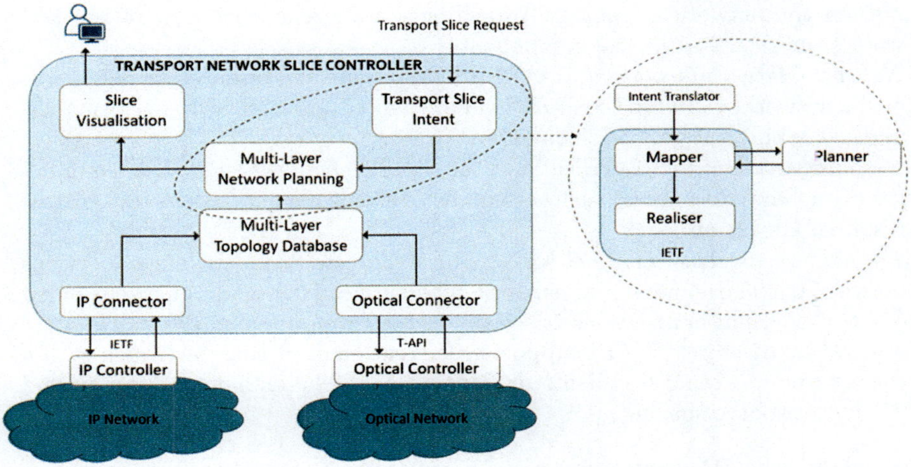


**Fig. 1.** A diagram showing the different modules to provision network slices in multi-layer networks.

## 3 Description of Modules

This section discusses the various modules that make up the functional architecture of the SDN network slice orchestrator described in Sect. 2 more in detail.

### 3.1 IP over Optical Transport Network

The network infrastructure comprises a layered architecture integrating IP over optical (DWDM) technologies.

The **optical connector** is responsible for obtaining information about the optical network layer. Transport-API (T-API) was introduced by the Open Networking Foundation (ONF) in 2014 as an initiative to develop an open and standardised interface for effectively managing SDN optical transport networks. The ONF provided a functional architecture and recommendations to guide the development and implementation of T-API [2].

The primary goal of T-API is to enable efficient management of transport networks that consist of multiple domains with diverse hardware and specifications. It establishes a standardised communication framework between the T-API client, typically a network operator, and the T-API server, typically a SDN controller.

At its core, the T-API interface facilitates control over network resources at a desired level of abstraction selected by the client. It does so by offering a range of services that enable the management and provision of the transport network. These services include:

1. *Obtaining network topology*: T-API allows the T-API client to retrieve information the network's topology, providing a comprehensive view of the network's physical and logical components.
2. *Network status information*: T-API provides real-time updates on the status of the network, enabling the T-API client to monitor the health, performance, and availability of network elements.
3. *Resource availability*: T-API allows the T-API client to query the availability of network resources, such as bandwidth or capacity, to ensure efficient resource allocation.
4. *Connection establishment*: T-API enables the establishment of connections between two endpoints in the network. This service facilitates the provisioning of point-to-point connections based on customer requirements.
5. *Virtual partitioning*: T-API supports the creation of virtual partitions within the network to separate clients or applications. This enhances security and efficient resource sharing.

Similarly, the **IP connector** plays a crucial role in gathering information belonging to the IP network layer. By interfacing with the network infrastructure, the IP connector retrieves essential details about the IP layer's topology and associated features. To facilitate standardised representation and interoperability, the Internet Engineering Task Force (IETF) has defined a generic YANG data model for expressing this network topology. This data model serves as a common framework for capturing the IP layer's characteristics.

### 3.2 Network Slice Visualisation

To test the functionality and usability of these standard interfaces, work is in progress to develop an interactive, dynamic, real-time graphical user interface (GUI) written in Python. This GUI will serve as a visualisation tool for network operators, enabling them to view the different layers of the transport network and the established connectivity services. The ultimate goal of this tool is to also provide network planning capabilities. With a visual representation of the network, operators can gain insights into the overall network architecture and make informed decisions regarding resource allocation, capacity planning, and network optimisation.

In the first test phase, we want to connect to the optical network controller of 'Centre Tecnològic de Telecomunicacions de Catalunya' (CTTC)[3]. The development of this tool involves the next steps. We utilise a REST-based API to connect to CTTC to retrieve real-time data and interact with their server and

[3] https://www.cttc.cat/.

retrieve real-time data. Through the REST-based API, we obtain the "context", or network information, in JSON format according to the T-API interface specifications. From this JSON file, we extract the topological information of the optical transport network. This information is processed and transformed into a visual representation, which is displayed in the GUI, as illustrated in the example in Fig. 2. The topological view illustrates the network elements, consisting of four optical nodes connected by unidirectional links. Similarly, the tool collects information on the IP network layer. By leveraging the extracted topological information, the GUI should help network operators to understand the network layout, identify potential bottlenecks or vulnerabilities, and effectively manage the network resources. In a next step, we also want to integrate a more dynamic mapping between both network layers in the tool and decide a suitable configuration of the IP layer. Furthermore, as part of our long-term objectives, we envision the GUI evolving into a comprehensive network planning tool. The tool could incorporate features such as connection service creation wizards, drag-and-drop functionalities, or the ability to simulate and evaluate network changes before their actual implementation.

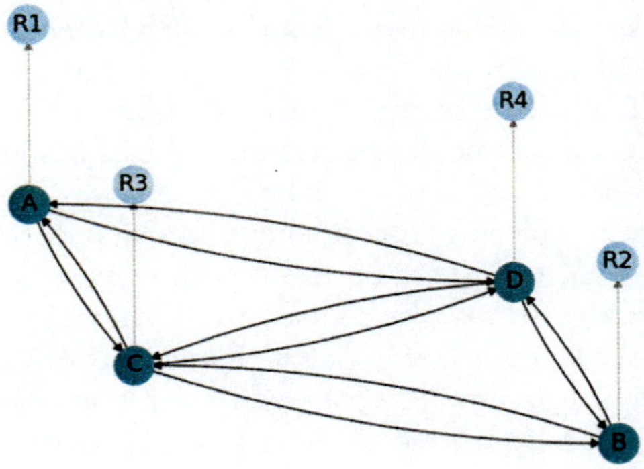


**Fig. 2.** Visualisation of CTTC optical network topology and mapping to IP layer.

### 3.3 Transport Slice Intent Modelling

The concept of an "Intent", as defined in [RFC9315], refers to a declaration of operational goals and desired outcomes for a network service without specifying the implementation details. It serves as an instruction for managing and monitoring network elements and addressing inconsistencies in the service level objective (SLO) targets [3]. In the context of B5G networks, there are three major types of network segments: the radio access network (RAN), the mobile core network and the transport network. To effectively manage the diverse requirements and topologies of each E2E slice, a network slice controller (NSC) is required. The NSC handles different intents across the network and provides the necessary resources accordingly [4].

The IETF proposed an internal structure for modelling the controller that retrieves the client's request and performs the slice mapping. Focusing on the transport network, the right part of the diagram in Fig. 1 illustrates a conceptual scheme of the interactions within the NSC, adapted to our architecture. The key building blocks are the mapper, planner and realiser [5]. The module receives an intent in 3GPP format, which is modelled using YANG. Before being passed to the mapper, the intent is modified to translate it to IETF format. The connection endpoints are extracted from the service delivery points (SDP) in the 3GPP intent. The mapper periodically receives the updated topology from the realiser and uses it to create a network resource partition (NRP) that is sent to the planner. The planner is responsible for path computation. Once the planner confirms the feasibility of the slice request, it notifies the mapper, which, in turn, provides the realiser with the list of slices to be created. Finally, the realiser informs controller of the respective connection type, and the network slices are established in the network.

### 3.4 Multi-layer Planning and Path Computation

An important aspect of our research deals with multi-layer path computation. In particular, we make use of Segment Routing (SR) technology to establish a desired path. Segment routing is a flexible and scalable approach to routing that enables efficient Traffic Engineering (TE) and path calculation. By utilising segment routing, we can dynamically steer traffic through specific paths in the network based on predefined segments or labels.

Furthermore, we are exploring how artificial intelligence techniques can be applied in segment routing to further optimise path computation. For instance, reinforcement learning algorithms can be used to dynamically adapt path selection based on previously learned traffic patterns and network data. In this way, the network controller learns how to optimise resource allocation in real-time. Looking ahead, some future topics we aim to address in the realm of multi-layer network planning include [6]:

1. *Path optimisation across hybrid networks*: To date, our study has focused on single-domain IP networks. We aim to extend our study to IP over optical networks comprising multiple vendors. With a focus on SR, a key consideration is how to leverage the SR policies within each domain to optimise multi-domain TE. While network service providers have specific requirements for multi-domain traffic engineering paths, there is still a need for further exploration of segment list computations across multiple domains. Defining methods for exchanging Segment IDentifier (SID) information is important here. A future research directions could be to identify the best concatenation of binding SIDs of SR policies spanning multiple domains. Such approach would reduce complexity and overhead in network operations [7].
2. *Service-level path computation*: We plan to investigate methods for providing service-level path computation, where paths are determined based on specific service requirements, such as latency, bandwidth, or quality of service. Such

approach enables a more granular control and customisation of paths for different services.

3. *Resilience and fault tolerance*: To ensure fault tolerance in the network, path planning should take into account potential failures or disruptions in the network. So far, our simulations focused on Fast ReRoute (FRR). We have carried out a study which analyses the performance of Topology Independent - Loop Free Alternate (TI-LFA), a FRR algorithm using segment routing [8]. Long-term failure resilience is still a topic for further research.
4. *Dynamic path reconfiguration*: We intend to develop mechanisms for dynamic path reconfiguration, allowing paths to be adjusted adaptively in response to changing network conditions or traffic demands. This flexibility will facilitate responsiveness to varying traffic patterns.

## 4 Conclusion and Future Work

In this paper, we discussed a functional architecture for a SDN network slice orchestrator that addresses the challenges of managing and provisioning resources in IP over optical transport networks. We discussed ongoing efforts in multi-layer network planning, transport slice intent modelling, and multi-layer path computation. By leveraging standardised interfaces such as the Transport-API (T-API), we enable efficient management and provisioning of network resources among different vendors. We also emphasised the benefit of developing a dynamic visualisation tool for network operators. Such tool can provide a complete picture of the network, allowing operators to make informed decisions on resource allocation and capacity planning. In our future work, we will explore optimisation techniques for multi-domain traffic engineering using segment routing. Additionally, we aim to develop service-level path computation methods that consider specific service requirements such as latency, bandwidth, quality of service, and resilience.

**Acknowledgements.** This work is part of the IoTalentum project, which has received funding from the EU H2020 research and innovation programme under the MSCA grant agreement No 953442.

## References

1. Shafi, M., et al.: 5G: a tutorial overview of standards, trials, challenges, deployment, and practice. IEEE J. Sel. Areas Commun. **35**(6), 1201–1221 (2017). https://doi.org/10.1109/JSAC.2017.2692307
2. TAPI v2.1.3 Reference Implementation Agreement. https://opennetworking.org/wp-content/uploads/2020/08/TR-547-TAPI-v2.1.3-Reference-Implementation-Agreement-1.pdf. Accessed 4 May 2023
3. Contreras, L.M., Demestichas, P., Tantsura, J.: «IETF Network Slice Intent», Internet Engineering Task Force, Internet Draft draft-contreras-nmrg-transport-slice-intent-06 (2022). https://datatracker.ietf.org/doc/draft-contreras-nmrg-transport-slice-intent-06. Accessed 4 May 2023

4. Geng, X., Contreras, L.M., Rokui, R., Dong, J., Bykov, I.: «IETF Network Slice Application in 3GPP 5G End-to-End Network Slice», Internet Engineering Task Force, Internet Draft draft-ietf-teas-5g-network-slice-application-00 (2023). https://datatracker.ietf.org/doc/draft-ietf-teas-5g-network-slice-application. Accessed 4 May 2023
5. Contreras, L.M., et al.: «IETF Network Slice Controller and its associated data models», Internet Engineering Task Force, Internet Draft draft-contreras-teas-slice-controller-models-05 (2023). https://datatracker.ietf.org/doc/draft-contreras-teas-slice-controller-models-05. Accessed 4 May 2023
6. Wu, D., Cui, L.: A comprehensive survey on segment routing traffic engineering. Digit. Commun. Netw. https://doi.org/10.1016/j.dcan.2022.02.006
7. Giorgetti, A., Sgambelluri, A., Paolucci, F., Cugini, F., Castoldi, P.: Segment routing for effective recovery and multi-domain traffic engineering. J. Opt. Commun. Network. **9**(2), A223–A232 (2017). https://doi.org/10.1364/JOCN.9.00A223
8. Roelens, L., de Dios, Ó.G., de Miguel, I., Echeverry, E., Barroso, R.J.D.: Performance evaluation of TI-LFA in traffic-engineered segment routing-based networks. In: 2023 19th International Conference on the Design of Reliable Communication Networks (DRCN), Vilanova i la Geltru, Spain, pp. 1–8 (2023). https://doi.org/10.1109/DRCN57075.2023.10108280

# Exploring the Cutting-Edge of Energy Aggregation Approaches and Business Models

Danny Bwalya[1,3], Manuel Azevedo[1,2(✉)], and Emilio S. Corchado[3]

[1] PH Energía, 4150-044 Porto, Portugal
{Danny.bwalya,manuel.azevedo}@energiasimples.pt
[2] Institute of Engineering – Polytechnic of Porto, 4249-015 Porto, Portugal
[3] Department of Computer Science and Automation, University of Salamanca, Salamanca, Spain
escorchado@usal.es

**Abstract.** Energy aggregation business models play a crucial role in maximizing the benefits of renewable energy generation and consumption. In this context, this paper presents an examination of energy system models and their correlation with aggregation business models, emphasizing algorithmic approaches. The distinct characteristics of energy system models in comparison to traditional business models are elucidated. The study encompasses diverse model types employed in the energy sector, including optimization, classification, clustering, and integrated assessment models. Additionally, it investigates the algorithms utilized in aggregation business models, such as demand response, virtual power plants, and peer-to-peer energy trading, highlighting both their advantages and challenges. The paper concludes that the advancement of energy aggregation business models holds significant promise in addressing the intricate complexities of contemporary power systems and the ongoing energy transition. Future research endeavors involve the exploration of Electroencephalography (EEG) techniques within energy aggregation models, inspired by the information processing mechanisms of the human brain.

**Keywords:** Aggregation · Energy Communities · Smart Grid · Business Model

## 1 Introduction

With access to secure, affordable, and sustainable energy being a hot topic, and the world's increasing focus on transitioning towards a carbon-neutral economy, climate communities and citizens are playing a critical role in driving this transition forward. In pursuit of solutions to address this and ensuing problems of the conventional electricity grid network, smart grids, and energy communities (ECs) have emerged and are showing great potential [1]. Historically, energy consumers have been viewed as passive recipients of energy and price takers, but as individuals and communities become more aware of the need to reduce carbon emissions and address climate change, they are becoming more active thereby directly participating in shaping the energy landscape [2].


R. Mehmood et al. (Eds.): DCAI 2023, LNNS 741, pp. 497–505, 2023.
https://doi.org/10.1007/978-3-031-38318-2_50

Citizens are adopting sustainable practices such as reducing energy consumption, using renewable energy sources (RES), and choosing environmentally friendly products. They are also advocating for policy changes and encouraging businesses to adopt sustainable practices. As a result, there is a growing trend towards energy democracy, where individuals and communities have greater control over their energy consumption, production, and pricing [3]. This shift towards active energy and price making by citizens is critical for achieving a carbon-neutral economy and mitigating the impacts of climate change. By empowering individuals and communities to take an active role in shaping the energy landscape, the transition towards a sustainable future can be accelerated.

Energy aggregation business models such as Peer-to-peer (P2P) sharing, and ECs are some innovative solutions for managing and maximizing the benefits of renewable energy generation and consumption. These models allow community members to share excess energy generated from their renewable assets with each other, creating a more flexible and sustainable energy system. P2P energy trading enables community members to directly trade energy with each other, reducing reliance on traditional centralized energy systems and providing greater control over energy consumption and production. This type of trading also allows for a more efficient use of renewable energy resources within the community, leading to a reduction in carbon emissions and an increase in renewable energy adoption [4]. ECs can also actively manage their flexible assets to enhance the benefits from their renewable assets. By pooling their resources and actively managing their flexible assets, these communities can generate value creation opportunities and participate in a range of market alternatives, such as selling excess energy back to the grid or participating in demand response (DR) programs [5].

Business models therefore play a crucial role in driving the transition towards a sustainable energy future. Nonetheless, implementing innovative aggregation business models in the energy sector can be challenging due to the complex environment of current power systems. The energy transition requires a significant shift in the way we generate, distribute, and consume energy, and this requires a collaborative effort among policymakers, regulators, and industry stakeholders. The introduction of new stakeholders such as prosumers, and incorporation of distributed energy resources (DER) with high levels of uncertainty and variability adds an extra layer of complexity to this process. It is therefore imperative that innovation in these business models occurs simultaneously with regulation and policy to achieve a successful transition. Policies and system regulations should be designed to facilitate the integration of innovative business models into the existing energy system and to ensure that the benefits of these models are realized and maximized.

The objectives of this paper are therefore to provide an overview of energy system models and related business models; explaining their characteristics and the differences from traditional business models; and to discuss some algorithms/approaches regarding aggregation business models.

The paper has four sections. Firstly, an introduction to the topic, followed by a brief state-of-the-art narration of energy systems models. The next section then addresses algorithms in energy aggregation business models, and finally Section 4 gives conclusions and indicates the direction of future studies.

## 2 State-of-the-Art: Energy Communities' Business Models

### 2.1 Background of Energy System Models

The development of energy system models (ESMs) has been aimed at addressing the issues arising from anthropogenic CO2 emissions and the need for a significant transformation in the energy sector. These ESMs have facilitated the integration of renewable energy resources into existing energy systems and the coupling of energy forms and sectors. Early attempts to predict future energy demands date back to the 1950s and were based on simple scenarios [6]. ESMs have since evolved based on the principle of peak-load-pricing, which distinguishes between the capacity and operating costs of facilities producing non-storable goods, such as electricity [7]. Their development has been driven by the need for supply reliability [8], advancements in computational resources [9], and the integration of non-dispatchable technologies, such as RES [10]. In general, ESM developers have had two options: focusing on economic mechanisms (top-down approach) or the technical dimension (bottom-up approach), including simulations and optimizations [11]. The complexity of ESMs is influenced by the spatiotemporal and techno-economic dimensions. Spatiotemporal resolution plays a role in the techno-economic dimension, whereby the temporal resolution imposes limitations on the operational accuracy of energy system components. These dimensions interact and contribute to the overall intricacy of ESMs, shaping their modeling and analysis capabilities in capturing the spatiotemporal dynamics and techno-economic interactions within energy systems.

### 2.2 Legal Framework for Energy Aggregation

**EU Context.** Directive (EC) 2019/944 [12] and Regulation (EC) 2019/943 [13] are important because they establish common rules for the internal market for electricity in the EU. These regulations are essential for creating a level playing field for energy market actors and promoting competition in the sector. The Renewable Energy Directive (RED II) [14] is another important directive aimed at increasing the share of RES in the EU's energy mix. This directive sets binding renewable energy targets for each member state and outlines measures to support the development of RESs.

**EC context.** Directive (EC) 2019/944 also sets common rules for the internal market for electricity within energy communities. It is essential for enabling ECs to participate in the wider electricity market and benefit from the opportunities offered by energy aggregation. The directive aims to create integrated, competitive, flexible, and transparent electricity markets that prioritize consumer needs and protection. It seeks to empower consumers, DR, and ECs, while also recognizing aggregators as market participants. In addition to its focus on promoting consumer engagement in the electricity market, the directive strongly advocates for the integration of RES.

### 2.3 Select Existing EC Business Models

There are several existing business models for ECs, including P2P energy trading, virtual power plants (VPP), energy cooperatives, microgrids, energy as a service (EaaS), DR programs, green tariffs, and community energy storage (CES). These models can be combined or adapted to meet the specific needs of different communities, as shown in Table 1—allowing them to take an active role in the transition to a more sustainable energy system. Various projects have implemented different energy trading designs. Under the P2P model for example, the projects can be categorized as either physical layer-oriented or virtual layer-oriented, depending on their focus. The SOLshare [15], Lumenaza [16], and Brooklyn Microgrid [17] projects focus on electricity trading among peers and utilize blockchain-based solutions. On the other hand, projects such as EMPOWER [18], and P2P-Smart Test [19] focus on aggregator business models with grid services offered by the distribution system operator (DSO) and/or the transmission system operator (TSO). Moreover, they employed cloud-based management solutions.

The services of the above business models can either be demand-based or supply-based, with P2P and VPP tending to a more neutral position in this regard. Based on the number of agents involved in a particular business model, the complexity of these models varies i.e., the more agents involved, the more complex the business model tends to be. Fig. 1 below is a graphical representation of the complexity levels of some aggregation models and their inclination on the supply or demand-based services scale.

### 2.4 Peculiarities: EC Business Models vs Traditional Business Models

Though EC business models are in principle similar to other conventional business models, they need to be innovated in a tailored fashion due to their five peculiarities from a business point of view [21].

**Not Profit-Oriented:** Unlike other business models, where companies are objectively committed to profit-making, their focus is to address energy-related issues, and to meet the social, environmental, and economic needs of the end users [22].

**Multisided Platform:** ECs tend to operate as multisided platforms [21] because of their emphasis on taking advantage of co-benefits and the distinct traits of their members. Multisided platforms are known for having multiple agents that directly engage with the platform they are associated with [23]. For example, energy producers may need to be incentivized to generate renewable energy, while energy consumers may require access to affordable and reliable energy. Energy service providers, on the other hand, may be interested in providing services that can help optimize energy usage or integrate RES. To successfully function as a multisided platform, ECs must carefully design their business models and develop strategies that enable different sides to interact effectively. This may involve the development of new technologies or services that can facilitate interactions between different sides or the creation of new business models that enable value creation for all parties involved.

**Influenced by Network Effect:** Network effects are a crucial factor in ECs. They refer to a situation where the value or usefulness of a product or service increases as more people use it or as more complementary products are added [24]. In the context of ECs, network effects can be observed when the benefits of being part of the community increase as more members join or as more complementary services are offered. This

**Table 1.** Overview of Sustainable Energy Business Models

| Business Model | Characteristics |
|---|---|
| P2P | • Community members trade excess renewable energy |
| | • Decentralized platform for energy trading |
| | • Facilitates local renewable energy production and consumption |
| VPP | • Aggregates DERs |
| | • Provides grid services or participates in energy markets |
| | • Enables the integration of variable RES |
| Energy cooperatives | • Pool resources to develop and operate renewable energy projects |
| | • Members share benefits and risks |
| | • Enables community engagement and empowerment |
| Microgrids | • Can operate independently from main grid |
| | • Can be integrated with RES |
| | • Offers increased energy resilience and reliability |
| EaaS | • Offers access to energy services rather than ownership of assets |
| | • Provides flexibility and scalability |
| | • Can reduce upfront costs for consumers |
| DR | • Incentivizes energy reduction during peak demand |
| | • Offers economic benefits for both utilities and consumers |
| | • Helps to balance the energy grid |
| CES | • Enables management of renewable energy generation & energy demand |
| | • Supports grid stability and resilience |
| | • Can increase the value of local renewable energy resources. |
| Green tariffs | • Direct purchase of renewable energy from utilities |
| | • Provides a price premium for renewable energy |
| | • Enables greater consumer choice and support for renewable energy |

can create a virtuous cycle where more members and more services lead to even greater benefits for everyone involved.

**Co-Benefits Leveraging:** Traditionally, energy providers have focused on selling a single service, such as electricity or natural gas. However, in an EC, there is an opportunity to leverage the co-benefits between different energy services and offer a more comprehensive suite of services to members [25]. For example, an EC could offer services that combine renewable energy generation [26], energy storage [27], and DR [28]. By

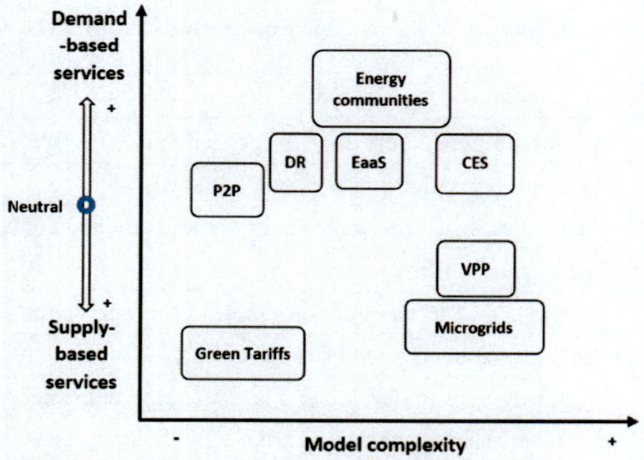


**Fig. 1.** Characteristics and complexity of aggregation business models (modified from source [20])

doing so, the community could optimize energy usage, reduce energy costs, and improve energy security.

**Community-Driven:** In an EC, the traditional model of a main firm innovating and implementing a business model is replaced by a community-driven model. The community itself becomes the operator and consumer in the business model [29], with each member playing an active role in the energy system. This community-driven approach empowers consumers to take an active role in shaping the energy landscape and to participate in the decision-making process. The success of an EC depends on the willingness of its members to participate actively in the community [30]. Therefore, it is essential that each member perceives a net positive value proposition from their participation in the community [21].

## 3 Algorithms in Energy Aggregation Business Model

The application of various algorithms in energy aggregation business models is a dynamic and active area of research. In [6] for example, time series analysis was used for predicting energy demand and electricity prices, thereby offering valuable insights for decision-making by the involved stakeholders. Additionally, classification algorithms are employed to segment energy market participants, enabling a better understanding of their diverse characteristics and behaviors.

In order to identify and analyze community members with similar ESS profiles, the authors in [31] used clustering, which is typically used in the context of energy aggregation models to help group similar participants. In optimization problems involving DR, DG, and ESSs, mixed-integer linear programming techniques have proven to be effective [31]. These mathematical optimization models facilitate optimal decision-making by considering discrete variables and constraints inherent in energy systems.

Artificial Neural Networks (ANNs) have also gained prominence in such models due to their capacity for pattern recognition. ANNs are employed for various tasks such as load forecasting, price prediction, and anomaly detection. By leveraging historical data, ANNs can identify patterns and trends, enabling more accurate predictions and facilitating proactive decision-making.

To address the uncertainty associated with the intermittence of RES and optimize resource allocation, stochastic optimization techniques and linear programming methods are employed, respectively. Stochastic optimization considers probabilistic scenarios to make decisions robust to uncertainties, while linear programming optimizes resource allocation based on linear constraints and objectives.

Genetic algorithms provide an alternative approach to solving optimization problems like DR and resource allocation. These algorithms mimic natural evolution processes, exploring different combinations of variables to identify optimal solutions. Similarly, ant colony optimization and particle swarm optimization algorithms offer alternative optimization approaches, providing efficient and scalable solutions for complex energy system dynamics.

Ongoing research efforts in this field are dedicated to enhancing the efficiency, scalability, and adaptability of these algorithms. Researchers aim to develop advanced optimization algorithms that can effectively handle the complexities inherent in energy aggregation business models. This involves improving computational efficiency, exploring parallel computing techniques, and addressing the challenges posed by real-time decision-making in dynamic energy systems.

## 4 Conclusion and Future Work

In conclusion, the reviewed literature gives the indication that innovation in energy aggregation business models provides a promising avenue for addressing the complex challenges of the modern power systems and energy transition. By leveraging insights from existing business models, lessons learned from already implemented projects, incentives, and the opportunities derived from regulations and legal frameworks, novel approaches can be explored, and new models developed to improve the state-of-the-art.

Moving forward, future work related to this paper would be to explore the possibility of integrating Brain-Computer Interface technology like Electroencephalography (EEG) techniques in energy aggregation business models. EEG is a non-invasive technique for measuring brain activity by recording electrical signals. This technology has the potential for application in energy aggregation business models by using insights of input data such as consumer characteristics, preferences, market prices, community assets etc., to recommend the most ideal business model for the datasets. This idea originates from the parallels between the brain's information processing mechanism and the dynamics of energy systems, which may offer some valuable insights for designing innovative energy aggregation solutions. The human brain, simply put, can be understood as a complex system with input and output functions, much like an energy aggregation model. Another interesting parallel that can be drawn between the two systems is their randomness—unpredictability for the brain and intermittence for RES.

Drawing from these parallels and more, the idea to approach new energy system modeling problems using a time-tested technique like EEG seems interesting. The hypothesis is that by adopting EEG-inspired techniques, decision-making processes in energy aggregation can be enhanced, leading to more efficient and tailored solutions that promote sustainable energy practices.

**Acknowledgements.** This work is part of the IoTalentum project, which has received funding from the EU H2020 research and innovation programme under the MSCA grant agreement No 953442.

## References

1. Rohde, F., Hielscher, S.: Smart grids and institutional change: emerging contestations between organisations over smart energy transitions. Ene. Res. Soc. Sci. **74**, 101974 (2021)
2. Gržanić, M., Capuder, T., Zhang, N., Huang, W.: Prosumers as active market participants: a systematic review of evolution of opportunities, models and challenges. Renew. Sustain. Ener. Rev. **154**, 111859 (2022)
3. Wahlund, M., Palm, J.: The role of energy democracy and energy citizenship for participatory energy transitions: a comprehensive review. Ener. Res. Soc. Sci. **87**, 102482 (2022)
4. Kong, K.G.H., et al.: Fuzzy optimization for peer-to-peer (P2P) multi-period renewable energy trading planning. J. Clean Prod. 368 (2022)
5. Okur, Ö., Heijnen, P., Lukszo, Z.: Aggregator's business models in residential and service sectors: A review of operational and financial aspects. Renew. Sustain. Ener. Rev. **139**, 110702 (2021)
6. Hoffmann, M., Kotzur, L., Stolten, D., Robinius, M.: A review on time series aggregation methods for energy system models. Energies (Basel) 13 (2020)
7. Chao, H.-p: Peak load pricing and capacity planning with demand and supply uncertainty. Bell J. Econ. **14**, 179–190 (1983)
8. Fodstad, M., et al.: Next frontiers in energy system modelling: A review on challenges and the state of the art. Renew. Sustain. Ener. Rev. 160 (2022)
9. Subramanian, A.S.R., Gundersen, T., Adams, T.A.: Modeling and simulation of energy systems: a review. Processes **6**, 238 (2018)
10. Panda, A., Dauda, A.K., Chua, H., Tan, R.R., Aviso, K.B.: Recent advances in the integration of renewable energy sources and storage facilities with hybrid power systems. Clean Eng. Technol. **12**, 100598 (2023)
11. Ringkjøb, H.K., Haugan, P.M., Solbrekke, I.M.: A review of modelling tools for energy and electricity systems with large shares of variable renewables. Renew. Sustain. Ener. Rev. **96**, 440–459 (2018)
12. Directive (EU) 2019/ 944 of the European Parliament and of the Council - of 5 June 2019 - on common rules for the internal market for electricity and amending Directive 2012/27/EU (2019)
13. Regulation (EU) 2019/943 of the European Parliament and of the Council - of 5 June 2019 - on the internal market for electricity (2019)
14. Directive (EU) 2018/2001 of the European Parliament and of the Council of 11 December 2018 on the promotion of the use of energy from renewable sources (recast) (Text with EEA relevance) (2018)
15. Home - SOLshare, https://solshare.com/, last accessed 5 May 2023
16. International Renewable Energy Agency, T.: Peer-to-peer electricity trading - Innovation Landscape Brief (2020)

17. Brooklyn Microgrid | Community Powered Energy, https://www.brooklyn.energy/, last accessed 5 May 2023
18. Local Electricity retail Markets for Prosumer smart grid pOWER services | EMPOWER Project | Fact Sheet | H2020 | CORDIS | European Commission, https://cordis.europa.eu/project/id/646476, last accessed 5 May 2023
19. EU H2020 Project "P2P-SMARTEST (Peer to Peer Smart Energy Distribution Networks (P2P-SmartTest))": description, participants, costs and EC-fundings, https://www.fabiodisconzi.com/open-h2020/projects/194444/index.html, last accessed 5 May 2023
20. Larrea, M., Maider, B., Ozamiz, B.: Modelos de negocio en recursos distribuidos de electricidad (2020)
21. Kubli, M., Puranik, S.: A typology of business models for energy communities: Current and emerging design options. Renew. Sustain. Ener. Rev. **176**, 113165 (2023)
22. Caramizaru, A., Uihlein, A.: Energy communities: an overview of energy and social innovation (2020)
23. Multisided Platforms and Markets: A Literature Review, https://www.researchgate.net/publication/325225786_Multisided_Platforms_and_Markets_A_Literature_Review, last accessed 26 April 2023
24. What Are Network Effects? | HBS Online, https://online.hbs.edu/blog/post/what-are-network-effects, last accessed 26 April 2023
25. Lowitzsch, J., Hoicka, C.E., van Tulder, F.J.: Renewable energy communities under the 2019 European Clean Energy Package – Governance model for the energy clusters of the future? Renew. Sustain. Ener. Rev. **122**, 109489 (2020)
26. Cai, Q., Xu, Q., Qing, J., Shi, G., Liang, Q.M.: Promoting wind and photovoltaics renewable energy integration through demand response: Dynamic pricing mechanism design and economic analysis for smart residential communities. Energy **261**, 125293 (2022)
27. The Role of Storage and Demand Response — Greening the Grid, https://greeningthegrid.org/resources/factsheets/the-role-of-storage-and-demand-response/view, last accessed 30 May 2023
28. Barnes, J., Hansen, P., Kamin, T., Golob, U., Musolino, M., Nicita, A.: Energy communities as demand-side innovators? Assessing the potential of European cases to reduce demand and foster flexibility. Ener. Res. Soc. Sci. **93**, 102848 (2022)
29. Hanke, F., Lowitzsch, J.: Empowering Vulnerable Consumers to Join Renewable Energy Communities—Towards an Inclusive Design of the Clean Energy Package. Energies **13**, 1615 (2020)
30. Koirala, B.P., Koliou, E., Friege, J., Hakvoort, R.A., Herder, P.M.: Energetic communities for community energy: A review of key issues and trends shaping integrated community energy systems. Renew. Sustain. Ener. Rev. **56**, 722–744 (2016)
31. Silva, C., Faria, P., Fernandes, A., Vale, Z.: Clustering distributed Energy Storage units for the aggregation of optimized local solar energy. Energy Reports **8**, 405–410 (2022)

# Author Index





R. Mehmood et al. (Eds.): DCAI 2023, LNNS 741, pp. 507–509, 2023.
https://doi.org/10.1007/978-3-031-38318-2